Intelligent Control and Learning Systems

Volume 18

Series Editor

Dong Shen, School of Mathematics, Renmin University of China, Beijing, China

The Springer book series Intelligent Control and Learning Systems addresses the emerging advances in intelligent control and learning systems from both mathematical theory and engineering application perspectives. It is a series of monographs and contributed volumes focusing on the in-depth exploration of learning theory in control such as iterative learning, machine learning, deep learning, and others sharing the learning concept, and their corresponding intelligent system frameworks in engineering applications. This series is featured by the comprehensive understanding and practical application of learning mechanisms. This book series involves applications in industrial engineering, control engineering, and material engineering, etc.

The Intelligent Control and Learning System book series promotes the exchange of emerging theory and technology of intelligent control and learning systems between academia and industry. It aims to provide a timely reflection of the advances in intelligent control and learning systems. This book series is distinguished by the combination of the system theory and emerging topics such as machine learning, artificial intelligence, and big data. As a collection, this book series provides valuable resources to a wide audience in academia, the engineering research community, industry and anyone else looking to expand their knowledge in intelligent control and learning systems.

Jun Hu · Zidong Wang · Chaoqing Jia

Variance-Constrained Filtering for Stochastic Complex Systems

Theories and Algorithms

Springer

Jun Hu
Department of Applied Mathematics
Harbin University of Science
and Technology
Harbin, Heilongjiang, China

Zidong Wang
Department of Computer Science
Brunel University London
Uxbridge, Middlesex, UK

Chaoqing Jia
School of Automation
Harbin University of Science
and Technology
Harbin, Heilongjiang, China

ISSN 2662-5458 ISSN 2662-5466 (electronic)
Intelligent Control and Learning Systems
ISBN 978-981-96-2636-6 ISBN 978-981-96-2637-3 (eBook)
https://doi.org/10.1007/978-981-96-2637-3

This Springer imprint is published by the registered company Springer Nature Singapore Pte Ltd.
The registered company address is: 152 Beach Road, #21-01/04 Gateway East, Singapore 189721, Singapore

Preface

Time-varying characteristic, stochasticity and nonlinearity are ubiquitous features in most of practical systems, which significantly enhance to the challenges of system analysis and inevitably affect the system performance. Consequently, the analysis and synthesis problems for time-varying nonlinear dynamical systems have long been the mainstream of research topics and a large number of schemes have been developed accordingly. Over the past decade, with the rapid developments of the networked systems equipped with advanced communication techniques, the filtering/estimation algorithm design problems for time-varying dynamical systems with incomplete information have become hot topics with increasing research interests.

This book is concerned with the variance-constrained filtering and its application problems for several classes of time-varying dynamical systems with incomplete information. The content of this book can be conceptually divided into three parts. In Part I, we focus on the robust variance-constrained filtering problems for some classes of time-varying nonlinear stochastic systems subject to modeling errors, stochastic nonlinearity, incomplete measurements, filter gain perturbations as well as event-triggered communication schemes. In particular, some new filtering methods are developed in terms of the solutions to Riccati-like difference equations, which are suitable for online implementations and applications. In Part II, based on the proposed variance-constrained filtering approaches, the joint state and fault estimation problems are discussed for several classes of time-varying nonlinear stochastic systems with randomly occurring nonlinearities (RONs), randomly occurring deception attacks (RODAs), randomly occurring faults (ROFs), stochastic nonlinearities and packet dropouts. In this part, some sufficient conditions are established to provide satisfactory joint estimation schemes and performance evaluation criterion. In Part III, the presented variance-constrained filtering/estimation methods are extended to provide the variance-constrained state estimation approaches for time-varying complex networks with uncertain inner coupling, switching topology, packet dropouts and communication protocols.

The compendious framework and description of this book are given as follows. Chapter 1 introduces the recent advances in variance-constrained filtering and estimation for time-varying dynamical systems subject to incomplete information. Chapter 2 is concerned with the robust variance-constrained filtering problem for a class of time-varying networked systems influenced by stochastic nonlinearity, randomly occurring quantization (ROQ) and randomly occurring uncertainties (ROUs). In Chap. 3, the resilient variance-constrained filtering scheme is given for time-varying nonlinear stochastic systems, in which the stochastic uncertainties, incomplete measurements as well as filter gain perturbations are all addressed. With the purpose of saving network resources, the proposed variance-constrained filtering schemes are further extended and the event-triggering condition is introduced in Chaps. 4–5, where both the stochastic uncertainties and missing measurements are examined. Chapter 6 is concerned with the co-design problem of the joint state and fault estimator for time-varying stochastic systems involving RONs and RODAs. In Chap. 7, the joint state and fault estimation problem is handled for time-varying systems with ROUs, stochastic nonlinearities and packet dropouts. In Chap. 8, the joint state and fault estimation issue is tackled for time-varying nonlinear stochastic systems with ROFs and sensor saturations. Next, the presented variance-constrained filtering/estimation methods are employed to address the state estimation problems for time-varying complex networks. In Chap. 9, the recursive state estimation issue is solved for time-varying coupled stochastic complex networks with missing measurements. In Chap. 10, the state estimation problem is discussed for time-varying complex networks with uncertain inner coupling and signal quantization. In order to decrease the communication burden and improve the transmission efficiency, the event-triggered mechanism (ETM) has been utilized in Chap. 11 to tackle the state estimation problem for time-varying complex networks under the variance constraint index. In Chap. 12, the state estimation issues are addressed for time-varying uncertain complex networks with RONs, switching topology and packet dropouts. In Chap. 13, the state estimation issue is handled for time-varying nonlinear complex networks through the communication scheduling in the variance-constrained framework. Chapter 14 draws conclusions on this book and points out some potential research directions.

This book is a research monograph whose intended audience is graduate and postgraduate students as well as researchers, serving as both a summary of the recent research results and a source of further research directions.

Harbin, China — Jun Hu
Uxbridge, UK — Zidong Wang
Harbin, China — Chaoqing Jia

Acknowledgements

We would like to acknowledge the successive help of many people who have been directly involved in various aspects of the research leading to this book. Special thanks go to Prof. Guo-Ping Liu from Southern University of Science and Technology of China, Prof. Raquel Caballero-Águila from Universidad de Jaén of Spain and Prof. Dongyan Chen from Harbin University of Science and Technology of China. We also extend our thanks to many colleagues who have offered support and encouragement throughout this research effort. In particular, we would like to acknowledge the contributions and friendly support from Hongli Dong, Lei Zou, Bo Shen, Hang Geng, Yang Liu, Wei Chen, Hongxu Zhang and Jiaxing Li. Finally, we are deeply indebted to our families for their never-ending understanding, unfailing encouragement and strong support when it was most required.

The writing of this book was supported in part by the National Natural Science Foundation of China under Grants 12171124, 61933007 and 12301567, the Natural Science Foundation of Heilongjiang Province of China under Grant ZD2022F003, the Course Ideological and Political Demonstration Project of Ministry of Education of China under Grant 2021-0038, the Fundamental Research Foundation for Universities of Heilongjiang Province of China under Grant 2022-KYYWF-0141, and the Alexander von Humboldt Foundation of Germany. The support of these organizations is much acknowledged.

Contents

Abbreviations

CKF	Cubature Kalman filtering
CSTR	Continuous stirred tank reactor
DPDs	Data packet dropouts
EBSE	Event-based state estimation
EEC	Estimation error covariance
EKF	Extended Kalman filtering
ETM	Event-triggered mechanism
ETMs	Event-triggered mechanisms
FEC	Filtering error covariance
i.i.d.	Independent and identically distributed
KF	Kalman filtering
MSE	Mean-square error
MSEs	Mean-square errors
OETE	Optimized event-triggered estimation
RAP	Random access protocol
RAPBSE	RAP-based state estimation
RCS	Random coupling strength
RFE	Recursive fault estimation
RHS	Right-hand side
RODAs	Randomly occurring deception attacks
ROFs	Randomly occurring faults
RONs	Randomly occurring nonlinearities
ROQ	Randomly occurring quantization
ROUs	Randomly occurring uncertainties
RSDs	Random sensor delays
RVCF	Robust variance-constrained filtering
SCS	Stochastic coupling strength
SPDs	Successive packet dropouts
UKF	Unscented Kalman filtering
UOPs	Uncertain occurrence probabilities

Acronyms

$\mathbb{R}^p$	The p-dimensional Euclidean space
$\mathbb{R}^{p \times q}$	The set of all $p \times q$ real-valued matrices
$\mathrm{tr}(X)$	The trace of the matrix X
X^T	The transpose of the matrix X
$\mathrm{sym}(X)$	$X + X^T$
$\|X\|$	The norm of the matrix X defined by $\|X\| = \sqrt{\mathrm{tr}(X^T X)}$
X^{-1}	The inverse of the matrix X
I	An identity matrix with appropriate dimension
I_p	The p-dimensional identity matrix
$1_{p \times p}$	The all-ones matrix of size $p \times p$
0	A zero matrix with proper dimension
$\mathrm{Prob}\{\cdot\}$	The occurrence probability of the event "·"
$\mathbb{E}\{x\}$	The expectation of the stochastic variable x
$\mathbb{E}\{x\|y\}$	The expectation of the stochastic variable x conditional on y
$\mathrm{Var}\{x\}$	The variance of the stochastic variable x
$\lambda_{\max}(X)$	The largest eigenvalue of a square matrix X
$\mathrm{diag}\{\cdots\}$	The block-diagonal matrix
$\otimes$	The Kronecker product
$\circ$	The Hadamard product defined by $[A \circ B]_{ij} = A_{ij} \cdot B_{ij}$
$A > B$	The $A - B$ is positive definite, where A and B are real-valued symmetric matrices
$A \geq B$	The $A - B$ is positive semi-definite, where A and B are real-valued symmetric matrices

Chapter 1
Introduction

During the past two decades, the variance-constrained filtering and estimation issues have been one of the dominant research topics on account of their great application potentials in a variety of research areas, such as imagine processing, target tracking and environmental monitoring. Along with the rapid advancements of network communication techniques, compared with the traditional control systems, the networked control systems possess prominent merits including flexible architecture, decreased maintenance cost, high reliability and so on. While the network communication brings some convenience, a series of network-induced phenomena generally emerge in the process of signal transmission due to constrained network bandwidth, unreliable communication link and complicated network environment, such as missing measurement, fading measurement, sensor saturation, sensor delay, network attacks, and so on. In addition, it is worth emphasizing that because of the modeling error or implementation error, some imperfect internal features regarding the dynamical systems might appear including stochastic uncertainty, stochastic nonlinearity and so on. On the other hand, it is noted that the systems are always time-varying in practical engineering applications on account of the complicated network media. As a result, it is crucial to tackle the filtering and estimation issues for time-varying dynamical systems affected by the incomplete information (networked-induced phenomena, modeling uncertainty and so on) under variance-constrained sense, which serves as a meaningful yet challenging topic.

1.1 Research Background, Motivations and Research Problems

In this section, the research background, motivations as well as research problems of this book are introduced from the following research route.

J. Hu et al., *Variance-Constrained Filtering for Stochastic Complex Systems*, Intelligent Control and Learning Systems 18,
https://doi.org/10.1007/978-981-96-2637-3_1

1.1.1 *Time-Varying Dynamical Systems*

In practical engineering situations, on account of the external environment changes, the internal property of the module as well as complicated interaction behaviors among subsystems, the time-varying characteristic is commonly considered during the system modeling, which can better describe the system behaviors from the reality perspective. Moreover, it is noted that the nonlinearity and engineering-oriented complexity are two representative features in most practical time-varying systems. If the nonlinearity and engineering-oriented complexity cannot be effectively addressed, the system performance might be deteriorated. As such, it is valuable to develop the filtering and estimation algorithms in the context of time-varying dynamical systems due to their practical applications in a variety of fields, including the dynamic positioning, target tracking, and so forth (Caballero-Águila et al. 2017; Hu et al. 2018; Priel and Zelazo 2023; Yang et al. 2023; Wang and Sun 2023; Park et al. 2016). Next, we aim to introduce the time-varying dynamical systems from the following two aspects.

1.1.1.1 Time-Varying Nonlinear Systems

In real-world scenarios, many practical dynamical systems exhibit nonlinear properties. The existence of the nonlinearities would unavoidably deteriorate the system performance and even give rise to the instability of the control systems. In this regard, the classical Kalman filtering (KF), which can be viewed as an optimal estimation scheme for linear systems affected by Gaussian noise, will not guarantee satisfactory performance. As a result, when it comes to the time-varying nonlinear systems, some variants with respect to the KF have been proposed to estimate the system states, such as extended Kalman filtering (EKF), unscented Kalman filtering (UKF), particle filtering, cubature Kalman filtering (CKF) and so on, see Hafez et al. (2023), Duan et al. (2021) for more details. To be more specific, in view of the unscented transform technique, the UKF aims to employ a sequence of sigma points with different weights to approximate the posterior mean and covariance associated with the system states. In Liu et al. (2020), a new UKF algorithm has been proposed for stochastic nonlinear systems under communication scheduling strategies. Further, a new unscented Tobit KF scheme has been presented in Li et al. (2022) for switched nonlinear systems in the presence of censored measurements, in which the variational Bayesian algorithm has been adopted to generate the marginal distributions. Different from the former, the CKF adopts a set of cubature points with the same weights to approximate the nonlinear function by resorting to the three-order spherical-radial cubature principle. In Ju et al. (2022), a distributed CKF problem has been addressed for nonlinear systems under the stochastic communication scheduling strategy, where the filter gain has been obtained through realizing the minimization of the upper bound regarding the filtering error covariance (FEC). By utilizing the Monte Carlo method, the particle filtering is utilized to estimate the state of the nonlinear systems that would encounter the calculational burden. For instance, a distributed particle filtering scheme has been proposed in Song et al. (2020) for nonlinear systems under the dynamic event-triggered scheduling strategy.

It is worthwhile to emphasize that the EKF has stirred much research interest owing to its convenience on implementation. Specifically, the primary idea of the EKF is to linearize the nonlinear function by resorting to the Taylor expansion method and the higher-order terms are omitted or further estimated. For instance, the EKF problem has been explored in Liu et al. (2016) for stochastic nonlinear systems influenced by cyber attacks. For the sake of avoiding data collisions and reducing communication burden, an EKF scheme has been developed in Xu et al. (2022) for nonlinear uncertain systems under round-robin protocol. To sum up, it is valuable yet urgent to investigate the filtering issues in the nonlinear context especially in the networked environment.

1.1.1.2 Time-Varying Complex Networks

Complex networks characterize a class of dynamical networks, which are composed of a great number of interacted nodes. Generally, the complex networks can be utilized to describe many practical networks, such as technological networks, power grids, social networks, airport networks and so on Boccaletti et al. (2006), Liang et al. (2024). During the past few decades, considerable research interest has been focused on the analysis issues with respect to the time-varying complex networks (Park et al. 2018; Chen et al. 2023). As a matter of fact, it is crucial to obtain the state information regarding the network nodes with hope to analyze the intrinsic structures as well as dynamical behaviors concerning on complex networks (Zhao et al. 2022; Ding et al. 2022; Rezaei et al. 2023). Nevertheless, when the massive network nodes become concern, it is generally common in real world that the exact state information is difficult to obtain owing to some unavoidable reasons, such as constrained network resources and complicated communication media, and thus only partial measurement signals of the network nodes can be availably obtained (Basin et al. 2013, 2018). Accordingly, the state estimation issues within the context of time-varying complex networks have recently stirred increasing research interest. It is noted that, because of the inherent characteristics of complex networks, more difficulties should be handled when addressing the state estimation issue in the context of time-varying complex networks in comparison with the time-invariant cases.

On account of the complicated network environment or dynamical interaction of practical networked environment, the connection relationship or strength cases among network nodes might be randomly changed. As such, some additional difficulties and complexities might occur when analyzing the time-varying complex networks. Specifically, according to different connection ways of nodes, the coupling cases among nodes can be generally categorized into the following types, i.e., normal coupling, random inner coupling, random outer coupling, stochastic coupling and switching topology, where the specific parameter definitions can be respectively found in the subsequent chapters. In addition, with the increasing developments of communication network, the incomplete information also generally emerges in the process of data transmission. Consequently, it is valuable to deal with the estimation problems for time-varying complex networks involving incomplete information.

1.1.2 Incomplete Information

Over the past decades, the networked control systems have been successfully applied in a diverse range of practical domains owing to their distinctive merits. Meanwhile, the introduction of the communication network also brings some additional challenges for the filter/estimator design (Li and Yang 2023; Wang et al. 2023). To be specific, in the process of transmitting information via the communication network, the data signals might undergo signal quantization, fading measurement, measurement outliers, sensor saturation, data drifting, sensor delays, cyber attacks and so forth. On the other hand, due to the limited network bandwidth and constrained network resources, the data congestion would inevitably appear in the transmission process. With the purpose of avoiding such a case and enhancing communication efficiency, some effective yet active communication scheduling strategies are commonly adopted to arrange the data transmission, such as event-triggered mechanism (ETM), random access protocol (RAP) and so on. Here, the above mentioned phenomena are called as the incomplete information. Now, we endeavor to introduce and discuss the incomplete information from the following aspects in detail.

1.1.2.1 Stochastic Nonlinearities and Stochastic Uncertainties

As it is well recognized, the nonlinearities and uncertainties pose certain complexities to the dynamical system modeling. If they cannot be effectively handled, the system performance will be deteriorated, and thus it is necessary to cope with them carefully for analyzing the time-varying dynamical systems (Ding et al. 2015a, b; Dong et al. 2014; Hou et al. 2016). To be specific, the so-called stochastic nonlinearities are encountered in practical engineering applications due to random system failures or sudden environment changes. In particular, the stochastic nonlinearities satisfying certain statistical properties could contain many known nonlinearities. For instance, it could characterize the functions in some linear systems involving the state multiplicative noises and the nonlinearities in some nonlinear systems possessing random disturbances. Roughly speaking, the statistical characteristic of the stochastic nonlinearities is related to the real system states and brings great burden for the filter design. Recently, the stochastic nonlinearities have received ongoing research interests in control and signal processing communities. To name a few, by resorting to the semi-definite programming approach, a filtering strategy has been proposed in Wei et al. (2009) for time-varying systems subject to stochastic nonlinearities. By virtue of the variance-constrained method, a recursive filtering algorithm has been proposed in Hu et al. (2013) for time-varying systems involving stochastic nonlinearities.

On the other hand, considering the constrained computational capacity, the modeling uncertainties should be introduced with hope to better reflect the dynamical behaviors of the practical engineering systems. Up to now, the commonly used modeling ways for the uncertainties primarily contain the norm-bounded uncertainties and the convex polytopic uncertainties, and much research effort has been devoted to addressing the analysis issues for time-varying dynamical systems with uncertainties, see e.g. Dong et al. (2015), Xiong et al. (2010), Basin et al. (2013), Karimi (2011) for more details. Moreover, in the networked world nowadays, the network transmissions are prevalent owing to their distinguished merits (Basin et al. 2008; Gao et al. 2008; Zhang et al. 2013). It is quite common that the uncertainties occur in a random way rather than a deterministic way especially within a networked setting. Therefore, there is a great requirement to present new filtering methods to address the estimation issue for nonlinear systems with stochastic uncertainties. For instance, a novel robust EKF scheme has been developed in Xiong et al. (2010) for time-varying nonlinear systems involving stochastic uncertainties, in which an optimized upper bound regarding the FEC has been derived.

1.1.2.2 Randomly Occurring Nonlinearities and Randomly Occurring Uncertainties

As mentioned above, the existence of the nonlinearities would inevitably deteriorate the system performance. In the networked setting, the nonlinearities may occur in a probabilistic manner and are randomly changeable with specified types that are probably caused by the sudden environment changes. To date, the so-called randomly occurring nonlinearities (RONs) have attained certain research attention (Hu et al. 2014, 2012, 2013). To name a few, the stochastic synchronization issue has been handled in Wang et al. (2010) for coupled stochastic complex networks subject to RONs and mixed time-delays. Motivated by the particular engineering importance of the RONs, the envelope-constrained H_∞ filtering algorithm has been proposed in Ding et al. (2015c) within the context of time-varying stochastic systems impacted by the RONs and fading measurements, where the H_∞ performance requirement and envelope constraint have been guaranteed simultaneously by providing sufficient conditions.

On the other hand, the presence of uncertainties would worsen the whole system performance (Liu et al. 2018; Min et al. 2018; Hu et al. 2018; Kao et al. 2015). Meanwhile, it is noted that the uncertainties might occur in a random way in some cases. Consequently, it is essential to put forward some proper ways to attenuate the impact from randomly occurring uncertainties (ROUs) onto the filtering performance. So far, plenty of results regarding the filtering issues have been published for uncertain time-varying systems (Ding et al. 2018; Ma et al. 2010; Hu et al. 2016; Zha et al. 2024). For instance, a state estimation strategy has been proposed in Hu et al. (2014) for uncertain nonlinear systems, in which the ROUs have been described by the Bernoulli distributed random variables. In Ma et al. (2010), a robust filter in a recursive manner has been constructed for uncertain systems, in which a sufficient criterion has been provided to guarantee the exponential mean-square boundedness with respect to filtering error dynamics from the theoretical viewpoint. It is noteworthy to see that it is essential to compensate the adverse effects resulted from ROUs and provide more efficient filtering algorithms with improved estimation precision.

1.1.2.3 Missing Measurements

Except considering the imperfect internal features regarding the dynamical systems, owing to the fact that the network bandwidth is constrained, a great deal of data information cannot be allowed to transmit simultaneously, otherwise the network collisions would frequently occur in the communication channel. For such a case, the filter/estimator side can only receive incomplete data signals rather than all data signals collected by sensors, which will worsen the estimation precision to a certain degree (Hu et al. 2018, 2023; Mao et al. 2019). Consequently, with the purpose of improving the estimation performance, it is vitally significant to take missing measurements into consideration when constructing the state estimator, especially in the networked environment (Gao and Chen 2007; Hounkpevi and Yaz 2007; Wang et al. 2003). For example, an optimized filtering scheme has been designed in Sun et al. (2016) within the context of linear stochastic systems with correlated noises and missing measurements. Further, the state estimation method has been proposed in Hu et al. (2016) for time-varying nonlinear complex networks involving missing measurements, in which a compensation strategy has been provided to reduce the influences from the missing measurements onto the state estimation performance.

It is highlighted that, among the existing literature regarding the state estimation schemes influenced by missing measurements, the Bernoulli distributed random variables with deterministic probability are usually employed to characterize the transmission cases of the measurements. Nevertheless, owing to the situation for extensive statistical tests and the potential bias in these tests, such a need cannot be always maintained, especially in some cases that the missing measurements may be changeable with time and the missing probability cannot be obtained precisely. Accordingly, it is meaningful to investigate the filtering and state estimation issues in the presence of missing measurements with uncertain occurrence probabilities (UOPs). For example, a distributed filtering problem has been examined in Hu et al. (2022) for saturated systems subject to missing measurements with uncertain missing probability. Nonetheless, when it comes to the variance-constrained filtering problems involving missing measurements with uncertain probabilities, the corresponding investigations are scared and more efficient methods are desirable.

1.1.2.4 Fading Measurements

Because of the external environment variation and internal characteristics of components, the fading measurements might occur in practical engineering applications (Han et al. 2023). From a mathematical perspective, the fading measurements can be regarded as a more general form of the missing measurements. In particular, when the fading factor is chosen as 1, the measurements can be normally transmitted, and the measurements are missed when the fading factor is set as 0. Recently, some investigations regarding the state estimation issues involving fading measurements have been published (Liu et al. 2023; Shu et al. 2016; Song et al. 2019; Wang et al. 2017; Wen et al. 2016; Yan et al. 2017). For instance, a new recursive filtering method has been designed in Wang et al. (2017) for 2-D systems impacted by fading measurements, where the monotonicity analysis has been given between the measurement degradation and the estimation performance. What's more, the distributed filtering problems have been addressed in Song et al. (2019) and Wen et al. (2016) to handle the imperfect measurements and the expected filter gains have been designed to guarantee the minimum upper bound of FEC, in which the boundedness of the filtering error systems and the monotonicity between the trace of the minimum upper bound and the fading probability have been carefully examined. It is imperative to mention that the occurrence of fading measurements certainly deteriorates the filtering performance, and thus it is valuable to conduct the investigations concerning on the filtering problems in the context of time-varying dynamical systems affected by fading measurements.

1.1.2.5 Successive Packet Dropouts

During the process of signal transmissions, the successive packet dropouts (SPDs) would occur because of various reasons (e.g., limited bandwidth, network collision and sudden environment variation). Briefly speaking, the SPDs are described by a set of random variables obeying the Markov chains (Wu et al. 2015; You et al. 2011) or a binary switching sequence (Ding et al. 2015d; Hu et al. 2016). It is worthwhile to note that if the SPDs cannot be adequately addressed, the system performance might be deteriorated. Up to now, a large number of filtering approaches have been proposed to handle the phenomenon of the packet dropouts/loss (Hu et al. 2023; Shen et al. 2011). For instance, the H_∞ filtering issue has been tackled in Shen et al. (2011) for uncertain time-varying systems subject to SPDs, in which a set of Bernoulli distributed random variables has been employed to describe the SPDs. In Wang et al. (2015), a novel state estimation strategy in a recursive way has been proposed in the context of nonlinear stochastic systems subject to SPDs, random delays as well as correlated noises. So far, the variance-constrained state estimation issues have not been enough investigated within the context of time-varying dynamical systems subject to SPDs.

1.1.2.6 Sensor Saturation

It is noted that sensors may not always deliver data signals with unlimited amplitudes on account of physical or technological restrictions, that is to say, the sensor saturation would occur. It is pointed out that the sensor saturation could destroy the system performance if not addressed suitably (Dong et al. 2012; Zuo et al. 2017). As a result, the filtering problems have stirred considerable research attention by taking the sensor saturation into consideration and some initial results have been reported in the existing literature (Zhang and Sun 2023; Yang and Li 2009; Basit et al. 2022). The chief difficulty faced in this topic can be listed on how to establish an effective filtering algorithm via using the available information affected by the sensor saturation under certain performance requirements. For instance, the fault detection problem has been addressed in Dong et al. (2012) for an array of discrete-time Markovian jump systems in the presence of sensor saturations and randomly varying nonlinearities. It is noteworthy to find that most existing literature has been focused on time-invariant systems only and there are few investigations with respect to the filtering issue for time-varying dynamical systems with sensor saturations in the variance-constrained sense despite the fact that most of practical systems have certain time-varying characteristics.

1.1.2.7 Random Sensor Delays

As it is well known, on account of sensor link failure or environmental variations, the sensor delays might appear in the communication process, in which the traditional estimation methods cannot be used to cope with such a case (Hu et al. 2022). Moreover, considering the constrained network bandwidth, it is noted that in practical engineering, the sensor delays might occur in a random way, i.e., the so-called random sensor delays (RSDs). A great deal of effort has been devoted to investigating the filtering/estimation problems influenced by RSDs. To name a few, a distributed fusion filtering problem has been explored in Hu et al. (2022) for multi-sensor systems involving RSDs and round-robin protocol, where a set of Bernoulli random variables has been adopted to describe the RSDs. Further, a novel resilient filtering scheme has been proposed in Hu et al. (2023), in which an array of random variables with known statistical properties has been employed to describe the RSDs and the boundedness analysis associated with the proposed filtering algorithm has been discussed from a theoretical perspective.

When it comes to the complex networks, it is worthwhile to point out that the occurrence of RSDs of each node is mutually independent, namely, some nodes may encounter the sensor delays while other nodes receive the measurements normally. Recently, the analysis problems have been discussed for complex networks with RSDs, see e.g. Liang et al. (2014), Ding et al. (2012), Chen (2010) for more details. In Na and Sun (2011), a novel state estimation scheme has been presented for discrete linear stochastic dynamical systems affected by RSDs. However, it is crucial to highlight that most of existing results depend on an explicit assumption that the occurrence probabilities of sensor delays are deterministic, which might constrain the corresponding applications especially in the networked environment. Thus, it is essential to describe the RSDs with the uncertain occurrence probability and evaluate the influences from RSDs onto the filtering and estimation performance.

1.1.2.8 Signal Quantization

With pervasive networked environments, the measurement signals should be quantized before transmitted to the receiver side due primarily to the constrained abilities of the network channels, and the quantization errors should be properly handled to attenuate the effects on the filtering algorithm performance. Briefly speaking, the quantization mechanisms can be categorized as the logarithmic quantization, uniform quantization, dynamic quantization, probabilistic quantization and so on Li and Xiong (2022), Zhang and Yang (2023), Zou et al. (2017). So far, considerable effort has been dedicated to studying the filtering and control issues with signal quantization (Tan et al. 2024; Lyu and Bo 2017; Liu et al. 2017; Fu and Xie 2005). For example, the sector bound method has been used in Fu and Xie (2005) to convert the quantization errors into the sector bound uncertainties, and it is noted that this approach has been widely employed for addressing the control and filtering problems in the context of time-varying networked systems influenced by the quantization. An H_∞ filtering scheme has been proposed in Hu et al. (2023) for time-varying systems involving signal measurements and non-Gaussian noises, where the applicability of the developed filtering algorithm has been validated by a mobile robot localization system. Next, a new robust H_∞ filtering scheme has been given in Lyu and Bo (2017) in the context of time-varying nonlinear systems with the randomly varying gain perturbations and quantized measurements, in which both the estimation error variance constraint and H_∞ performance have been fully discussed and ensured by providing sufficient conditions.

On the other hand, it is noted that most of filtering methods are primarily focused on handling the deterministic quantization effects. Due to the inaccuracy of statistical experiment results, considering the uncertainty can better reflect the unreliable network media. Accordingly, it is imperative to highlight that there is a requirement to consider the randomly occurring quantization (ROQ) effects onto the filtering performance. Recently, some initial investigations regarding signal quantization have been reported. To name a few, a finite-time H_∞ asynchronous filtering problem has been explored in Zhu et al. (2020) for Markov jump systems affected by ROQ, where a novel filter has been designed such that the H_∞ performance and the finite-time boundedness can be simultaneously guaranteed. By resorting to T-S fuzzy method, an asynchronous filtering scheme has been proposed in Tao et al. (2018) for Markov jump systems involving the ROQ, where the filter gain has been derived in view of solving a set of matrix inequalities. So far, there are few results to investigate the filtering problem for time-varying dynamical systems subject to ROQ in the variance-constrained sense.

1.1.2.9 Randomly Occurring Deception Attacks

With the rapid developments of the communication networks, there have been increasing research interests on analyzing the security of the networked systems (Ge et al. 2023; Wang et al. 2024; Liu and Yang 2022; Ren et al. 2023; Shang et al. 2022). Accordingly, the malicious attacks in the cyber world become a practical yet important concern for the deployment of networked systems (Li et al. 2022; Li et al. 2023). The malicious cyber attacks commonly addressed in the literature contain, but are not limited to, replay attack, denial-of-service attack and deception attack (Amin et al. 2009; Bedi et al. 2014; Fouchal et al. 2015; Long et al. 2005; Amin et al. 2013; Pang and Liu 2012). Among them, the replay attack indicates that a set of data signals would be repeatedly sent to the transmission channel to damage the system performance. The denial-of-service attack is implemented by preventing the signal transmission to deteriorate the system performance. The main purpose of deception attacks is to viciously inject a bias into the transmitted information by attackers. Recently, considerable research attention has been focused on the investigations with cyber attacks. For example, the networked predictive control scheme in a recursive manner has been explored in Pang and Liu (2012) to compensate the negative influences caused by the deception attacks.

Due to the fact that the occurrence of the cyber attacks relies on the network conditions, the deception attacks may occur in a probabilistic manner, i.e., the randomly occurring deception attacks (RODAs). Nevertheless, it is crucial to highlight that the RODAs have been vastly overlooked in spite of their practical significance in networked media. Consequently, it is practical yet significant to address the filtering problems influenced by RODAs and enhance the security concerning on the networked systems.

1.1.2.10 Event-Triggered Mechanism

In order to actively avoid data congestion and alleviate communication burden, the ETM, as an effective communication scheduling strategy, has received increasing research attention around the world (Basit et al. 2022; Cheng et al. 2022; Li and Ye 2023; Zou et al. 2017; Hu et al. 2017; Ma et al. 2019; Chen and Ho 2023). Unlike the time-triggered mechanism, whether the data transmission can be successfully is determined by a pre-defined triggered condition, that is to say, only the triggered condition is satisfied and then the data information can be delivered to the receiver side, thus reducing the transmission frequency and decreasing communication consumption (Yi and Xu 2024; Hu et al. 2018; Muehlebach and Trimpe 2018; Yao and Wei 2023). Until now, much research attention has been endeavored on the filtering and estimation problems with ETM (Gu and Yang 2017; Wang et al. 2017; Hou et al. 2018). For example, an ETM has been used in Zhang et al. (2017) for delayed dynamical networks, in which the event generator function relies on the measurement and two different triggered thresholds have been established. The event generator function in Hu et al. (2015) depends on a single threshold, which can be viewed a special case in Zhang et al. (2017) when another threshold is equal to zero. Furthermore, the event generator function has been constructed in Wang et al. (2016) for each node and the event-based state estimation (EBSE) strategy has been presented for complex networks involving mixed delays and nonlinear disturbances, in which the estimator gain has been derived for all nodes but possessing more operation burdens.

According to different triggered conditions, the event-triggered mechanisms (ETMs) can be generalized as the static ETM, dynamic ETM, memory ETM, and so on, see Zha et al. (2024), Sun et al. (2022), Wang et al. (2024), Song et al. (2023), Chen et al. (2024) for more details. To be more specific, in the static ETM, the triggered threshold is static, that is to say, the triggered threshold cannot be changed according to the practical case. For the dynamic ETM, the triggered threshold can be dynamically adjusted according to the practical engineering case. For instance, by introducing a dynamic auxiliary variable, a dynamic ETM has been constructed in Wang et al. (2024) to arrange the data transmission, where a non-fragile quadratic filtering problem has been examined for non-Gaussian systems. Besides, noting the memory characteristics of some practical systems, the memory ETM is adopted to arrange the signal transmission, which can better reflect the reality. In Sun et al. (2022), a new filtering scheme has been developed for nonlinear systems under memory ETM, in which the filtering error has been proved to be input-to-state stable in the mean-square sense. Nevertheless, when it comes to the event-triggered variance-constrained filtering problems, there are few investigations for time-varying dynamical systems.

1.1.2.11 Random Access Protocol

In order to further improve data transmission efficiency and reduce signal transmission frequency, some communication protocols have been proposed to regulate signal transmissions (Zhao et al. 2023; Li et al. 2022, 2023; Zou et al. 2017; Zhou and Sun 2024; Meng et al. 2022; Wang et al. 2023). Actually, the commonly employed communication protocols include the round-robin protocol, weighted try-once-discard protocol and RAP. Specifically, the round-robin protocol can be viewed as a static scheduling strategy, which arranges the node transmission in a fixed circular sequence. The weighted try-once-discard protocol is a kind of dynamic scheduling scheme, which is based on the maximum-error-first rule to select the transmission node. Different from the above two scheduling strategies, the node choosing way of the RAP is according to a stochastic process, which is easy to implement and realize. Recently, the RAP has been widely applied in various fields, such as satellite networks and cellular wireless networks. A great number of analysis and synthesis issues for networked systems under RAP have stirred increasing research attention (Zou et al. 2018; Liu et al. 2022; Wan et al. 2022; Ju et al. 2022).

Briefly speaking, there are two ways to characterize the scheduling principle of the RAP, i.e., independent and identically distributed (i.i.d.) random variables and Markov chain. For instance, a novel recursive filtering scheme has been presented in Zou et al. (2018) for time-varying systems under RAP, where a set of i.i.d. random variables has been utilized to describe the RAP and the boundedness of the upper bound of the FEC has been examined. In Zou et al. (2021), a moving horizon estimation scheme has been proposed for nonlinear systems affected by the RAP, in which the Markov chain has been adopted to model the scheduling behavior of the RAP and the ultimate boundedness of the estimation errors has been testified in the mean-square sense. Nonetheless, it can be observed that the variance-constrained state estimation problems for time-varying dynamical systems influenced by the RAP have not received sufficient research attention.

1.1.3 Variance-Constrained Filtering and Estimation

As discussed in Sect. 1.1.1, it is a significant yet challenging research topic to study the variance-constrained filtering problems owing to the nonlinearities and complexities of the practical dynamical systems.

1.1.3.1 Recursive Filter Design

The variance-constrained filtering issues have obtained the sustained research attention and considerable effort has been dedicated to dealing with the variance-constrained filtering problems within the context of time-varying dynamical systems, see Shen and Sun (2023), Tan et al. (2022), Gao et al. (2022, 2023) for more details. Taking the intermittent network collisions, random failures and sudden environment exchanges into account, the stochastic nonlinearities and RONs inevitably occur in the system. Much effort has been endeavored to develop the recursive filtering algorithm in the context of time-varying nonlinear systems. To be specific, for time-varying nonlinear systems, the common handling way is to linearize the nonlinear function by means of some effective approaches and then design the filtering algorithm according to the recursive principle, which is suitable for online computations. To name a few, by resorting to the Taylor series expansion approach, the robust extended Kalman filter has been constructed in Xiong et al. (2011) for the nonlinear systems with stochastic disturbances. Compared with the traditional extended Kalman filter, it can be observed that the new filter possesses better filtering performance. With the aid of the known statistical properties of the stochastic nonlinearity, the filtering problem has been addressed in Hu et al. (2013) for systems with stochastic nonlinearities in the variance-constrained sense.

On the other hand, due to some physical restrictions such as the finite word length, resistance errors, roundoff errors in numerical computation, finite resolution of the instrumentation and imprecision in analogue-digital conversion (Feng et al. 2011), it is of vital importance to consider the filter gain perturbations when constructing the filter. It is noted that the non-fragile or resilient filtering problems have attracted increasing attention in recent years (Ge et al. 2023; Ding et al. 2022; Yang and Wang

2001; Mahmoud 2004; Chang and Yang 2014; Zheng et al. 2023). To name a few, a non-fragile state estimation algorithm has been proposed in Wang et al. (2019) for 2-D systems under redundant channels, where the statistical properties of the filter gain variation have been exhibited including the mean and the bound of the covariance. Further, the issue of resilient recursive filtering for complex networks has been explored in Ding et al. (2022), where the gain variations have been characterized using a set of random variables satisfying a uniform distribution. So far, the resilient filtering issues have been vastly ignored for time-varying dynamical systems influenced by incomplete information on account of the mathematical complexities and analysis difficulties.

1.1.3.2 Fault Estimator Design

It has been well recognized that the faults commonly occur in many practical engineering systems and much effort has been focused on addressing the problems of fault detection, fault estimation and fault-torelat control (Dong et al. 2014; Li et al. 2023; Yan and Edwards 2007; Wang et al. 2014; Ma and Li 2023). It is imperative to acknowledge that the corresponding results concerning on the fault estimation problems within the context of time-varying nonlinear systems are scarce owing to the mathematical complexities compared with time-invariant nonlinear systems. To name a few, the fault estimation problem has been examined in Dong et al. (2014) by the solutions of the coupled backward recursive Riccati-like difference equations. It should be pointed out that, until now, little research effort has been made to cope with the co-design issue of the state estimator and fault estimator for time-varying dynamical systems. Besides, accompanying with the ever-increasing development of the networked control systems, the fault signals as a kind of network-induced nature have attracted plentiful attention. It has been revealed that the appearance of the faults could be inconsecutive or even random in networked settings because of unpredictable parameter exchanges or structural changes through the networks. For instance, an effective H_∞ fault estimation scheme has been proposed in Dong et al. (2014) with the randomly occurring faults (ROFs), in which a backward recursive Riccati-like difference equation has been solved. Up to now, as far as we know, the joint state and fault estimation issue has not been thoroughly handled yet within the context of time-varying dynamical systems with incomplete information. Moreover, it is worth emphasizing that most of existing approaches cannot provide the performance analysis concerning on the estimation strategy for designed time-varying dynamical systems with certain complexities.

1.1.3.3 State Estimator Design

In the real-world conditions, the states of dynamical systems are not often available because of physical limitations or technological constraints (Basin and Calderon-Alvarez 2010; Haughton and Heydt 2013; Liu and Cao 2010; Liang et al. 2010;

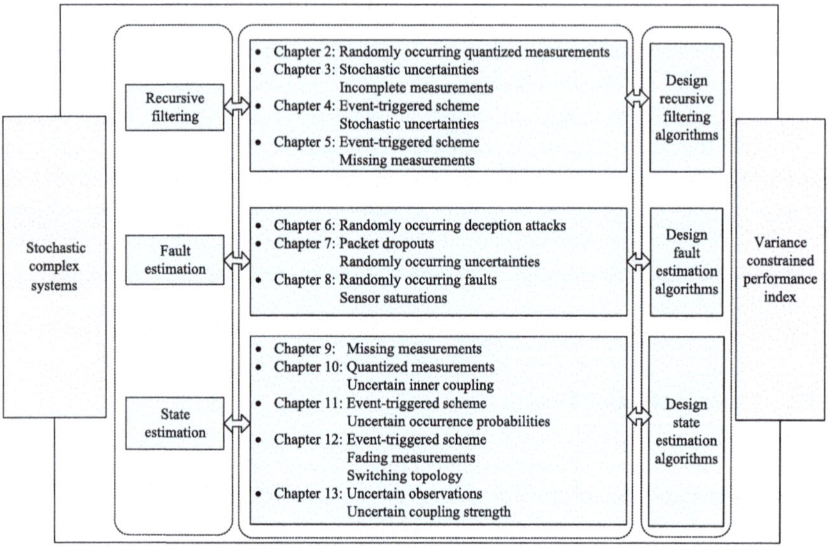

Fig. 1.1 Organization structure of the book

Li et al. 2022). Accordingly, it is essential to conduct the investigation concerning on the state estimation. To better analyze the dynamical behaviors of the complex networks and reach satisfactory engineering requirements, it is important to provide the state estimations of the network nodes based on the available measurement signals. As a result, the state estimation issues under different performance indices have attracted considerable research interest (Liang et al. 2024; Park et al. 2018; Shen et al. 2011). For example, for the sake of attenuating the effects induced by energy-bounded disturbances, an H_∞ state estimation algorithm has been presented in Shen et al. (2011) in light of available measurement information. Under the sense of variance constraint, a novel state estimation strategy has been put forward in Hu et al. (2016) for time-varying networked complex networks in the presence of stochastic noises, in which the influences from the missing measurements onto the estimation performance have been comprehensively evaluated. In what follows, on the basis of the above result given in Hu et al. (2016), the variance-constrained state estimator has been constructed in Li et al. (2018) for time-varying nonlinear complex networks subject to stochastic noises. Along with the rapid developments of networked systems, the incomplete information induced by the network is usually occurred in the transmission process and it is of significance to establish novel estimation schemes for time-varying dynamical systems under incomplete information.

1.2 Outline

In this book, we aim to investigate the variance-constrained filtering and estimation issues for several classes of time-varying dynamical systems affected by incomplete information. The organization structure of this book can be found in Fig. 1.1 and the specific outline of this book is presented as follows:

- In this chapter, the research background, motivations and research problems are firstly introduced, which primarily contain the time-varying dynamical systems, incomplete information, variance-constrained filtering and estimation, then the outline of the book is exhibited.
- Chapter 2 studies the robust optimal filtering issue in the context of time-varying networked stochastic nonlinear systems influenced by ROQ by virtue of the variance-constrained method. The phenomenon of quantized measurements randomly occurs and the occurrence probability is described by an array of Bernoulli random variables with known probability, in which the logarithmic quantizer is employed to schedule the measurements. Attention is focused on the construction of a recursive filtering method such that an upper bound of the FEC is solved and an expected filter gain is established under the premise of minimizing the proposed upper bound. Moreover, by resorting to the matrix theory, a sufficient condition is provided to guarantee the exponential boundedness regarding the filtering error in the mean-square sense.
- Chapter 3 investigates the resilient recursive filtering issue for time-varying networked nonlinear systems involving stochastic uncertainties and incomplete measurements. The incomplete measurements simultaneously include the ROQ and missing measurements. Meanwhile, the multiplicative noise with known mean and variance is used to describe the stochastic perturbations of the filter gain. On the basis of the available measurement information, a novel filtering method is constructed for time-varying nonlinear systems such that, for both stochastic uncertainties and incomplete measurements, there is an upper bound of the FEC that can be optimized by appropriately constructing a filter gain in the minimum variance sense.
- In Chap. 4, the recursive filtering issues are addressed for two classes of time-varying dynamical systems involving stochastic uncertainties. Firstly, the stochastic uncertainties are proposed to characterize the modeling errors. In order to improve the communication efficiency, the ETM is used to arrange the node transmissions. And the filter gain perturbation is discussed and described through the norm-bounded uncertainties. In light of the available information, a novel event-based filtering scheme is presented for time-varying dynamical systems in the presence of stochastic uncertainties and filter gain perturbations. Secondly, the designed filtering algorithm is extended to address the design problem of the event-triggered resilient filtering for time-varying systems involving stochastic uncertainties and SPDs. Here, the Bernoulli distributed random variables are utilized to describe the behavior of the SPDs. Based on the available measurements, a

resilient event-triggered filtering algorithm is developed such that an upper bound of the FEC can be minimized.

- In Chap. 5, the event-triggered filtering issues are studied for two classes of time-varying nonlinear systems with missing measurements. Firstly, the descriptions of the RONs and the missing measurements are described by the Bernoulli distributed random variables with UOPs. Particularly, for the sake of reducing the network resource consumption, the ETM is utilized to schedule the information transmission. Moreover, the filter gain perturbations are described by utilizing the norm bounded uncertainties. A robust event-triggered resilient filter is designed against the RONs and missing measurements. Secondly, the conducted filtering method is extended to handle the event-triggered filtering problem for time-varying dynamical systems impacted by multiple missing measurements with uncertain missing probabilities. The multiple missing measurements are described by a sequence of mutually independent random variables satisfying Bernoulli distribution and the missing probabilities could be uncertain. What's more, the ETM is utilized to decrease the network communication burden. A recursive filter is constructed and an upper bound on the FEC is derived in view of the solution to a set of matrix equations. By employing the similar techniques, a desired filter gain is obtained under the variance-constrained sense.
- Chapter 6 deals with the fault estimation problem for time-varying nonlinear stochastic systems involving RONs and RODAs. In particular, both the RONs and RODAs are characterized by the mutually independent random variables obeying the Bernoulli distribution. By virtue of the augmentation method, the co-design problem of the estimator and fault estimator is converted into the recursive filter design problem. Based on the available measurements, a novel estimation scheme is proposed against RONs and RODAs. After that, the estimator gain is designed through solving two sets of Riccati-like difference equations and related joint estimation method is proposed accordingly.
- Chapter 7 addresses the fault estimation issue for time-varying dynamical systems influenced by the ROUs, stochastic nonlinearities and packet dropouts. The stochastic nonlinearities are considered which include many known nonlinearities as special cases. Both the ROUs and packet dropouts are depicted by the mutually independent random variables with given occurrence probabilities. The purpose of this chapter is to propose a fault estimation algorithm under the premise of ensuring the minimization of the upper bound concerning on the estimation error covariance (EEC). Additionally, the monotonicity of the developed estimation scheme is discussed and revealed from the perspective of mathematical analysis. Moreover, it is observed that the designed recursive filtering method is suitable for online implementations.
- Chapter 8 copes with the fault estimation issue for time-varying uncertain nonlinear stochastic systems subject to ROFs and sensor saturations. Specifically, a set of Bernoulli distributed random variables is adopted to depict the ROFs. The measurements might undergo the saturation, which can be described by the combination of signum function and minimum function by comparing the saturation

threshold and real value. The purpose of this chapter is to construct a locally optimal estimation approach such that, for both the ROFs and sensor saturations, the EEC has an upper bound and the estimator gain is constructed via minimizing the trace of the presented upper bound at each time instant. Moreover, a sufficient criterion is provided to prove the boundedness of the presented estimation algorithm.

- In Chap. 9, the recursive state estimation issue is studied for time-varying complex networks with missing measurements. The random variables satisfying known probabilistic distributions are adopted to describe the missing measurements. The nonlinearities are addressed by adopting the Taylor series expansion approach and the high-order terms with regard to the linearization errors are estimated. Attention is focused on the establishment of the state estimation scheme against missing measurements and the random disturbances such that the minimum error variance performance requirement can be guaranteed. By solving the Riccati-like difference equations, the expected estimator gain matrix is derived and it is revealed that the presented recursive state estimation scheme is applicable for online implementations.
- In Chap. 10, a recursive state estimation issue is addressed for time-varying stochastic complex networks in the presence of uncertain inner coupling and signal quantization. It is noted that the coupling strengths are always varying that belong to given intervals due to the intricate network structure and environment exchanges. Additionally, the data signals are needed to be quantized to accommodate the communication requirements. The purpose of this chapter is to design a recursive state estimation scheme to guarantee the existence of a locally minimized upper bound of the EEC. In addition, the boundedness analysis of the estimation error dynamics is provided and a sufficient criterion is built to guarantee exponential boundedness of the estimation error dynamics in the mean-square sense.
- In Chap. 11, the state estimation issue is discussed for time-varying delayed complex networks involving ETM, stochastic coupling and missing measurements with UOPs. Two sequences of mutually independent Bernoulli random variables are adopted to characterize the stochastic coupling and missing measurements, in which the uncertainties of the occurrence probabilities are discussed and modeled. Besides, a robust state estimation approach is presented for time-varying complex networks such that sufficient conditions are given to guarantee the presence of an upper bound regarding the EEC. Moreover, the monotonicity evaluation between the proposed upper bound and the deterministic missing probability is further conducted by means of the matrix theory.
- In Chap. 12, the EBSE problems have been tackled for two classes of time-varying complex networks with fading measurements and uncertain switching topology. Firstly, for the sake of enhancing the communication efficiency, the ETM is given to regulate the data transmissions. Moreover, the RONs and data packet dropouts (DPDs) are described by two sets of Bernoulli distributed random variables. In addition, the gain perturbations are considered. The main purpose of this chapter is to establish a novel state estimation scheme for considered time-varying complex networks such that an optimized upper bound of the EEC can be derived via

selecting a desirable estimator gain. Secondly, taking the fading measurements into account, a set of random variables obeying given probability distribution is used to model the fading measurements. The focus is on the design of the state estimator under the influences of fading measurements, stochastic coupling strength (SCS) as well as ETM. Additionally, the monotonicity relationship between the trace of the upper bound concerning on the EEC and the fading probability is discussed and shown from a theoretical perspective.

- In Chap. 13, the variance-constrained state estimation problems have been tackled for two classes of time-varying nonlinear complex networks with stochastic coupling. Firstly, the stochastic inner coupling is modeled by the multiplicative noise. The missing measurements are depicted by a sequence of mutually independent Bernoulli distributed random variables with UOPs. With the aim of avoiding the data congestions, the RAP is adopted to arrange the node transmission. Through solving the recursive matrix equations, an upper bound of the EEC is proposed and the estimator gain is constructed by minimizing the presented upper bound. What's more, the monotonicity analysis in relation to the deterministic occurrence probability of missing measurements and the algorithm performance are adequately discussed. Secondly, the variance-constrained optimized state estimation issue is addressed for time-varying nonlinear complex networks subject to RSDs, random coupling strength (RCS) and ETM. A set of Bernoulli distributed random variables is given to model the RSDs. For the sake of avoiding potential data collisions, the ETM is used to arrange the data communication. After that, a new robust state estimation strategy is presented by fully taking the RSDs and RCS into consideration. Furthermore, a sufficient condition is provided to guarantee the boundedness of the estimation error systems.
- Chapter 14 provides the conclusions and discussions of this book and exhibits some future research topics regarding the current work.

References

Amin S, Cárdenas AA, Sastry SS (2009) Safe and secure networked control systems under denial-of-service attacks. Hybrid Syst Comput Control 5469:31–45

Amin S, Litrico X, Sastry S, Bayen AM (2013) Cyber security of water SCADA systems-Part I: analysis and experimentation of stealthy deception attacks. IEEE Trans Control Syst Technol 21(5):1963–1970

Basin MV, Calderon-Alvarez D (2010) Optimal filtering over linear observations with unknown parameters. J Frankl Inst-Eng Appl Math 347(6):988–1000

Basin MV, Alcorta-Garcia MA, Alanis-Duran A (2008) Optimal filtering for linear systems with state and multiple observation delays. Int J Syst Sci 39(5):547–555

Basin MV, Loukianov AG, Hernandez-Gonzalez M (2013) Joint state and parameter estimation for uncertain stochastic nonlinear polynomial systems. Int J Syst Sci 44(7):1200–1208

Basin MV, Yu P, Shtessel YB (2018) Hypersonic missile adaptive sliding mode control using finite- and fixed-time observers. IEEE Trans Ind Electron 65(1):930–941

Basit A, Tufail M, Rehan M (2022) Event-triggered distributed state estimation under unknown parameters and sensor saturations over wireless sensor networks. IEEE Trans Circuits Syst II-Express Briefs 69(3):1772–1776
Basit A, Tufail M, Rehan M (2022) Event-triggered distributed state estimation under unknown parameters and sensor saturations over wireless sensor networks. IEEE Trans Circuits Syst II-Express Briefs 69(3):1772–1776
Bedi H, Shiva S, Roy S (2014) A game inspired defense mechanism against distributed denial of service attacks. Secur Commun Netw 7(12):2389–2404

Boccaletti S, Latora V, Moreno Y, Chavez M, Hwang DU (2006) Complex networks: structure and dynamics. Phys Rep 424(4–5):175–308

Caballero-Águila R, Hermoso-Carazo A, Linares-Pérez J (2017) Distributed fusion filters from uncertain measured outputs in sensor networks with random packet losses. Inf Fusion 34:70–79

Chang X, Yang G (2014) Non-fragile H_∞ filter design for discrete-time fuzzy systems with multiplicative gain variations. Inf Sci 266:171–185

Chen M (2010) Synchronization in complex dynamical networks with random sensor delay. IEEE Trans Circuits Syst II-Express Briefs 57(1):46–50

Chen S, Ho DWC (2023) Edge-based sender-receiver event-triggered schemes for distributed filtering. IEEE Trans Circuits Syst I-Regul Pap 70(5):2143–2155

Chen Y, Meng X, Wang Z, Dong H (2023) Event-triggered recursive state estimation for stochastic complex dynamical networks under hybrid attacks. IEEE Trans Neural Netw Learn Syst 34(3):1465–1477

Chen G, Zhou Q, Li H, Yao D, Ahn CK (2024) Event-triggered state estimation and control for networked nonlinear systems under dynamic sparse attacks. IEEE Trans Netw Sci Eng 11(2):1947–1958

Cheng J, Park JH, Yan H, Wu Z (2022) An event-triggered round-robin protocol to dynamic output feedback control for nonhomogeneous Markov switching systems. Automatica 145:110525

Ding D, Wang Z, Shen B, Shu H (2012) H_∞ state estimation for discrete-time complex networks with randomly occurring sensor saturations and randomly varying sensor delays. IEEE Trans Neural Netw Learn Syst 23(5):725–736

Ding D, Wang Z, Lam J, Shen B (2015) Finite-horizon H_∞ control for discrete time-varying systems with randomly occurring nonlinearities and fading measurements. IEEE Trans Autom Control 60(9):2488–2493

Ding D, Wang Z, Alsaadi FE, Shen B (2015) Receding horizon filtering for a class of discrete time-varying nonlinear systems with multiple missing measurements. Int J Gen Syst 44(2):198–211

Ding D, Wang Z, Shen B, Dong H (2015) Envelope-constrained H_∞ filtering with fading measurements and randomly occurring nonlinearities: the finite horizon case. Automatica 55:37–45

Ding D, Wang Z, Shen B, Dong H (2015) H_∞ state estimation with fading measurements, randomly varying nonlinearities and probabilistic distributed delays. Int J Robust Nonlinear Control 25(13):2180–2195

Ding D, Han QL, Xiang Y, Ge X, Zhang XM (2018) A survey on security control and attack detection for industrial cyber-physical systems. Neurocomputing 275:1674–1683

Ding D, Liu H, Dong H, Liu H (2022) Resilient filtering of nonlinear complex dynamical networks under randomly occurring faults and hybrid cyber-attacks. IEEE Trans Netw Sci Eng 9(4):2341–2352

Ding D, Liu H, Dong H, Liu H (2022) Resilient filtering of nonlinear complex dynamical networks under randomly occurring faults and hybrid cyber-attacks. Trans Netw Sci Eng 9(4):2341–2352

Dong H, Wang Z, Gao H (2012) Fault detection for markovian jump systems with sensor saturations and randomly varying nonlinearities. IEEE Trans Circuits Syst I-Regul Pap 59(10):2354–2362

Dong H, Wang Z, Ding SX, Gao H (2014) Finite-horizon estimation of randomly occurring faults for a class of nonlinear time-varying systems. Automatica 50(12):3182–3189
Dong H, Wang Z, Ding SX, Gao H (2014) Finite-horizon estimation of randomly occurring faults for a class of nonlinear time-varying systems. Automatica 50(12):3182–3189

Dong H, Wang Z, Ding SX, Gao H (2015) Finite-horizon reliable control with randomly occurring uncertainties and nonlinearities subject to output quantization. Automatica 52:355–362

Duan P, Duan Z, Lv Y, Chen G (2021) Distributed finite-horizon extended Kalman filtering for uncertain nonlinear systems. IEEE Trans Cybern 51(2):512–520

Feng J, Wang Z, Zeng M (2011) Optimal robust non-fragile Kalman-type recursive filtering with finite-step autocorrelated noises and multiple packet dropouts. Aerosp Sci Technol 15(6):486–494

Fouchal S, Mansouri D, Mokdad L, Iouallalen M (2015) Recursive-clustering-based approach for denial of service (DoS) attacks in wireless sensors networks. Int J Commun Syst 28(2):309–324

Fu M, Xie L (2005) The sector bound approach to quantized feedback control. IEEE Trans Autom Control 50(11):1698–1711

Gao H, Chen T (2007) H_∞ estimation for uncertain systems with limited communication capacity. IEEE Trans Autom Control 52(11):2070–2084

Gao H, Chen T, Lam J (2008) A new delay system approach to network-based control. Automatica 44(1):39–52

Gao H, Dong H, Wang Z, Han F (2022) Recursive minimum-variance filter design for state-saturated complex networks with uncertain coupling strengths subject to deception attacks. IEEE Trans Cybern 52(10):11121–11132

Gao H, Li Y, Yu L, Yu H (2023) Collaborative-prediction-based recursive filtering for nonlinear systems with sensor saturation under duty cycle scheduling. Syst Sci Control Eng 11(1):2247007

Ge X, Han QL, Wu Q, Zhang XM (2023) Resilient and safe platooning control of connected automated vehicles against intermittent denial-of-service attacks. IEEE-CAA J Automatica Sin 10(5):1234–1251

Gu Y, Yang G (2017) Event-triggered fault detection for discrete-time Lipschitz nonlinear networked systems in finite-frequency domain. Neurocomputing 260:245–256

Hafez I, Wadi A, Abdel-Hafez MF, Hussein AA (2023) Variational Bayesian-based maximum correntropy cubature Kalman filter method for state-of-charge estimation of Li-ion battery cells. IEEE Trans Veh Technol 72(3):3090–3104

Han F, Liu J, Li J, Song J, Wang M, Zhang Y (2023) Consensus control for multi-rate multi-agent systems with fading measurements: the dynamic event-triggered case. Syst Sci Control Eng 11(1):2158959

Haughton DA, Heydt GT (2013) A linear state estimation formulation for smart distribution systems. IEEE Trans Power Syst 28(2):1187–1195

Hou N, Dong H, Wang Z, Ren W, Alsaadi FE (2016) Non-fragile state estimation for discrete Markovian jumping neural networks. Neurocomputing 179:238–245

Hou N, Dong H, Zhang W, Liu Y, Alsaadi FE (2018) Event-triggered state estimation for time-delayed complex networks with gain variations based on partial nodes. Int J Gen Syst 47(5):408–421

Hounkpevi FO, Yaz E (2007) Robust minimum variance linear state estimators for multiple sensors with different failure rates. Automatica 43(7):1274–1280

Hu J, Wang Z, Gao H, Stergioulas LK (2012) Probability-guaranteed H_∞ finite-horizon filtering for a class of nonlinear time-varying systems with sensor saturations. Syst Control Lett 61(4):477–484

Hu J, Wang Z, Gao H (2013) Recursive filtering with random parameter matrices, multiple fading measurements and correlated noises. Automatica 49(11):3440–3448

Hu J, Wang Z, Shen B, Gao H (2013) Gain-constrained recursive filtering with stochastic nonlinearities and probabilistic sensor delays. IEEE Trans Signal Process 61(5):1230–1238

Hu J, Chen D, Du J (2014) State estimation for a class of discrete nonlinear systems with randomly occurring uncertainties and distributed sensor delays. Int J Gen Syst 43(3–4):387–401

Hu J, Wang Z, Liang J, Dong H (2015) Event-triggered distributed state estimation with randomly occurring uncertainties and nonlinearities over sensor networks: a delay-fractioning approach. J Frankl Inst-Eng Appl Math 352(9):3750–3763

Hu J, Wang Z, Chen D, Alsaadi FE (2016) Estimation, filtering and fusion for networked systems with network-induced phenomena: new progress and prospects. Inf Fusion 31:65–75

Hu J, Wang Z, Liu S, Gao H (2016) A variance-constrained approach to recursive state estimation for time-varying complex networks with missing measurements. Automatica 64:155–162

Hu J, Wang Z, Alsaadi FE, Hayat T (2017) Event-based filtering for time-varying nonlinear systems subject to multiple missing measurements with uncertain missing probabilities. Inf Fusion 38:74–83

Hu L, Wang Z, Han QL, Liu X (2018) Event-based input and state estimation for linear discrete time-varying systems. Int J Control 91(1):101–113

Hu L, Wang Z, Han Q-L, Liu X (2018) State estimation under false data injection attacks: security analysis and system protection. Automatica 87:176–183

Hu J, Wang Z, Gao H (2018) Joint state and fault estimation for uncertain time-varying nonlinear systems with randomly occurring faults and sensor saturations. Automatica 97:150–160

Hu J, Wang Z, Gao H (2018) Joint state and fault estimation for uncertain time-varying nonlinear systems with randomly occurring faults and sensor saturations. Automatica 97:150–160

Hu Z, Hu J, Tan H, Huang J, Cao Z (2022) Distributed resilient fusion filtering for nonlinear systems with random sensor delay under round-robin protocol. Int J Syst Sci 53(13):2786–2799

Hu J, Hu Z, Dong H, Liu H (2023) Distributed resilient fusion filtering for nonlinear systems with random sensor delays: optimized algorithm design and boundedness analysis. J Syst Sci Complex 36(4):1423–1442

Hu J, Hu Z, Caballero-Águila R, Chen C, Fan S, Yi X (2023) Distributed resilient fusion filtering for nonlinear systems with multiple missing measurements via dynamic event-triggered mechanism. Inf Sci 637:118950

Hu J, Li J, Kao Y, Chen D (2022) Optimal distributed filtering for nonlinear saturated systems with random access protocol and missing measurements: The uncertain probabilities case. Appl Math Comput 418:126844

Hu J, Wang C, Caballero-Águila R, Liu H (2023) Distributed optimal fusion filtering for singular systems with random transmission delays and packet dropout compensations. Commun Nonlinear Sci Numer Simul 119:107093

Ju Y, Liu D, Ding D, Wei G (2022) Distributed cubature Kalman filtering for nonlinear systems with stochastic communication protocol. Asian J Control 24(6):3566–3579

Ju Y, Liu D, Ding D, Wei G (2022) Distributed cubature Kalman filtering for nonlinear systems with stochastic communication protocol. Asian J Control 24(6):3566–3579

Kao Y, Xie J, Wang C, Karimi HR (2015) A sliding mode approach to H_∞ non-fragile observer-based control design for uncertain Markovian neutral-type stochastic systems. Automatica 52:218–226

Karimi HR (2011) Robust delay-dependent H_∞ control of uncertain time-delay systems with mixed neutral, discrete, and distributed time-delays and Markovian switching parameters. IEEE Trans Circuits Syst I-Regul Pap 58(8):1910–1923

Li Z, Xiong J (2022) Event-triggered fuzzy filtering for nonlinear networked systems with dynamic quantization and stochastic cyber attacks. ISA Trans 121:53–62

Li X, Ye D (2023) Dynamic event-triggered distributed filtering design for interval type-2 fuzzy systems over sensor networks under deception attacks. Int J Syst Sci 54(15):2875–2890

Li W, Jia Y, Du J (2018) Variance-constrained state estimation for nonlinearly coupled complex networks. IEEE Trans Cybern 48(2):818–824

Li M, Liang J, Wang F (2022) Robust set-membership filtering for two-dimensional systems with sensor saturation under the round-robin protocol. Int J Syst Sci 53(13):2773–2785

Li J, Dong H, Shen Y, Hou N (2022) Encoding-decoding strategy based resilient state estimation for bias-corrupted stochastic nonlinear systems. ISA Trans 127:80–87

Li X, Wei G, Ding D, Liu S (2022) Recursive filtering for time-varying discrete sequential systems subject to deception attacks: weighted try-once-discard protocol. IEEE Trans Syst Man Cybern-Syst 52(6):3704–3713

Li M, Liang J, Qiu J (2023) Set-membership filtering for 2-D systems under uniform quantization and weighted try-once-discard protocol. IEEE Trans Circuits Syst II-Express Briefs 70(9):3474–3478

Li Q, Zhi Y, Tan H, Sheng W (2023) Protocol-based zonotopic state and fault estimation for communication-constrained industrial cyber-physical systems. Inf Sci 634:730–743

Liang Y, Chen T, Pan Q (2010) Optimal linear state estimator with multiple packet dropouts. IEEE Trans Autom Control 55(6):1428–1433

Liang J, Wang Z, Liu Y, Liu X (2014) State estimation for two-dimensional complex networks with randomly occurring nonlinearities and randomly varying sensor delays. Int J Robust Nonlinear Control 24(1):18–38

Liang L, Cheng J, Yan H, Cao J, Park JH (2024) Protocol-based state estimation for complex networks with nonhomogeneous sojourn probabilities. IEEE Trans Syst Man Cybern-Syst. https://doi.org/10.1109/TSMC.2024.3378693

Li X, Song J, Hou N, Dai D, Yang F (2023) Finite-horizon distributed set-membership filtering with dynamical bias and DoS attacks under binary encoding schemes. Inf Sci 641:119084

Liu X, Cao J (2010) Robust state estimation for neural networks with discontinuous activations. IEEE Trans Syst Man Cybern B Cybern 40(6):1425–1437

Liu Y, Yang G (2022) Resilient event-triggered distributed state estimation for nonlinear systems against DoS attacks. IEEE Trans Cybern 52(9):9076–9089

Liu S, Wei G, Song Y, Liu Y (2016) Extended Kalman filtering for stochastic nonlinear systems with randomly occurring cyber attacks. Neurocomputing 207:708–716

Liu S, Wei G, Song Y, Ding D (2017) Set-membership state estimation subject to uniform quantization effects and communication constraints. J Frankl Inst-Eng Appl Math 354(15):7012–7027

Liu G, Xu S, Wei Y, Qi Z, Zhang Z (2018) New insight into reachable set estimation for uncertain singular time-delay systems. Appl Math Comput 320:769–780

Liu S, Wang Z, Chen Y, Wei G (2020) Protocol-based unscented Kalman filtering in the presence of stochastic uncertainties. IEEE Trans Autom Control 65(3):1303–1309

Liu S, Zhao X, Tian E, Wei G (2022) Distributed recursive filtering under random access protocols: A multirate strategy. Int J Robust Nonlinear Control 32(12):7132–7148

Liu J, Gong E, Zha L, Xie X, Tian E (2023) Outlier-resistant recursive security filtering for multirate networked systems under fading measurements and round-robin protocol. IEEE Trans Control Netw Syst 10(4):1962–1974

Li J, Wei G, Li W (2022) Unscented Tobit Kalman filtering for switched nonlinear systems with censored measurement. Appl Math Comput 431:127327

Li W, Yang F (2023) Information fusion over network dynamics with unknown correlations: an overview. Int J Netw Dyn Intell 2(2):100003

Long M, Wu C, Hung JY (2005) Denial of service attacks on network-based control systems: impact and mitigation. IEEE Trans Ind Inf 1(2):85–96

Lyu M, Bo Y (2017) Variance-constrained resilient H_∞ filtering for time-varying nonlinear networked systems subject to quantization effects. Neurocomputing 267:283–294

Ma S, Li Y (2023) Adaptive fuzzy fault-tolerant control for active seat suspension systems with full-state constraints. Syst Sci Control Eng 11(1):2153391

Ma L, Wang Z, Hu J, Bo Y, Guo Z (2010) Robust variance-constrained filtering for a class of nonlinear stochastic systems with missing measurements. Signal Process 90(6):2060–2071

Ma H, Li H, Liang H, Dong G (2019) Adaptive fuzzy event-triggered control for stochastic nonlinear systems with full state constraints and actuator faults. IEEE Trans Fuzzy Syst 27(11):2242–2254

Mahmoud MS (2004) Resilient linear filtering for uncertain systems. Automatica 40(10):1797–1802

Mao J, Ding D, Wei G, Liu H (2019) Networked recursive filtering for time-delayed nonlinear stochastic systems with uniform quantisation under round-robin protocol. Int J Syst Sci 50:871–884

Meng X, Chen Y, Ma L, Liu H (2022) Protocol-based variance constrained distributed secure filtering with measurement censoring. Int J Syst Sci 53(15):3322–3338

Min H, Xu S, Li Y, Chu Y, Wei Y, Zhang Z (2018) Adaptive finite-time control for stochastic nonlinear systems subject to unknown covariance noise. J Frankl Inst-Eng Appl Math 355:2645–2661

Muehlebach M, Trimpe S (2018) Distributed event-based state estimation for networked systems: an LMI approach. IEEE Trans Autom Control 63(1):269–276

Na J, Sun S (2011) Optimal linear estimations for systems with random sensor delays, multiple packet dropouts and uncertain observations. IEEE Trans Signal Process 59(11):5181–5192
Pang Z, Liu G (2012) Design and implementation of secure networked predictive control systems under deception attacks. IEEE Trans Control Syst Technol 20(5):1334–1342
Park JH, Mathiyalagan K, Sakthivel R (2016) Fault estimation for discrete-time switched nonlinear systems with discrete and distributed delays. Int J Robust Nonlinear Control 26(17):3755–3771
Park M, Lee SH, Kwon OM, Seuret A (2018) Closeness-centrality-based synchronization criteria for complex dynamical networks with interval time-varying coupling delays. IEEE Trans Cybern 48(7):2192–2202
Priel A, Zelazo D (2023) Event-triggered consensus Kalman filtering for time-varying networks and intermittent observations. Int J Robust Nonlinear Control 33(13):7430–7451
Ren H, Cheng Z, Qin J, Lu R (2023) Deception attacks on event-triggered distributed consensus estimation for nonlinear systems. Automatica 154:111100
Rezaei H, Farnam A, Crevecoeur G (2023) Scalable recursive state estimation for nonlinear two-dimensional complex networks with switching topology. Int J Robust Nonlinear Control 33(12):6639–6654
Shang H, Zong G, Qi W (2022) Security control for networked discrete-time semi-Markov jump systems with round-robin protocol. IEEE Trans Circuits Syst II-Express Briefs 69(6):2812–2816
Shen Y, Sun S (2023) Distributed recursive filtering for multi-rate uniform sampling systems with packet losses in sensor networks. Int J Syst Sci 54(8):1729–1745
Shen B, Wang Z, Liu X (2011) Bounded H_∞ synchronization and state estimation for discrete time-varying stochastic complex networks over a finite horizon. IEEE Trans Neural Netw 22(1):145–157
Shen B, Wang Z, Shu H, Wei G (2011) H_∞ filtering for uncertain time-varying systems with multiple randomly occurred nonlinearities and successive packet dropouts. Int J Robust Nonlinear Control 21(14):1693–1709
Shu H, Zhang S, Shen B, Liu Y (2016) Unknown input and state estimation for linear discrete-time systems with missing measurements and correlated noises. Int J Gen Syst 45:648–661
Song W, Wang J, Wang C, Shan J (2019) A variance-constrained approach to event-triggered distributed extended Kalman filtering with multiple fading measurements. Int J Robust Nonlinear Control 29:1558–1576
Song W, Wang Z, Wang J, Alsaadi FE, Shan J (2020) Distributed auxiliary particle filtering with diffusion strategy for target tracking: a dynamic event-triggered approach. IEEE Trans Signal Process 69:328–340
Song H, Ding D, Shen B, Dong H (2023) Jointly distributed filtering based on generalized maximum correntropy criterion: memory-based event-triggered cases. IEEE Trans Ind Inform 19(10):10024–10033
Sun S, Tian T, Lin H (2016) Optimal linear estimators for systems with finite-step correlated noises and packet dropout compensations. IEEE Trans Signal Process 64(21):5672–5681
Sun Y, Xiao H, Ding D, Liu S (2022) Secure filtering under adaptive event-triggering protocols with memory mechanisms. ISA Trans 127:13–21
Tan H, Shen B, Li Q, Qian W (2022) Recursive filtering for nonlinear systems with self-interferences over full-duplex relay networks. IEEE-CAA J Automatica Sin 9(11):2037–2040
Tan H, Shen B, Li Q, Huang T (2024) Zonotopic set-membership estimation for time-varying systems subject to dynamical biases and quantization effects. Inf Sci 654:119869
Tao J, Lu R, Su H, Shi P, Wu Z (2018) Asynchronous filtering of nonlinear Markov jump systems with randomly occurred quantization via TCS fuzzy models. IEEE Trans Fuzzy Syst 26(4):1866–1877
Wang R, Sun S (2023) Distributed matrix-weighted fusion consensus filtering with two-stage filtering for sensor networks. IEEE Sens J 23(5):5003–5013
Wang Z, Ho DWC, Liu X (2003) Variance-constrained filtering for uncertain stochastic systems with missing measurements. IEEE Trans Autom Control 48(7):1254–1258

Wang Z, Wang Y, Liu Y (2010) Global synchronization for discrete-time stochastic complex networks with randomly occurred nonlinearities and mixed time-delays. IEEE Trans Neural Netw 21(1):11–25

Wang Y, Ding SX, Xu D, Shen B (2014) An H_∞ fault estimation scheme of wireless networked control systems for industrial real-time applications. IEEE Trans Control Syst Technol 22(6):2073–2086

Wang S, Fang H, Tian X (2015) Recursive estimation for nonlinear stochastic systems with multi-step transmission delays, multiple packet dropouts and correlated noises. Signal Process 115:164–175

Wang L, Wang Z, Huang T, Wei G (2016) An event-triggered approach to state estimation for a class of complex networks with mixed time delays and nonlinearities. IEEE Trans Cybern 46(11):2497–2508

Wang S, Fang H, Tian X (2017) Event-based robust state estimator for linear time-varying system with uncertain observations and randomly occurring uncertainties. J Frankl Inst-Eng Appl Math 354(3):1403–1420

Wang F, Liang J, Wang Z, Liu X (2017) A variance-constrained approach to recursive filtering for nonlinear 2-D systems with measurement degradations. IEEE Trans Cybern 46:1877–1887

Wang F, Wang Z, Liang J, Liu X (2019) Resilient state estimation for 2-D time-varying systems with redundant channels: A variance-constrained approach. IEEE Trans Cybern 49(7):2479–2489

Wang F, Liang J, Lam J, Yang J, Zhao C (2023) Robust filtering for 2-D systems with uncertain-variance noises and weighted try-once-discard protocols. IEEE Trans Syst Man Cybern-Syst 53(5):2914–2924

Wang S, Wang Z, Dong H, Chen Y, Lu G (2024) Dynamic event-triggered quadratic nonfragile filtering for non-Gaussian systems: tackling multiplicative noises and missing measurements. IEEE-CAA J Automatica Sin 11(5):1127–1138

Wang X, Tian E, Zheng W, Xie X (2024) Important-data-based DoS attack mechanism and resilient H_∞ filter design for networked T-S fuzzy systems. IEEE Trans Cybern 54(5):3352–3362

Wang Y, Shen B, Zou L, Han Q-L (2023) A survey on recent advances in distributed filtering over sensor networks subject to communication constraints. Int J Netw Dyn Intell 2(2):100007

Wan X, Li Y, Li Y, Wu M (2022) Finite-time H_∞ state estimation for two-time-scale complex networks under stochastic communication protocol. IEEE Trans Neural Netw Learn Syst 33(1):25–36

Wei G, Wang Z, Shu H (2009) Robust filtering with stochastic nonlinearities and multiple missing measurements. Automatica 45(3):836–841

Wen C, Wang Z, Liu Q, Alsaadi FE (2016) Recursive distributed filtering for a class of state-saturated systems with fading measurements and quantization effects. IEEE Trans Syst Man Cybern-Syst 48:930–941

Wu J, Shi L, Xie L, Johansson KH (2015) An improved stability condition for Kalman filtering with bounded Markovian packet losses. Automatica 62:32–38

Xiong K, Wei C, Liu L (2010) Robust extended Kalman filtering for nonlinear systems with stochastic uncertainties. IEEE Trans Syst Man Cybern Part A-Syst Hum 40(2):399–405

Xiong K, Liu L, Liu Y (2011) Robust extended Kalman filtering for nonlinear systems with multiplicative noises. Optim Control Appl Methods 32(1):47–63

Xu Y, Lv W, Lin W, Lu R, Quevedo DE (2022) On extended state estimation for nonlinear uncertain systems with round-robin protocol. Automatica 138:110154

Yan X, Edwards C (2007) Nonlinear robust fault reconstruction and estimation using a sliding mode observer. Automatica 43(9):1605–1614

Yan L, Zhang S, Ding D, Liu Y, Alsaadi FE (2017) H_∞ state estimation for memristive neural networks with multiple fading measurements. Neurocomputing 230:23–29

Yang F, Li Y (2009) Set-membership filtering for systems with sensor saturation. Automatica 45(8):1896–1902

Yang G, Wang J (2001) Robust nonfragile Kalman filtering for uncertain linear systems with estimator gain uncertainty. IEEE Trans Autom Control 46(2):343–348

Yang L, Gao Y, Feng Z, Wu L (2023) Distributed filtering with dynamic event-triggered communication for networked periodic systems. IEEE Trans Control Netw Syst 10(1):26–37
Yao M, Wei G (2023) Dynamic event-triggered control of continuous time systems with random impulses. IEEE-CAA J Automatica Sin 10(12):2292–2299
Yi X, Xu T (2024) Distributed event-triggered estimation for dynamic average consensus: a perturbation-injected privacy-preservation scheme. Inf Fusion 108:102396
You K, Fu M, Xie L (2011) Mean square stability for Kalman filtering with Markovian packet losses. Automatica 47(12):2647–2657
Zha L, Guo Y, Liu J, Xie X, Tian E (2024) Protocol-based distributed security fusion estimation for time-varying uncertain systems over sensor networks: tackling DoS attacks. IEEE Trans Signal Inf Proc Netw 10:119–130
Zha L, Liao R, Liu J, Xie X, Tian E, Cao J (2024) Outlier-resistant distributed filtering over sensor networks under dynamic event-triggered schemes and DoS attacks. IEEE Trans Autom Sci Eng. https://doi.org/10.1109/TASE.2024.3360718
Zhang D, Yang H (2023) Ultimately bounded filtering for nonlinear systems subject to integral-type event-triggered mechanism and dynamic quantization. IEEE Sens J 23(22):27574–27580
Zhang L, Gao H, Kaynak O (2013) Network-induced constraints in networked control systems-a survey. IEEE Trans Ind Inf 9(1):403–426
Zhang W, Wang Z, Liu Y, Ding D, Alsaadi FE (2017) Event-based state estimation for a class of complex networks with time-varying delays: a comparison principle approach. Phys Lett A 381(1):10–18
Zhang L, Sun S (2023) Distributed H_∞ fusion filtering for multi-sensor networked systems with DoS attacks and sensor saturations. Digit Signal Prog 134:103908
Zhao Y, He X, Ma L, Liu H (2022) Unbiasedness-constrained least squares state estimation for time-varying systems with missing measurements under round-robin protocol. Int J Syst Sci 53(9):1925–1941
Zhao Z, Wang Z, Zou L, Liu H, Alsaadi FE (2023) Zonotopic multi-sensor fusion estimation with mixed delays under try-once-discard protocol: a set-membership framework. Inf Fusion 91:681–693
Zheng X, Kang Y, Li H, Li J (2023) Multiple description encoding-decoding-based resilient filtering for complex networks under the round-robin protocol. Appl Math Comput 458:128250
Zhou H, Sun S (2024) Distributed filtering for multi-sensor networked systems with stochastic communication protocol and correlated noises. Inf Fusion 104:102121
Zhu Y, Song X, Wang M, Lu J (2020) Finite-time asynchronous H_∞ filtering design of Markovian jump systems with randomly occurred quantization. Int J Control Autom Syst 18(2):450–461
Zou L, Wang Z, Zhou DH (2017) Event-based control and filtering of networked systems: a survey. Int J Autom Comput 14(3):239–253
Zou L, Wang Z, Han QL, Zhou D (2017) Ultimate boundedness control for networked systems with try-once-discard protocol and uniform quantization effects. IEEE Trans Autom Control 62(12):6582–6588
Zou L, Wang Z, Han QL, Zhou DH (2018) Recursive filtering for time-varying systems with random access protocol. IEEE Trans Autom Control 64(2):720–727
Zou L, Wang Z, Han QL, Zhou D (2021) Moving horizon estimation of networked nonlinear systems with random access protocol. IEEE Trans Syst Man Cybern-Syst 51(5):2937–2948
Zuo Z, Luo Q, Wang Y (2017) Stabilization of linear systems with direct feedthrough term in the presence of output saturation. Automatica 77:254–258

Chapter 2
Recursive Filtering and Boundedness Analysis with Randomly Occurring Quantization

In this chapter, the optimized recursive filtering issue is discussed for networked time-varying nonlinear systems with ROQ. A sequence of binary random variables is used to describe the ROQ, in which the logarithmic quantizer is employed to describe the phenomenon of quantized measurements. The purpose of this chapter is to construct a desirable filter of recursive form such that, for all ROUs, ROQ and stochastic nonlinearity, an upper bound regarding the FEC is solved and the expected filter gain with expression form is derived and given. Furthermore, a sufficient condition is provided to guarantee the boundedness of the filtering error dynamics. Lastly, some comparative simulation examples are provided to validate the usefulness of established recursive filtering algorithm.

2.1 Problem Formulation

In this section, we consider the following time-varying nonlinear systems with ROUs and stochastic nonlinearity:

$$x_{k+1} = (A_k + \alpha_k \Delta A_k)x_k + f(x_k, \xi_k) + B_k \omega_k, \tag{2.1}$$

$$y_k = C_k x_k + \nu_k, \tag{2.2}$$

where $x_k \in \mathbb{R}^n$ is the state and its initial value x_0 has mean $\bar{x}_0$ and covariance $P_0 > 0$, $y_k \in \mathbb{R}^m$ is the measurement signal, $\xi_k \in \mathbb{R}$ is a zero-mean Gaussian white noise, $\omega_k \in \mathbb{R}^l$ and $\nu_k \in \mathbb{R}^m$ denote the zero-mean Gaussian noises with covariance $Q_k > 0$ and $R_k > 0$. A_k, B_k and C_k represent known matrices of appropriate dimensions.

The uncertain matrix ΔA_k satisfies

$$\Delta A_k = H_k F_k M_k \tag{2.3}$$

J. Hu et al., *Variance-Constrained Filtering for Stochastic Complex Systems*, Intelligent Control and Learning Systems 18,
https://doi.org/10.1007/978-981-96-2637-3_2

with H_k and M_k denoting known matrices, and F_k is an uncertain matrix satisfying

$$F_k^T F_k \leq I.$$

To describe the ROUs, the Bernoulli distributed random variable α_k is introduced and satisfies

$$\begin{aligned} \text{Prob}\{\alpha_k = 1\} &= \bar{\alpha}_k, \\ \text{Prob}\{\alpha_k = 0\} &= 1 - \bar{\alpha}_k, \end{aligned} \tag{2.4}$$

where $\bar{\alpha}_k \in [0, 1]$ denotes a given constant.

The stochastic nonlinear function $f(x_k, \xi_k)$ with $f(0, \xi_k) = 0$ meets the following characteristics:

$$\mathbb{E}\{f(x_k, \xi_k)|x_k\} = 0, \tag{2.5}$$

$$\mathbb{E}\{f(x_k, \xi_k)f^T(x_l, \xi_l)|x_k\} = 0, \ k \neq l, \tag{2.6}$$

$$\mathbb{E}\{f(x_k, \xi_k)f^T(x_k, \xi_k)|x_k\} = \sum_{i=1}^{s} \Pi_i x_k^T \Gamma_i x_k \tag{2.7}$$

with $s > 0$ being a given integer, Π_i and Γ_i being known matrices of suitable dimensions for $i = 1, 2, \ldots, s$.

Remark 2.1 As a matter of fact, it is impossible to obtain an accurate system model in certain cases, thus it is necessary to deal with the modeling errors and analyze their influences onto the filtering performance. In addition, it is noted that the modeling errors encounter the random changes, thereby the ROUs are considered via introducing a random variable α_k with hope to meet the engineering reality. On the other hand, the stochastic nonlinearity $f(\cdot)$ could include many known nonlinearities. To name a few, it can characterize the functions in general linear systems with the state-multiplicative noises $\xi_k x_k$, in which ξ_k denotes a zero-mean noise with bounded covariance and the nonlinearities in nonlinear systems with random disturbances. As a result, it is meaningful to investigate the filtering problems with ROUs as well as stochastic nonlinearity, and the different filtering methods are desirable to handle the above mentioned phenomena.

Because of the constrained channel bandwidth, the signal quantization may appear in a random manner. To better characterize this phenomenon, the quantization process is modeled as

$$q(y_k) = \left[q_1(y_k^1) \; q_2(y_k^2) \; \cdots \; q_m(y_k^m) \right]^T,$$

where $y_k^1, y_k^2, \ldots, y_k^m$ are the components of y_k.

The following set of quantization levels of each $q_j(\cdot)$ $(j = 1, 2, \ldots, m)$ are considered:

$$\mathfrak{U}_j = \left\{ \pm u_i^{(j)},\ u_i^{(j)} = \left(\chi^{(j)}\right)^i u_0^{(j)},\ i = 0, \pm 1, \pm 2, \ldots \right\} \cup \{0\}$$

with $\chi^{(j)} \in (0, 1)$ denoting the quantization density and $u_0^{(j)} > 0$. Furthermore, by resorting to Fu and Xie (2005), Elia and Mitter (2001), the following logarithmic quantizer is used:

$$q_j(y_k^j) = \begin{cases} u_i^{(j)}, & \frac{1}{1+\delta_j} u_i^{(j)} < y_k^j \leq \frac{1}{1-\delta_j} u_i^{(j)} \\ 0, & y_k^j = 0 \\ -q_j(-y_k^j), & y_k^j < 0, \end{cases}$$

where $\delta_j = \frac{1-\chi^{(j)}}{1+\chi^{(j)}}$. It is obvious to find that $q_j(y_k^j) = \left(1 + \Delta_k^{(j)}\right) y_k^j$ with $|\Delta_k^{(j)}| \leq \delta_j$. Then, the effects caused by the signal quantization are transformed into the sector-bounded uncertainties as mentioned in Fu and Xie (2005). The transformation employed here is a feasible yet effective way to simplify the filter design under signal quantization. Denoting

$$\begin{aligned} \mathscr{F}_k &= \Delta_k \Upsilon^{-1},\ \Upsilon = \text{diag}\{\delta_1, \delta_2, \ldots, \delta_m\}, \\ \Delta_k &= \text{diag}\{\Delta_k^{(1)}, \Delta_k^{(2)}, \ldots, \Delta_k^{(m)}\}, \end{aligned}$$

it is seen that $\mathscr{F}_k$ is an unknown matrix and satisfies $\mathscr{F}_k \mathscr{F}_k^T = \mathscr{F}_k^T \mathscr{F}_k \leq I$.

The *real* measurements received can be modeled as

$$\tilde{y}_k = \Lambda_k y_k + (I - \Lambda_k) q(y_k), \tag{2.8}$$

where $\Lambda_k = \text{diag}\{\lambda_{k,1}, \lambda_{k,2}, \ldots, \lambda_{k,m}\}$ and $\lambda_{k,i}$ $(i = 1, 2, \ldots, m)$ are random variables satisfying

$$\text{Prob}\left\{\lambda_{k,i} = 1\right\} = \bar{\lambda}_{k,i},\ \text{Prob}\left\{\lambda_{k,i} = 0\right\} = 1 - \bar{\lambda}_{k,i} \tag{2.9}$$

with $\bar{\lambda}_{k,i}$ denoting given scalars. In the meantime, assume that random variables ξ_k, α_k, ω_k, $\lambda_{k,i}$, ν_k and x_0 are mutually independent throughout this chapter.

In this chapter, we design the following two-step recursive filter:

$$\hat{x}_{k+1|k} = A_k \hat{x}_{k|k}, \tag{2.10}$$

$$\hat{x}_{k+1|k+1} = \hat{x}_{k+1|k} + K_{k+1}(\tilde{y}_{k+1} - \bar{\Lambda}_{k+1} C_{k+1} \hat{x}_{k+1|k}), \tag{2.11}$$

where $\hat{x}_{k|k}$ represents the state estimate, $\hat{x}_{k+1|k}$ is the one-step prediction, K_{k+1} stands for the filter gain and $\bar{\Lambda}_{k+1} = \mathbb{E}\{\Lambda_{k+1}\}$.

The objectives of this chapter primarily include the following three aspects: (1) solve an upper bound of the FEC based on matrix theory and stochastic analysis method; (2) design a proper filter gain by minimizing the obtained upper bound; and

(3) provide a sufficient condition to ensure the boundedness of the filtering error dynamics in the mean-square sense.

Before proceeding, the following lemmas are introduced to facilitate the proof of main results.

Lemma 2.1 (Wang et al. 1992) *Given matrices E, H, G and F with suitable dimensions such that $FF^T \leq I$, let X be a symmetric positive definite matrix and γ be an arbitrary positive constant such that $\gamma^{-1}I - GXG^T > 0$, one has*

$$(E + HFG)X(E + HFG)^T \leq E\left(X^{-1} - \gamma G^T G\right)^{-1} E^T + \gamma^{-1} HH^T.$$

Lemma 2.2 (Hardy et al. 1988) *For two real vectors c and d, the following inequality holds:*

$$cd^T + dc^T \leq \varepsilon cc^T + \varepsilon^{-1} dd^T$$

with $\varepsilon > 0$ being a scalar.

Lemma 2.3 *For matrices L, N, K and X with appropriate dimensions, the following properties:*

$$\frac{\partial \mathrm{tr}(LKN)}{\partial K} = L^T N^T, \quad \frac{\partial \mathrm{tr}(LK^T N)}{\partial K} = NL,$$
$$\frac{\partial \mathrm{tr}(LKNK^T X)}{\partial K} = L^T X^T K N^T + XLKN$$

hold.

Lemma 2.4 (Horn and Johnson 1991) *For a matrix $A = [a_{ij}]_{n\times n}$ and a stochastic diagonal matrix $E = \mathrm{diag}\{e_1, e_2, \ldots, e_n\}$, one has*

$$\mathbb{E}\{EAE^T\} = \begin{bmatrix} \mathbb{E}\{e_1^2\} & \mathbb{E}\{e_1 e_2\} & \cdots & \mathbb{E}\{e_1 e_n\} \\ \mathbb{E}\{e_2 e_1\} & \mathbb{E}\{e_2^2\} & \cdots & \mathbb{E}\{e_2 e_n\} \\ \vdots & \vdots & \ddots & \vdots \\ \mathbb{E}\{e_n e_1\} & \mathbb{E}\{e_n e_2\} & \cdots & \mathbb{E}\{e_n^2\} \end{bmatrix} \circ A$$

with $\circ$ representing the Hadamard product.

2.2 Design of Optimal Filtering Algorithm

In this section, a minimized upper bound concerning on the FEC is established by means of the stochastic analysis approach. Furthermore, the expected filter gain is constructed by solving a set of matrix equations within the optimization framework.

Firstly, let's compute the one-step prediction error and filtering error. Denote $\tilde{x}_{k+1|k} = x_{k+1} - \hat{x}_{k+1|k}$ and $\tilde{x}_{k+1|k+1} = x_{k+1} - \hat{x}_{k+1|k+1}$, respectively. Subtracting (2.10) from (2.1) results in

$$\tilde{x}_{k+1|k} = A_k \tilde{x}_{k|k} + \bar{\alpha}_k \Delta A_k x_k + \tilde{\alpha}_k \Delta A_k x_k + f(x_k, \xi_k) + B_k \omega_k, \tag{2.12}$$

where $\tilde{\alpha}_k = \alpha_k - \bar{\alpha}_k$. Along the same line with (2.12), one has

$$\begin{aligned} \tilde{x}_{k+1|k+1} = & (I - K_{k+1}\bar{\Lambda}_{k+1}C_{k+1})\tilde{x}_{k+1|k} - K_{k+1}\tilde{\Lambda}_{k+1}C_{k+1}x_{k+1} - K_{k+1}\breve{\Lambda}_{k+1} \\ & \times (I + \Delta_{k+1})C_{k+1}x_{k+1} + K_{k+1}\tilde{\Lambda}_{k+1}(I + \Delta_{k+1})C_{k+1}x_{k+1} \\ & - K_{k+1}\breve{\Lambda}_{k+1}\Delta_{k+1}\nu_{k+1} + K_{k+1}\tilde{\Lambda}_{k+1}\Delta_{k+1}\nu_{k+1} - K_{k+1}\nu_{k+1}, \end{aligned} \tag{2.13}$$

where $\tilde{\Lambda}_{k+1} = \Lambda_{k+1} - \bar{\Lambda}_{k+1}$ and $\breve{\Lambda}_{k+1} = I - \bar{\Lambda}_{k+1}$.

Next, the expected expression forms of the one-step prediction error covariance and FEC are calculated.

Theorem 2.1 *The one-step prediction error covariance $P_{k+1|k}$ satisfies*

$$\begin{aligned} P_{k+1|k} = & A_k P_{k|k} A_k^T + \bar{\alpha}_k \Delta A_k \mathbb{E}\{x_k x_k^T\} \Delta A_k^T + B_k Q_k B_k^T + \bar{\alpha}_k A_k \mathbb{E}\{\tilde{x}_{k|k} x_k^T\} \Delta A_k^T \\ & + \bar{\alpha}_k \Delta A_k \mathbb{E}\{x_k \tilde{x}_{k|k}^T\} A_k^T + \sum_{i=1}^{s} \Pi_i \mathrm{tr}(\mathbb{E}\{x_k x_k^T\} \Gamma_i). \end{aligned} \tag{2.14}$$

Proof With the help of (2.12) and the independent properties of random variables, it can be derived that (2.14) holds readily.

Theorem 2.2 *The FEC $P_{k+1|k+1}$ is computed as*

$$\begin{aligned} P_{k+1|k+1} = & (I - K_{k+1}\bar{\Lambda}_{k+1}C_{k+1})P_{k+1|k}(I - K_{k+1}\bar{\Lambda}_{k+1}C_{k+1})^T + K_{k+1}R_{k+1} \\ & \times K_{k+1}^T + \mathscr{L}_1 + \mathscr{L}_1^T + K_{k+1}\breve{\Lambda}_{k+1}(I + \Delta_{k+1})C_{k+1}\mathbb{E}\{x_{k+1}x_{k+1}^T\} \\ & \times C_{k+1}^T(I + \Delta_{k+1})^T \breve{\Lambda}_{k+1} K_{k+1}^T + \mathscr{L}_2 + \mathscr{L}_2^T + K_{k+1}\breve{\Lambda}_{k+1}\Delta_{k+1} \\ & \times R_{k+1}\Delta_{k+1}^T \breve{\Lambda}_{k+1} K_{k+1}^T + K_{k+1}\Big\{ \breve{\Xi}_{k+1} \circ [C_{k+1}\mathbb{E}\{x_{k+1}x_{k+1}^T\}C_{k+1}^T \\ & + (I + \Delta_{k+1})C_{k+1}\mathbb{E}\{x_{k+1}x_{k+1}^T\}C_{k+1}^T(I + \Delta_{k+1})^T + \Delta_{k+1}R_{k+1} \\ & \times \Delta_{k+1}^T + \mathscr{L}_3 + \mathscr{L}_3^T] \Big\} K_{k+1}^T, \end{aligned} \tag{2.15}$$

where

$$\begin{aligned} \mathscr{L}_1 = & K_{k+1}\mathbb{E}\{\nu_{k+1}\nu_{k+1}^T\}\Delta_{k+1}^T \breve{\Lambda}_{k+1} K_{k+1}^T, \\ \mathscr{L}_2 = & -(I - K_{k+1}\bar{\Lambda}_{k+1}C_{k+1})\mathbb{E}\{\tilde{x}_{k+1|k}x_{k+1}^T\}C_{k+1}^T(I + \Delta_{k+1})^T \breve{\Lambda}_{k+1} K_{k+1}^T, \\ \mathscr{L}_3 = & -C_{k+1}\mathbb{E}\{x_{k+1}x_{k+1}^T\}C_{k+1}^T(I + \Delta_{k+1})^T, \\ \breve{\Xi}_{k+1} = & \mathrm{diag}\{\bar{\lambda}_{k+1,1}(1 - \bar{\lambda}_{k+1,1}), \bar{\lambda}_{k+1,2}(1 - \bar{\lambda}_{k+1,2}), \ldots, \bar{\lambda}_{k+1,m}(1 - \bar{\lambda}_{k+1,m})\}. \end{aligned}$$

Proof According to (2.13) and Lemma 2.4, one has

$$
\begin{aligned}
P_{k+1|k+1} = &\ (I - K_{k+1}\bar{\Lambda}_{k+1}C_{k+1})P_{k+1|k}(I - K_{k+1}\bar{\Lambda}_{k+1}C_{k+1})^T + K_{k+1}R_{k+1} \\
&\times K_{k+1}^T + \mathscr{L}_1 + \mathscr{L}_1^T + K_{k+1}\breve{\Lambda}_{k+1}(I + \Delta_{k+1})C_{k+1}\mathbb{E}\{x_{k+1}x_{k+1}^T\} \\
&\times C_{k+1}^T(I + \Delta_{k+1})^T\breve{\Lambda}_{k+1}K_{k+1}^T + \mathscr{L}_2 + \mathscr{L}_2^T + K_{k+1}\breve{\Lambda}_{k+1}\Delta_{k+1} \\
&\times R_{k+1}\Delta_{k+1}^T\breve{\Lambda}_{k+1}K_{k+1}^T + K_{k+1}\Big\{\breve{\Xi}_{k+1} \circ [C_{k+1}\mathbb{E}\{x_{k+1}x_{k+1}^T\}C_{k+1}^T \\
&+(I + \Delta_{k+1})C_{k+1}\mathbb{E}\{x_{k+1}x_{k+1}^T\}C_{k+1}^T(I + \Delta_{k+1})^T + \Delta_{k+1}R_{k+1} \\
&\times \Delta_{k+1}^T + \mathscr{L}_3 + \mathscr{L}_3^T]\Big\}K_{k+1}^T + \sum_{l=1}^{18}(\mathscr{W}_l + \mathscr{W}_l^T),
\end{aligned}
$$

where

$$
\begin{aligned}
\mathscr{W}_1 &= -\mathbb{E}\{(I - K_{k+1}\bar{\Lambda}_{k+1}C_{k+1})\tilde{x}_{k+1|k}x_{k+1}^T C_{k+1}^T\tilde{\Lambda}_{k+1}K_{k+1}^T\}, \\
\mathscr{W}_2 &= \mathbb{E}\{(I - K_{k+1}\bar{\Lambda}_{k+1}C_{k+1})\tilde{x}_{k+1|k}x_{k+1}^T C_{k+1}^T(I + \Delta_{k+1})^T\tilde{\Lambda}_{k+1}K_{k+1}^T\}, \\
\mathscr{W}_3 &= -\mathbb{E}\{(I - K_{k+1}\bar{\Lambda}_{k+1}C_{k+1})\tilde{x}_{k+1|k}v_{k+1}^T\Delta_{k+1}^T\breve{\Lambda}_{k+1}K_{k+1}^T\}, \\
\mathscr{W}_4 &= \mathbb{E}\{(I - K_{k+1}\bar{\Lambda}_{k+1}C_{k+1})\tilde{x}_{k+1|k}v_{k+1}^T\Delta_{k+1}^T\tilde{\Lambda}_{k+1}K_{k+1}^T\}, \\
\mathscr{W}_5 &= -\mathbb{E}\{(I - K_{k+1}\bar{\Lambda}_{k+1}C_{k+1})\tilde{x}_{k+1|k}v_{k+1}^T K_{k+1}^T\}, \\
\mathscr{W}_6 &= \mathbb{E}\{K_{k+1}\tilde{\Lambda}_{k+1}C_{k+1}x_{k+1}x_{k+1}^T C_{k+1}^T(I + \Delta_{k+1})^T\breve{\Lambda}_{k+1}K_{k+1}^T\}, \\
\mathscr{W}_7 &= \mathbb{E}\{K_{k+1}\tilde{\Lambda}_{k+1}C_{k+1}x_{k+1}v_{k+1}^T\Delta_{k+1}^T\breve{\Lambda}_{k+1}K_{k+1}^T\}, \\
\mathscr{W}_8 &= -\mathbb{E}\{K_{k+1}\tilde{\Lambda}_{k+1}C_{k+1}x_{k+1}v_{k+1}^T\Delta_{k+1}^T\tilde{\Lambda}_{k+1}K_{k+1}^T\}, \\
\mathscr{W}_9 &= \mathbb{E}\{K_{k+1}\tilde{\Lambda}_{k+1}C_{k+1}x_{k+1}v_{k+1}^T K_{k+1}^T\}, \\
\mathscr{W}_{10} &= -\mathbb{E}\{K_{k+1}\breve{\Lambda}_{k+1}(I + \Delta_{k+1})C_{k+1}x_{k+1}x_{k+1}^T C_{k+1}^T(I + \Delta_{k+1})^T\tilde{\Lambda}_{k+1}K_{k+1}^T\}, \\
\mathscr{W}_{11} &= \mathbb{E}\{K_{k+1}\breve{\Lambda}_{k+1}(I + \Delta_{k+1})C_{k+1}x_{k+1}v_{k+1}^T\Delta_{k+1}^T\breve{\Lambda}_{k+1}K_{k+1}^T\}, \\
\mathscr{W}_{12} &= -\mathbb{E}\{K_{k+1}\breve{\Lambda}_{k+1}(I + \Delta_{k+1})C_{k+1}x_{k+1}v_{k+1}^T\Delta_{k+1}^T\tilde{\Lambda}_{k+1}K_{k+1}^T\}, \\
\mathscr{W}_{13} &= \mathbb{E}\{K_{k+1}\breve{\Lambda}_{k+1}(I + \Delta_{k+1})C_{k+1}x_{k+1}v_{k+1}^T K_{k+1}^T\}, \\
\mathscr{W}_{14} &= -\mathbb{E}\{K_{k+1}\tilde{\Lambda}_{k+1}(I + \Delta_{k+1})C_{k+1}x_{k+1}v_{k+1}^T\Delta_{k+1}^T\breve{\Lambda}_{k+1}K_{k+1}^T\}, \\
\mathscr{W}_{15} &= \mathbb{E}\{K_{k+1}\tilde{\Lambda}_{k+1}(I + \Delta_{k+1})C_{k+1}x_{k+1}v_{k+1}^T\Delta_{k+1}^T\tilde{\Lambda}_{k+1}K_{k+1}^T\}, \\
\mathscr{W}_{16} &= -\mathbb{E}\{K_{k+1}\tilde{\Lambda}_{k+1}(I + \Delta_{k+1})C_{k+1}x_{k+1}v_{k+1}^T K_{k+1}^T\}, \\
\mathscr{W}_{17} &= -\mathbb{E}\{K_{k+1}\breve{\Lambda}_{k+1}\Delta_{k+1}v_{k+1}v_{k+1}^T\Delta_{k+1}^T\tilde{\Lambda}_{k+1}K_{k+1}^T\}, \\
\mathscr{W}_{18} &= -\mathbb{E}\{K_{k+1}\tilde{\Lambda}_{k+1}\Delta_{k+1}v_{k+1}v_{k+1}^T K_{k+1}^T\}.
\end{aligned}
$$

Note that v_{k+1} and Λ_{k+1} are mutually independent and $\mathbb{E}\{\tilde{\Lambda}_{k+1}\} = 0$, then it can be clearly derived that $\mathscr{W}_i = 0$ $(i = 1, 2, \ldots, 18)$. Accordingly, the result in (2.15) is obtained easily.

Remark 2.2 Generally speaking, the global optimal filtering approach is best objective. However, it is impossible to achieve such a purpose because of the existences of the parameter uncertainties, nonlinearity and ROQ. To overcome these potential difficulties, the alternative way is adopted to calculate an upper bound of FEC and then minimize such upper bound through selecting an appropriate filter gain matrix at each sampling time.

Up to now, the specific forms of the one-step prediction error covariance and the FEC have been solved. In what follows, the purpose is to search for an upper bound on the FEC and design a filter gain matrix suitably.

Theorem 2.3 *Let* $\gamma_{k+1,1} > 0$ *and* $\varepsilon_i > 0$ $(i = 1, 2, \ldots, 6)$ *be known real numbers. If the following matrix difference equations:*

$$\begin{aligned}\Sigma_{k+1|k} = &\ (1+\bar{\alpha}_k\varepsilon_1)A_k\Sigma_{k|k}A_k^T + \Omega_k + B_kQ_kB_k^T \\ &+(1+\varepsilon_1^{-1})\bar{\alpha}_k\mathrm{tr}(M_k\bar{\mathscr{L}}_kM_k^T)H_kH_k^T \end{aligned} \tag{2.16}$$

and

$$\begin{aligned}\Sigma_{k+1|k+1} = &\ (1+\varepsilon_5)(I-K_{k+1}\bar{\Lambda}_{k+1}C_{k+1})\Sigma_{k+1|k}(I-K_{k+1}\bar{\Lambda}_{k+1}C_{k+1})^T \\ &+(1+\varepsilon_4)K_{k+1}R_{k+1}K_{k+1}^T + (1+\varepsilon_5^{-1})\mathrm{tr}(C_{k+1}\bar{\Pi}_{k+1}C_{k+1}^T) \\ &\times K_{k+1}\breve{\Lambda}_{k+1}[(I-\gamma_{k+1,1}\Upsilon\Upsilon)^{-1}+\gamma_{k+1,1}^{-1}I]\breve{\Lambda}_{k+1}K_{k+1}^T + K_{k+1} \\ &\times\Psi_{k+1}K_{k+1}^T + (1+\varepsilon_4^{-1})\mathrm{tr}(\Upsilon R_{k+1}\Upsilon)K_{k+1}\breve{\Lambda}_{k+1}^2K_{k+1}^T \end{aligned} \tag{2.17}$$

under $\gamma_{k+1,1}^{-1}I - \Upsilon\Upsilon > 0$ *and* $\Sigma_{0|0} = P_{0|0} > 0$, *have solutions* $\Sigma_{k+1|k} > 0$ *and* $\Sigma_{k+1|k+1} > 0$, *then* $P_{k+1|k+1} \le \Sigma_{k+1|k+1}$. *Besides, if the filter gain* K_{k+1} *is designed as*

$$\begin{aligned}K_{k+1} = &\ (1+\varepsilon_5)\Sigma_{k+1|k}C_{k+1}^T\bar{\Lambda}_{k+1}\Big\{(1+\varepsilon_4)R_{k+1} + (1+\varepsilon_5^{-1})\mathrm{tr}(C_{k+1}\bar{\Pi}_{k+1}C_{k+1}^T) \\ &\times\breve{\Lambda}_{k+1}[(I-\gamma_{k+1,1}\Upsilon\Upsilon)^{-1}+\gamma_{k+1,1}^{-1}I]\breve{\Lambda}_{k+1} + (1+\varepsilon_5)\bar{\Lambda}_{k+1}C_{k+1} \\ &\times\Sigma_{k+1|k}C_{k+1}^T\bar{\Lambda}_{k+1} + (1+\varepsilon_4^{-1})\mathrm{tr}(\Upsilon R_{k+1}\Upsilon)\breve{\Lambda}_{k+1}^2 + \Psi_{k+1}\Big\}^{-1}, \end{aligned} \tag{2.18}$$

it is revealed that $\mathrm{tr}(\Sigma_{k+1|k+1})$ *is minimized, where*

$$\begin{aligned}\Omega_k = &\ \sum_{i=1}^{s}\Pi_i\mathrm{tr}(\bar{\mathscr{L}}_k\Gamma_i), \\ \bar{\Pi}_{k+1} = &\ (1+\varepsilon_3)\Sigma_{k+1|k} + (1+\varepsilon_3^{-1})\hat{x}_{k+1|k}\hat{x}_{k+1|k}^T, \\ \Psi_{k+1} = &\ \breve{\Xi}_{k+1}\circ\big\{(1+\varepsilon_6^{-1})\mathrm{tr}(C_{k+1}\bar{\Pi}_{k+1}C_{k+1}^T)[(I-\gamma_{k+1,1}\Upsilon\Upsilon)^{-1}+\gamma_{k+1,1}^{-1}I] \\ &+\mathrm{tr}(\Upsilon R_{k+1}\Upsilon) + (1+\varepsilon_6)C_{k+1}\bar{\Pi}_{k+1}C_{k+1}^T\big\}, \\ \bar{\mathscr{L}}_k = &\ (1+\varepsilon_2)\Sigma_{k|k} + (1+\varepsilon_2^{-1})\hat{x}_{k|k}\hat{x}_{k|k}^T. \end{aligned} \tag{2.19}$$

Proof The mathematical induction method is adopted to prove this theorem. By means of (2.14) and Lemma 2.2, it is deduced that

$$\begin{aligned}&\bar{\alpha}_k A_k \mathbb{E}\{\tilde{x}_{k|k} x_k^T\} \Delta A_k^T + \bar{\alpha}_k \Delta A_k \mathbb{E}\{x_k \tilde{x}_{k|k}^T\} A_k^T \\ \leq\ & \bar{\alpha}_k \varepsilon_1 A_k P_{k|k} A_k^T + \bar{\alpha}_k \varepsilon_1^{-1} \Delta A_k \mathbb{E}\{x_k x_k^T\} \Delta A_k^T\end{aligned} \tag{2.20}$$

with $\varepsilon_1 > 0$ being a scalar. Thus, one obtains

$$\begin{aligned}P_{k+1|k} \leq\ & (1 + \bar{\alpha}_k \varepsilon_1) A_k P_{k|k} A_k^T + \sum_{i=1}^{s} \Pi_i \mathrm{tr}(\mathbb{E}\{x_k x_k^T\} \Gamma_i) + (1 + \varepsilon_1^{-1}) \bar{\alpha}_k \Delta A_k \\ & \times \mathbb{E}\{x_k x_k^T\} \Delta A_k^T + B_k Q_k B_k^T.\end{aligned} \tag{2.21}$$

Subsequently, we have

$$\begin{aligned}\mathbb{E}\{x_k x_k^T\} \leq\ & (1 + \varepsilon_2) P_{k|k} + (1 + \varepsilon_2^{-1}) \hat{x}_{k|k} \hat{x}_{k|k}^T \\ :=\ & \tilde{\mathscr{L}}_k\end{aligned} \tag{2.22}$$

with $\varepsilon_2 > 0$ denoting a constant. Noting (2.3), we have

$$\Delta A_k \mathbb{E}\{x_k x_k^T\} \Delta A_k^T \leq \mathrm{tr}(M_k \tilde{\mathscr{L}}_k M_k^T) H_k H_k^T. \tag{2.23}$$

It follows from (2.21)–(2.23) that

$$\begin{aligned}P_{k+1|k} \leq\ & (1 + \bar{\alpha}_k \varepsilon_1) A_k P_{k|k} A_k^T + \sum_{i=1}^{s} \Pi_i \mathrm{tr}(\tilde{\mathscr{L}}_k \Gamma_i) + (1 + \varepsilon_1^{-1}) \bar{\alpha}_k \mathrm{tr}(M_k \tilde{\mathscr{L}}_k M_k^T) \\ & \times H_k H_k^T + B_k Q_k B_k^T.\end{aligned} \tag{2.24}$$

Next, it is obvious to find that

$$\begin{aligned}\mathbb{E}\{x_{k+1} x_{k+1}^T\} \leq\ & (1 + \varepsilon_3) P_{k+1|k} + (1 + \varepsilon_3^{-1}) \hat{x}_{k+1|k} \hat{x}_{k+1|k}^T \\ :=\ & \check{\Pi}_{k+1}\end{aligned} \tag{2.25}$$

with $\varepsilon_3 > 0$ being a scalar.

By resorting to Lemma 2.2 and (2.25), we have

$$\begin{aligned}\mathscr{L}_1 + \mathscr{L}_1^T \leq\ & \varepsilon_4 K_{k+1} R_{k+1} K_{k+1}^T + \varepsilon_4^{-1} K_{k+1} \check{\Lambda}_{k+1} \Delta_{k+1} R_{k+1} \Delta_{k+1}^T \check{\Lambda}_{k+1} K_{k+1}^T, \\ \mathscr{L}_2 + \mathscr{L}_2^T \leq\ & \varepsilon_5 (I - K_{k+1} \bar{\Lambda}_{k+1} C_{k+1}) P_{k+1|k} (I - K_{k+1} \bar{\Lambda}_{k+1} C_{k+1})^T + \varepsilon_5^{-1} \\ & \times K_{k+1} \check{\Lambda}_{k+1} (I + \Delta_{k+1}) C_{k+1} \check{\Pi}_{k+1} C_{k+1}^T (I + \Delta_{k+1})^T \check{\Lambda}_{k+1} K_{k+1}^T, \\ \mathscr{L}_3 + \mathscr{L}_3^T \leq\ & \varepsilon_6 C_{k+1} \check{\Pi}_{k+1} C_{k+1}^T + \varepsilon_6^{-1} (I + \Delta_{k+1}) C_{k+1} \check{\Pi}_{k+1} C_{k+1}^T \\ & \times (I + \Delta_{k+1})^T\end{aligned} \tag{2.26}$$

with $\varepsilon_i > 0$ $(i = 4, 5, 6)$ being scalars. It follows from (2.26) that

$$\begin{aligned} P_{k+1|k+1} \leq\ & (1+\varepsilon_5)(I - K_{k+1}\bar{\Lambda}_{k+1}C_{k+1})P_{k+1|k}(I - K_{k+1}\bar{\Lambda}_{k+1}C_{k+1})^T \\ & +(1+\varepsilon_4)K_{k+1}R_{k+1}K_{k+1}^T + (1+\varepsilon_5^{-1})K_{k+1}\breve{\Lambda}_{k+1}(I+\Delta_{k+1})C_{k+1} \\ & \times\breve{\Pi}_{k+1}C_{k+1}^T(I+\Delta_{k+1})^T\breve{\Lambda}_{k+1}K_{k+1}^T + (1+\varepsilon_4^{-1})K_{k+1}\breve{\Lambda}_{k+1}\Delta_{k+1} \\ & \times R_{k+1}\Delta_{k+1}^T\breve{\Lambda}_{k+1}K_{k+1}^T + K_{k+1}\Big\{\breve{\Xi}_{k+1}\circ[(1+\varepsilon_6)C_{k+1}\breve{\Pi}_{k+1}C_{k+1}^T \\ & +(1+\varepsilon_6^{-1})(I+\Delta_{k+1})C_{k+1}\breve{\Pi}_{k+1}C_{k+1}^T(I+\Delta_{k+1})^T \\ & +\Delta_{k+1}R_{k+1}\Delta_{k+1}^T]\Big\}K_{k+1}^T. \end{aligned} \tag{2.27}$$

Considering $\Delta_{k+1} = \mathscr{F}_{k+1}\Upsilon$ $(\mathscr{F}_{k+1}\mathscr{F}_{k+1}^T \leq I)$ and Lemma 2.1, one has

$$\begin{aligned} & (I+\Delta_{k+1})C_{k+1}\breve{\Pi}_{k+1}C_{k+1}^T(I+\Delta_{k+1})^T \\ \leq\ & \mathrm{tr}(C_{k+1}\breve{\Pi}_{k+1}C_{k+1}^T)[(I - \gamma_{k+1,1}\Upsilon\Upsilon)^{-1} + \gamma_{k+1,1}^{-1}I], \end{aligned} \tag{2.28}$$

$$\Delta_{k+1}R_{k+1}\Delta_{k+1}^T \leq \mathrm{tr}(\Upsilon R_{k+1}\Upsilon)I, \tag{2.29}$$

where $\gamma_{k+1,1} > 0$ is a scalar. Taking (2.28)–(2.29) into consideration, one gets

$$\begin{aligned} P_{k+1|k+1} \leq\ & (1+\varepsilon_5)(I - K_{k+1}\bar{\Lambda}_{k+1}C_{k+1})P_{k+1|k}(I - K_{k+1}\bar{\Lambda}_{k+1}C_{k+1})^T \\ & +(1+\varepsilon_4)K_{k+1}R_{k+1}K_{k+1}^T + (1+\varepsilon_5^{-1})\mathrm{tr}(C_{k+1}\breve{\Pi}_{k+1}C_{k+1}^T)K_{k+1} \\ & \times\breve{\Lambda}_{k+1}[(I - \gamma_{k+1,1}\Upsilon\Upsilon)^{-1} + \gamma_{k+1,1}^{-1}I]\breve{\Lambda}_{k+1}K_{k+1}^T + K_{k+1}\Big\{\breve{\Xi}_{k+1} \\ & \circ\big\{(1+\varepsilon_6^{-1})\mathrm{tr}(C_{k+1}\breve{\Pi}_{k+1}C_{k+1}^T)[(I - \gamma_{k+1,1}\Upsilon\Upsilon)^{-1} + \gamma_{k+1,1}^{-1}I] \\ & +\mathrm{tr}(\Upsilon R_{k+1}\Upsilon)I + (1+\varepsilon_6)C_{k+1}\breve{\Pi}_{k+1}C_{k+1}^T\big\}\Big\}K_{k+1}^T \\ & +(1+\varepsilon_4^{-1})\mathrm{tr}(\Upsilon R_{k+1}\Upsilon)K_{k+1}\breve{\Lambda}_{k+1}^2K_{k+1}^T. \end{aligned} \tag{2.30}$$

Then, it can be obtained from (2.16) (2.17) (2.24) and (2.30) that $P_{k+1|k+1} \leq \Sigma_{k+1|k+1}$.

In what follows, the main purpose is to minimize $\mathrm{tr}(\Sigma_{k+1|k+1})$ and design the filter gain matrix. Noting Lemma 2.3 and computing the partial derivative of $\mathrm{tr}(\Sigma_{k+1|k+1})$ concerning on K_{k+1} yield

$$\begin{aligned} & \frac{\partial\mathrm{tr}(\Sigma_{k+1|k+1})}{\partial K_{k+1}} \\ =\ & -2(1+\varepsilon_5)(I - K_{k+1}\bar{\Lambda}_{k+1}C_{k+1})\Sigma_{k+1|k}C_{k+1}^T\bar{\Lambda}_{k+1} + 2(1+\varepsilon_4)K_{k+1}R_{k+1} \\ & +2(1+\varepsilon_5^{-1})\mathrm{tr}(C_{k+1}\bar{\Pi}_{k+1}C_{k+1}^T)K_{k+1}\breve{\Lambda}_{k+1}[(I - \gamma_{k+1,1}\Upsilon\Upsilon)^{-1} + \gamma_{k+1,1}^{-1}I] \\ & \times\breve{\Lambda}_{k+1} + 2(1+\varepsilon_4^{-1})\mathrm{tr}(\Upsilon R_{k+1}\Upsilon)K_{k+1}\breve{\Lambda}_{k+1}^2 + 2K_{k+1}\Psi_{k+1}, \end{aligned} \tag{2.31}$$

where $\bar{\Pi}_{k+1}$ and Ψ_{k+1} are given in (2.19). Letting (2.31) be zero, one has

$$\begin{aligned} K_{k+1} = & (1+\varepsilon_5)\Sigma_{k+1|k}C_{k+1}^T\bar{\Lambda}_{k+1}\Big\{(1+\varepsilon_4)R_{k+1} + (1+\varepsilon_5^{-1})\mathrm{tr}(C_{k+1} \\ & \times\bar{\Pi}_{k+1}C_{k+1}^T)\breve{\Lambda}_{k+1}[(I-\gamma_{k+1,1}\Upsilon\Upsilon)^{-1} + \gamma_{k+1,1}^{-1}I]\breve{\Lambda}_{k+1} \\ & +(1+\varepsilon_5)\bar{\Lambda}_{k+1}C_{k+1}\Sigma_{k+1|k}C_{k+1}^T\bar{\Lambda}_{k+1} \\ & +(1+\varepsilon_4^{-1})\mathrm{tr}(\Upsilon R_{k+1}\Upsilon)\breve{\Lambda}_{k+1}^2 + \Psi_{k+1}\Big\}^{-1}, \end{aligned} \tag{2.32}$$

which is same with (2.18). Consequently, the proof is finished now.

Remark 2.3 As depicted in Theorem 2.3, the upper bound of the FEC is optimized by constructing the filter gain K_{k+1}. It is crucial to emphasize that $\gamma_{k+1,1}$ can be selected based on the constraint condition $\gamma_{k+1,1}^{-1}I - \Upsilon\Upsilon > 0$. After that, $\gamma_{k+1,1}$ can be adjusted to enhance the solvability of the filtering algorithm on the premise of guaranteeing certain filtering accuracy. Hence, the proposed filtering method has certain flexibility with the loss of certain estimation accuracy. Besides, it is crucial to point out that the newly designed filtering algorithm is suitable for online implementations due to its recursive way.

According to Theorem 2.3, the robust variance-constrained filtering (RVCF) algorithm is provided to facilitate the implementation.

RVCF Algorithm:
Step 1 : Set $k = 0$ and select the initial values.
Step 2 : Compute the one-step prediction $\hat{x}_{k+1|k}$ based on (2.10).
Step 3 : Calculate the value of $\Sigma_{k+1|k}$ by (2.16).
Step 4 : Solve the estimator gain matrix K_{k+1} by (2.18).
Step 5 : Compute the filtering update equation $\hat{x}_{k+1|k+1}$ by (2.11).
Step 6 : Obtain $\Sigma_{k+1|k+1}$ by (2.17).
Step 7 : Set $k = k + 1$, and go to *Step* 2.

2.3 Boundedness Analysis

In this section, the boundedness of the filtering error is discussed and proved by proposing the sufficient condition. To proceed, the definition of exponential boundedness of stochastic process is provided.

Definition 2.1 (Reif et al. 1999) If there are scalars $\varrho > 0$, $\nu > 0$, and $0 < \vartheta < 1$ such that

$$\mathbb{E}\{\|\zeta_k\|^2\} \leq \varrho\|\zeta_0\|^2\vartheta^k + \nu \tag{2.33}$$

holds, then the stochastic process ζ_k is exponentially mean-square bounded.

For the sake of conducting the boundedness analysis with regard to the filtering error, the following assumption is provided.

Assumption 2.1 For every $i \in \{1, 2, \ldots, m\}$ and $k \geq 0$, there are positive scalars $\overline{a}$, $\underline{c}$, $\overline{c}$, $\overline{h}$, $\overline{m}$, $\overline{l}_1$, $\overline{l}_2$ $\overline{f}$, $\underline{b}_1$, $\overline{b}_1$, $\underline{\omega}$, $\overline{\omega}$, $\overline{\nu}$, $\underline{\lambda}$ and $\overline{\lambda}$ such that

$$\begin{aligned}&\|A_k\| \leq \overline{a}, \quad \|H_k\| \leq \overline{h}, \quad \|M_k\| \leq \overline{m}, \quad \underline{c} \leq \|C_k\| \leq \overline{c}, \quad \underline{\lambda} \leq \bar{\lambda}_{k,i} \leq \overline{\lambda}, \\ &\mathrm{tr}(\bar{\mathscr{L}}_k) \leq \overline{l}_1, \quad \mathrm{tr}(\Omega_k) \leq \overline{f}, \quad \mathrm{tr}(\bar{\Pi}_{k+1}) \leq \overline{l}_2, \quad \underline{b}_1 I \leq B_k B_k^T \leq \overline{b}_1 I, \\ &\underline{\omega} I \leq Q_k \leq \overline{\omega} I, \quad R_k \leq \overline{\nu} I.\end{aligned}$$

Besides, the following inequality:

$$\overline{a}\left(1 + \frac{\overline{c}^2}{\underline{c}^2}\right) < 1 \tag{2.34}$$

holds.

Theorem 2.4 *For the time-varying systems (2.1)–(2.2) and the filter (2.10)–(2.11), under Assumption 2.1, the filtering error $\tilde{x}_{k|k}$ is exponentially mean-square bounded.*

Proof Substituting (2.12) into (2.13) yields

$$\tilde{x}_{k+1|k+1} = \check{A}_{k+1}\tilde{x}_{k|k} + r_{k+1} + z_{k+1}, \tag{2.35}$$

where

$$\begin{aligned}\check{A}_{k+1} &= \Xi_{k+1} A_k, \\ \Xi_{k+1} &= I - K_{k+1}\bar{\Lambda}_{k+1}C_{k+1}, \\ r_{k+1} &= \bar{\alpha}_k \Xi_{k+1}\Delta A_k x_k - K_{k+1}\check{\Lambda}_{k+1}(I + \Delta_{k+1})C_{k+1}x_{k+1}, \\ z_{k+1} &= \tilde{\alpha}_k \Xi_{k+1}\Delta A_k x_k + \Xi_{k+1} f(x_k, \xi_k) + \Xi_{k+1}B_k\omega_k - K_{k+1}\tilde{\Lambda}_{k+1}C_{k+1}x_{k+1} \\ &\quad - K_{k+1}\nu_{k+1} + K_{k+1}\tilde{\Lambda}_{k+1}(I + \Delta_{k+1})C_{k+1}x_{k+1} - K_{k+1}\check{\Lambda}_{k+1}\Delta_{k+1} \\ &\quad \times \nu_{k+1} + K_{k+1}\tilde{\Lambda}_{k+1}\Delta_{k+1}\nu_{k+1}.\end{aligned}$$

Based on (2.18) and Assumption 2.1, it is obvious to derive that

$$\begin{aligned}\|K_{k+1}\| &< \|(1 + \varepsilon_5)\Sigma_{k+1|k}C_{k+1}^T\bar{\Lambda}_{k+1}[(1 + \varepsilon_5)\bar{\Lambda}_{k+1}C_{k+1}\Sigma_{k+1|k}C_{k+1}^T\bar{\Lambda}_{k+1}]^{-1}\| \\ &\leq \frac{\overline{c}}{\underline{c}^2\underline{\lambda}} := \overline{k}\end{aligned}$$

and

$$\begin{aligned}\|\Xi_{k+1}\| &< \|I - (1 + \varepsilon_5)\Sigma_{k+1|k}C_{k+1}^T\bar{\Lambda}_{k+1}[(1 + \varepsilon_5)\bar{\Lambda}_{k+1}C_{k+1}\Sigma_{k+1|k}C_{k+1}^T \\ &\quad \times \bar{\Lambda}_{k+1}]^{-1}\bar{\Lambda}_{k+1}C_{k+1}\| \\ &\leq 1 + \frac{\overline{c}^2}{\underline{c}^2} := \overline{\varsigma}_1.\end{aligned}$$

Then, one has

$$\|\breve{A}_{k+1}\| = \|\Xi_{k+1} A_k\| \le \|\Xi_{k+1}\| \|A_k\| \le \overline{\varsigma}_1 \overline{a} := \overline{a}_1.$$

With the aid of Lemma 2.2 and Assumption 2.1, one obtains

$$\begin{aligned} \mathbb{E}\{r_{k+1}^T r_{k+1}\} \le\ & \mathbb{E}\{(1+\sigma_1)\bar{\alpha}_k^2 x_k^T \Delta A_k^T \Xi_{k+1}^T \Xi_{k+1} \Delta A_k x_k + (1+\sigma_1^{-1}) x_{k+1}^T C_{k+1}^T \\ & \times (I+\Delta_{k+1})^T \breve{A}_{k+1} K_{k+1}^T K_{k+1} \breve{A}_{k+1} (I+\Delta_{k+1}) C_{k+1} x_{k+1}\} \\ \le\ & (1+\sigma_1)\bar{\alpha}_k^2 \overline{\varsigma}_1^2 \overline{h}^2 \overline{m}^2 \overline{l}_1 + (1+\sigma_1^{-1})(1-\underline{\lambda})^2 (1+\delta)^2 \overline{c}^2 \overline{k}^2 \overline{l}_2 \\ :=\ & \overline{r}^2, \end{aligned}$$

where $\sigma_1 > 0$ is a scalar and $\delta = \max\{\delta_1, \delta_2, \ldots, \delta_m\}$. Further, one gets

$$\begin{aligned} \mathbb{E}\{z_{k+1}^T z_{k+1}\} \le\ & (\bar{\alpha}_k - \bar{\alpha}_k^2)\overline{h}^2 \overline{\varsigma}_1^2 \overline{m}^2 \overline{l}_1 + \overline{\varsigma}_1^2 \overline{f} + \overline{\varsigma}_1^2 l \overline{b}_1 \overline{\omega} + (1+\sigma_2)\overline{k}^2 \hat{\lambda}^2 \overline{c}^2 \overline{l}_2 \\ & + (1+\sigma_3)\overline{k}^2 m \overline{v} + (1+\sigma_3^{-1})\overline{k}^2 (1-\underline{\lambda})^2 \delta^2 m \overline{v} + (1+\sigma_2^{-1}) \\ & \times (1+\delta)^2 \overline{k}^2 \hat{\lambda}^2 \overline{c}^2 \overline{l}_2 + m \overline{k}^2 \hat{\lambda}^2 \delta^2 \overline{v} \\ :=\ & \overline{z}^2, \end{aligned}$$

where σ_2 and σ_3 are positive scalars and $\hat{\lambda} = \max\{1-\underline{\lambda}, \overline{\lambda}\}$.

In what follows, we consider the following recursive equation about Θ_k as:

$$\Theta_{k+1} = \breve{A}_{k+1} \Theta_k \breve{A}_{k+1}^T + B_k Q_k B_k^T, \tag{2.36}$$

under the initial condition $\Theta_0 = B_0 Q_0 B_0^T$. It is easy to see that

$$\begin{aligned} \|\Theta_{k+1}\| \le\ & \|\Theta_k\| \|\breve{A}_{k+1}\|^2 + \|B_k Q_k B_k^T\| \\ \le\ & \overline{a}_1^2 \|\Theta_k\| + \overline{\omega} \overline{b}_1. \end{aligned}$$

Moreover, it is clear that

$$\|\Theta_k\| \le \overline{a}_1^{2k} \|\Theta_0\| + \overline{\omega} \overline{b}_1 \sum_{i=0}^{k-1} \overline{a}_1^{2i}.$$

It follows from (2.34) that $0 < \overline{a}_1 < 1$ and we have

$$\begin{aligned} \|\Theta_k\| \le\ & \|\Theta_0\| + \overline{\omega} \overline{b}_1 \sum_{i=0}^{\infty} \overline{a}_1^{2i} \\ =\ & \|\Theta_0\| + \frac{\overline{b}_1 \overline{\omega}}{1-\overline{a}_1^2}. \end{aligned} \tag{2.37}$$

On account of the positive definite property of matrix Θ_k, it is immediate to know that

$$\Theta_{k+1} \geq B_k Q_k B_k^T \geq \underline{b}_1 \underline{\omega} I. \tag{2.38}$$

By virtue of (2.37) and (2.38), it indicates that there are $\underline{\theta} > 0$ and $\overline{\theta} > 0$ with $\underline{\theta} I \leq \Theta_k \leq \overline{\theta} I$ for each $k \geq 0$.

In terms of (2.36) and the matrix inversion lemma, it can be easily known that

$$\begin{aligned}
& \check{A}_{k+1}^T \Theta_{k+1}^{-1} \check{A}_{k+1} - \Theta_k^{-1} \\
= \ & \check{A}_{k+1}^T (\check{A}_{k+1} \Theta_k \check{A}_{k+1}^T + B_k Q_k B_k^T)^{-1} \check{A}_{k+1} - \Theta_k^{-1} \\
= \ & (\Theta_k + \check{A}_{k+1}^{-1} B_k Q_k B_k^T \check{A}_{k+1}^{-T})^{-1} - \Theta_k^{-1} \\
= \ & -\Theta_k^{-1} \check{A}_{k+1}^{-1} \big[(B_k Q_k B_k^T)^{-1} + \check{A}_{k+1}^{-T} \Theta_k^{-1} \check{A}_{k+1}^{-1}\big]^{-1} \check{A}_{k+1}^{-T} \Theta_k^{-1} \\
= \ & -\big[\check{A}_{k+1}^T (B_k Q_k B_k^T)^{-1} \check{A}_{k+1} \Theta_k + I\big]^{-1} \Theta_k^{-1} \\
\leq \ & -\left(\frac{\overline{a}_1^2 \overline{\theta}}{\underline{b}_1 \underline{\omega}} + 1\right)^{-1} \Theta_k^{-1}.
\end{aligned}$$

Let $\eta_0 = \left(\frac{\overline{a}_1^2 \overline{\theta}}{\underline{b}_1 \underline{\omega}} + 1\right)^{-1}$ and $V_k(\tilde{x}_{k|k}) = \tilde{x}_{k|k}^T \Theta_k^{-1} \tilde{x}_{k|k}$. Next, it is easy to observe that $\eta_0 \in (0, 1)$, and there exists $\beta > 0$ satisfying $\eta = (1 - \eta_0)(1 + \beta) < 1$. As a result, it can be derived from Lemma 2.2 and (2.35) that

$$\begin{aligned}
& \mathbb{E}\{V_{k+1}(\tilde{x}_{k+1|k+1}) | x_{k|k}\} - (1 + \beta) V_k(\tilde{x}_{k|k}) \\
= \ & \mathbb{E}\{\tilde{x}_{k+1|k+1}^T \Theta_{k+1}^{-1} \tilde{x}_{k+1|k+1} | \tilde{x}_{k|k}\} - (1 + \beta) V_k(\tilde{x}_{k|k}) \\
= \ & \mathbb{E}\{(\check{A}_{k+1} \tilde{x}_{k|k} + r_{k+1} + z_{k+1})^T \Theta_{k+1}^{-1} (\check{A}_{k+1} x_{k|k} + r_{k+1} + z_{k+1}) | \tilde{x}_{k|k}\} \\
& -(1 + \beta) V_k(\tilde{x}_{k|k}) \\
\leq \ & \mathbb{E}\{(1 + \beta) \tilde{x}_{k|k}^T \check{A}_{k+1}^T \Theta_{k+1}^{-1} \check{A}_{k+1} \tilde{x}_{k|k} - (1 + \beta) \tilde{x}_{k|k}^T \Theta_k^{-1} \tilde{x}_{k|k} | \tilde{x}_{k|k}\} \\
& +(1 + \beta^{-1}) \mathbb{E}\{r_{k+1}^T \Theta_{k+1}^{-1} r_{k+1} | \tilde{x}_{k|k}\} + \mathbb{E}\{z_{k+1}^T \Theta_{k+1}^{-1} z_{k+1} | \tilde{x}_{k|k}\} \\
= \ & (1 + \beta) \mathbb{E}\{\tilde{x}_{k|k}^T [\check{A}_{k+1}^T \Theta_{k+1}^{-1} \check{A}_{k+1} - \Theta_k^{-1}] \tilde{x}_{k|k} | \tilde{x}_{k|k}\} + \mathbb{E}\{z_{k+1}^T \Theta_{k+1}^{-1} z_{k+1} | \tilde{x}_{k|k}\} \\
& +(1 + \beta^{-1}) \mathbb{E}\{r_{k+1}^T \Theta_{k+1}^{-1} r_{k+1} | \tilde{x}_{k|k}\} \\
\leq \ & -\eta_0 (1 + \beta) V_k(\tilde{x}_{k|k}) + \upsilon,
\end{aligned}$$

where $\upsilon = \frac{(1+\beta^{-1})\overline{r}^2 + \overline{z}^2}{\underline{\theta}}$. Consequently, we have

$$\mathbb{E}\{V_{k+1}(\tilde{x}_{k+1|k+1}) | \tilde{x}_{k|k}\} \leq \eta V_k(\tilde{x}_{k|k}) + \upsilon.$$

By iteration and $\frac{1}{\overline{\theta}} I \leq \Theta_k^{-1} \leq \frac{1}{\underline{\theta}} I$, we arrive at

$$\begin{aligned}\mathbb{E}\{\|\tilde{x}_{k|k}\|^2\} &\leq \frac{\overline{\theta}}{\underline{\theta}}\|\tilde{x}_{0|0}\|^2\eta^k + \upsilon\overline{\theta}\sum_{i=0}^{\infty}\eta^i \\ &= \frac{\overline{\theta}}{\underline{\theta}}\|\tilde{x}_{0|0}\|^2\eta^k + \frac{\upsilon\overline{\theta}}{1-\eta},\end{aligned}$$

under $0 < \eta < 1$. Next, it can be obtained from Definition 2.1 that $\tilde{x}_{k|k}$ is exponentially bounded in the mean-square sense, which completes the proof.

Remark 2.4 Notice that some effective filtering approaches have been reported in Ma et al. (2017, 2018) for networked systems subject to energy bounded noises. Unlike (Ma et al. 2017, 2018), a novel RVCF scheme has been designed for uncertain time-varying nonlinear systems with ROQ and stochastic noises under variance-constrained sense. To be specific, it is crucial to point out that the merits of the developed filtering scheme include its local optimality under the minimum variance sense and online applications. In addition, the developed filtering approach can be extended to deal with the mean-square consensus issue for time-varying multi-agent systems (Ma et al. 2017).

2.4 An Illustrative Example

In this section, the comparative simulation experiment is used to illustrate the effectiveness of the developed filtering scheme.

The corresponding system parameters in (2.1) and (2.2) are given by

$$A_k = \begin{bmatrix} 0.6 - 0.6\cos(k) & -0.36 \\ 0.5 - \sin(k)\cos(k) & 0.65 + 0.4\cos(k) \end{bmatrix}, \quad B_k = \begin{bmatrix} 0.095 \\ 0.095 - 1.5\sin(k) \end{bmatrix},$$
$$F_k = \sin(2k), \quad H_k = \begin{bmatrix} 0.012 \; 0.024 \end{bmatrix}^T, \quad M_k = \begin{bmatrix} 0.042 \; 0.014 \end{bmatrix},$$
$$Q_k = 0.04, \; R_k = 0.05, \; C_k = \begin{bmatrix} 1 \; 0.9 \end{bmatrix}.$$

The stochastic nonlinearity $f(x_k, \xi_k)$ is modeled by

$$f(x_k, \xi_k) = \begin{bmatrix} 0.3 \\ 0.2 \end{bmatrix} [0.2\text{sign}\left(x_{1,k}\right) x_{1,k}\xi_{1,k} + 0.3\text{sign}\left(x_{2,k}\right) x_{2,k}\xi_{2,k}],$$

where $x_k = \begin{bmatrix} x_{1,k} \; x_{2,k} \end{bmatrix}^T$ and $\xi_{i,k}$ $(i = 1, 2)$ denote zero-mean noises with unity covariances. It is not difficult to see that

$$\Pi_1 = \begin{bmatrix} 0.09 \; 0.06 \\ 0.06 \; 0.06 \end{bmatrix}, \quad \Gamma_1 = \begin{bmatrix} 0.04 & 0 \\ 0 & 0.09 \end{bmatrix}.$$

The related parameters regarding the quantizer are selected as $u_0^1 = 0.6$ and $\chi^{(1)} = 0.03$. Other parameters are chosen as $\varepsilon_1 = 0.01$, $\varepsilon_2 = 0.08$, $\varepsilon_3 = 0.06$, $\varepsilon_4 =$

0.02, $\varepsilon_5 = 0.03$, $\varepsilon_6 = 0.8$, $\gamma_{k+1,1} = [\lambda_{\max}(\Upsilon\Upsilon) + 0.5]^{-1}$, $\bar{\alpha}_k = 0.7$, $\bar{\Lambda}_k = 0.4$, $x_0 = \hat{x}_{0|0} = \begin{bmatrix} 2 & 2 \end{bmatrix}^T$ and $\Sigma_{0|0} = 3I_2$. Here, MSEi ($i = 1, 2$) represent the mean-square errors (MSEs) for the estimations of the states $x_{i,k}$ ($i = 1, 2$).

Figure 2.1 depicts the measurements with and without ROQ. For the purpose of reflecting the superiority of the proposed scheme, the system states and their estimations are exhibited in Figs. 2.2 and 2.3 on the basis of the established recursive variance-constrained filtering approach and the traditional Kalman filter (KF) scheme. The obtained upper bound and log(MSEi) ($i = 1, 2$) are plotted in Figs. 2.4 and 2.5, which can be obviously seen that the MSEs are indeed below the upper bound. The log(MSEi) ($i = 1, 2$) of the filtering scheme proposed in this chapter and the traditional KF method are given in Figs. 2.6 and 2.7, where we can find that the developed filtering algorithm has the smaller error than the traditional KF approach.

In order to analyze the impacts from the ROQ onto the filtering algorithm, the traces of the proposed upper bounds are given in Fig. 2.8 with $\bar{\Lambda}_k = 0.2$, $\bar{\Lambda}_k = 0.7$, $\bar{\Lambda}_k = 0.8$ and $\bar{\Lambda}_k = 0.9$. It can be observed from this figure that the filtering algorithm performance is enhanced if less quantized measurements are employed, namely, more original measurements are delivered to the filter side and then the better filtering precision is obtained.

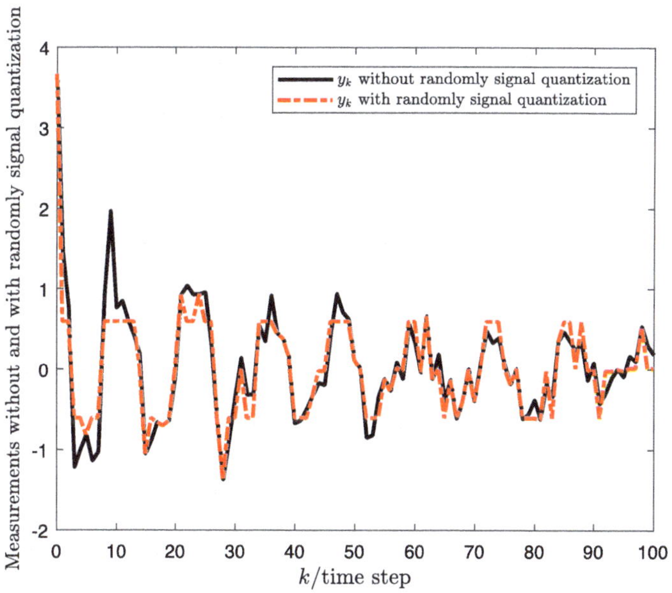

Fig. 2.1 y_k without and with randomly occurring signal quantization

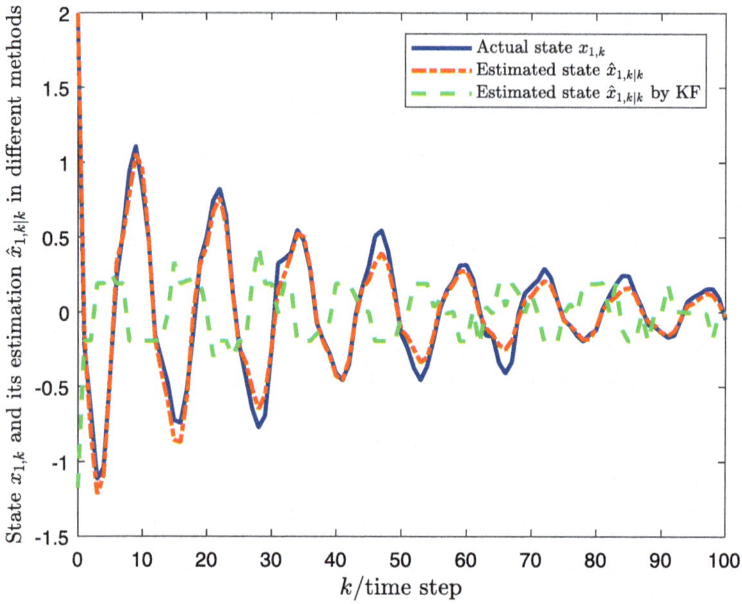

Fig. 2.2 State $x_{1,k}$ and its estimation $\hat{x}_{1,k|k}$

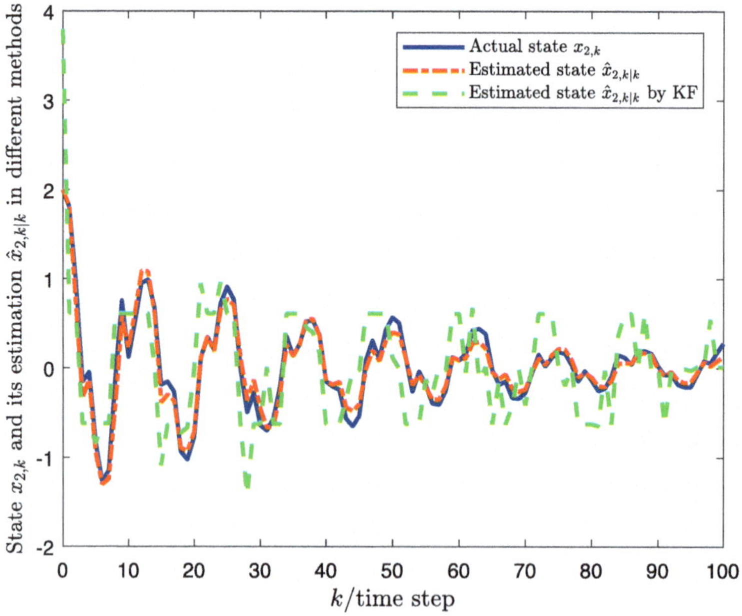

Fig. 2.3 State $x_{2,k}$ and its estimation $\hat{x}_{2,k|k}$

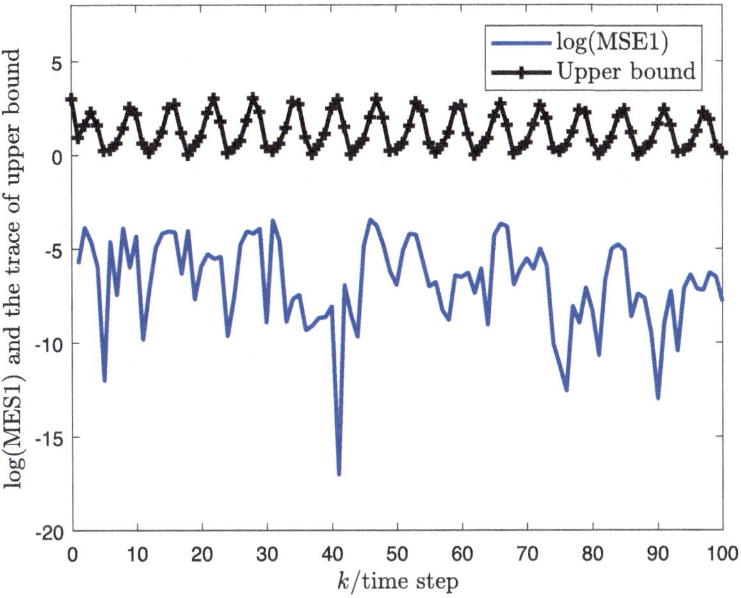

Fig. 2.4 log(MSE1) and its upper bound

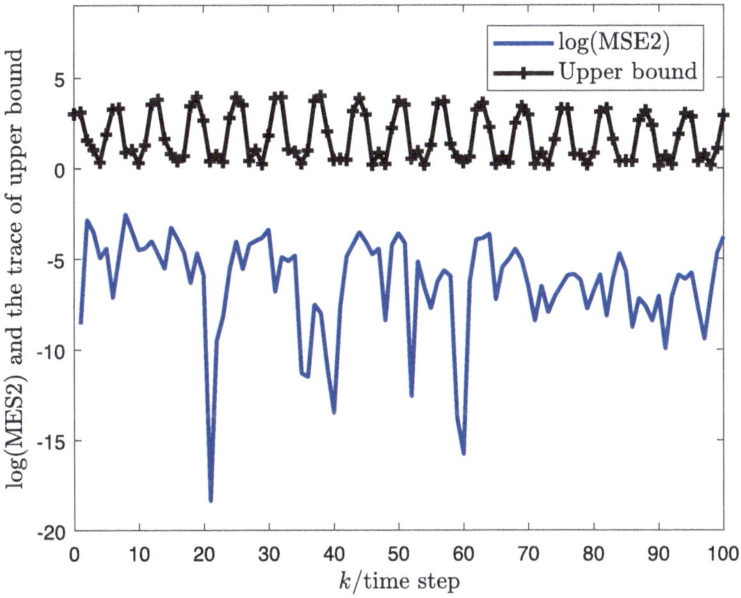

Fig. 2.5 log(MSE2) and its upper bound

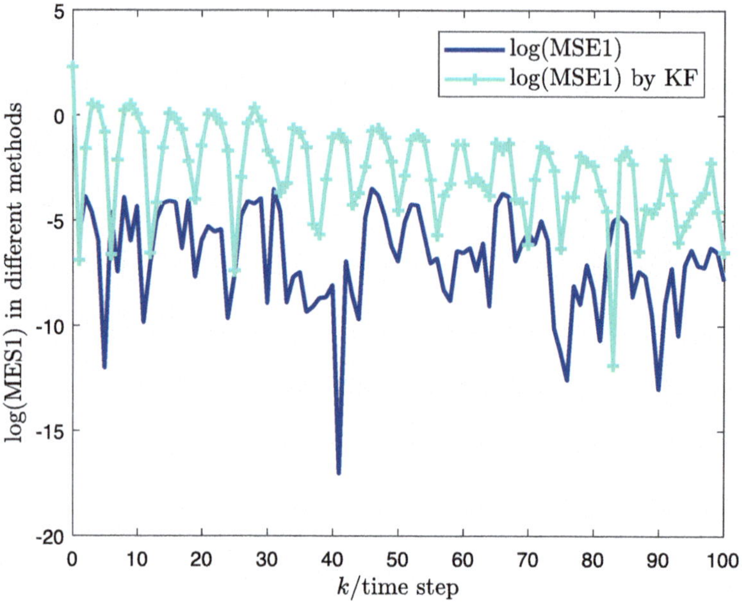

Fig. 2.6 log(MSE1) in different methods

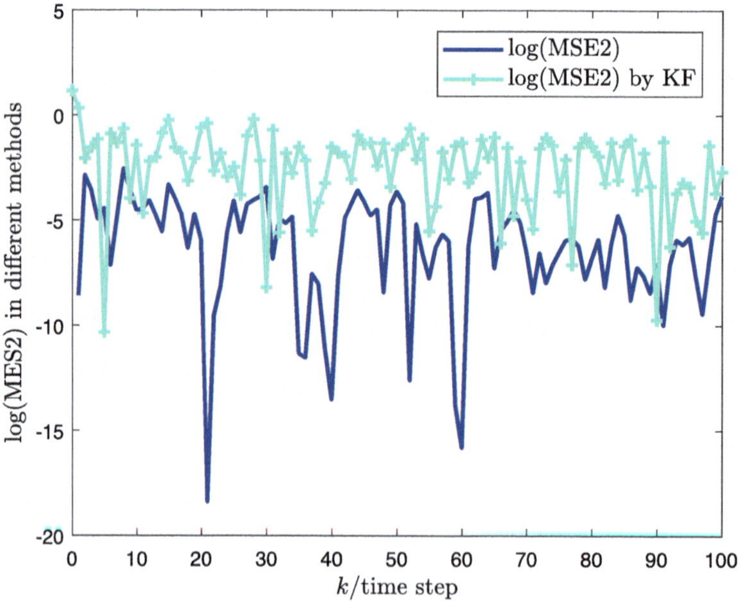

Fig. 2.7 log(MSE2) in different methods

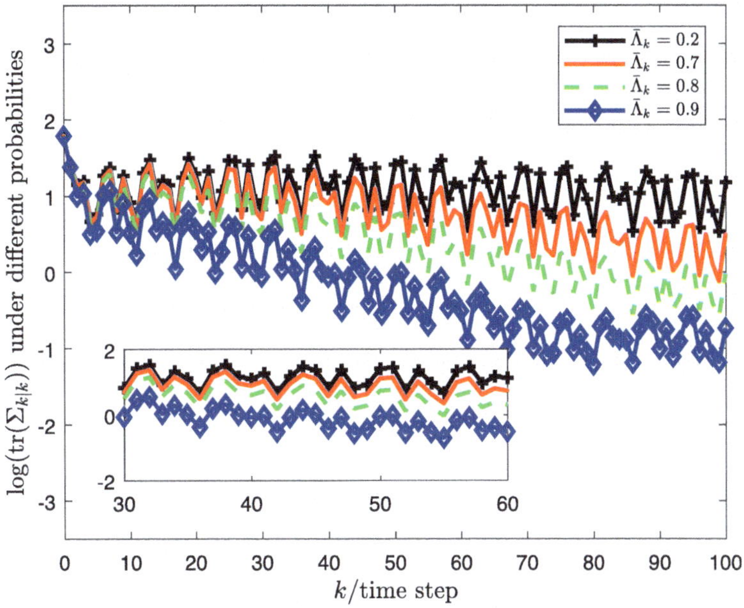

Fig. 2.8 $\log(\mathrm{tr}(\Sigma_{k|k}))$ under different occurrence probabilities

2.5 Summary

In this chapter, the robust filtering issue under variance-constraint has been handled for time-varying nonlinear stochastic systems with stochastic nonlinearity, ROUs and ROQ. In particular, the ROUs and ROQ have been described through the mutually independent random variables satisfying the Bernoulli distribution. A variance-constrained filtering strategy has been presented, in which the desirable filter gain matrix has been constructed to optimize the obtained upper bound concerning on the FEC. Besides, a sufficient criterion has been given to guarantee the exponential mean square boundedness of the filtering error dynamics. At last, some simulation experiments with comparisons have been provided to demonstrate the advantages of the designed filtering method.

References

Elia N, Mitter SK (2001) Stabilization of linear systems with limited information. IEEE Trans Autom Control 46(9):1384–1400

Fu M, Xie L (2005) The sector bound approach to quantized feedback control. IEEE Trans Autom Control 50(11):1698–1711

Hardy GH, Littlewood JE, Pólya G (1988) Inequalities, 2nd edn. Cambridge University Press

Horn RA, Johnson CR (1991) Topics in matrix analysis. Cambridge University Press, New York

Ma L, Wang Z, Lam HK, Kyriakoulis N (2017) Distributed event-based set-membership filtering for a class of nonlinear systems with sensor saturations over sensor networks. IEEE Trans Cybern 47(11):3772–3783

Ma L, Wang Z, Han QL, Lam HK (2018) Envelope-constrained H_∞ filtering for nonlinear systems with quantization effects: The finite horizon case. Automatica 93:527–534

Ma L, Wang Z, Han QL, Liu Y (2017) Consensus control of stochastic multi-agent systems: a survey. Sci China-Inf Sci 60:120201

Reif K, Günther S, Yaz E, Unbehauen R (1999) Stochastic stability of the discrete-time extended Kalman filter. IEEE Trans Autom Control 44(4):714–728

Wang Y, Xie L, de Souza CE (1992) Robust control of a class of uncertain nonlinear systems. Syst Control Lett 19(2):139–149

Chapter 3
Resilient Filtering with Stochastic Uncertainties and Incomplete Measurements

In this chapter, the resilient filtering problem is addressed for networked time-varying nonlinear systems with stochastic uncertainties and incomplete measurements. In particular, the incomplete measurements including ROQ and missing measurements are described by resorting to the Kronecker delta function in a unified framework. Moreover, the filter gain perturbations are modeled and discussed in order to reflect the practical situation. In light of the available measurements, a novel filtering scheme with compensation ability is proposed to guarantee that an upper bound of the FEC is ensured and this upper bound is minimized by constructing an appropriate filter gain. It is revealed that the expression form of the expected filter gain can be derived via solving two sets of matrix difference equations, and the designed filtering scheme of recursive way is suitable for online calculations and applications. Finally, a simulation experiment is provided to validate the effectiveness of developed resilient filtering method.

3.1 Problem Formulation

The resilient filter design problem is discussed, where the stochastic uncertainties, incomplete measurements and stochastic perturbations regarding the filter gain are all involved and considered. Specifically, the stochastic uncertainties are considered to characterize the random modelling errors, and the incomplete measurements contain the ROQ and missing measurements simultaneously. Besides, the filtering gain perturbations are also considered in order to better cater the engineering reality with certain robustness requirement when designing the filtering method. Subsequently, we model the physical plant, stochastic uncertainties as well as incomplete measurements in a mathematical manner.

J. Hu et al., *Variance-Constrained Filtering for Stochastic Complex Systems*, Intelligent Control and Learning Systems 18,
https://doi.org/10.1007/978-981-96-2637-3_3

The physical plant and the *ideal* measurement are given by

$$x_{k+1} = f(x_k) + \eta_k B_k \phi(x_k) + D_k \omega_k, \tag{3.1}$$

$$y_k = C_k x_k, \tag{3.2}$$

where $x_k \in \mathbb{R}^n$ is the system state, x_0 is the initial value with mean $\bar{x}_0$ and covariance $P_0 > 0$, $y_k \in \mathbb{R}^m$ is the ideal measurement, $\eta_k \in \mathbb{R}$ denotes a zero-mean multiplicative noise with unity variance, and $\omega_k \in \mathbb{R}^r$ represents the process noise with zero-mean and covariance $Q_k > 0$. The nonlinear function $f(x_k)$ is analytic everywhere with continuous and bounded second-order partial derivative. $\phi(x_k)$ is an unknown nonlinear function with $\mathbb{E}\{\phi(x_k)\phi^T(x_k)\} \leq \Theta_k$, where Θ_k is a known matrix. B_k, C_k and D_k are known matrices of proper dimensions.

In the real-world cases, the quantization commonly occurs on account of the limited transmission capacity of the network communication channels. Hence, it is essential to consider the signal quantization when designing the filter for the networked systems. In this chapter, the quantization process can be characterized as

$$q(y_k) = \left[q_1(y_k^{(1)}) \; q_2(y_k^{(2)}) \; \cdots \; q_m(y_k^{(m)}) \right]^T .$$

The set of quantization levels of each $q_j(\cdot)$ $(j = 1, 2, \ldots, m)$ is defined as

$$\mathfrak{U}_j = \left\{ \pm u_i^{(j)}, \; u_i^{(j)} = \left(\chi^{(j)}\right)^i u_0^{(j)}, \; i = 0, \pm 1, \pm 2, \ldots \right\} \cup \{0\},$$

where $\chi^{(j)} \in (0, 1)$ are the quantization densities and $u_i^{(j)}$ represent the quantization levels with $u_0^{(j)} > 0$.

In what follows, the quantizer is supposed to be of the logarithmic form

$$q_j(y_k^{(j)}) = \begin{cases} u_i^{(j)}, & \frac{1}{1+\sigma_j} u_i^{(j)} < y_k^{(j)} \leq \frac{1}{1-\sigma_j} u_i^{(j)} \\ 0, & y_k^{(j)} = 0 \\ -q_j(-y_k^{(j)}), & y_k^{(j)} < 0 \end{cases}$$

where $\sigma_j = \frac{1-\chi^{(j)}}{1+\chi^{(j)}}$. From the mentioned definition, it is easy to observe that

$$q_j(y_k^{(j)}) = \left(1 + \Im_k^{(j)}\right) y_k^{(j)},$$

where $|\Im_k^{(j)}| \leq \sigma_j$.

The *actual* measurement received by the filter is of following mathematical form:

$$\tilde{y}_k = \delta_{\alpha_k,1} y_k + \delta_{\alpha_k,2} q(y_k) + \zeta_k E_k \psi(x_k) + \nu_k, \tag{3.3}$$

where $\zeta_k \in \mathbb{R}$ is a zero-mean multiplicative noise with unity variance, E_k is a known matrix of proper dimension, $\psi(x_k)$ is an unknown nonlinear function satisfying $\mathbb{E}\{\psi(x_k)\psi^T(x_k)\} \leq \Psi_k$ with Ψ_k being a known matrix, ν_k represents the measurement noise with zero-mean and covariance $R_k > 0$, and $\delta_{k,l}$ is a Kronecker delta function satisfying

$$\delta_{k,l} = \begin{cases} 1, & k = l \\ 0, & k \neq l. \end{cases} \tag{3.4}$$

The random variable α_k obeys the following probability distribution:

$$\text{Prob}\{\alpha_k = j_1\} = \lambda_{j_1}, \quad j_1 = 1, 2, 3, \tag{3.5}$$

where $\lambda_{j_1} \in [0, 1]$ are known scalars with $\sum_{j_1=1}^{3} \lambda_{j_1} = 1$.

Setting $\Im_k = \text{diag}\left\{\Im_k^{(1)}, \Im_k^{(2)}, \ldots, \Im_k^{(m)}\right\}$, together with (3.2) and (3.3), the measurements transmitted to the remote filter can be rewritten as

$$\tilde{y}_k = \delta_{\alpha_k,1} C_k x_k + \delta_{\alpha_k,2}(I + \Im_k) C_k x_k + \zeta_k E_k \psi(x_k) + \nu_k. \tag{3.6}$$

Letting $\Lambda = \text{diag}\{\sigma_1, \sigma_2, \ldots, \sigma_m\}$ and setting $F_k = \Im_k \Lambda^{-1}$, we have F_k satisfying $F_k F_k^T = F_k^T F_k \leq I$.

Remark 3.1 As given in Shen et al. (2013), a new measurement model is employed in (3.3) to model the phenomenon of incomplete measurements. If $\alpha_k = 1$, it means that the sensors can normally work; if $\alpha_k = 2$, the signal quantization occurs; and if $\alpha_k = 3$, the measurements contain the noises and stochastic uncertainties only. The proposed model (3.3) is fairly comprehensive, which can better reflect the engineering practice.

In this chapter, we design the following resilient filter:

$$\hat{x}_{k+1|k} = f(\hat{x}_{k|k}), \tag{3.7}$$

$$\hat{x}_{k+1|k+1} = \hat{x}_{k+1|k} + (K_{k+1} + \Delta K_{s,k+1})(\tilde{y}_{k+1} - \lambda_1 C_{k+1}\hat{x}_{k+1|k}), \tag{3.8}$$

where $\hat{x}_{k|k}$ is the state estimate, $\hat{x}_{k+1|k}$ is the one-step prediction, K_{k+1} is the filter gain, and $\Delta K_{s,k+1}$ represents the stochastic perturbations of the filter gain characterized by

$$\Delta K_{s,k+1} = \varrho_{k+1} K_{s,k+1}. \tag{3.9}$$

Here, $K_{s,k+1}$ is a known real matrix with suitable dimensions, and $\varrho_{k+1} \in \mathbb{R}$ is a zero-mean multiplicative noise with unity covariance. Throughout this chapter, we assume that all noise signals are mutually independent.

Remark 3.2 In Eqs. (3.7)–(3.8), the stochastic perturbations of the filter gain are considered when designing the filter for time-varying networked systems. In particular, the parameter ϱ_{k+1} is introduced to reflect the stochastic perturbations in the filter gains. As discussed in Yaz et al. (2006), such perturbations may be caused by the finite word length, finite resolution of the instrumentation and manual tuning to achieve the desired filtering performance. During the implementation, a zero-mean multiplicative noise is employed to represent the parameter ϱ_{k+1}. The "zero-mean" property of the multiplicative noise can describe the physical situation where the perturbations can be positive or negative in an equally likely manner for the components of the filter gains during the operation. On the other hand, it is worth mentioning that the occurrence probability λ_1 is introduced in Eq. (3.8) with hope to compensate the influences of the incomplete measurements.

Let the filtering error be $\tilde{x}_{k+1|k+1} = x_{k+1} - \hat{x}_{k+1|k+1}$ and the FEC be $P_{k+1|k+1} = \mathbb{E}\{\tilde{x}_{k+1|k+1}\tilde{x}_{k+1|k+1}^T\}$. The objective of this chapter is to construct the filter gain K_{k+1} in Eq. (3.8) such that the following two objectives are achieved.

(R1) There is a set of matrices $\Sigma_{k+1|k+1} > 0$ satisfying the following inequality:

$$P_{k+1|k+1} \leq \Sigma_{k+1|k+1}. \tag{3.10}$$

(R2) Such an upper bound $\Sigma_{k+1|k+1}$ is minimized by appropriately constructing the filter gain K_{k+1}.

3.2 Main Results

In this section, for all stochastic uncertainties, incomplete measurement as well as stochastic perturbations of the filter gain, an upper bound regarding the FEC is calculated and then minimized at each sampling instant by appropriately constructing the filter gain.

Define the one-step prediction error as $\tilde{x}_{k+1|k} = x_{k+1} - \hat{x}_{k+1|k}$. Subtracting Eq. (3.7) from Eq. (3.1), one obtains

$$\tilde{x}_{k+1|k} = f(x_k) - f(\hat{x}_{k|k}) + \eta_k B_k \phi(x_k) + D_k \omega_k. \tag{3.11}$$

By employing the Taylor series expansion approach around $\hat{x}_{k|k}$, the nonlinear function $f(x_k)$ is addressed as

$$f(x_k) = f(\hat{x}_{k|k}) + A_k \tilde{x}_{k|k} + o(|\tilde{x}_{k|k}^2|), \tag{3.12}$$

where $A_k = \frac{\partial f(x_k)}{\partial x_k}|_{x_k=\hat{x}_{k|k}}$. Based on Hu et al. (2013), Xiong et al. (2010), Calafiore (2005), the high-order term $o(|\tilde{x}_{k|k}^2|)$ is approximated by

$$o(|\tilde{x}_{k|k}^2|) = M_k \Omega_k N_k \tilde{x}_{k|k}^2, \tag{3.13}$$

where M_k is a problem-dependent scaling matrix, N_k is a tuning matrix, and Ω_k is an unknown matrix satisfying $\Omega_k\Omega_k^T \leq I$.

According to Eqs. (3.11)–(3.13), the one-step prediction error is rewritten as

$$\tilde{x}_{k+1|k} = (A_k + M_k\Omega_k N_k)\,\tilde{x}_{k|k} + \eta_k B_k\phi(x_k) + D_k\omega_k. \tag{3.14}$$

Moreover, it follows from Eqs. (3.6) and (3.8) that

$$\begin{aligned}\tilde{x}_{k+1|k+1} = {} & [I - \lambda_1(K_{k+1} + \varrho_{k+1}K_{s,k+1})C_{k+1}]\tilde{x}_{k+1|k} \\ & -(\delta_{\alpha_{k+1},1} - \lambda_1)(K_{k+1} + \varrho_{k+1}K_{s,k+1})C_{k+1}x_{k+1} \\ & -\lambda_2(K_{k+1} + \varrho_{k+1}K_{s,k+1})(I + \Im_{k+1})C_{k+1}x_{k+1} \\ & -(\delta_{\alpha_{k+1},2} - \lambda_2)(K_{k+1} + \varrho_{k+1}K_{s,k+1})(I + \Im_{k+1})C_{k+1}x_{k+1} \\ & -\zeta_{k+1}(K_{k+1} + \varrho_{k+1}K_{s,k+1})E_{k+1}\psi(x_{k+1}) \\ & -(K_{k+1} + \varrho_{k+1}K_{s,k+1})\nu_{k+1}.\end{aligned} \tag{3.15}$$

To facilitate the filtering scheme developments, the nonlinearity $f(x_k)$ is linearized by utilizing the Taylor series expansion method, where the linearization errors are estimated and will be handled in the subsequent derivations. Based on the obtained one-step prediction error and the filtering error, the next step is to derive the concrete forms of the one-step prediction error covariance and FEC according to Eqs. (3.14) and (3.15).

Theorem 3.1 *The one-step prediction error covariance $P_{k+1|k} = \mathbb{E}\{\tilde{x}_{k+1|k}\tilde{x}_{k+1|k}^T\}$ and the FEC $P_{k+1|k+1} = \mathbb{E}\{\tilde{x}_{k+1|k+1}\tilde{x}_{k+1|k+1}^T\}$ obey the following recursions:*

$$\begin{aligned}P_{k+1|k} = {} & (A_k + M_k\Omega_k N_k)P_{k|k}(A_k + M_k\Omega_k N_k)^T \\ & +B_k\mathbb{E}\{\phi(x_k)\phi^T(x_k)\}B_k^T + D_kQ_kD_k^T\end{aligned} \tag{3.16}$$

and

$$\begin{aligned}P_{k+1|k+1} = {} & (I - \lambda_1K_{k+1}C_{k+1})P_{k+1|k}(I - \lambda_1K_{k+1}C_{k+1})^T \\ & +\lambda_1^2K_{s,k+1}C_{k+1}P_{k+1|k}C_{k+1}^TK_{s,k+1}^T - \mathscr{A}_{k+1} - \mathscr{A}_{k+1}^T \\ & +\lambda_1(1 - \lambda_1)K_{k+1}C_{k+1}\mathbb{E}\{x_{k+1}x_{k+1}^T\}C_{k+1}^TK_{k+1}^T \\ & +\lambda_1(1 - \lambda_1)K_{s,k+1}C_{k+1}\mathbb{E}\{x_{k+1}x_{k+1}^T\}C_{k+1}^TK_{s,k+1}^T + \mathscr{B}_{k+1} + \mathscr{B}_{k+1}^T \\ & +\lambda_2K_{k+1}(I + \Im_{k+1})C_{k+1}\mathbb{E}\{x_{k+1}x_{k+1}^T\}C_{k+1}^T(I + \Im_{k+1})^TK_{k+1}^T \\ & +\lambda_2K_{s,k+1}(I + \Im_{k+1})C_{k+1}\mathbb{E}\{x_{k+1}x_{k+1}^T\}C_{k+1}^T(I + \Im_{k+1})^TK_{s,k+1}^T \\ & +K_{k+1}E_{k+1}\mathbb{E}\{\psi(x_{k+1})\psi^T(x_{k+1})\}E_{k+1}^TK_{k+1}^T \\ & +K_{s,k+1}E_{k+1}\mathbb{E}\{\psi(x_{k+1})\psi^T(x_{k+1})\}E_{k+1}^TK_{s,k+1}^T \\ & +K_{k+1}R_{k+1}K_{k+1}^T + K_{s,k+1}R_{k+1}K_{s,k+1}^T,\end{aligned} \tag{3.17}$$

where

$$\mathscr{A}_{k+1} = \lambda_2 \mathbb{E}\Big\{[I - \lambda_1(K_{k+1} + \varrho_{k+1} K_{s,k+1})C_{k+1}]\tilde{x}_{k+1|k} x_{k+1}^T C_{k+1}^T \times (I + \Im_{k+1})^T (K_{k+1} + \varrho_{k+1} K_{s,k+1})^T\Big\},$$

$$\mathscr{B}_{k+1} = \mathbb{E}\Big\{(\delta_{\alpha_{k+1},1} - \lambda_1)(\delta_{\alpha_{k+1},2} - \lambda_2)(K_{k+1} + \varrho_{k+1} K_{s,k+1})C_{k+1} x_{k+1} \times x_{k+1}^T C_{k+1}^T (I + \Im_{k+1})^T (K_{k+1} + \varrho_{k+1} K_{s,k+1})^T\Big\}. \tag{3.18}$$

Proof According to the definition $P_{k+1|k} = \mathbb{E}\{\tilde{x}_{k+1|k}\tilde{x}_{k+1|k}^T\}$ and (3.14), it is not difficult to show that Eq. (3.16) is true.

Secondly, it follows from the definition $P_{k+1|k+1} = \mathbb{E}\{\tilde{x}_{k+1|k+1}\tilde{x}_{k+1|k+1}^T\}$ and (3.15) that we can obtain the below results

$$\begin{aligned}
& P_{k+1|k+1} \\
={}& \mathbb{E}\Big\{[I - \lambda_1(K_{k+1} + \varrho_{k+1} K_{s,k+1})]C_{k+1}\tilde{x}_{k+1|k}\tilde{x}_{k+1|k}^T C_{k+1}^T \\
& \times [I - \lambda_1(K_{k+1} + \varrho_{k+1} K_{s,k+1})]^T\Big\} - \mathscr{A}_{k+1} - \mathscr{A}_{k+1}^T + \lambda_1(1 - \lambda_1) \\
& \times \mathbb{E}\Big\{(K_{k+1} + \varrho_{k+1} K_{s,k+1})C_{k+1} x_{k+1} x_{k+1}^T C_{k+1}^T (K_{k+1} + \varrho_{k+1} K_{s,k+1})^T\Big\} \\
& + \mathscr{B}_{k+1} + \mathscr{B}_{k+1}^T + \lambda_2^2 \mathbb{E}\Big\{(K_{k+1} + \varrho_{k+1} K_{s,k+1})(I + \Im_{k+1})C_{k+1} x_{k+1} x_{k+1}^T C_{k+1}^T \\
& \times (I + \Im_{k+1})^T (K_{k+1} + \varrho_{k+1} K_{s,k+1})^T\Big\} + \lambda_2(1 - \lambda_2)\mathbb{E}\Big\{(K_{k+1} + \varrho_{k+1} K_{s,k+1}) \\
& \times (I + \Im_{k+1})C_{k+1} x_{k+1} x_{k+1}^T C_{k+1}^T (I + \Im_{k+1})^T (K_{k+1} + \varrho_{k+1} K_{s,k+1})^T\Big\} \\
& + \mathbb{E}\Big\{(K_{k+1} + \varrho_{k+1} K_{s,k+1})E_{k+1}\psi(x_{k+1})\psi^T(x_{k+1})E_{k+1}^T (K_{k+1} + \varrho_{k+1} K_{s,k+1})^T\Big\} \\
& + \mathbb{E}\Big\{(K_{k+1} + \varrho_{k+1} K_{s,k+1})R_{k+1}(K_{k+1} + \varrho_{k+1} K_{s,k+1})^T\Big\} \\
={}& (I - \lambda_1 K_{k+1} C_{k+1})P_{k+1|k}(I - \lambda_1 K_{k+1} C_{k+1})^T + \lambda_1^2 K_{s,k+1} C_{k+1} P_{k+1|k} C_{k+1}^T K_{s,k+1}^T \\
& - \mathscr{A}_{k+1} - \mathscr{A}_{k+1}^T + \lambda_1(1 - \lambda_1) K_{k+1} C_{k+1} \mathbb{E}\{x_{k+1} x_{k+1}^T\} C_{k+1}^T K_{k+1}^T \\
& + \lambda_1(1 - \lambda_1) K_{s,k+1} C_{k+1} \mathbb{E}\{x_{k+1} x_{k+1}^T\} C_{k+1}^T K_{s,k+1}^T + \mathscr{B}_{k+1} + \mathscr{B}_{k+1}^T \\
& + \lambda_2 K_{k+1}(I + \Im_{k+1})C_{k+1}\mathbb{E}\{x_{k+1} x_{k+1}^T\} C_{k+1}^T (I + \Im_{k+1})^T K_{k+1}^T \\
& + \lambda_2 K_{s,k+1}(I + \Im_{k+1})C_{k+1}\mathbb{E}\{x_{k+1} x_{k+1}^T\} C_{k+1}^T (I + \Im_{k+1})^T K_{s,k+1}^T \\
& + K_{k+1} E_{k+1}\mathbb{E}\{\psi(x_{k+1})\psi^T(x_{k+1})\} E_{k+1}^T K_{k+1}^T \\
& + K_{s,k+1} E_{k+1}\mathbb{E}\{\psi(x_{k+1})\psi^T(x_{k+1})\} E_{k+1}^T K_{s,k+1}^T \\
& + K_{k+1} R_{k+1} K_{k+1}^T + K_{s,k+1} R_{k+1} K_{s,k+1}^T,
\end{aligned} \tag{3.19}$$

where $\mathscr{A}_{k+1}$ and $\mathscr{B}_{k+1}$ are defined in (3.18). Thus, it is concluded that the specific form of FEC in (3.17) holds. Based on the above derivations, the proof of this theorem is complete.

Remark 3.3 It is crucial to highlight that there are some unknown terms Ω_k, $\mathscr{A}_{k+1}$, $\mathscr{B}_{k+1}$, $\mathbb{E}\left\{x_{k+1}x_{k+1}^T\right\}$, $\Im_{k+1}$, $\mathbb{E}\left\{\phi(x_{k+1})\phi^T(x_{k+1})\right\}$ and $\mathbb{E}\left\{\psi(x_{k+1})\psi^T(x_{k+1})\right\}$ in the one-step prediction error covariance $P_{k+1|k}$ and the FEC $P_{k+1|k+1}$. Hence, it is impossible to obtain exact value of $P_{k+1|k+1}$. In what follows, an alternative scheme is adopted to derive the filter gain, which has the easy-to-implement form. In other words, an upper bound concerning on the FEC is firstly solved and the filter gain with recursive expression is constructed to minimize the trace of this obtained upper bound.

Theorem 3.2 *Let ε_1, ε_2, ε_3, $\gamma_{1,k}$ and $\gamma_{2,k+1}$ denote positive real numbers. If the following Riccati-like difference equations:*

$$\begin{aligned}\Sigma_{k+1|k} = {} & A_k(\Sigma_{k|k}^{-1} - \gamma_{1,k}N_k^T N_k)^{-1}A_k^T + \gamma_{1,k}^{-1}M_k M_k^T \\ & + B_k \Theta_k B_k^T + D_k Q_k D_k^T \end{aligned} \tag{3.20}$$

and

$$\begin{aligned}\Sigma_{k+1|k+1} = {} & (1+\varepsilon_1)(I - \lambda_1 K_{k+1}C_{k+1})\Sigma_{k+1|k}(I - \lambda_1 K_{k+1}C_{k+1})^T \\ & + K_{s,k+1}\Bigg\{(1+\varepsilon_1)\lambda_1^2 C_{k+1}\Sigma_{k+1|k}C_{k+1}^T + \lambda_1(1-\lambda_1+\varepsilon_2\lambda_1)C_{k+1} \\ & \times \Xi_{k+1|k}C_{k+1}^T + R_{k+1} + E_{k+1}\Psi_{k+1}E_{k+1}^T + \lambda_2(1+\varepsilon_1^{-1}\lambda_2+\varepsilon_2^{-1}\lambda_2) \\ & \times \mathrm{tr}(\Phi_{k+1|k})\Big[(I - \gamma_{2,k+1}\Lambda\Lambda)^{-1} + \gamma_{2,k+1}^{-1}I\Big]\Bigg\}K_{s,k+1}^T \\ & + K_{k+1}\Bigg\{\lambda_1(1-\lambda_1+\varepsilon_2\lambda_1)C_{k+1}\Xi_{k+1|k}C_{k+1}^T + R_{k+1} \\ & + E_{k+1}\Psi_{k+1}E_{k+1}^T + \lambda_2(1+\varepsilon_1^{-1}\lambda_2^2+\varepsilon_2^{-1}\lambda_2^2)\mathrm{tr}(\Phi_{k+1|k}) \\ & \times\Big[(I - \gamma_{2,k+1}\Lambda\Lambda)^{-1} + \gamma_{2,k+1}^{-1}I\Big]\Bigg\}K_{k+1}^T \end{aligned} \tag{3.21}$$

under the initial condition $\Sigma_{0|0} = P_{0|0} > 0$ as well as

$$\gamma_{1,k}^{-1}I - N_k\Sigma_{k|k}N_k^T > 0, \tag{3.22}$$

$$\gamma_{2,k+1}^{-1}I - \Lambda\Lambda > 0 \tag{3.23}$$

have solutions $\Sigma_{k+1|k} > 0$ and $\Sigma_{k+1|k+1} > 0$, then the matrix $P_{k+1|k+1}$ has an upper bound $\Sigma_{k+1|k+1}$, i.e.,

$$P_{k+1|k+1} \le \Sigma_{k+1|k+1}, \tag{3.24}$$

where

$$\Lambda = \text{diag}\{\sigma_1, \sigma_2, \ldots, \sigma_m\}, \tag{3.25}$$

$$\Xi_{k+1|k} = (1+\varepsilon_3)\Sigma_{k+1|k} + (1+\varepsilon_3^{-1})\hat{x}_{k+1|k}\hat{x}_{k+1|k}^T, \tag{3.26}$$

$$\Phi_{k+1|k} = C_{k+1}\Xi_{k+1|k}C_{k+1}^T. \tag{3.27}$$

Moreover, if we design the filter gain K_{k+1} *as*

$$\begin{aligned} K_{k+1} = {} & \lambda_1(1+\varepsilon_1)\Sigma_{k+1|k}C_{k+1}^T\Big\{\lambda_1^2(1+\varepsilon_1)C_{k+1}\Sigma_{k+1|k}C_{k+1}^T \\ & +\lambda_1(1-\lambda_1+\varepsilon_2\lambda_1)C_{k+1}\Xi_{k+1|k}C_{k+1}^T + E_{k+1}\Psi_{k+1}E_{k+1}^T + R_{k+1} \\ & +\lambda_2(1+\varepsilon_1^{-1}\lambda_2+\varepsilon_2^{-1}\lambda_2)\text{tr}(\Phi_{k+1|k}) \\ & \times\Big[(I-\gamma_{2,k+1}\Lambda\Lambda)^{-1}+\gamma_{2,k+1}^{-1}I\Big]\Big\}^{-1}, \end{aligned} \tag{3.28}$$

the obtained upper bound $\Sigma_{k+1|k+1}$ *can be minimized.*

Proof The mathematical induction is utilized here. Firstly, noting the initial condition, one has $P_{0|0} \le \Sigma_{0|0}$. Then, assuming $P_{k|k} \le \Sigma_{k|k}$, the aim is to testify that $P_{k+1|k+1} \le \Sigma_{k+1|k+1}$.

Now, let us show that $P_{k+1|k} \le \Sigma_{k+1|k}$. Considering the property of the nonlinearity $\phi(x_k)$ and using Lemma 2.1, we have

$$\begin{aligned} P_{k+1|k} & \le (A_k + M_k\Omega_kN_k)P_{k|k}(A_k+M_k\Omega_kN_k)^T + B_k\Theta_kB_k^T + D_kQ_kD_k^T \\ & \le A_k(\Sigma_{k|k}^{-1} - \gamma_{1,k}N_k^TN_k)^{-1} + \gamma_{1,k}^{-1}M_kM_k^T + B_k\Theta_kB_k^T + D_kQ_kD_k^T \\ & = \Sigma_{k+1|k}, \end{aligned} \tag{3.29}$$

where $\gamma_{1,k}$ is a positive scalar.

By using Lemma 2.2, we have

$$\begin{aligned} & -\mathscr{A}_{k+1} - \mathscr{A}_{k+1}^T \\ \le & \varepsilon_1(I-\lambda_1K_{k+1}C_{k+1})P_{k+1|k}(I-\lambda_1K_{k+1}C_{k+1})^T \\ & + \varepsilon_1\lambda_1^2K_{s,k+1}C_{k+1}P_{k+1|k}C_{k+1}^TK_{s,k+1}^T \\ & + \varepsilon_1^{-1}\lambda_2^2K_{k+1}(I+\Im_{k+1})C_{k+1}\mathbb{E}\{x_{k+1}x_{k+1}^T\}C_{k+1}^T(I+\Im_{k+1})^TK_{k+1}^T \\ & + \varepsilon_1^{-1}\lambda_2^2K_{s,k+1}(I+\Im_{k+1})C_{k+1}\mathbb{E}\{x_{k+1}x_{k+1}^T\}C_{k+1}^T(I+\Im_{k+1})^TK_{s,k+1}^T \end{aligned} \tag{3.30}$$

and

$$
\begin{aligned}
&\mathscr{B}_{k+1} + \mathscr{B}_{k+1}^T \\
\leq &\varepsilon_2 \lambda_1^2 K_{k+1} C_{k+1} \mathbb{E}\{x_{k+1} x_{k+1}^T\} C_{k+1}^T K_{k+1}^T \\
&+ \varepsilon_2 \lambda_1^2 K_{s,k+1} C_{k+1} \mathbb{E}\{x_{k+1} x_{k+1}^T\} C_{k+1}^T K_{s,k+1}^T \\
&+ \varepsilon_2^{-1} \lambda_2^2 K_{k+1} (I + \Im_{k+1}) C_{k+1} \mathbb{E}\{x_{k+1} x_{k+1}^T\} C_{k+1}^T (I + \Im_{k+1})^T K_{k+1}^T \\
&+ \varepsilon_2^{-1} \lambda_2^2 K_{s,k+1} (I + \Im_{k+1}) C_{k+1} \mathbb{E}\{x_{k+1} x_{k+1}^T\} C_{k+1}^T (I + \Im_{k+1})^T K_{s,k+1}^T,
\end{aligned} \tag{3.31}
$$

where ε_1 and ε_2 are two positive scalars. Subsequently, it follows from the property of the nonlinearity $\psi(x_k)$ that

$$
\begin{aligned}
&K_{k+1} E_{k+1} \mathbb{E}\{\psi(x_{k+1}) \psi^T(x_{k+1})\} E_{k+1}^T K_{k+1}^T \\
\leq &K_{k+1} E_{k+1} \Psi_{k+1} E_{k+1}^T K_{k+1}^T
\end{aligned} \tag{3.32}
$$

and

$$
\begin{aligned}
&K_{s,k+1} E_{k+1} \mathbb{E}\{\psi(x_{k+1}) \psi^T(x_{k+1})\} E_{k+1}^T K_{s,k+1}^T \\
\leq &K_{s,k+1} E_{k+1} \Psi_{k+1} E_{k+1}^T K_{s,k+1}^T.
\end{aligned} \tag{3.33}
$$

Substituting the inequalities (3.30)–(3.33) into (3.17), we have

$$
\begin{aligned}
&P_{k+1|k+1} \\
\leq &(1 + \varepsilon_1)(I - \lambda_1 K_{k+1} C_{k+1}) P_{k+1|k} (I - \lambda_1 K_{k+1} C_{k+1})^T \\
&+ (1 + \varepsilon_1) \lambda_1^2 K_{s,k+1} C_{k+1} P_{k+1|k} C_{k+1}^T K_{s,k+1}^T \\
&+ (\lambda_1 - \lambda_1^2 + \varepsilon_2 \lambda_1^2) K_{k+1} C_{k+1} \mathbb{E}\{x_{k+1} x_{k+1}^T\} C_{k+1}^T K_{k+1}^T \\
&+ (\lambda_1 - \lambda_1^2 + \varepsilon_2 \lambda_1^2) K_{s,k+1} C_{k+1} \mathbb{E}\{x_{k+1} x_{k+1}^T\} C_{k+1}^T K_{s,k+1}^T \\
&+ (\lambda_2 + \varepsilon_1^{-1} \lambda_2^2 + \varepsilon_2^{-1} \lambda_2^2) K_{k+1} (I + \Im_{k+1}) \\
&\times C_{k+1} \mathbb{E}\{x_{k+1} x_{k+1}^T\} C_{k+1}^T (I + \Im_{k+1})^T K_{k+1}^T \\
&+ (\lambda_2 + \varepsilon_1^{-1} \lambda_2^2 + \varepsilon_2^{-1} \lambda_2^2) K_{s,k+1} (I + \Im_{k+1}) \\
&\times C_{k+1} \mathbb{E}\{x_{k+1} x_{k+1}^T\} C_{k+1}^T (I + \Im_{k+1})^T K_{s,k+1}^T \\
&+ K_{k+1} E_{k+1} \Psi_{k+1} E_{k+1}^T K_{k+1}^T + K_{s,k+1} E_{k+1} \Psi_{k+1} E_{k+1}^T K_{s,k+1}^T \\
&+ K_{k+1} R_{k+1} K_{k+1}^T + K_{s,k+1} R_{k+1} K_{s,k+1}^T.
\end{aligned} \tag{3.34}
$$

Subsequently, we obtain

$$
\tilde{x}_{k+1|k} \hat{x}_{k+1|k}^T + \hat{x}_{k+1|k} \tilde{x}_{k+1|k}^T \leq \varepsilon_3 \tilde{x}_{k+1|k} \tilde{x}_{k+1|k}^T + \varepsilon_3^{-1} \hat{x}_{k+1|k} \hat{x}_{k+1|k}^T
$$

with $\varepsilon_3 > 0$ being a constant. Then, $\mathbb{E}\left\{x_{k+1} x_{k+1}^T\right\}$ is expressed as

$$
\begin{aligned}
\mathbb{E}\left\{x_{k+1} x_{k+1}^T\right\} &= \mathbb{E}\left\{(\tilde{x}_{k+1|k} + \hat{x}_{k+1|k})(\tilde{x}_{k+1|k} + \hat{x}_{k+1|k})^T\right\} \\
&\leq (1 + \varepsilon_3) P_{k+1|k} + (1 + \varepsilon_3^{-1}) \hat{x}_{k+1|k} \hat{x}_{k+1|k}^T.
\end{aligned} \tag{3.35}
$$

Setting $\Pi_{k+1|k} = (1+\varepsilon_3)P_{k+1|k} + (1+\varepsilon_3^{-1})\hat{x}_{k+1|k}\hat{x}_{k+1|k}^T$, we have

$$\begin{aligned}&(\lambda_1 - \lambda_1^2 + \varepsilon_2\lambda_1^2)K_{k+1}C_{k+1}\mathbb{E}\left\{x_{k+1}x_{k+1}^T\right\}C_{k+1}^T K_{k+1}^T\\ \leq&(\lambda_1 - \lambda_1^2 + \varepsilon_2\lambda_1^2)K_{k+1}C_{k+1}\Pi_{k+1|k}C_{k+1}^T K_{k+1}^T\end{aligned} \tag{3.36}$$

and

$$\begin{aligned}&(\lambda_1 - \lambda_1^2 + \varepsilon_2\lambda_1^2)K_{s,k+1}C_{k+1}\mathbb{E}\left\{x_{k+1}x_{k+1}^T\right\}C_{k+1}^T K_{s,k+1}^T\\ \leq&(\lambda_1 - \lambda_1^2 + \varepsilon_2\lambda_1^2)K_{s,k+1}C_{k+1}\Pi_{k+1|k}C_{k+1}^T K_{s,k+1}^T.\end{aligned} \tag{3.37}$$

Adopting Lemma 2.1, the fifth term of the right-hand side (RHS) of inequality (3.34) can be tackled as

$$\begin{aligned}&(\lambda_2 + \varepsilon_1^{-1}\lambda_2^2 + \varepsilon_2^{-1}\lambda_2^2)K_{k+1}(I+\Im_{k+1})C_{k+1}\mathbb{E}\left\{x_{k+1}x_{k+1}^T\right\}C_{k+1}^T(I+\Im_{k+1})^T K_{k+1}^T\\ \leq&(\lambda_2 + \varepsilon_1^{-1}\lambda_2^2 + \varepsilon_2^{-1}\lambda_2^2)K_{k+1}(I+F_{k+1}\Lambda)C_{k+1}\Pi_{k+1|k}C_{k+1}^T(I+F_{k+1}\Lambda)^T K_{k+1}^T\\ \leq&(\lambda_2 + \varepsilon_1^{-1}\lambda_2^2 + \varepsilon_2^{-1}\lambda_2^2)\mathrm{tr}(\Upsilon_{k+1|k})K_{k+1}(I+F_{k+1}\Lambda)(I+F_{k+1}\Lambda)^T K_{k+1}^T\\ \leq&(\lambda_2 + \varepsilon_1^{-1}\lambda_2^2 + \varepsilon_2^{-1}\lambda_2^2)\mathrm{tr}(\Upsilon_{k+1|k})K_{k+1}\left[(I-\gamma_{2,k+1}\Lambda\Lambda)^{-1} + \gamma_{2,k+1}^{-1}I\right]K_{k+1}^T,\end{aligned} \tag{3.38}$$

where $\gamma_{2,k+1} > 0$ and $\Upsilon_{k+1|k} = C_{k+1}\Pi_{k+1|k}C_{k+1}^T$. Similarly, the sixth term of the RHS of inequality (3.34) can be calculated as

$$\begin{aligned}&(\lambda_2 + \varepsilon_1^{-1}\lambda_2^2 + \varepsilon_2^{-1}\lambda_2^2)K_{s,k+1}(I+\Im_{k+1})C_{k+1}\\ &\times\mathbb{E}\left\{x_{k+1}x_{k+1}^T\right\}(I+\Im_{k+1})C_{k+1}^T K_{s,k+1}^T\\ \leq&(\lambda_2 + \varepsilon_1^{-1}\lambda_2^2 + \varepsilon_2^{-1}\lambda_2^2)\mathrm{tr}(\Upsilon_{k+1|k})K_{s,k+1}\\ &\times\left[(I-\gamma_{2,k+1}\Lambda\Lambda)^{-1} + \gamma_{2,k+1}^{-1}I\right]K_{s,k+1}^T.\end{aligned} \tag{3.39}$$

Then, it follows from inequalities (3.34), (3.36)–(3.39) that

$$\begin{aligned}P_{k+1|k+1} \leq&(1+\varepsilon_1)(I-\lambda_1K_{k+1}C_{k+1})P_{k+1|k}(I-\lambda_1K_{k+1}C_{k+1})^T\\ &+(1+\varepsilon_1)\lambda_1^2K_{s,k+1}C_{k+1}P_{k+1|k}C_{k+1}^TK_{s,k+1}^T\\ &+(\lambda_1 - \lambda_1^2 + \varepsilon_2\lambda_1^2)K_{k+1}C_{k+1}\Pi_{k+1|k}C_{k+1}^T K_{k+1}^T\\ &+(\lambda_1 - \lambda_1^2 + \varepsilon_2\lambda_1^2)K_{s,k+1}C_{k+1}\Pi_{k+1|k}C_{k+1}^T K_{s,k+1}^T\\ &+(\lambda_2 + \varepsilon_1^{-1}\lambda_2^2 + \varepsilon_2^{-1}\lambda_2^2)\mathrm{tr}(\Upsilon_{k+1|k})K_{k+1}\\ &\times\left[(I-\gamma_{2,k+1}\Lambda\Lambda)^{-1} + \gamma_{2,k+1}^{-1}I\right]K_{k+1}^T\\ &+(\lambda_2 + \varepsilon_1^{-1}\lambda_2^2 + \varepsilon_2^{-1}\lambda_2^2)\mathrm{tr}(\Upsilon_{k+1|k})K_{s,k+1}\end{aligned}$$

$$
\begin{aligned}
&\times\left[(I-\gamma_{2,k+1}\Lambda\Lambda)^{-1}+\gamma_{2,k+1}^{-1}I\right]K_{s,k+1}^{T}\\
&+K_{k+1}E_{k+1}\Psi_{k+1}E_{k+1}^{T}K_{k+1}^{T}+K_{s,k+1}E_{k+1}\Psi_{k+1}E_{k+1}^{T}K_{s,k+1}^{T}\\
&+K_{k+1}R_{k+1}K_{k+1}^{T}+K_{s,k+1}R_{k+1}K_{s,k+1}^{T}. \qquad (3.40)
\end{aligned}
$$

According to inequality (3.40), it is obvious to show that $P_{k+1|k+1} \leq \Sigma_{k+1|k+1}$.

Finally, we devote to deriving the explicit form of the filter gain such that the calculated upper bound of the FEC is minimized. On the basis of Lemma 2.3, taking the partial derivative of the trace of this upper bound regarding K_{k+1} and letting the derivative be zero, it can be derived that

$$
\begin{aligned}
&\frac{\partial \mathrm{tr}(\Sigma_{k+1|k+1})}{\partial K_{k+1}}\\
&=-2\lambda_1(1+\varepsilon_1)(I-\lambda_1K_{k+1}C_{k+1})\Sigma_{k+1|k}C_{k+1}^{T}\\
&\quad+2K_{k+1}\Big\{\lambda_1(1-\lambda_1+\varepsilon_2\lambda_1)C_{k+1}\Xi_{k+1|k}C_{k+1}^{T}\\
&\quad+\lambda_2(1+\varepsilon_1^{-1}\lambda_2+\varepsilon_2^{-1}\lambda_2)\mathrm{tr}(\Phi_{k+1|k})\left[(I-\gamma_{2,k+1}\Lambda\Lambda)^{-1}+\gamma_{2,k+1}^{-1}I\right]\\
&\quad+E_{k+1}\Psi_{k+1}E_{k+1}^{T}+R_{k+1}\Big\}\\
&=0, \qquad (3.41)
\end{aligned}
$$

where $\Xi_{k+1|k}$ and $\Phi_{k+1|k}$ are defined as in (3.26) and (3.27), respectively. Through the straightforward algebraic manipulations, it is obvious to verify that the filter gain K_{k+1} can be given in (3.28). Then, the proof of this theorem is complete.

Remark 3.4 It is worth mentioning that the addressed time-varying nonlinear stochastic system is more comprehensive (including the stochastic uncertainties and incomplete measurements) than the existed ones. In particular, a compensation strategy is proposed, where the missing probability is introduced to compensate the influence of the incomplete measurements when constructing the filter. Moreover, the stochastic perturbations of the filter gain are considered to better reflect the engineering practice. These two aspects constitute the main differences compared with the existing filtering approaches (e.g. KF and EKF).

Remark 3.5 So far, a novel recursive filter with certain robustness is constructed in the context of time-varying networked systems subject to stochastic uncertainties and incomplete measurements. By solving the Riccati-like difference equations, an optimized upper bound concerning on the FEC is calculated and the filter gain is derived by minimizing the trace of the obtained upper bound. Note that $\gamma_{1,k}$ and $\gamma_{2,k+1}$ are involved in inequalities (3.22) and (3.23), $\gamma_{1,k}$ and $\gamma_{2,k+1}$ could be given *a prior* and adjusted to ensure the feasibility of inequalities (3.22) and (3.23) in Theorem 3.2. Moreover, it is worth pointing out that we make the first attempt to construct the recursive filter for networked *time-varying* systems with signal quantization and missing measurements in a random way. Due to the recursive nature of the developed filtering strategy, the design procedure is suitable for online real-time applications.

3.3 An Illustrative Example

In this section, a simulation experiment is adopted to validate the feasibility of the presented filtering algorithm.

The physical plant with stochastic uncertainties and incomplete measurements is considered and described in (3.1)–(3.3), where

$$
\begin{aligned}
f(x_k) &= \begin{bmatrix} 0.75x_{1,k} + \sin(x_{1,k}x_{2,k}) \\ 1.5x_{2,k} - \sin(x_{1,k}x_{2,k}) \end{bmatrix}, \quad D_k = \begin{bmatrix} 0.06 \\ 0.3 + 0.5e^{-2k} \end{bmatrix}, \\
C_k &= \begin{bmatrix} 0.85 & 0 \\ 0 & -1.5 \end{bmatrix}, \quad B_k = \begin{bmatrix} 0.3 \\ 0.1 \end{bmatrix}, \quad E_k = \begin{bmatrix} 0.1 \\ 0.45 \end{bmatrix}, \\
\phi(x_k) &= \sqrt{0.75}\sin(x_{1,k}), \quad \psi(x_k) = \sqrt{1.2}\sin(x_{2,k}),
\end{aligned}
$$

and $x_k = \begin{bmatrix} x_{1,k} \; x_{2,k} \end{bmatrix}^T$ with $x_{i,k}$ $(i = 1, 2)$ denoting the i-th element of the system state, $\omega_k \in \mathbb{R}$ and $\nu_k \in \mathbb{R}^2$ are zero-mean Gaussian white noises with covariances 0.4 and $0.3I_2$, respectively.

The initial conditions are set as follows:

$$
\hat{x}_{0|0} = \bar{x}_0 = \begin{bmatrix} 0.35 \; 0.25 \end{bmatrix}^T, \quad P_{0|0} = 10I_2.
$$

Suppose that $\lambda_1 = 0.9$, $\lambda_2 = 0.05$ and $\lambda_3 = 0.05$. Let the parameters of the logarithmic quantizer be $u_0^1 = 0.2$, $u_0^2 = 0.3$, $\chi^{(1)} = 0.6$ and $\chi^{(2)} = 0.4$. Other parameters are selected as $M_k = \text{diag}\{0.2, 0.2\}$, $N_k = 0.5I_2$, $\Theta_k = 0.75$, $\Psi_k = 1.2$, $\varepsilon_1 = 0.6$, $\varepsilon_2 = 0.6$, $\varepsilon_3 = 0.1$, $\gamma_{1,k} = [3\lambda_{\max}(N_k\Sigma_{k|k}N_k^T) + 0.5]^{-1}$, and $\gamma_{2,k+1} = 0.02$. The stochastic perturbation of the filter gain is chosen as $\Delta K_{s,k+1} = \varrho_{k+1}K_{s,k+1}$ with $K_{s,k+1} = \begin{bmatrix} 0.04 & 0.02 \\ 0.06 & 0.1 \end{bmatrix}$ and $\varrho_{k+1} \in \mathbb{R}$ being a zero mean Gaussian white noise with covariance 1.

Let MSE1 be the mean-square error (MSE) for the estimation of $x_{1,k}$,

$$
(1/M)\sum_{j=1}^{M}\left(x_{1,k}^{(j)} - \hat{x}_{1,k|k}^{(j)}\right)^2,
$$

where M denotes the number of experiment test. $x_{1,k}^{(j)}$ and $\hat{x}_{1,k|k}^{(j)}$ denote the values of $x_{1,k}$ and $\hat{x}_{1,k|k}$ in the j-th experiment test. Likewise, MSE2 can be defined by $(1/M)\sum_{j=1}^{M}\left(x_{2,k}^{(j)} - \hat{x}_{2,k|k}^{(j)}\right)^2$. On the basis of Theorem 3.2, the filter gains can be designed recursively and the experiment results are depicted in Figs. 3.1, 3.2, 3.3 and

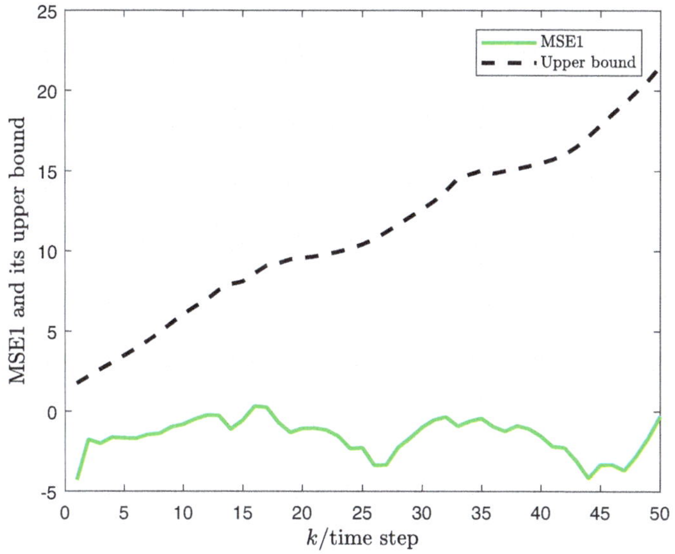

Fig. 3.1 MSE1 and its upper bound

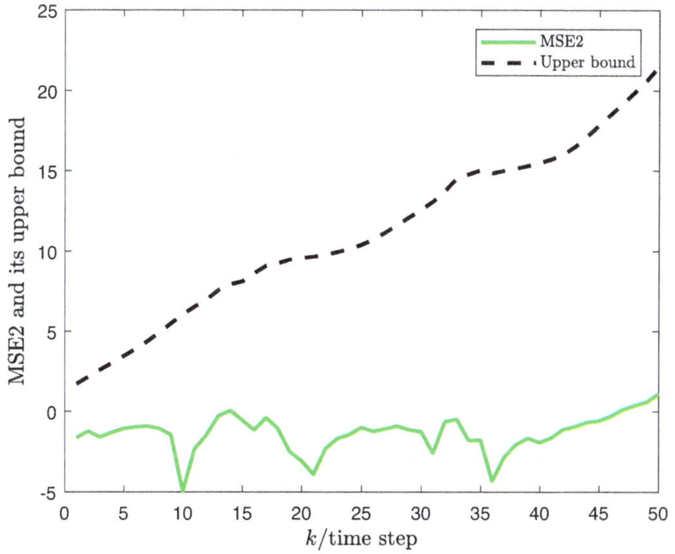

Fig. 3.2 MSE2 and its upper bound

3.4. Among them, Figs. 3.1 and 3.2 represent the upper bounds $\Sigma^{11}_{k|k}$ and $\Sigma^{22}_{k|k}$ as well as the MSE for the states $x_{1,k}$ and $x_{2,k}$, which reflect that the MSEs keep below their upper bounds. The trajectories of the actual states $x_{i,k}$ and their estimates $\hat{x}_{i,k|k}$ $(i = 1, 2)$ are displayed in Figs. 3.3 and 3.4. It is obvious to find that the established

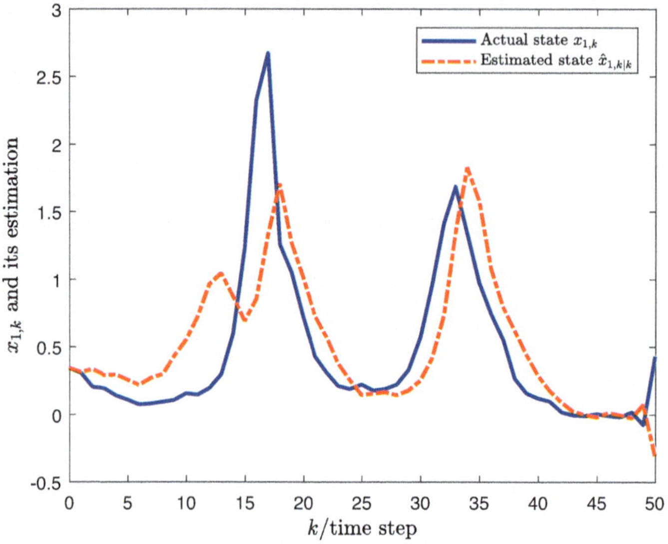

Fig. 3.3 The real state $x_{1,k}$ and its estimation $\hat{x}_{1,k|k}$

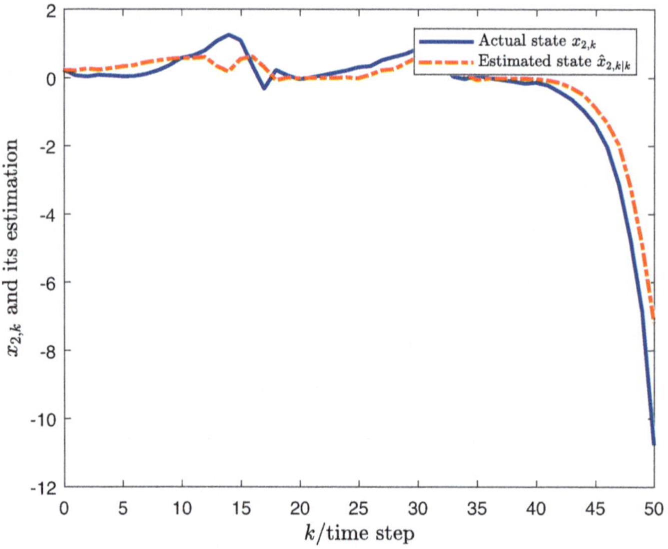

Fig. 3.4 The real state $x_{2,k}$ and its estimation $\hat{x}_{2,k|k}$

filtering scheme can perform well to estimate the system states in spite of taking the stochastic uncertainties, incomplete measurements and stochastic perturbations of the filter gain into consideration. Moreover, with the aim to demonstrate the effect from the incomplete measurements onto the filtering performance, the related

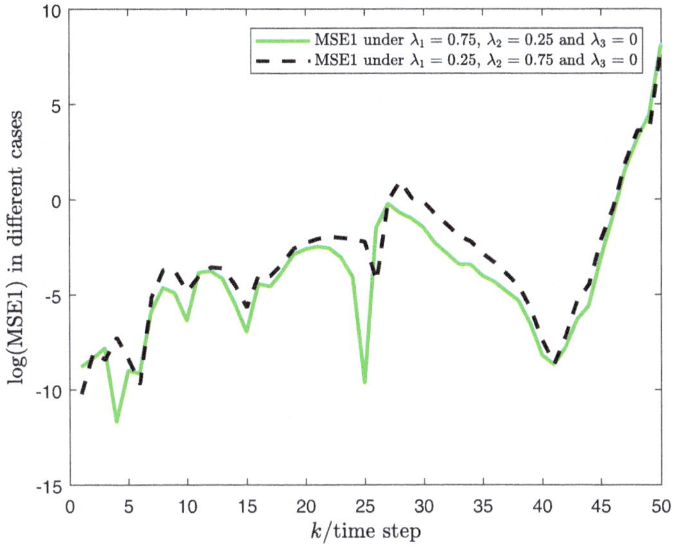

Fig. 3.5 MSE1 under different λ_{j_1} ($j_1 = 1, 2, 3$)

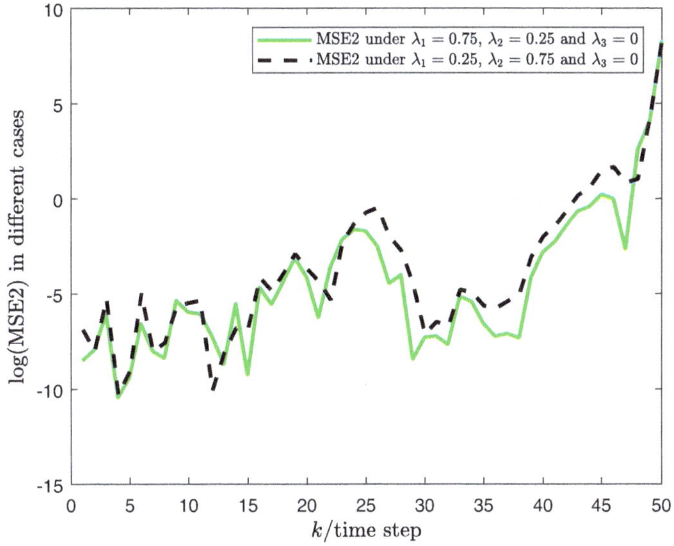

Fig. 3.6 MSE2 under different λ_{j_1} ($j_1 = 1, 2, 3$)

simulation results are provided in Figs. 3.5 and 3.6. It can be obviously seen that the MSE1 and MSE2 become smaller when more measurements can be received.

3.4 Summary

In this chapter, the resilient recursive filtering issue has been tackled for networked time-varying nonlinear systems affected by stochastic uncertainties and incomplete measurements. By resorting to a stochastic Kronecker delta function, the incomplete measurements have been described and discussed. The Taylor series expansion method has been adopted to handle the nonlinearity and the linearization errors have been estimated. Based on the available information, a new filtering scheme has been proposed such that an optimized upper bound associated with the FEC has been computed and the filter gain has been derived through minimizing the trace of this upper bound. Finally, the usefulness of the developed filtering method has been validated by a simulation example.

References

Calafiore G (2005) Reliable localization using set-valued nonlinear filters. IEEE Trans Syst Man Cybern Part A-Syst Hum 35(2):189–197

Hu J, Wang Z, Shen B, Gao H (2013) Quantised recursive filtering for a class of nonlinear systems with multiplicative noises and missing measurements. Int J Control 86(4):650–663

Shen B, Wang Z, Ding D, Shu H (2013) H_∞ state estimation for complex networks with uncertain inner coupling and incomplete measurements. IEEE Trans Neural Netw Learn Syst 24(12):2027–2037

Xiong K, Wei C, Liu L (2010) Robust extended Kalman filtering for nonlinear systems with stochastic uncertainties. IEEE Trans Syst Man Cybern Part A-Syst Hum 40(2):399–405

Yaz EE, Jeong CS, Yaz YI (2006) An LMI approach to discrete-time observer design with stochastic resilience. J Comput Appl Math 188(2):246–255

Chapter 4
Event-Triggered Resilient Filtering with Stochastic Uncertainties and Successive Packet Dropouts

In this chapter, we aim to discuss the effects from the stochastic uncertainties and limited communication capability onto the whole filtering algorithm accuracy. Firstly, the event-based resilient filtering issue is tackled for time-varying systems with stochastic uncertainties. With the objective to decrease communication frequency, the ETM is adopted to manage the signal transmissions. The stochastic uncertainties are addressed to characterize the modelling errors and the filter gain perturbation is modeled by norm-bounded uncertainties. An event-based filtering method is given such that, for both stochastic uncertainties and filter gain perturbations, an optimized upper bound on the FEC is guaranteed via constructing an appropriate filter gain. Secondly, the event-triggered resilient filtering issue is addressed for time-varying systems affected by stochastic uncertainties and SPDs. Here, the SPDs are described through a set of binary random variables. Subsequently, a resilient event-based filtering method is proposed for time-varying systems such that, in the presence of stochastic uncertainties, SPDs and filter gain perturbation, an optimized upper bound concerning on the FEC is solved via appropriately constructing the filter gain. Lastly, some simulation experiments are given to demonstrate the validity of the designed filtering algorithm.

4.1 Event-Triggered Resilient Filtering with Stochastic Uncertainties

4.1.1 Problem Formulation

Consider the resilient filter design problem for time-varying systems subject to stochastic uncertainties under the ETM. The specific schematic diagram of this chapter is showed in Fig. 4.1, in which the measurements are scheduled by the ETM. The system model is described as

J. Hu et al., *Variance-Constrained Filtering for Stochastic Complex Systems*, Intelligent Control and Learning Systems 18,
https://doi.org/10.1007/978-981-96-2637-3_4

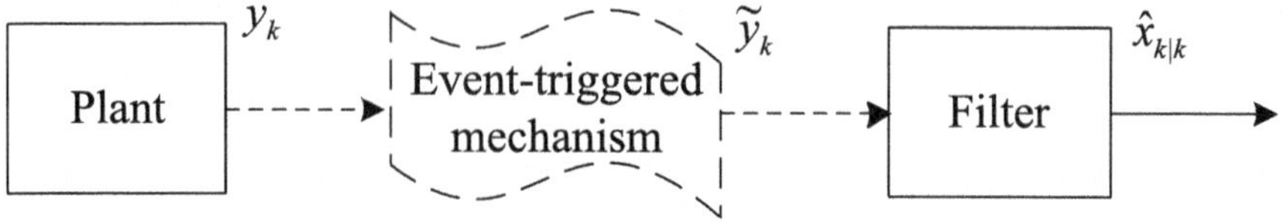

Fig. 4.1 Filtering under the event-triggered scheduling strategy

$$x_{k+1} = A_k x_k + \eta_k B_k \varphi(x_k) + D_k \omega_k, \tag{4.1}$$

$$y_k = C_k x_k + \xi_k E_k \psi(x_k) + \nu_k, \tag{4.2}$$

where $x_k \in \mathbb{R}^n$ is the state and its initial value x_0 has the mean $\bar{x}_0$ and covariance $P_0 > 0$, $y_k \in \mathbb{R}^m$ stands for the measurement. The process noise $\omega_k \in \mathbb{R}^p$ is zero-mean Gaussian white noise with covariance $Q_k > 0$ and the measurement noise $\nu_k \in \mathbb{R}^m$ is zero-mean Gaussian white noise with covariance $R_k > 0$, $\eta_k \in \mathbb{R}$ and $\xi_k \in \mathbb{R}$ are zero-mean Gaussian white noises with unity covariances. A_k, B_k, C_k, D_k and E_k are all known matrices of proper dimensions. $\varphi(x_k)$ and $\psi(x_k)$ characterize uncertain functions satisfying $\mathbb{E}\{\varphi(x_k)\varphi^T(x_k)\} \leq \Phi_k$ and $\mathbb{E}\{\psi(x_k)\psi^T(x_k)\} \leq \Psi_k$ with Φ_k and Ψ_k being known matrices. Without loss of generality, it is supposed that η_k, ξ_k, ω_k, ν_k and x_0 are mutually independent.

In this chapter, the event-based scheduling scheme is used to decrease the network communication costs. Specifically, the ETM is constructed as

$$(y_{k+j} - y_{k_i})^T (y_{k+j} - y_{k_i}) > \delta \tag{4.3}$$

with $\delta > 0$ denoting a given scalar, y_{k+j} representing the measurement at time instant $k + j$ and y_{k_i} standing for the latest delivered measurement. At the triggering time k_i with $k_0 = 0$, the released data $\tilde{y}_k$ received by the filter side is described as

$$\tilde{y}_k = y_{k_i}, \ k \in \{k_i, k_i + 1, \ldots, k_{i+1} - 1\}. \tag{4.4}$$

In this chapter, we design the following recursive filter:

$$\hat{x}_{k+1|k} = A_k \hat{x}_{k|k}, \tag{4.5}$$

$$\hat{x}_{k+1|k+1} = \hat{x}_{k+1|k} + \check{K}_{k+1}(\tilde{y}_{k+1} - C_{k+1}\hat{x}_{k+1|k}), \tag{4.6}$$

where $\hat{x}_{k|k}$ and $\hat{x}_{k+1|k}$ denote the filtering and the one-step prediction, respectively. $\check{K}_{k+1} = K_{k+1} + \Delta K_{k+1}$, K_{k+1} stands for the filter gain and ΔK_{k+1} denotes the filter gain perturbation satisfying

$$\Delta K_{k+1} = M_{k+1} F_{k+1} N_{k+1}, \tag{4.7}$$

where M_{k+1} and N_{k+1} are given matrices and $F_{k+1}^T F_{k+1} \leq I$.

Let the one-step prediction error be $\tilde{x}_{k+1|k} = x_{k+1} - \hat{x}_{k+1|k}$ and the filtering error be $\tilde{x}_{k+1|k+1} = x_{k+1} - \hat{x}_{k+1|k+1}$. Moreover, $P_{k+1|k} = \mathbb{E}\{\tilde{x}_{k+1|k}\tilde{x}^T_{k+1|k}\}$ and $P_{k+1|k+1} = \mathbb{E}\{\tilde{x}_{k+1|k+1}\tilde{x}^T_{k+1|k+1}\}$ characterize the one-step prediction error covariance and the FEC, respectively. We aim to construct an event-based resilient filter as in (4.5) and (4.6) such that, for the stochastic uncertainties and filter gain perturbation, an upper bound $\Sigma_{k+1|k+1}$ of $P_{k+1|k+1}$ is guaranteed. Moreover, we decide to minimize $\mathrm{tr}(\Sigma_{k+1|k+1})$ and obtain the filter gain K_{k+1} with easy-to-implement form.

4.1.2 Design of Filter Gain

In this section, the one-step prediction error covariance and FEC are computed and an optimized upper bound of the FEC is solved through constructing an appropriate filter gain.

Subtracting (4.5) from (4.1), one has

$$\tilde{x}_{k+1|k} = A_k\tilde{x}_{k|k} + \eta_k B_k\varphi(x_k) + D_k\omega_k. \tag{4.8}$$

Next, subtracting (4.6) from (4.1) yields

$$\begin{aligned}\tilde{x}_{k+1|k+1} = &(I - \check{K}_{k+1}C_{k+1})\tilde{x}_{k+1|k} + \check{K}_{k+1}(y_{k+1} - \tilde{y}_{k+1})\\ &-\xi_{k+1}\check{K}_{k+1}E_{k+1}\psi(x_{k+1}) - \check{K}_{k+1}v_{k+1}.\end{aligned} \tag{4.9}$$

Theorem 4.1 *$P_{k+1|k}$ satisfies the following recursive form:*

$$P_{k+1|k} = A_kP_{k|k}A_k^T + B_k\mathbb{E}\{\varphi(x_k)\varphi^T(x_k)\}B_k^T + D_kQ_kD_k^T. \tag{4.10}$$

Proof The proof procedure is omitted owing to its briefness.

Theorem 4.2 *$P_{k+1|k+1}$ satisfies the following recursive form:*

$$\begin{aligned}P_{k+1|k+1} = &(I - \check{K}_{k+1}C_{k+1})P_{k+1|k}(I - \check{K}_{k+1}C_{k+1})^T\\ &+\check{K}_{k+1}\mathbb{E}\{(y_{k+1} - \tilde{y}_{k+1})(y_{k+1} - \tilde{y}_{k+1})^T\}\check{K}^T_{k+1}\\ &+\check{K}_{k+1}E_{k+1}\mathbb{E}\{\psi(x_{k+1})\psi^T(x_{k+1})\}E^T_{k+1}\check{K}^T_{k+1} + \check{K}_{k+1}R_{k+1}\check{K}^T_{k+1}\\ &+\mathcal{N}_1 + \mathcal{N}_1^T - \mathcal{N}_2 - \mathcal{N}_2^T - \mathcal{N}_3 - \mathcal{N}_3^T,\end{aligned} \tag{4.11}$$

where

$$\begin{aligned}\mathcal{N}_1 &= \mathbb{E}\{(I - \check{K}_{k+1}C_{k+1})\tilde{x}_{k+1|k}(y_{k+1} - \tilde{y}_{k+1})^T\check{K}^T_{k+1}\},\\ \mathcal{N}_2 &= \mathbb{E}\{\xi_{k+1}\check{K}_{k+1}(y_{k+1} - \tilde{y}_{k+1})\psi^T(x_{k+1})E^T_{k+1}\check{K}^T_{k+1}\},\\ \mathcal{N}_3 &= \mathbb{E}\{\check{K}_{k+1}(y_{k+1} - \tilde{y}_{k+1})v^T_{k+1}\check{K}^T_{k+1}\}.\end{aligned}$$

Proof According to (4.9), we have

$$\begin{aligned}P_{k+1|k+1} &= (I - \check{K}_{k+1}C_{k+1})P_{k+1|k}(I - \check{K}_{k+1}C_{k+1})^T \\ &\quad + \check{K}_{k+1}\mathbb{E}\{(y_{k+1} - \tilde{y}_{k+1})(y_{k+1} - \tilde{y}_{k+1})^T\}\check{K}_{k+1}^T \\ &\quad + \check{K}_{k+1}E_{k+1}\mathbb{E}\{\psi(x_{k+1})\psi^T(x_{k+1})\}E_{k+1}^T\check{K}_{k+1}^T + \check{K}_{k+1}R_{k+1}\check{K}_{k+1}^T \\ &\quad + \mathscr{N}_1 + \mathscr{N}_1^T - \mathscr{N}_2 - \mathscr{N}_2^T - \mathscr{N}_3 - \mathscr{N}_3^T - \mathscr{M}_1 \\ &\quad - \mathscr{M}_1^T - \mathscr{M}_2 - \mathscr{M}_2^T + \mathscr{M}_3 + \mathscr{M}_3^T,\end{aligned}$$

where

$$\begin{aligned}\mathscr{M}_1 &= \mathbb{E}\{\xi_{k+1}(I - \check{K}_{k+1}C_{k+1})\tilde{x}_{k+1|k}\psi^T(x_{k+1})E_{k+1}^T\check{K}_{k+1}^T\}, \\ \mathscr{M}_2 &= \mathbb{E}\{(I - \check{K}_{k+1}C_{k+1})\tilde{x}_{k+1|k}v_{k+1}^T\check{K}_{k+1}^T\}, \\ \mathscr{M}_3 &= \mathbb{E}\{\xi_{k+1}\check{K}_{k+1}E_{k+1}\psi(x_{k+1})v_{k+1}^T\check{K}_{k+1}^T\}.\end{aligned}$$

In terms of $\mathbb{E}\{\xi_{k+1}\} = 0$ and $\mathbb{E}\{v_{k+1}\} = 0$, we can get $\mathscr{M}_i = 0\,(i = 1, 2, 3)$. Then, (4.11) can be obtained and the proof is finished.

Remark 4.1 In (4.10) and (4.11), there exist some uncertain terms due to the simultaneous consideration of event-based communication protocol, stochastic uncertainties as well as filter gain perturbation. Accordingly, it is tough to get the precise value of the FEC. In the next step, a suboptimal approach is adopted to derive the filter gain K_{k+1}, where a locally optimal upper bound regarding the FEC can be guaranteed.

In the following theorem, an optimized upper bound concerning on the FEC is provided and the filter gain is constructed accordingly.

Theorem 4.3 *Let* $\varepsilon_i > 0\,(i = 1, 2, 3)$ *and* $\gamma_{j,k+1} > 0\,(j = 1, 2)$ *be constants. If the following Riccati-like difference equations:*

$$\Sigma_{k+1|k} = A_k\Sigma_{k|k}A_k^T + B_k\Phi_kB_k^T + D_kQ_kD_k^T, \tag{4.12}$$

$$\begin{aligned}\Sigma_{k+1|k+1} &= (1+\varepsilon_1)[(I - K_{k+1}C_{k+1})(\Sigma_{k+1|k}^{-1} - \gamma_{1,k+1}C_{k+1}^TN_{k+1}^TN_{k+1}C_{k+1})^{-1} \\ &\quad \times(I - K_{k+1}C_{k+1})^T + \gamma_{1,k+1}^{-1}M_{k+1}M_{k+1}^T] + K_{k+1}(\Pi_{k+1}^{-1} - \gamma_{2,k+1}N_{k+1}^T \\ &\quad \times N_{k+1})^{-1}K_{k+1}^T + \gamma_{2,k+1}^{-1}M_{k+1}M_{k+1}^T,\end{aligned} \tag{4.13}$$

under the initial condition $\Sigma_{0|0} = P_{0|0}$ *and*

$$\begin{aligned}&\gamma_{1,k+1}^{-1}I - N_{k+1}C_{k+1}\Sigma_{k+1|k}C_{k+1}^TN_{k+1}^T > 0, \\ &\gamma_{2,k+1}^{-1}I - N_{k+1}\Pi_{k+1}N_{k+1}^T > 0,\end{aligned}$$

have solutions $\Sigma_{k+1|k} > 0$ *and* $\Sigma_{k+1|k+1} > 0$*, where*

$$\begin{aligned} \Pi_{k+1} = &(1+\varepsilon_1^{-1}+\varepsilon_2+\varepsilon_3)\delta I + (1+\varepsilon_2^{-1})E_{k+1}\Psi_{k+1}E_{k+1}^T \\ &+(1+\varepsilon_3^{-1})R_{k+1}, \end{aligned} \tag{4.14}$$

then one has $P_{k+1|k+1} \le \Sigma_{k+1|k+1}$*. Moreover, if the filter gain* K_{k+1} *is designed by*

$$\begin{aligned} K_{k+1} = &(1+\varepsilon_1)(\Sigma_{k+1|k}^{-1} - \gamma_{1,k+1}C_{k+1}^T N_{k+1}^T N_{k+1} C_{k+1})^{-1} \\ &\times C_{k+1}^T\Big[(\Pi_{k+1}^{-1} - \gamma_{2,k+1}N_{k+1}^T N_{k+1})^{-1} + (1+\varepsilon_1)C_{k+1} \\ &\times(\Sigma_{k+1|k}^{-1} - \gamma_{1,k+1}C_{k+1}^T N_{k+1}^T N_{k+1}C_{k+1})^{-1}C_{k+1}^T\Big]^{-1}, \end{aligned} \tag{4.15}$$

then $\mathrm{tr}(\Sigma_{k+1|k+1})$ *can be minimized.*

Proof Firstly, we are devoted to handling the uncertain term $\mathbb{E}\{\varphi(x_k)\varphi^T(x_k)\}$ in (4.10). It is obvious to find that

$$P_{k+1|k} \le A_k P_{k|k} A_k^T + B_k \Phi_k B_k^T + D_k Q_k D_k^T. \tag{4.16}$$

According to Lemma 2.2, we have

$$\mathcal{N}_1 + \mathcal{N}_1^T \le \varepsilon_1 (I - \check{K}_{k+1}C_{k+1})P_{k+1|k}(I - \check{K}_{k+1}C_{k+1})^T + \varepsilon_1^{-1}\check{K}_{k+1}\mathbb{E}\{(y_{k+1}-\tilde{y}_{k+1})(y_{k+1}-\tilde{y}_{k+1})^T\}\check{K}_{k+1}^T, \tag{4.17}$$

$$-\mathcal{N}_2 - \mathcal{N}_2^T \le \varepsilon_2\check{K}_{k+1}\mathbb{E}\{(y_{k+1}-\tilde{y}_{k+1})(y_{k+1}-\tilde{y}_{k+1})^T\}\check{K}_{k+1}^T + \varepsilon_2^{-1}\check{K}_{k+1}E_{k+1}\mathbb{E}\{\psi(x_{k+1})\psi^T(x_{k+1})\}E_{k+1}^T\check{K}_{k+1}^T, \tag{4.18}$$

$$-\mathcal{N}_3 - \mathcal{N}_3^T \le \varepsilon_3\check{K}_{k+1}\mathbb{E}\{(y_{k+1}-\tilde{y}_{k+1})(y_{k+1}-\tilde{y}_{k+1})^T\}\check{K}_{k+1}^T + \varepsilon_3^{-1}\check{K}_{k+1}R_{k+1}\check{K}_{k+1}^T, \tag{4.19}$$

where $\varepsilon_i > 0$ $(i = 1, 2, 3)$. Substituting (4.17)–(4.19) into (4.11), we obtain

$$\begin{aligned} P_{k+1|k+1} \le &(1+\varepsilon_1)(I - \check{K}_{k+1}C_{k+1})P_{k+1|k}(I - \check{K}_{k+1}C_{k+1})^T \\ &+ (1+\varepsilon_1^{-1}+\varepsilon_2+\varepsilon_3)\check{K}_{k+1}\mathbb{E}\{(y_{k+1}-\tilde{y}_{k+1})(y_{k+1}-\tilde{y}_{k+1})^T\}\check{K}_{k+1}^T \\ &+ (1+\varepsilon_2^{-1})\check{K}_{k+1}E_{k+1}\mathbb{E}\{\psi(x_{k+1})\psi^T(x_{k+1})\}E_{k+1}^T\check{K}_{k+1}^T \\ &+ (1+\varepsilon_3^{-1})\check{K}_{k+1}R_{k+1}\check{K}_{k+1}^T. \end{aligned} \tag{4.20}$$

Based on (4.3), one obtains

$$(\tilde{y}_k - y_k)(\tilde{y}_k - y_k)^T \le \delta I. \tag{4.21}$$

Taking (4.21) into consideration and together with $\mathbb{E}\{\psi(x_{k+1})\psi^T(x_{k+1})\} \le \Psi_{k+1}$, we have

$$
\begin{aligned}
P_{k+1|k+1} \le\ & (1+\varepsilon_1)(I-\check{K}_{k+1}C_{k+1})P_{k+1|k}(I-\check{K}_{k+1}C_{k+1})^T \\
& + \check{K}_{k+1}\Pi_{k+1}\check{K}_{k+1}^T,
\end{aligned}
\tag{4.22}
$$

where Π_{k+1} is defined in (4.14).

Recalling Lemma 2.1 and $\Delta K_{k+1} = M_{k+1}F_{k+1}N_{k+1}$, we have

$$
\begin{aligned}
& (I-\check{K}_{k+1}C_{k+1})P_{k+1|k}(I-\check{K}_{k+1}C_{k+1})^T \\
& \le (I-K_{k+1}C_{k+1})(P_{k+1|k}^{-1}-\gamma_{1,k+1}C_{k+1}^T N_{k+1}^T N_{k+1}C_{k+1})^{-1} \\
& \times (I-K_{k+1}C_{k+1})^T + \gamma_{1,k+1}^{-1}M_{k+1}M_{k+1}^T, \\
& \check{K}_{k+1}\Pi_{k+1}\check{K}_{k+1}^T \\
& \le K_{k+1}(\Pi_{k+1}^{-1}-\gamma_{2,k+1}N_{k+1}^T N_{k+1})^{-1}K_{k+1}^T + \gamma_{2,k+1}^{-1}M_{k+1}M_{k+1}^T.
\end{aligned}
$$

Then, it concludes that

$$
\begin{aligned}
P_{k+1|k+1} \le\ & (1+\varepsilon_1)[(I-K_{k+1}C_{k+1})(P_{k+1|k}^{-1}-\gamma_{1,k+1}C_{k+1}^T N_{k+1}^T N_{k+1}C_{k+1})^{-1} \\
& \times (I-K_{k+1}C_{k+1})^T + \gamma_{1,k+1}^{-1}M_{k+1}M_{k+1}^T] + K_{k+1} \\
& \times (\Pi_{k+1}^{-1}-\gamma_{2,k+1}N_{k+1}^T N_{k+1})^{-1}K_{k+1}^T + \gamma_{2,k+1}^{-1}M_{k+1}M_{k+1}^T.
\end{aligned}
\tag{4.23}
$$

According to the mathematical induction approach, (4.12), (4.13), (4.16) and (4.23), it concludes that $P_{k+1|k+1} \le \Sigma_{k+1|k+1}$ holds.

In what follows, in light of Lemma 2.3, taking the derivative of $\mathrm{tr}(\Sigma_{k+1|k+1})$ in relation to K_{k+1} and letting the result be zero, one has

$$
\begin{aligned}
& \frac{\partial \mathrm{tr}(\Sigma_{k+1|k+1})}{\partial K_{k+1}} \\
&= -2(1+\varepsilon_1)(I-K_{k+1}C_{k+1})(\Sigma_{k+1|k}^{-1}-\gamma_{1,k+1}C_{k+1}^T N_{k+1}^T N_{k+1}C_{k+1})^{-1}C_{k+1}^T \\
& \quad +2K_{k+1}(\Pi_{k+1}^{-1}-\gamma_{2,k+1}N_{k+1}^T N_{k+1})^{-1} \\
&= 0.
\end{aligned}
$$

In terms of the above equation, it is not difficult to get the following filter gain:

$$
\begin{aligned}
K_{k+1} =\ & (1+\varepsilon_1)(\Sigma_{k+1|k}^{-1}-\gamma_{1,k+1}C_{k+1}^T N_{k+1}^T N_{k+1}C_{k+1})^{-1}C_{k+1}^T \\
& \times\Big[(\Pi_{k+1}^{-1}-\gamma_{2,k+1}N_{k+1}^T N_{k+1})^{-1} + (1+\varepsilon_1)C_{k+1} \\
& \times(\Sigma_{k+1|k}^{-1}-\gamma_{1,k+1}C_{k+1}^T N_{k+1}^T N_{k+1}C_{k+1})^{-1}C_{k+1}^T\Big]^{-1},
\end{aligned}
$$

and the proof is complete.

Remark 4.2 In Theorem 4.3, all uncertain terms included in the one-step prediction error covariance and FEC have been effectively tackled in light of the matrix analysis technology, and a recursive filtering scheme has been established. Particularly, an optimized upper bound concerning on the FEC has been calculated, and the specific expression form of the filter gain has been derived in a recursive manner.

4.2 Event-Triggered Resilient Filtering with SPDs

4.2.1 Problem Formulation

In this section, we consider the time-varying systems with stochastic uncertainties and SPDs. It can be clearly seen from Fig. 4.2 that the SPDs might appear in the channel and then the ETM is adopted to schedule the signal transmission.

The time-varying nonlinear system model is given by

$$\breve{x}_{k+1} = \breve{A}_k\breve{x}_k + \eta_k\breve{B}_k\varphi(\breve{x}_k) + \breve{D}_k\breve{\omega}_k, \tag{4.24}$$

$$\breve{y}_k = \breve{C}_k\breve{x}_k + \xi_k\breve{E}_k\psi(\breve{x}_k) + \breve{\nu}_k, \tag{4.25}$$

$$y_k = \lambda_k\breve{y}_k + (1 - \lambda_k)y_{k-1}, \tag{4.26}$$

where $\breve{x}_k \in \mathbb{R}^n$ is the state and the initial value $\breve{x}_0$ has the mean $\bar{\breve{x}}_0$ and covariance $\breve{P}_0 > 0$, $\breve{y}_k \in \mathbb{R}^m$ is the ideal measurement and $y_k \in \mathbb{R}^m$ is the measurement received by the relay. $\breve{\omega}_k \in \mathbb{R}^p$ and $\breve{\nu}_k \in \mathbb{R}^m$ stand for zero mean Gaussian noises with covariances $\breve{Q}_k > 0$ and $\breve{R}_k > 0$. $\eta_k \in \mathbb{R}$ and $\xi_k \in \mathbb{R}$ are random noises with zero mean and unity covariances. $\breve{A}_k$, $\breve{B}_k$, $\breve{C}_k$, $\breve{D}_k$ and $\breve{E}_k$ represent given matrices. $\varphi(\breve{x}_k)$ and $\psi(\breve{x}_k)$ denote unknown nonlinear functions satisfying $\mathbb{E}\{\varphi(\breve{x}_k)\varphi^T(\breve{x}_k)\} \leq \breve{\Phi}_k$ and $\mathbb{E}\{\psi(\breve{x}_k)\psi^T(\breve{x}_k)\} \leq \breve{\Psi}_k$ with $\breve{\Phi}_k$ and $\breve{\Psi}_k$ denoting given matrices.

The Bernoulli distributed random variable λ_k is utilized to describe the SPDs, where the statistical properties of λ_k are given as

$$\text{Prob}\{\lambda_k = 1\} = \bar{\lambda}_k,$$
$$\text{Prob}\{\lambda_k = 0\} = 1 - \bar{\lambda}_k,$$

here $\bar{\lambda}_k \in [0, 1]$ denotes a given constant scalar. Moreover, we assume that random variables λ_k, η_k, ξ_k, $\breve{\omega}_k$, $\breve{\nu}_k$ and $\breve{x}_0$ are mutually independent.

Remark 4.3 With the purpose of depicting the model uncertainties with *stochastic characteristics* as in Kai et al. (2010), the stochastic uncertainties are introduced here, i.e., η_k, ξ_k, $\varphi(\breve{x}_k)$ and $\psi(\breve{x}_k)$ in (4.24) and (4.25). Specifically, the nonlinear functions

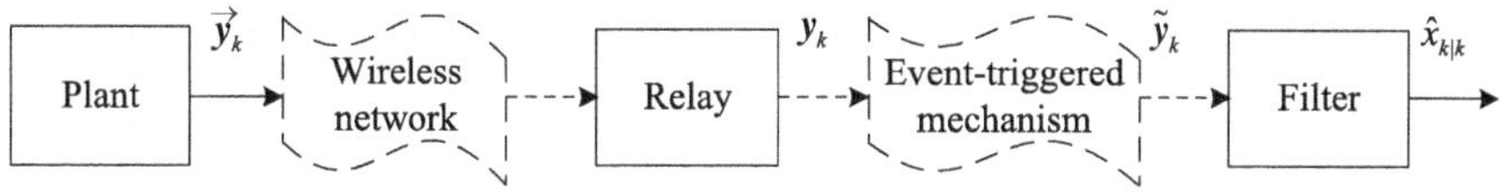

Fig. 4.2 Filtering with SPDs and ETM

$\varphi(\breve{x}_k)$ and $\psi(\breve{x}_k)$ are bounded, the noise sequences η_k and ξ_k have known statistical characteristics, and $\breve{B}_k$ and $\breve{E}_k$ can be attained through using the prior information regarding practical systems. Moreover, it is crucial to highlight that the relays are considered in the process of remote transmission and then different types of networks can be cooperatively used. When the measurements are delivered to the relay via the open yet shared communication networks, the SPDs would be occurred. As it is recognized, the filtering algorithm precision will be damaged if the data packets are lost due to the fact that the updated information cannot be obtained, and thus it is essential to enhance the filtering performance via compensating the negative impacts from the SPDs.

Taking into account the constrained transmission capacity of communication network, the ETM is used to attenuate the communication burden, where the event generator function is constructed as

$$(y_{k+j} - y_{k_i})^T (y_{k+j} - y_{k_i}) > \sigma \tag{4.27}$$

with $\sigma > 0$ being a pre-determined scalar, y_{k+j} being the measurement at step $k + j$ and y_{k_i} denoting the latest transmitted measurement. In light of (4.27), at the triggering time k_i with $k_0 = 0$, the released measurement $\tilde{y}_k$ is modeled by

$$\tilde{y}_k = y_{k_i}, \ k \in \{k_i, k_i + 1, \ldots, k_{i+1} - 1\}. \tag{4.28}$$

Next, a compact form of (4.24)–(4.26) is described as follows:

$$x_{k+1} = A_k x_k + B_k \phi(x_k) + D_k \omega_k, \tag{4.29}$$

$$y_k = C_k x_k + \lambda_k \xi_k E_k \psi(x_k) + \lambda_k \nu_k, \tag{4.30}$$

where

$$\begin{aligned} x_k &= \begin{bmatrix} \breve{x}_k \\ y_{k-1} \end{bmatrix}, \quad A_k = \begin{bmatrix} \breve{A}_k & 0 \\ \lambda_k \breve{C}_k & (1-\lambda_k) I_m \end{bmatrix}, \quad B_k = \begin{bmatrix} \eta_k \breve{B}_k & 0 \\ 0 & \lambda_k \xi_k \breve{E}_k \end{bmatrix}, \\ \phi(x_k) &= \begin{bmatrix} \varphi(\breve{x}_k) \\ \psi(\breve{x}_k) \end{bmatrix}, \quad D_k = \begin{bmatrix} \breve{D}_k & 0 \\ 0 & \lambda_k I_m \end{bmatrix}, \quad \omega_k = \begin{bmatrix} \breve{\omega}_k \\ \breve{\nu}_k \end{bmatrix}, \\ C_k &= \begin{bmatrix} \lambda_k \breve{C}_k & (1-\lambda_k) I_m \end{bmatrix}, \quad E_k = \breve{E}_k, \quad \nu_k = \breve{\nu}_k, \quad \psi(x_k) = \psi(\breve{x}_k). \end{aligned} \tag{4.31}$$

Define the following notations:

$$\begin{aligned} \bar{A}_k &= \mathbb{E}\{A_k\} = \begin{bmatrix} \breve{A}_k & 0 \\ \bar{\lambda}_k \breve{C}_k & (1-\bar{\lambda}_k) I_m \end{bmatrix}, \quad \bar{D}_k = \mathbb{E}\{D_k\} = \begin{bmatrix} \breve{D}_k & 0 \\ 0 & \bar{\lambda}_k I_m \end{bmatrix}, \\ \bar{C}_k &= \mathbb{E}\{C_k\} = \begin{bmatrix} \bar{\lambda}_k \breve{C}_k & (1-\bar{\lambda}_k) I_m \end{bmatrix}, \quad \bar{B}_k = \mathbb{E}\{B_k\} = 0, \end{aligned}$$

and denote $\tilde{A}_k = A_k - \bar{A}_k$, $\tilde{C}_k = C_k - \bar{C}_k$ and $\tilde{D}_k = D_k - \bar{D}_k$.

Based on above discussions, the following recursive filter is constructed as:

$$\hat{x}_{k+1|k} = \bar{A}_k \hat{x}_{k|k}, \tag{4.32}$$

$$\hat{x}_{k+1|k+1} = \hat{x}_{k+1|k} + \bar{K}_{k+1}(\tilde{y}_{k+1} - \bar{C}_{k+1}\hat{x}_{k+1|k}), \tag{4.33}$$

where $\hat{x}_{k|k}$ is the state estimate with $\hat{x}_{0|0} = \begin{bmatrix} \bar{x}_0^T & 0^T \end{bmatrix}^T$, $\hat{x}_{k+1|k}$ is the one-step prediction at step k, $\bar{K}_{k+1} = K_{k+1} + \Delta K_{k+1}$, K_{k+1} is the filter gain, ΔK_{k+1} denotes the filter gain perturbation with $\Delta K_{k+1} = M_{k+1} F_{k+1} N_{k+1}$, M_{k+1} and N_{k+1} are pre-defined matrices and F_{k+1} satisfies $F_{k+1}^T F_{k+1} \leq I$.

Set the one-step prediction error and filtering error as

$$\tilde{x}_{k+1|k} = x_{k+1} - \hat{x}_{k+1|k},$$

$$\tilde{x}_{k+1|k+1} = x_{k+1} - \hat{x}_{k+1|k+1}.$$

As a result, $P_{k+1|k} = \mathbb{E}\{\tilde{x}_{k+1|k}\tilde{x}_{k+1|k}^T\}$ is the one-step prediction error covariance and $P_{k+1|k+1} = \mathbb{E}\{\tilde{x}_{k+1|k+1}\tilde{x}_{k+1|k+1}^T\}$ is the FEC. Subsequently, we aim to construct a resilient filter (4.32) and (4.33) such that an upper bound concerning on the FEC $\Sigma_{k+1|k+1}$ is obtained. Moreover, $\mathrm{tr}(\Sigma_{k+1|k+1})$ can be minimized via properly selecting the filter gain K_{k+1}.

4.2.2 Main Results

In this subsection, by resorting to the matrix theory, an upper bound associated with the FEC is calculated and the proper filter gain is devised.

Subtracting (4.32) from (4.29) yields

$$\tilde{x}_{k+1|k} = \tilde{A}_k x_k + \bar{A}_k \tilde{x}_{k|k} + B_k \phi(x_k) + D_k \omega_k. \tag{4.34}$$

Based on (4.33), we have

$$\begin{aligned}\tilde{x}_{k+1|k+1} = {} & (I - \bar{K}_{k+1}\bar{C}_{k+1})\tilde{x}_{k+1|k} - \bar{K}_{k+1}\tilde{C}_{k+1}x_{k+1} - \bar{K}_{k+1}(\tilde{y}_{k+1} - y_{k+1}) \\ & -\lambda_{k+1}\xi_{k+1}\bar{K}_{k+1}E_{k+1}\psi(x_{k+1}) - \lambda_{k+1}\bar{K}_{k+1}\nu_{k+1}.\end{aligned} \tag{4.35}$$

It is clear that

$$\mathbb{E}\{\omega_k\omega_k^T\} = \mathrm{diag}\{\breve{Q}_k, \breve{R}_k\} := Q_k,$$

$$\tilde{A}_k = A_k - \bar{A}_k = \begin{bmatrix} 0 & 0 \\ (\lambda_k - \bar{\lambda}_k)\breve{C}_k & (\bar{\lambda}_k - \lambda_k)I_m \end{bmatrix} = (\lambda_k - \bar{\lambda}_k)\mathscr{A}_k,$$

$$\tilde{C}_k = C_k - \bar{C}_k = \begin{bmatrix} (\lambda_k - \bar{\lambda}_k)\breve{C}_k & (\bar{\lambda}_k - \lambda_k)I_m \end{bmatrix} = (\lambda_k - \bar{\lambda}_k)\mathscr{C}_k,$$

$$B_k = \begin{bmatrix} \eta_k \breve{B}_k & 0 \\ 0 & \lambda_k \xi_k \breve{E}_k \end{bmatrix} = \Lambda_{1,k} \mathscr{B}_k, \quad D_k = \begin{bmatrix} \breve{D}_k & 0 \\ 0 & \lambda_k I_m \end{bmatrix} = \Lambda_{2,k} \mathscr{D}_k,$$

where

$$\mathscr{A}_k = \begin{bmatrix} 0 & 0 \\ \breve{C}_k & -I_m \end{bmatrix}, \quad \mathscr{B}_k = \begin{bmatrix} \breve{B}_k & 0 \\ 0 & \breve{E}_k \end{bmatrix}, \quad \mathscr{D}_k = \begin{bmatrix} \breve{D}_k & 0 \\ 0 & I_m \end{bmatrix},$$
$$\Lambda_{1,k} = \begin{bmatrix} \eta_k I_n & 0 \\ 0 & \lambda_k \xi_k I_m \end{bmatrix}, \quad \Lambda_{2,k} = \begin{bmatrix} I_n & 0 \\ 0 & \lambda_k I_m \end{bmatrix}, \quad \mathscr{C}_k = \begin{bmatrix} \breve{C}_k & -I_m \end{bmatrix}.$$

Then, the recursive forms of one-step prediction error covariance and FEC can be obtained, and the corresponding proofs are omitted here.

Theorem 4.4 *$P_{k+1|k}$ satisfies the following recursive form:*

$$\begin{aligned} P_{k+1|k} = {} & \bar{A}_k P_{k|k} \bar{A}_k^T + \breve{E}_1 \circ [\mathscr{B}_k \mathbb{E}\{\phi(x_k)\phi^T(x_k)\}\mathscr{B}_k^T] + \breve{E}_2 \circ (\mathscr{D}_k Q_k \mathscr{D}_k^T) \\ & + \bar{\lambda}_k(1-\bar{\lambda}_k)\mathscr{A}_k \mathbb{E}\{x_k x_k^T\}\mathscr{A}_k^T, \end{aligned} \tag{4.36}$$

where

$$\breve{E}_1 = \begin{bmatrix} \breve{E}_{n\times n} & 0 \\ 0 & \bar{\lambda}_k \breve{E}_{m\times m} \end{bmatrix}, \quad \breve{E}_2 = \begin{bmatrix} \breve{E}_{n\times n} & \bar{\lambda}_k \breve{E}_{n\times m} \\ * & \bar{\lambda}_k \breve{E}_{m\times m} \end{bmatrix}, \quad \breve{E}_{m\times n} = \begin{bmatrix} 1 & 1 & \cdots & 1 \\ 1 & 1 & \cdots & 1 \\ \vdots & & \ddots & \vdots \\ 1 & 1 & \cdots & 1 \end{bmatrix}_{m\times n}.$$

Theorem 4.5 *$P_{k+1|k+1}$ satisfies the following recursive form:*

$$\begin{aligned} P_{k+1|k+1} = {} & (I - \bar{K}_{k+1}\bar{C}_{k+1})P_{k+1|k}(I - \bar{K}_{k+1}\bar{C}_{k+1})^T + \bar{K}_{k+1}\mathbb{E}\{(\tilde{y}_{k+1} - y_{k+1}) \\ & \times (\tilde{y}_{k+1} - y_{k+1})^T\}\bar{K}_{k+1}^T + \bar{\lambda}_{k+1}\bar{K}_{k+1}E_{k+1}\mathbb{E}\{\psi(x_{k+1})\psi^T(x_{k+1})\} \\ & \times E_{k+1}^T \bar{K}_{k+1}^T + \bar{\lambda}_{k+1}\bar{K}_{k+1}\breve{R}_{k+1}\bar{K}_{k+1}^T + \bar{\lambda}_{k+1}(1-\bar{\lambda}_{k+1})\bar{K}_{k+1}\mathscr{C}_{k+1} \\ & \times \mathbb{E}\{x_{k+1}x_{k+1}^T\}\mathscr{C}_{k+1}^T \bar{K}_{k+1}^T - \mathscr{N}_1 - \mathscr{N}_1^T + \mathscr{N}_2 + \mathscr{N}_2^T + \mathscr{N}_3 + \mathscr{N}_3^T \\ & + \mathscr{N}_4 + \mathscr{N}_4^T, \end{aligned} \tag{4.37}$$

where

$$\begin{aligned} \mathscr{N}_1 &= \mathbb{E}\{(I - \bar{K}_{k+1}\bar{C}_{k+1})\tilde{x}_{k+1|k}(\tilde{y}_{k+1} - y_{k+1})^T \bar{K}_{k+1}^T\}, \\ \mathscr{N}_2 &= \mathbb{E}\{\lambda_{k+1}\xi_{k+1}\bar{K}_{k+1}(\tilde{y}_{k+1} - y_{k+1})\psi^T(x_{k+1})E_{k+1}^T \bar{K}_{k+1}^T\}, \\ \mathscr{N}_3 &= \mathbb{E}\{\lambda_{k+1}\bar{K}_{k+1}(\tilde{y}_{k+1} - y_{k+1})v_{k+1}^T \bar{K}_{k+1}^T\}, \\ \mathscr{N}_4 &= \mathbb{E}\{\bar{K}_{k+1}(\tilde{y}_{k+1} - y_{k+1})x_{k+1}^T \tilde{C}_{k+1}^T \bar{K}_{k+1}^T\}. \end{aligned}$$

Theorem 4.6 *Let $\varepsilon_i > 0$ $(i = 1, 2, \ldots, 6)$ and $\gamma_{j,k+1} > 0$ $(j = 1, 2)$ be scalars. If the following matrix equations:*

$$\Sigma_{k+1|k} = \bar{A}_k \Sigma_{k|k} \bar{A}_k^T + \check{E}_1 \circ \left\{[\mathrm{tr}(\check{\Phi}_k) + \mathrm{tr}(\check{\Psi}_k)]\mathscr{B}_k \mathscr{B}_k^T\right\} + \check{E}_2 \circ (\mathscr{D}_k Q_k \mathscr{D}_k^T)$$
$$+\bar{\lambda}_k (1-\bar{\lambda}_k)\mathscr{A}_k \check{\mathscr{W}}_k \mathscr{A}_k^T, \tag{4.38}$$
$$\Sigma_{k+1|k+1} = (1+\varepsilon_3)[(I - K_{k+1}\bar{C}_{k+1})(\Sigma_{k+1|k}^{-1} - \gamma_{1,k+1}\bar{C}_{k+1}^T N_{k+1}^T N_{k+1}\bar{C}_{k+1})^{-1}$$
$$\times (I - K_{k+1}\bar{C}_{k+1})^T + \gamma_{1,k+1}^{-1} M_{k+1} M_{k+1}^T] + K_{k+1}$$
$$\times (\check{\Omega}_{k+1}^{-1} - \gamma_{2,k+1} N_{k+1}^T N_{k+1})^{-1} K_{k+1}^T + \gamma_{2,k+1}^{-1} M_{k+1} M_{k+1}^T \tag{4.39}$$

under the initial condition $\Sigma_{0|0} = P_{0|0}$ and

$$\gamma_{1,k+1}^{-1} I - N_{k+1}\bar{C}_{k+1}\Sigma_{k+1|k}\bar{C}_{k+1}^T N_{k+1}^T > 0,$$
$$\gamma_{2,k+1}^{-1} I - N_{k+1}\check{\Omega}_{k+1} N_{k+1}^T > 0,$$

have solutions $\Sigma_{k+1|k} > 0$ and $\Sigma_{k+1|k+1} > 0$, where

$$\check{\mathscr{W}}_k = (1+\varepsilon_1)\Sigma_{k|k} + (1+\varepsilon_1^{-1})\hat{x}_{k|k}\hat{x}_{k|k}^T,$$
$$\check{\mathscr{L}}_{k+1} = (1+\varepsilon_2)\Sigma_{k+1|k} + (1+\varepsilon_2^{-1})\hat{x}_{k+1|k}\hat{x}_{k+1|k}^T,$$
$$\check{\Omega}_{k+1} = (1+\varepsilon_4^{-1})\bar{\lambda}_{k+1} E_{k+1}\check{\Psi}_{k+1} E_{k+1}^T + (1+\varepsilon_5^{-1})\bar{\lambda}_{k+1}\check{R}_{k+1} + (1+\varepsilon_6^{-1})$$
$$\times \bar{\lambda}_{k+1}(1-\bar{\lambda}_{k+1})\mathscr{C}_{k+1}\check{\mathscr{L}}_{k+1}\mathscr{C}_{k+1}^T + (1+\varepsilon_3^{-1}+\varepsilon_4+\varepsilon_5+\varepsilon_6)\sigma I,$$

then one has $P_{k+1|k+1} \le \Sigma_{k+1|k+1}$. Besides, if the filter gain K_{k+1} is determined as

$$K_{k+1} = (1+\varepsilon_3)(\Sigma_{k+1|k}^{-1} - \gamma_{1,k+1}\bar{C}_{k+1}^T N_{k+1}^T N_{k+1}\bar{C}_{k+1})^{-1}\bar{C}_{k+1}^T$$
$$\times \Big[(\check{\Omega}_{k+1}^{-1} - \gamma_{2,k+1} N_{k+1}^T N_{k+1})^{-1} + (1+\varepsilon_3)\bar{C}_{k+1}$$
$$\times (\Sigma_{k+1|k}^{-1} - \gamma_{1,k+1}\bar{C}_{k+1}^T N_{k+1}^T N_{k+1}\bar{C}_{k+1})^{-1}\bar{C}_{k+1}^T\Big]^{-1}, \tag{4.40}$$

then the minimum value of $\mathrm{tr}(\Sigma_{k+1|k+1})$ *can be obtained.*

Proof Adopting Lemma 2.2, it is easy to obtain

$$\tilde{x}_{k|k}\hat{x}_{k|k}^T + \hat{x}_{k|k}\tilde{x}_{k|k}^T \le \varepsilon_1 \tilde{x}_{k|k}\tilde{x}_{k|k}^T + \varepsilon_1^{-1}\hat{x}_{k|k}\hat{x}_{k|k}^T. \tag{4.41}$$

Thus, we have

$$\mathbb{E}\{x_k x_k^T\} \le (1+\varepsilon_1)P_{k|k} + (1+\varepsilon_1^{-1})\hat{x}_{k|k}\hat{x}_{k|k}^T$$
$$:= \mathscr{W}_k, \tag{4.42}$$

where $\varepsilon_1 > 0$ is a constant. From (4.31), one has

$$\mathbb{E}\{\phi(x_k)\phi^T(x_k)\} \le [\mathrm{tr}(\check{\Phi}_k) + \mathrm{tr}(\check{\Psi}_k)]I. \tag{4.43}$$

Combing (4.42) and (4.43), we have

$$
\begin{aligned}
P_{k+1|k} \leq\; & \bar{A}_k P_{k|k} \bar{A}_k^T + \check{E}_1 \circ \left\{[\mathrm{tr}(\breve{\Phi}_k) + \mathrm{tr}(\breve{\Psi}_k)]\mathscr{B}_k \mathscr{B}_k^T\right\} \\
& + \check{E}_2 \circ (\mathscr{D}_k Q_k \mathscr{D}_k^T) + \bar{\lambda}_k (1 - \bar{\lambda}_k) \mathscr{A}_k \mathscr{W}_k \mathscr{A}_k^T. \qquad (4.44)
\end{aligned}
$$

Along the same line with (4.42), it is obvious to know that

$$
\begin{aligned}
\mathbb{E}\{x_{k+1} x_{k+1}^T\} &\leq (1 + \varepsilon_2) P_{k+1|k} + (1 + \varepsilon_2^{-1}) \hat{x}_{k+1|k} \hat{x}_{k+1|k}^T \\
&:= \mathscr{L}_{k+1}, \qquad (4.45)
\end{aligned}
$$

where $\varepsilon_2 > 0$ is a constant. Similarly, we obtain

$$
\begin{aligned}
-\mathscr{N}_1 - \mathscr{N}_1^T &\leq \varepsilon_3 (I - \bar{K}_{k+1} \bar{C}_{k+1}) P_{k+1|k} (I - \bar{K}_{k+1} \bar{C}_{k+1})^T \\
&\quad + \varepsilon_3^{-1} \bar{K}_{k+1} \mathbb{E}\{(\tilde{y}_{k+1} - y_{k+1})(\tilde{y}_{k+1} - y_{k+1})^T\} \bar{K}_{k+1}^T, \qquad (4.46) \\
\mathscr{N}_2 + \mathscr{N}_2^T &\leq \varepsilon_4 \bar{K}_{k+1} \mathbb{E}\{(\tilde{y}_{k+1} - y_{k+1})(\tilde{y}_{k+1} - y_{k+1})^T\} \bar{K}_{k+1}^T \\
&\quad + \varepsilon_4^{-1} \bar{\lambda}_{k+1} \bar{K}_{k+1} E_{k+1} \mathbb{E}\{\psi(x_{k+1}) \psi^T(x_{k+1})\} E_{k+1}^T \bar{K}_{k+1}^T, \qquad (4.47) \\
\mathscr{N}_3 + \mathscr{N}_3^T &\leq \varepsilon_5 \bar{K}_{k+1} \mathbb{E}\{(\tilde{y}_{k+1} - y_{k+1})(\tilde{y}_{k+1} - y_{k+1})^T\} \bar{K}_{k+1}^T \\
&\quad + \varepsilon_5^{-1} \bar{\lambda}_{k+1} \bar{K}_{k+1} \breve{R}_{k+1} \bar{K}_{k+1}^T, \qquad (4.48) \\
\mathscr{N}_4 + \mathscr{N}_4^T &\leq \varepsilon_6 \bar{K}_{k+1} \mathbb{E}\{(\tilde{y}_{k+1} - y_{k+1})(\tilde{y}_{k+1} - y_{k+1})^T\} \bar{K}_{k+1}^T \\
&\quad + \varepsilon_6^{-1} \bar{\lambda}_{k+1} (1 - \bar{\lambda}_{k+1}) \bar{K}_{k+1} \mathscr{C}_{k+1} \mathbb{E}\{x_{k+1} x_{k+1}^T\} \mathscr{C}_{k+1}^T \bar{K}_{k+1}^T \qquad (4.49)
\end{aligned}
$$

with $\varepsilon_i > 0$ $(i = 3, 4, 5, 6)$ denoting scalars. Substituting (4.45)–(4.49) into (4.37) yields

$$
\begin{aligned}
P_{k+1|k+1} \leq\; & (1 + \varepsilon_3)(I - \bar{K}_{k+1} \bar{C}_{k+1}) P_{k+1|k} (I - \bar{K}_{k+1} \bar{C}_{k+1})^T + (1 + \varepsilon_4^{-1}) \bar{\lambda}_{k+1} \bar{K}_{k+1} \\
& \times E_{k+1} \mathbb{E}\{\psi(x_{k+1}) \psi^T(x_{k+1})\} E_{k+1}^T \bar{K}_{k+1}^T + (1 + \varepsilon_5^{-1}) \bar{\lambda}_{k+1} \bar{K}_{k+1} \breve{R}_{k+1} \bar{K}_{k+1}^T \\
& + (1 + \varepsilon_6^{-1}) \bar{\lambda}_{k+1} (1 - \bar{\lambda}_{k+1}) \bar{K}_{k+1} \mathscr{C}_{k+1} \mathscr{L}_{k+1} \mathscr{C}_{k+1}^T \bar{K}_{k+1}^T + (1 + \varepsilon_3^{-1} + \varepsilon_4 \\
& + \varepsilon_5 + \varepsilon_6) \bar{K}_{k+1} \mathbb{E}\{(\tilde{y}_{k+1} - y_{k+1})(\tilde{y}_{k+1} - y_{k+1})^T\} \bar{K}_{k+1}^T. \qquad (4.50)
\end{aligned}
$$

By virtue of (4.27), one has

$$
(\tilde{y}_k - y_k)(\tilde{y}_k - y_k)^T \leq \sigma I. \qquad (4.51)
$$

Substituting (4.51) into (4.50), one has

$$
\begin{aligned}
P_{k+1|k+1} \leq\; & (1 + \varepsilon_3)(I - \bar{K}_{k+1} \bar{C}_{k+1}) P_{k+1|k} (I - \bar{K}_{k+1} \bar{C}_{k+1})^T \\
& + \bar{K}_{k+1} \Omega_{k+1} \bar{K}_{k+1}^T, \qquad (4.52)
\end{aligned}
$$

where

$$
\begin{aligned}
\Omega_{k+1} = &(1+\varepsilon_4^{-1})\bar{\lambda}_{k+1}E_{k+1}\breve{\Psi}_{k+1}E_{k+1}^T + (1+\varepsilon_5^{-1})\bar{\lambda}_{k+1}\breve{R}_{k+1} \\
&+ (1+\varepsilon_6^{-1})\bar{\lambda}_{k+1}(1-\bar{\lambda}_{k+1})\mathscr{C}_{k+1}\mathscr{L}_{k+1}\mathscr{C}_{k+1}^T \\
&+ (1+\varepsilon_3^{-1}+\varepsilon_4+\varepsilon_5+\varepsilon_6)\sigma I.
\end{aligned}
$$

In light of Lemma 2.1, one obtains

$$
\begin{aligned}
&(I-\bar{K}_{k+1}\bar{C}_{k+1})P_{k+1|k}(I-\bar{K}_{k+1}\bar{C}_{k+1})^T \\
&\le (I-K_{k+1}\bar{C}_{k+1})(P_{k+1|k}^{-1}-\gamma_{1,k+1}\bar{C}_{k+1}^T N_{k+1}^T N_{k+1}\bar{C}_{k+1})^{-1} \\
&\times (I-K_{k+1}\bar{C}_{k+1})^T + \gamma_{1,k+1}^{-1}M_{k+1}M_{k+1}^T, \\
&\bar{K}_{k+1}\Omega_{k+1}\bar{K}_{k+1}^T \\
&\le K_{k+1}(\Omega_{k+1}^{-1}-\gamma_{2,k+1}N_{k+1}^T N_{k+1})^{-1}K_{k+1}^T + \gamma_{2,k+1}^{-1}M_{k+1}M_{k+1}^T.
\end{aligned}
\tag{4.53}
$$

Then, it can be concluded that

$$
\begin{aligned}
P_{k+1|k+1} \le &(1+\varepsilon_3)[(I-K_{k+1}\bar{C}_{k+1})(P_{k+1|k}^{-1}-\gamma_{1,k+1}\bar{C}_{k+1}^T N_{k+1}^T N_{k+1}\bar{C}_{k+1})^{-1} \\
&\times (I-K_{k+1}\bar{C}_{k+1})^T + \gamma_{1,k+1}^{-1}M_{k+1}M_{k+1}^T] \\
&+ K_{k+1}(\Omega_{k+1}^{-1}-\gamma_{2,k+1}N_{k+1}^T N_{k+1})^{-1}K_{k+1}^T \\
&+ \gamma_{2,k+1}^{-1}M_{k+1}M_{k+1}^T.
\end{aligned}
\tag{4.54}
$$

On the basis of the mathematical induction method, (4.38), (4.39), (4.44) and (4.54), it can be obtained that $P_{k+1|k+1} \le \Sigma_{k+1|k+1}$ holds.

Finally, taking the derivative of $\mathrm{tr}(\Sigma_{k+1|k+1})$ regarding K_{k+1} and letting the final result be zero, we arrive at

$$
\begin{aligned}
&\frac{\partial \mathrm{tr}(\Sigma_{k+1|k+1})}{\partial K_{k+1}} \\
&= -2(1+\varepsilon_3)(I-K_{k+1}\bar{C}_{k+1})(\Sigma_{k+1|k}^{-1}-\gamma_{1,k+1}\bar{C}_{k+1}^T N_{k+1}^T N_{k+1}\bar{C}_{k+1})^{-1}\bar{C}_{k+1}^T \\
&\quad +2K_{k+1}(\breve{\Omega}_{k+1}^{-1}-\gamma_{2,k+1}N_{k+1}^T N_{k+1})^{-1} \\
&= 0.
\end{aligned}
$$

It is not difficult to testify that (4.40) is obtained and the proof is complete.

Remark 4.4 In Theorem 4.6, all uncertain terms reflected in the one-step prediction error covariance and FEC have been handled through adopting the matrix analysis method. Particularly, an upper bound regarding the FEC has been derived by adequately considering the available data information, and then minimized by selecting an appropriate filter gain. Finally, the desirable filter method with recursive feature has been given under variance-constrained index.

4.3 Two Illustrative Examples

In this section, two illustrative experiments with comparative results are provided to testify the validity and advantages of the proposed event-based filtering schemes.

Example 1 The system parameters in (4.1) and (4.2) are chosen as:

$$A_k = \begin{bmatrix} 0.5 & 0.6 \\ 0.3 & 0.5 \end{bmatrix}, \quad B_k = \begin{bmatrix} 0.2 \\ 0.6 \end{bmatrix}, \quad D_k = \begin{bmatrix} 0.4 \\ 0.6 \end{bmatrix},$$
$$N_k = \begin{bmatrix} 0.04 \\ 0.08 \end{bmatrix}, \quad M_k = \begin{bmatrix} 0.04 & 0.03 \\ 0.05 & 0.02 \end{bmatrix}^T, \quad E_k = 0.5,$$
$$F_k = \begin{bmatrix} 0.03\sin(k) & 0 \\ 0 & 0.07\cos(k) \end{bmatrix}, \quad C_k = \begin{bmatrix} 0.8 & 0.5 \end{bmatrix},$$

and $x_k = \begin{bmatrix} x_{1,k} \\ x_{2,k} \end{bmatrix}$ is the system state. The noise covariances are set as $Q_k = R_k = 0.2$.

Let the initial values be $\hat{x}_{0|0} = \begin{bmatrix} 2 & 2 \end{bmatrix}^T$ and $\Sigma_{0|0} = 3I_2$. Other known parameters are set as

$$\begin{aligned} &\varepsilon_1 = 0.8, \ \varepsilon_2 = 0.6, \ \varepsilon_3 = 0.4, \ \varphi(x_k) = \sqrt{0.6}\sin(x_{1,k}), \\ &\psi(x_k) = \sqrt{0.5}\sin(x_{2,k}), \ \Phi_k = 0.6, \ \Psi_k = 0.5, \\ &\gamma_{1,k+1} = [2\lambda_{\max}(N_{k+1}C_{k+1}\Sigma_{k+1|k}C_{k+1}^T N_{k+1}^T) + 1]^{-1}, \\ &\gamma_{2,k+1} = [2\lambda_{\max}(N_{k+1}\Pi_{k+1}N_{k+1}^T) + 1]^{-1}. \end{aligned}$$

The simulations are shown in Figs. 4.3, 4.4, 4.5, 4.6, 4.7, 4.8 and 4.9. Among them, Figs. 4.3 and 4.4 plot the state $x_{1,k}$ and its estimation $\hat{x}_{1,k|k}$ in the case of $\delta = 0.1$ and $\delta = 1$, respectively. Figures 4.5 and 4.6 depict the state $x_{2,k}$ and its estimation $\hat{x}_{2,k|k}$ in the case of $\delta = 0.1$ and $\delta = 1$, respectively. Figures 4.7 and 4.8 show the relationship between log(MSE) and the trace of $\Sigma_{k|k}$. Here, MSE stands for the mean-square error for the estimation of the system state. It can be seen that the trace of $\Sigma_{k|k}$ is always above log(MSE) and the estimation accuracy becomes better when the parameter δ decreases. Furthermore, for the sake of clarifying the effect of the triggered threshold onto the filtering performance, the curves of the trace of the upper bound under different triggered threshold δ are given in Fig. 4.9, which can observe that the trace of the upper bound becomes bigger when the triggered threshold is bigger, that is to say, the bigger triggered threshold is, the less data can be transmitted, and thus the worse filtering precision gets.

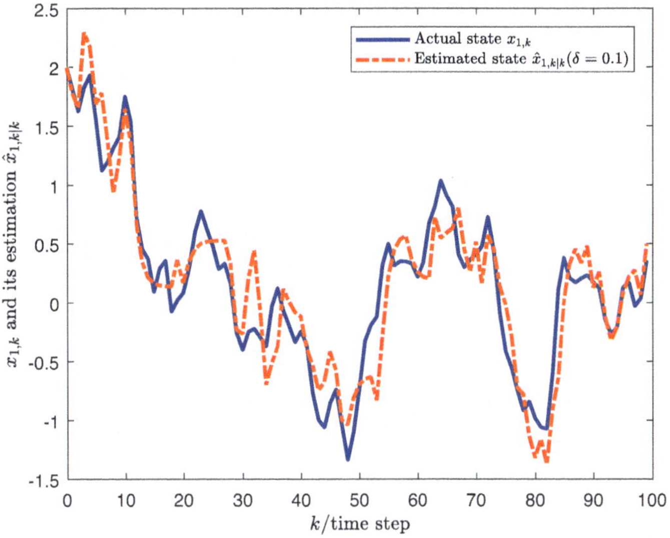

Fig. 4.3 State $x_{1,k}$ and its estimation $\hat{x}_{1,k|k}$ with $\delta = 0.1$

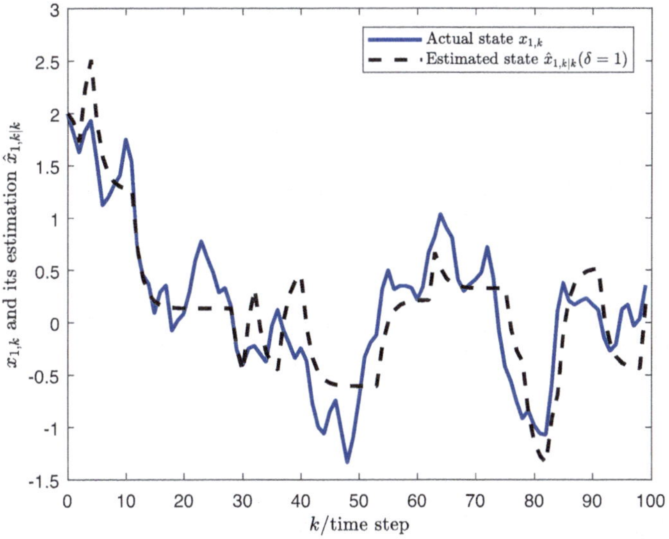

Fig. 4.4 State $x_{1,k}$ and its estimation $\hat{x}_{1,k|k}$ with $\delta = 1$

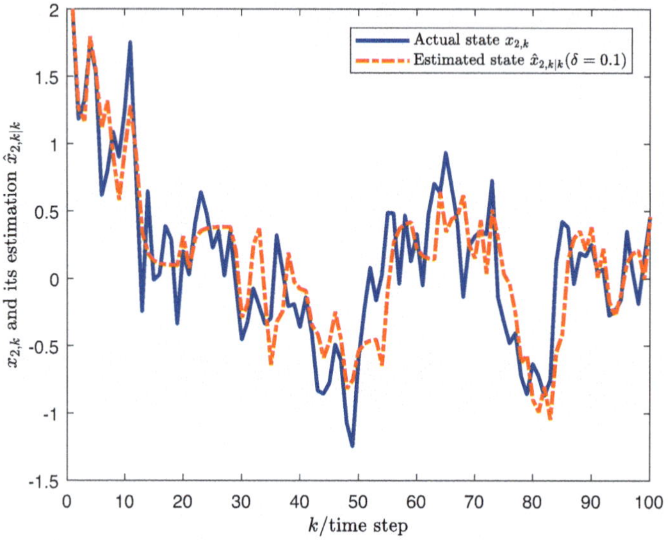

Fig. 4.5 State $x_{2,k}$ and its estimation $\hat{x}_{2,k|k}$ with $\delta = 0.1$

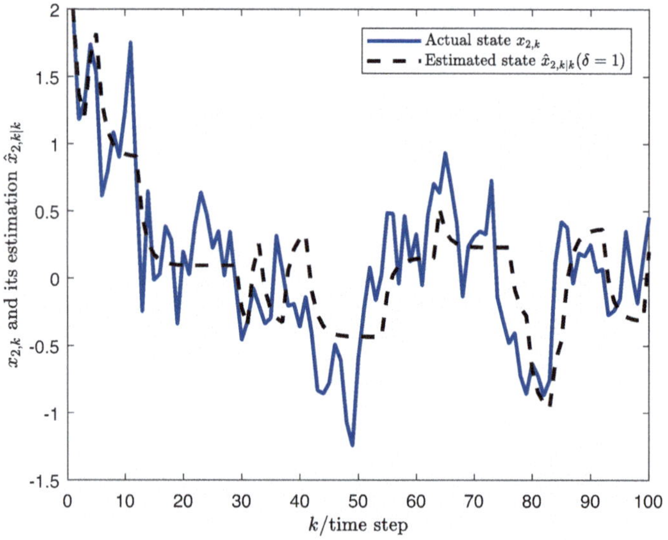

Fig. 4.6 State $x_{2,k}$ and its estimation $\hat{x}_{2,k|k}$ with $\delta = 1$

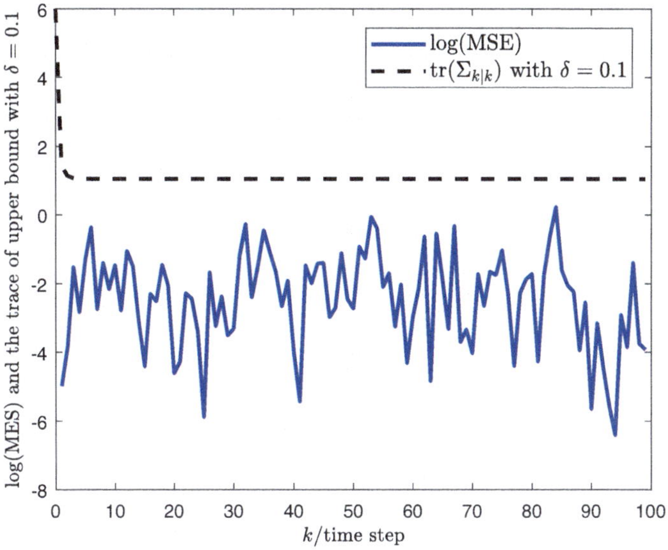

Fig. 4.7 log(MSE) and the trace of upper bound when $\delta = 0.1$

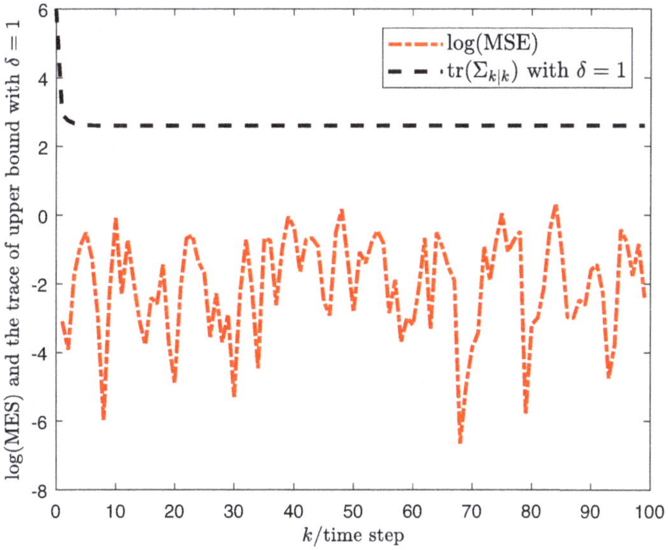

Fig. 4.8 log(MSE) and the trace of upper bound when $\delta = 1$

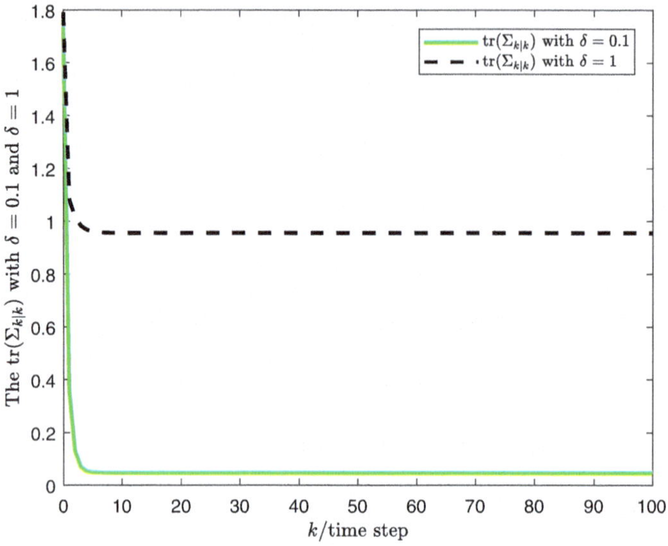

Fig. 4.9 The trace of upper bound under different δ

Example 2 The system parameters in (4.24)–(4.25) are given as

$$\breve{A}_k = \begin{bmatrix} 0.65 & 0.48 \\ 0.12 & 0.68 + 0.05\sin(k) \end{bmatrix}, \quad \breve{B}_k = \begin{bmatrix} 0.4 \\ 0.4 \end{bmatrix},$$
$$\breve{C}_k = \begin{bmatrix} 1 & 0.1 \\ 0.8 & 0.5 + 0.05\sin(k) \end{bmatrix}, \quad N_k = \begin{bmatrix} 0.04 & 0 \\ 0 & 0.08 \end{bmatrix},$$
$$M_k = \begin{bmatrix} 0.04 & 0.03 & 0.02 & 0.03 \\ 0.05 & 0.02 & 0.06 & 0.07 \end{bmatrix}^T, \quad \breve{D}_k = \begin{bmatrix} 0.3 \\ 0.4 \end{bmatrix},$$
$$F_k = \begin{bmatrix} 0.05\cos(k) & 0 \\ 0 & 0.04\sin(k) \end{bmatrix}, \quad \breve{E}_k = \begin{bmatrix} 0.5 \\ 0.6 \end{bmatrix},$$

and $\breve{x}_k = \begin{bmatrix} \breve{x}_{1,k} & \breve{x}_{2,k} \end{bmatrix}^T$ is the system state. The noise covariances are set as $\breve{Q}_k = 0.4$ and $\breve{R}_k = 0.2I_2$.

The initial conditions are chosen as $\hat{x}_{0|0} = \begin{bmatrix} 2 & 2 & 0 & 0 \end{bmatrix}^T$ and $\Sigma_{0|0} = 3I_4$. Other coefficients are set as

$$\begin{aligned}
&\varepsilon_1 = 0.8, \ \varepsilon_2 = 0.5, \ \varepsilon_3 = 0.2, \ \varepsilon_4 = 0.4, \\
&\varepsilon_5 = 0.2, \ \varepsilon_6 = 0.1, \ \varphi(\breve{x}_k) = \sqrt{0.2}\sin(\breve{x}_{1,k}), \\
&\psi(\breve{x}_k) = \sqrt{0.2}\sin(\breve{x}_{2,k}), \ \breve{\Phi}_k = 0.2, \ \breve{\Psi}_k = 0.2, \ \bar{\lambda}_k = 0.8, \\
&\gamma_{1,k+1} = [1.2\lambda_{\max}(N_{k+1}\bar{C}_{k+1}\Sigma_{k+1|k}\bar{C}_{k+1}^T\bar{N}_{k+1}^T) + 1]^{-1}, \\
&\gamma_{2,k+1} = [1.2\lambda_{\max}(N_{k+1}\breve{\Omega}_{k+1}N_{k+1}^T) + 1]^{-1}.
\end{aligned}$$

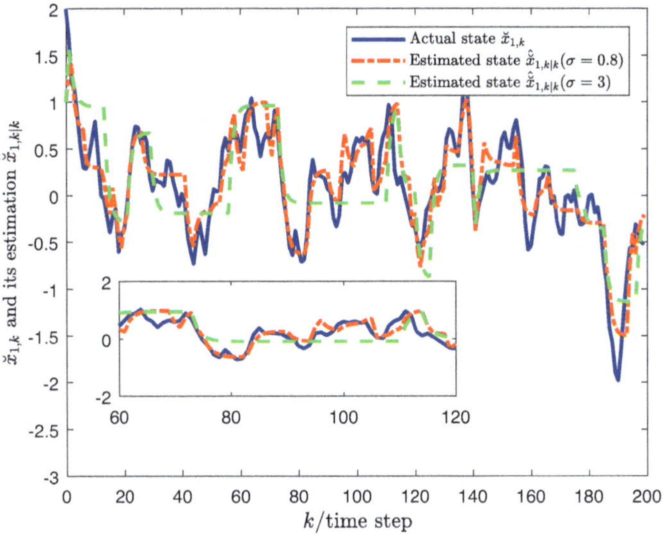

Fig. 4.10 State $\breve{x}_{1,k}$ and its estimation $\hat{\breve{x}}_{1,k|k}$

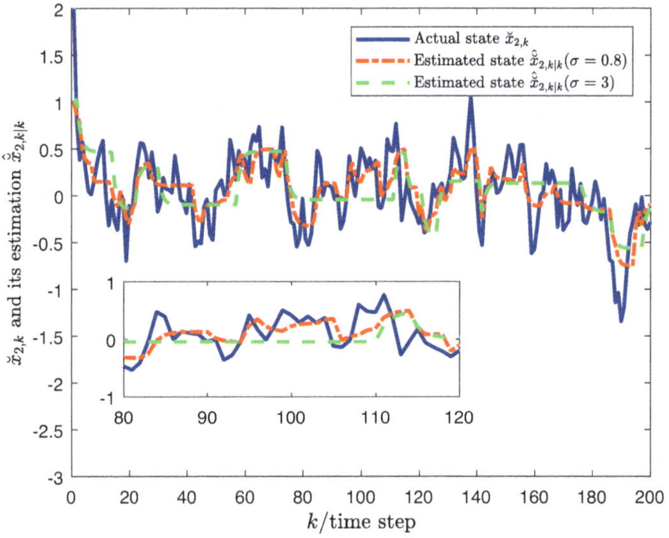

Fig. 4.11 State $\breve{x}_{2,k}$ and its estimation $\hat{\breve{x}}_{2,k|k}$

To clarify the influence of the event-triggered scheduling scheme, we set $\sigma = 0.8$ and $\sigma = 3$.

Figures 4.10 and 4.11 depict the system states $\breve{x}_{i,k}$ and their estimates $\hat{\breve{x}}_{i,k|k}$ $(i = 1, 2)$ under different triggered conditions. Figure 4.12 plots the log(MSE) of the state $\breve{x}_k$ and the trace of upper bound when $\sigma = 0.8$ and $\sigma = 3$. The value of λ_k is displayed

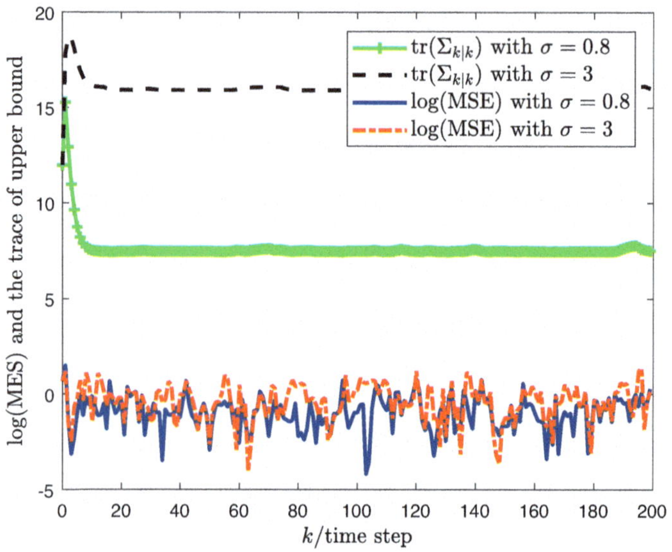

Fig. 4.12 log(MSE) and the trace of upper bound

in Fig. 4.13. In addition, the relation between the SPDs and the filtering performance is illustrated in Fig. 4.14. It is clearly observed that the bigger $\bar{\lambda}_k$ is, and then the lower MSE is, in other words, more measurements are obtained, thus the better filtering accuracy gets. Figure 4.15 presents the MESs with $\sigma = 8$ and $\sigma = 40$.

Remark 4.5 It follows from the simulation results given in Figs. 4.10, 4.11, 4.12 and 4.15 that the tracking accuracy remains better along with the decreasing triggered threshold σ. Moreover, it is pointed out that the purposes of introducing the ETM mainly include two aspects: (1) providing a better estimation performance; and (2) reducing the network transmission burden.

4.4 Summary

In this chapter, we have discussed the effects from stochastic uncertainties and SPDs onto the filtering performance under the limited communication capability. Firstly, a resilient filtering issue has been discussed for time-varying systems with stochastic uncertainties under ETM. A minimum upper bound of the FEC has been computed via constructing the filter gain. Secondly, a resilient filtering issue has been studied for time-varying systems affected by SPDs, ETM and stochastic uncertainties. An upper bound regarding the FEC has been derived through devising a proper filter

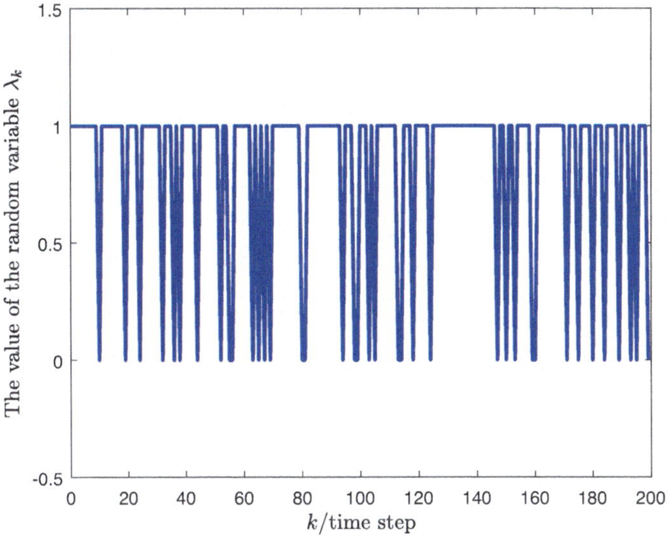

Fig. 4.13 The value of the random variable λ_k

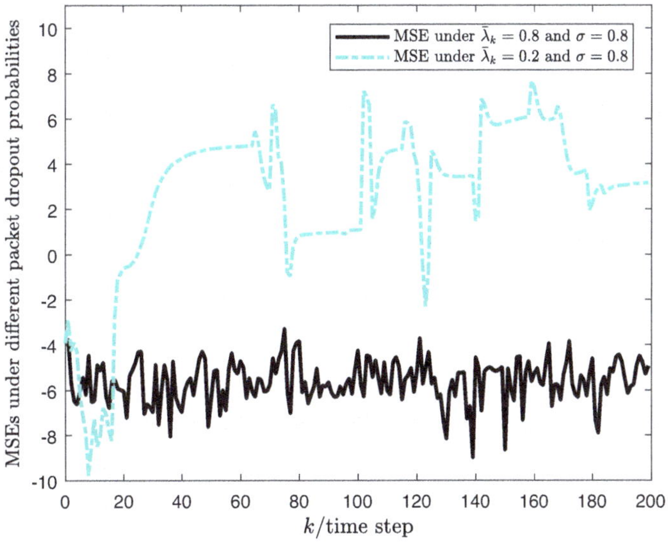

Fig. 4.14 MSEs under different $\bar{\lambda}_k$

gain. Lastly, two simulation experiments with comparative results have been provided to validate the effectiveness and applicability of the designed resilient filtering algorithms regulated by ETM.

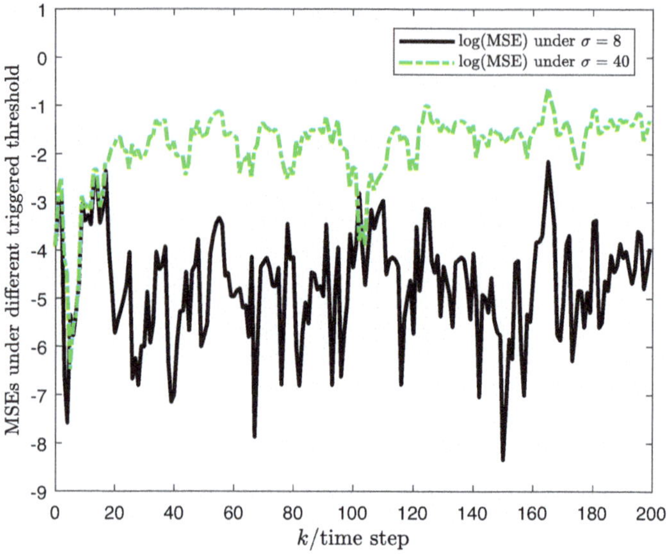

Fig. 4.15 MSEs under different σ

Reference

Kai X, Wei C, Liu L (2010) Robust extended Kalman filtering for nonlinear systems with stochastic uncertainties. IEEE Trans Syst Man Cybern Part A-Syst Hum 40(2):399–405

Chapter 5
Event-Triggered Filtering with Missing Measurements

In this chapter, the objective is to tackle the filtering problems for two classes of nonlinear systems in the presence of ETM and missing measurements. Firstly, the RONs and missing measurements are examined. Here, the binary random variables are employed to characterize the RONs and the missing measurements, in which the occurrence probabilities of them are assumed to be uncertain. With the aim to reduce resource consumptions, the ETM is utilized during the data transmissions through the network. Additionally, the filter gain perturbations are characterized by employing the norm bounded uncertainties. A robust event-triggered resilient filtering algorithm is given against the RONs and missing measurements. Note that the analytical expressions of the FEC cannot be computed directly, hence we derive its upper bound as an alternative way and subsequently minimize this upper bound via choosing an appropriate filter gain. Secondly, the event-triggered filtering issue is addressed for nonlinear stochastic systems influenced by multiple missing measurements. To be specific, the measurements from different sensors may encounter the missing phenomena that are described by a set of binary random variables and the missing probabilities are uncertain. Furthermore, the ETM is used to improve the transmission efficiency, where the current measurement is delivered to the remote filter only when it changes greatly compared with the previous one. A new recursive filter is designed and an optimized upper bound of the FEC is solved and then minimized by choosing the filter gain matrix. The specific expression form of the filter gain is presented by solving the matrix equations. Finally, three illustrative simulation experiments are employed to demonstrate the feasibility and validity of the proposed event-triggered filtering strategy.

J. Hu et al., *Variance-Constrained Filtering for Stochastic Complex Systems*, Intelligent Control and Learning Systems 18,
https://doi.org/10.1007/978-981-96-2637-3_5

5.1 Event-Triggered Resilient Filtering with Missing Measurements and Stochastic Uncertainties

5.1.1 Problem Formulation

Consider the following class of time-varying nonlinear stochastic systems:

$$x_{k+1} = (A_k + \Delta A_k)x_k + \alpha_k f(x_k) + B_k\omega_k, \tag{5.1}$$

$$y_k = \lambda_k C_k x_k + \nu_k, \tag{5.2}$$

where $x_k \in \mathbb{R}^n$ is the state of the target plant, the initial state value x_0 is with mean $\bar{x}_0$ and covariance P_0, $y_k \in \mathbb{R}^m$ is the measurement, $\omega_k \in \mathbb{R}^p$ and $\nu_k \in \mathbb{R}^m$ are zero mean Gaussian noises with covariances Q_k and R_k. The nonlinear function $f(x_k)$ has continuous and bounded second order partial derivatives, $\Delta A_k = M_1 F_{1,k} N_1$ satisfies $F_{1,k}F_{1,k}^T \leq I$, M_1 and N_1 are known matrices with compatible dimensions, A_k, B_k and C_k are pre-defined matrices of compatible dimensions.

The Bernoulli distributed random variables α_k and λ_k taking values 1 or 0 with UOPs are adopted to characterize the phenomenon of RONs and missing measurements respectively

$$\begin{aligned}\text{Prob}\{\alpha_k = 1\} &= \bar{\alpha}_k + \Delta\alpha_k,\\ \text{Prob}\{\lambda_k = 1\} &= \bar{\lambda}_k + \Delta\lambda_k,\end{aligned} \tag{5.3}$$

where $\bar{\alpha}_k + \Delta\alpha_k \in [0, 1]$ and $\bar{\lambda}_k + \Delta\lambda_k \in [0, 1]$, $\bar{\alpha}_k$ and $\bar{\lambda}_k$ are known constants, $\Delta\alpha_k$ and $\Delta\lambda_k$ are unknown constants satisfying $|\Delta\alpha_k| \leq \kappa_1$ and $|\Delta\lambda_k| \leq \kappa_2$ with κ_1 and κ_2 being known constants.

The following ETM is utilized to decrease the transmission frequency when the measurement information is delivered to the filter side through communication network. Specially, the event generator function is described as

$$(y_{k+l} - y_{k_t})^T(y_{k+l} - y_{k_t}) > \delta, \tag{5.4}$$

where $\delta > 0$ is a given scalar, y_{k+l} represents the measurement at time instant $k + l$, and y_{k_t} represents the latest transmitted measurement at time step k_t.

Based on (5.4), the released signal $\tilde{y}_k$ at the triggering time k_i is

$$\tilde{y}_k = y_{k_i}, \quad k \in \{k_i, k_i + 1, \ldots, k_{i+1} - 1\} \tag{5.5}$$

with $k_0 = 0$.

The following robust recursive filter is constructed:

$$\hat{x}_{k+1|k} = A_k\hat{x}_{k|k} + \bar{\alpha}_k f(\hat{x}_{k|k}), \tag{5.6}$$

$$\hat{x}_{k+1|k+1} = \hat{x}_{k+1|k} + (K_{k+1} + \Delta K_{k+1})(\tilde{y}_{k+1} - \bar{\lambda}_{k+1}C_{k+1}\hat{x}_{k+1|k}), \tag{5.7}$$

where $\hat{x}_{k|k}$ is the filtering, $\hat{x}_{k+1|k}$ is the one-step prediction, K_{k+1} is the filter gain, ΔK_{k+1} is the filter gain perturbation satisfying $\Delta K_{k+1} = M_3 F_{3,k+1} N_3$, M_3 and N_3 are known matrices of compatible dimensions and $F_{3,k+1}$ satisfies $F_{3,k+1} F_{3,k+1}^T \leq I$.

Define the one-step prediction error as $\tilde{x}_{k+1|k} = x_{k+1} - \hat{x}_{k+1|k}$ and the filtering error as $\tilde{x}_{k+1|k+1} = x_{k+1} - \hat{x}_{k+1|k+1}$. Subsequently, we aim to design the filter of (5.6)–(5.7) such that

$$\mathbb{E}\{\tilde{x}_{k+1|k+1}\tilde{x}_{k+1|k+1}^T\} \leq \Sigma_{k+1|k+1}, \tag{5.8}$$

where $\Sigma_{k+1|k+1}$ is a set of positive definite matrices. Furthermore, the filter gain K_{k+1} is designed via minimizing $\mathrm{tr}(\Sigma_{k+1|k+1})$.

5.1.2 New Optimized Filtering Method

In this subsection, a locally optimal filtering scheme is given for addressed time-varying systems by means of solving the matrix difference equations.

Subtracting (5.6) from (5.1), we obtain

$$\begin{aligned} \tilde{x}_{k+1|k} &= (A_k + \Delta A_k)x_k - A_k\hat{x}_{k|k} + (\tilde{\alpha}_k + \Delta\alpha_k)f(x_k) \\ &\quad + \bar{\alpha}_k[f(x_k) - f(\hat{x}_{k|k})] + B_k\omega_k, \end{aligned} \tag{5.9}$$

where $\tilde{\alpha}_k = \alpha_k - (\bar{\alpha}_k + \Delta\alpha_k)$ and $\mathbb{E}\{\tilde{\alpha}_k^2\} = (\bar{\alpha}_k + \Delta\alpha_k)(1 - \bar{\alpha}_k - \Delta\alpha_k) := \rho_k$. By employing the Taylor series expansion at $\hat{x}_{k|k}$, one has the following equation with the approximated linearization error estimation:

$$f(x_k) = f(\hat{x}_{k|k}) + D_k\tilde{x}_{k|k} + o(|\tilde{x}_{k|k}^2|), \tag{5.10}$$

where $D_k = \frac{\partial f(x_k)}{\partial x_k}|_{x_k=\hat{x}_{k|k}}$ and $o(|\tilde{x}_{k|k}^2|) = M_2 F_{2,k} N_2 \tilde{x}_{k|k}$ with M_2 and N_2 being bounded scaling matrices, and $F_{2,k}$ being an unknown time-varying matrix satisfying $F_{2,k}F_{2,k}^T \leq I$.

In light of (5.9) and (5.10), one has

$$\begin{aligned} \tilde{x}_{k+1|k} &= (A_k + \bar{\alpha}_k D_k + \bar{\alpha}_k M_2 F_{2,k} N_2)\tilde{x}_{k|k} + \Delta A_k x_k \\ &\quad + \tilde{\alpha}_k f(x_k) + \Delta\alpha_k f(x_k) + B_k\omega_k. \end{aligned} \tag{5.11}$$

Similarly, one obtains

$$\begin{aligned} \tilde{x}_{k+1|k+1} &= [I - \bar{\lambda}_{k+1}(K_{k+1} + \Delta K_{k+1})C_{k+1}]\tilde{x}_{k+1|k} \\ &\quad - (\tilde{\lambda}_{k+1} + \Delta\lambda_{k+1})(K_{k+1} + \Delta K_{k+1})C_{k+1}x_{k+1} \\ &\quad - (K_{k+1} + \Delta K_{k+1})(\tilde{y}_{k+1} - y_{k+1}) \\ &\quad - (K_{k+1} + \Delta K_{k+1})v_{k+1}, \end{aligned} \tag{5.12}$$

where $\tilde{\lambda}_{k+1} = \lambda_{k+1} - (\bar{\lambda}_{k+1} + \Delta\lambda_{k+1})$ and $\mathbb{E}\{\tilde{\lambda}_{k+1}^2\} = (\bar{\lambda}_{k+1} + \Delta\lambda_{k+1})(1 - \bar{\lambda}_{k+1} - \Delta\lambda_{k+1}) := \sigma_{k+1}$.

Next, the aim is to derive the one-step prediction error covariance and FEC according to (5.11) and (5.12).

Theorem 5.1 *The recursion of $P_{k+1|k}$ is calculated as*

$$\begin{aligned} P_{k+1|k} = &(A_k + \bar{\alpha}_k D_k + \bar{\alpha}_k M_2 F_{2,k} N_2) P_{k|k} (A_k + \bar{\alpha}_k D_k \\ &+ \bar{\alpha}_k M_2 F_{2,k} N_2)^T + \left(\Delta\alpha_k^2 + \rho_k\right) \mathbb{E}\{f(x_k) f^T(x_k)\} \\ &+ \Delta A_k \mathbb{E}\{x_k x_k^T\} \Delta A_k^T + B_k Q_k B_k^T + \mathscr{A}_1 + \mathscr{A}_1^T \\ &+ \mathscr{A}_2 + \mathscr{A}_2^T + \mathscr{A}_3 + \mathscr{A}_3^T, \end{aligned} \tag{5.13}$$

where

$$\begin{aligned} \mathscr{A}_1 &= \mathbb{E}\{(A_k + \bar{\alpha}_k D_k + \bar{\alpha}_k M_2 F_{2,k} N_2) \tilde{x}_{k|k} x_k^T \Delta A_k^T\}, \\ \mathscr{A}_2 &= \mathbb{E}\{(A_k + \bar{\alpha}_k D_k + \bar{\alpha}_k M_2 F_{2,k} N_2) \tilde{x}_{k|k} \Delta\alpha_k f^T(x_k)\}, \\ \mathscr{A}_3 &= \mathbb{E}\{\Delta\alpha_k f(x_k) x_k^T \Delta A_k^T\}. \end{aligned} \tag{5.14}$$

Proof Noting $\mathbb{E}\{\tilde{\alpha}_k\} = 0, \mathbb{E}\{\omega_k\} = 0$ and considering the expression form of (5.11), we can show that (5.13) is true.

Theorem 5.2 *$P_{k+1|k+1}$ is computed as*

$$\begin{aligned} P_{k+1|k+1} = &[I - \bar{\lambda}_{k+1}(K_{k+1} + \Delta K_{k+1}) C_{k+1}] P_{k+1|k} [I - \bar{\lambda}_{k+1} \\ &\times (K_{k+1} + \Delta K_{k+1}) C_{k+1}]^T + (\sigma_{k+1} + \Delta\lambda_{k+1}^2) \\ &\times (K_{k+1} + \Delta K_{k+1}) C_{k+1} \mathbb{E}\{x_{k+1} x_{k+1}^T\} C_{k+1}^T (K_{k+1} + \Delta K_{k+1})^T \\ &+ (K_{k+1} + \Delta K_{k+1}) \mathbb{E}\{(\tilde{y}_{k+1} - y_{k+1})(\tilde{y}_{k+1} - y_{k+1})^T\} \\ &\times (K_{k+1} + \Delta K_{k+1})^T + (K_{k+1} + \Delta K_{k+1}) R_{k+1} (K_{k+1} + \Delta K_{k+1})^T \\ &- \mathscr{B}_1 - \mathscr{B}_1^T - \mathscr{B}_2 - \mathscr{B}_2^T + \mathscr{B}_3 + \mathscr{B}_3^T + \mathscr{B}_4 + \mathscr{B}_4^T, \end{aligned} \tag{5.15}$$

where

$$\begin{aligned} \mathscr{B}_1 &= \mathbb{E}\{[I - \bar{\lambda}_{k+1}(K_{k+1} + \Delta K_{k+1}) C_{k+1}] \Delta\lambda_{k+1} \\ &\quad \times \tilde{x}_{k+1|k} x_{k+1}^T C_{k+1}^T (K_{k+1} + \Delta K_{k+1})^T\}, \\ \mathscr{B}_2 &= \mathbb{E}\{[I - \bar{\lambda}_{k+1}(K_{k+1} + \Delta K_{k+1}) C_{k+1}] \tilde{x}_{k+1|k} \\ &\quad \times (\tilde{y}_{k+1} - y_{k+1})^T (K_{k+1} + \Delta K_{k+1})^T\}, \\ \mathscr{B}_3 &= \mathbb{E}\{(\tilde{\lambda}_{k+1} + \Delta\lambda_{k+1})(K_{k+1} + \Delta K_{k+1}) C_{k+1} \\ &\quad \times x_{k+1} (\tilde{y}_{k+1} - y_{k+1})^T (K_{k+1} + \Delta K_{k+1})^T\}, \\ \mathscr{B}_4 &= (K_{k+1} + \Delta K_{k+1}) \mathbb{E}\{(\tilde{y}_{k+1} - y_{k+1}) v_{k+1}^T\} \\ &\quad \times (K_{k+1} + \Delta K_{k+1})^T. \end{aligned} \tag{5.16}$$

Proof Noticing $\mathbb{E}\{\tilde{\lambda}_{k+1}\} = 0$, $\mathbb{E}\{\nu_{k+1}\} = 0$ and considering (5.12), we can easily obtain (5.15).

Remark 5.1 Note that it is not possible to compute the exact value of $P_{k+1|k+1}$ due to the existence of some uncertain terms such as $\Delta\alpha_k$, ΔA_k, $\mathbb{E}\{x_k x_k^T\}$, σ_{k+1} in (5.13) and (5.15). Thus, we cannot derive the filter gain directly and minimize the FEC as made in conventional Kalman filter. Next, a suboptimal method is adopted to find the desirable filter gain.

Theorem 5.3 *For scalars* $\varepsilon_i > 0$ $(i = 1, 2, \ldots, 7)$, $\gamma_{1,k} > 0$, $\gamma_{j,k+1} > 0$ $(j = 2, 3, \ldots, 5)$ *and* $\eta_s > 0$ $(s = 1, 2, \ldots, 5)$, *if the following matrix equations:*

$$\begin{aligned}\Sigma_{k+1|k} &= \tau_1(A_k + \bar{\alpha}_k D_k)\bar{\Psi}_{1,k}^{-1}(A_k + \bar{\alpha}_k D_k)^T + \tau_1\gamma_{1,k}^{-1}\bar{\alpha}_k^2 M_2 M_2^T \\ &\quad + \tau_2\bar{\Upsilon}_{1,k} + \tau_3\bar{\Psi}_{5,k} + B_k Q_k B_k^T\end{aligned} \tag{5.17}$$

and

$$\begin{aligned}\Sigma_{k+1|k+1} &= \tau_4(I - \bar{\lambda}_{k+1}K_{k+1}C_{k+1})\bar{\Psi}_{2,k+1}^{-1}(I - \bar{\lambda}_{k+1}K_{k+1}C_{k+1})^T \\ &\quad + \tau_4\gamma_{2,k+1}^{-1}\bar{\lambda}_{k+1}^2 M_3 M_3^T + (\tau_5\bar{\sigma}_{k+1} + \tau_5\kappa_2^2 + \eta_1^{-1}\kappa_2^2) \\ &\quad \times [K_{k+1}C_{k+1}(\bar{\Psi}_{3,k+1}^{-1} - \gamma_{3,k+1}C_{k+1}^T N_3^T N_3 C_{k+1})^{-1}C_{k+1}^T K_{k+1}^T \\ &\quad + \gamma_{3,k+1}^{-1}M_3 M_3^T] + \tau_6\delta K_{k+1}(I - \gamma_{4,k+1}N_3^T N_3)^{-1}K_{k+1}^T \\ &\quad + \tau_6\gamma_{4,k+1}^{-1}\delta M_3 M_3^T + \tau_7 K_{k+1}(R_{k+1}^{-1} - \gamma_{5,k+1}N_3^T N_3)^{-1}K_{k+1}^T \\ &\quad + \tau_7\gamma_{5,k+1}^{-1}M_3 M_3^T\end{aligned}$$

with the initial condition $\Sigma_{0|0} = P_{0|0}$ *satisfying the following inequalities:*

$$\begin{aligned}&\gamma_{1,k}^{-1}I - N_2\Sigma_{k|k}N_2^T > 0, \\ &\gamma_{2,k+1}^{-1}I - N_3 C_{k+1}\Sigma_{k+1|k}C_{k+1}^T N_3^T > 0, \\ &\gamma_{3,k+1}^{-1}I - N_3 C_{k+1}\bar{\Psi}_{3,k+1}C_{k+1}^T N_3^T > 0, \\ &\gamma_{4,k+1}^{-1}I - N_3 N_3^T > 0, \\ &\gamma_{5,k+1}^{-1}I - N_3 R_{k+1}N_3^T > 0,\end{aligned} \tag{5.18}$$

have solutions $\Sigma_{k+1|k} > 0$ *and* $\Sigma_{k+1|k+1} > 0$, *where*

$$\begin{aligned}\tau_1 &= 1 + \varepsilon_1 + \varepsilon_2,\ \ \tau_2 = \bar{\rho}_k + (1 + \varepsilon_2^{-1} + \varepsilon_3)\kappa_1^2, \\ \tau_3 &= 1 + \varepsilon_1^{-1} + \varepsilon_3^{-1},\ \ \tau_4 = 1 + \eta_1 + \eta_2, \\ \tau_5 &= 1 + \eta_3,\ \ \tau_6 = 1 + \eta_2^{-1} + \eta_3^{-1} + \eta_4,\ \ \tau_7 = 1 + \eta_4^{-1}, \\ \bar{\Psi}_{1,k} &= \Sigma_{k|k}^{-1} - \gamma_{1,k}N_2^T N_2, \\ \bar{\Psi}_{2,k+1} &= \Sigma_{k+1|k}^{-1} - \gamma_{3,k+1}C_{k+1}^T N_3^T N_3 C_{k+1},\end{aligned}$$

$$
\begin{aligned}
\bar{\Psi}_{3,k+1} &= (1+\eta_5)\Sigma_{k+1|k} + (1+\eta_5^{-1})\hat{x}_{k+1|k}\hat{x}_{k+1|k}^T, \\
\bar{\Psi}_{4,k} &= (1+\varepsilon_7)\Sigma_{k|k} + (1+\varepsilon_7^{-1})\hat{x}_{k|k}\hat{x}_{k|k}^T, \\
\bar{\Psi}_{5,k} &= \mathrm{tr}(N_1\bar{\Psi}_{4,k}N_1^T)M_1M_1^T, \\
\bar{\Upsilon}_{1,k} &= (1+\varepsilon_4+\varepsilon_5)f(\hat{x}_{k|k})f^T(\hat{x}_{k|k}) + (1+\varepsilon_4^{-1}+\varepsilon_6)D_k\Sigma_{k|k}D_k^T \\
&\quad + (1+\varepsilon_5^{-1}+\varepsilon_6^{-1})\mathrm{tr}(N_2\Sigma_{k|k}N_2^T)M_2M_2^T, \\
\bar{\sigma}_{k+1} &= \min\{0.25, (\bar{\lambda}_{k+1}+\kappa_2)(1-\bar{\lambda}_{k+1}+\kappa_2)\}, \\
\bar{\rho}_k &= \min\{0.25, (\bar{\alpha}_k+\kappa_1)(1-\bar{\alpha}_k+\kappa_1)\},
\end{aligned} \tag{5.19}
$$

then one has $P_{k+1|k+1} \leq \Sigma_{k+1|k+1}$. *Moreover, if the filter gain* K_{k+1} *is designed as*

$$
\begin{aligned}
K_{k+1} &= \tau_4\bar{\lambda}_{k+1}\bar{\Psi}_{2,k+1}^{-1}C_{k+1}^T\big[\tau_4\bar{\lambda}_{k+1}^2C_{k+1}\bar{\Psi}_{2,k+1}^{-1}C_{k+1}^T + (\tau_5\bar{\sigma}_{k+1} + \tau_5\kappa_2^2 + \eta_1^{-1}\kappa_2^2) \\
&\quad \times C_{k+1}(\bar{\Psi}_{3,k+1}^{-1} - \gamma_{3,k+1}C_{k+1}^TN_3^TN_3C_{k+1})^{-1}C_{k+1}^T \\
&\quad + \tau_6\delta(I - \gamma_{4,k+1}N_3^TN_3)^{-1} + \tau_7(R_{k+1}^{-1} - \gamma_{5,k+1}N_3^TN_3)^{-1}\big]^{-1},
\end{aligned} \tag{5.20}
$$

then we can show that $\mathrm{tr}(\Sigma_{k+1|k+1})$ *is minimized.*

Proof Applying Lemma 2.2, we have

$$
\begin{aligned}
\mathscr{A}_1 + \mathscr{A}_1^T &\leq \varepsilon_1(A_k + \bar{\alpha}_kD_k + \bar{\alpha}_kM_2F_{2,k}N_2)P_{k|k}(A_k + \bar{\alpha}_kD_k + \bar{\alpha}_kM_2F_{2,k}N_2)^T \\
&\quad + \varepsilon_1^{-1}\Delta A_k\mathbb{E}\{x_kx_k^T\}\Delta A_k^T, && (5.21) \\
\mathscr{A}_2 + \mathscr{A}_2^T &\leq \varepsilon_2(A_k + \bar{\alpha}_kD_k + \bar{\alpha}_kM_2F_{2,k}N_2)P_{k|k}(A_k + \bar{\alpha}_kD_k + \bar{\alpha}_kM_2F_{2,k}N_2)^T \\
&\quad + \varepsilon_2^{-1}\Delta\alpha_k^2\mathbb{E}\{f(x_k)f^T(x_k)\}, && (5.22) \\
\mathscr{A}_3 + \mathscr{A}_3^T &\leq \varepsilon_3\Delta\alpha_k^2\mathbb{E}\{f(x_k)f^T(x_k)\} + \varepsilon_3^{-1}\Delta A_k\mathbb{E}\{x_kx_k^T\}\Delta A_k^T, && (5.23)
\end{aligned}
$$

where $\varepsilon_i > 0$ $(i = 1, 2, 3)$ are constants. Noting $\rho_k \leq \bar{\rho}_k$, $|\Delta\alpha_k| < \kappa_1$ and substituting (5.21)–(5.23) into (5.13) yield

$$
\begin{aligned}
P_{k+1|k} &\leq \tau_1(A_k + \bar{\alpha}_kD_k + \bar{\alpha}_kM_2F_{2,k}N_2)P_{k|k}(A_k + \bar{\alpha}_kD_k + \bar{\alpha}_kM_2F_{2,k}N_2)^T \\
&\quad + \tau_2\mathbb{E}\{f(x_k)f^T(x_k)\} + \tau_3\Delta A_k\mathbb{E}\{x_kx_k^T\}\Delta A_k^T + B_kQ_kB_k^T,
\end{aligned} \tag{5.24}
$$

where τ_i $(i = 1, 2, 3)$ are given in (5.19). Based on Lemma 2.1, the first term of the RHS of (5.24) is shown as

$$
\begin{aligned}
&(A_k + \bar{\alpha}_kD_k + \bar{\alpha}_kM_2F_{2,k}N_2)P_{k|k}(A_k + \bar{\alpha}_kD_k + \bar{\alpha}_kM_2F_{2,k}N_2)^T \\
&\leq (A_k + \bar{\alpha}_kD_k)\Psi_{1,k}^{-1}(A_k + \bar{\alpha}_kD_k)^T + \gamma_{1,k}^{-1}\bar{\alpha}_k^2M_2M_2^T,
\end{aligned} \tag{5.25}
$$

where $\Psi_{1,k} = P_{k|k}^{-1} - \gamma_{1,k} N_2^T N_2$ and $\gamma_{1,k}$ is a positive scalar. Moreover, the second term on the RHS of (5.24) is calculated as

$$\begin{aligned}
&\mathbb{E}\{f(x_k) f^T(x_k)\} \\
&\leq (1+\varepsilon_4+\varepsilon_5) f(\hat{x}_{k|k}) f^T(\hat{x}_{k|k}) + (1+\varepsilon_4^{-1}+\varepsilon_6) D_k P_{k|k} D_k^T \\
&+ (1+\varepsilon_5^{-1}+\varepsilon_6^{-1}) \mathrm{tr}(N_2 P_{k|k} N_2^T) M_2 M_2^T \\
&:= \Upsilon_{1,k},
\end{aligned} \tag{5.26}$$

where $\varepsilon_i > 0$ $(i = 4, 5, 6)$ are scalars. Next, it is easy to see that

$$\begin{aligned}
\mathbb{E}\{x_k x_k^T\} &= \mathbb{E}\{(\tilde{x}_{k|k} + \hat{x}_{k|k})(\tilde{x}_{k|k} + \hat{x}_{k|k})^T\} \\
&\leq \mathbb{E}\{(1+\varepsilon_7)\tilde{x}_{k|k}\tilde{x}_{k|k}^T + (1+\varepsilon_7^{-1})\hat{x}_{k|k}\hat{x}_{k|k}^T\} \\
&= (1+\varepsilon_7) P_{k|k} + (1+\varepsilon_7^{-1})\hat{x}_{k|k}\hat{x}_{k|k}^T \\
&:= \Psi_{4,k}
\end{aligned} \tag{5.27}$$

with ε_7 being a positive scalar. Thus, we obtain

$$\begin{aligned}
\Delta A_k \mathbb{E}\{x_k x_k^T\} \Delta A_k^T &\leq \mathrm{tr}(N_1 \Psi_{4,k} N_1^T) M_1 M_1^T \\
&:= \Psi_{5,k}.
\end{aligned} \tag{5.28}$$

Together with (5.25)–(5.28), one has

$$\begin{aligned}
P_{k+1|k} \leq\ & \tau_1 (A_k + \bar{\alpha}_k D_k) \Psi_{1,k}^{-1} (A_k + \bar{\alpha}_k D_k)^T + \tau_1 \gamma_{1,k}^{-1} \bar{\alpha}_k^2 M_2 M_2^T \\
&+ \tau_2 \Upsilon_{1,k} + \tau_3 \Psi_{5,k} + B_k Q_k B_k^T.
\end{aligned} \tag{5.29}$$

Subsequently, we aim to reveal that $P_{k+1|k+1} \leq \Sigma_{k+1|k+1}$. Considering the terms in (5.15) and similar in (5.21), we have

$$\begin{aligned}
&- \mathscr{B}_1 - \mathscr{B}_1^T \\
&\leq \eta_1 [I - \bar{\lambda}_{k+1}(K_{k+1} + \Delta K_{k+1}) C_{k+1}] P_{k+1|k} [I - \bar{\lambda}_{k+1}(K_{k+1} + \Delta K_{k+1}) C_{k+1}]^T \\
&+ \eta_1^{-1} \Delta\lambda_{k+1}^2 (K_{k+1} + \Delta K_{k+1}) C_{k+1} \mathbb{E}\{x_{k+1} x_{k+1}^T\} C_{k+1}^T (K_{k+1} + \Delta K_{k+1})^T, \quad (5.30) \\
&- \mathscr{B}_2 - \mathscr{B}_2^T \\
&\leq \eta_2 [I - \bar{\lambda}_{k+1}(K_{k+1} + \Delta K_{k+1}) C_{k+1}] P_{k+1|k} [I - \bar{\lambda}_{k+1}(K_{k+1} + \Delta K_{k+1}) C_{k+1}]^T \\
&+ \eta_2^{-1} (K_{k+1} + \Delta K_{k+1}) \mathbb{E}\{(\tilde{y}_{k+1} - y_{k+1})(\tilde{y}_{k+1} - y_{k+1})^T\} \\
&\times (K_{k+1} + \Delta K_{k+1})^T, \quad (5.31) \\
&\mathscr{B}_3 + \mathscr{B}_3^T
\end{aligned}$$

$$
\begin{aligned}
&\leq \eta_3(\sigma_{k+1} + \Delta\lambda_{k+1}^2)(K_{k+1} + \Delta K_{k+1})C_{k+1}\mathbb{E}\{x_{k+1}x_{k+1}^T\}C_{k+1}^T(K_{k+1} + \Delta K_{k+1})^T \\
&+ \eta_3^{-1}(K_{k+1} + \Delta K_{k+1})\mathbb{E}\{(\tilde{y}_{k+1} - y_{k+1})(\tilde{y}_{k+1} - y_{k+1})^T\} \\
&\times (K_{k+1} + \Delta K_{k+1})^T, \qquad (5.32) \\
&\mathscr{B}_4 + \mathscr{B}_4^T \\
&\leq \eta_4(K_{k+1} + \Delta K_{k+1})\mathbb{E}\{(\tilde{y}_{k+1} - y_{k+1})(\tilde{y}_{k+1} - y_{k+1})^T\}(K_{k+1} + \Delta K_{k+1})^T \\
&+ \eta_4^{-1}(K_{k+1} + \Delta K_{k+1})R_{k+1}(K_{k+1} + \Delta K_{k+1})^T, \qquad (5.33)
\end{aligned}
$$

where $\eta_i > 0$ $(i = 1, 2, 3, 4)$ are scalars. Noting $\sigma_{k+1} \leq \bar{\sigma}_{k+1}$, $|\Delta\lambda_{k+1}| < \kappa_2$ and substituting (5.30)–(5.33) into (5.15) yield

$$
\begin{aligned}
P_{k+1|k+1} \leq\ & \tau_4[I - \bar{\lambda}_{k+1}(K_{k+1} + \Delta K_{k+1})C_{k+1}]P_{k+1|k}[I - \bar{\lambda}_{k+1}(K_{k+1} + \Delta K_{k+1})C_{k+1}]^T \\
&+ (\tau_5\bar{\sigma}_{k+1} + \tau_5\kappa_2^2 + \eta_1^{-1}\kappa_2^2)(K_{k+1} + \Delta K_{k+1})C_{k+1}\mathbb{E}\{x_{k+1}x_{k+1}^T\}C_{k+1}^T \\
&\times (K_{k+1} + \Delta K_{k+1})^T + \tau_6(K_{k+1} + \Delta K_{k+1})\mathbb{E}\{(\tilde{y}_{k+1} - y_{k+1}) \\
&\times (\tilde{y}_{k+1} - y_{k+1})^T\}(K_{k+1} + \Delta K_{k+1})^T + \tau_7(K_{k+1} + \Delta K_{k+1}) \\
&\times R_{k+1}(K_{k+1} + \Delta K_{k+1})^T, \qquad (5.34)
\end{aligned}
$$

where $\tau_i > 0$ $(i = 4, 5, 6, 7)$ are constants.

Next, the first term of (5.34) is tackled as

$$
\begin{aligned}
&[I - \bar{\lambda}_{k+1}(K_{k+1} + \Delta K_{k+1})C_{k+1}]P_{k+1|k}[I - \bar{\lambda}_{k+1}(K_{k+1} + \Delta K_{k+1})C_{k+1}]^T \\
&\leq (I - \bar{\lambda}_{k+1}K_{k+1}C_{k+1})\Psi_{2,k+1}^{-1}(I - \bar{\lambda}_{k+1}K_{k+1}C_{k+1})^T \\
&+ \gamma_{2,k+1}^{-1}\bar{\lambda}_{k+1}^2 M_3 M_3^T, \qquad (5.35)
\end{aligned}
$$

where $\Psi_{2,k+1} = P_{k+1|k}^{-1} - \gamma_{2,k+1}C_{k+1}^T N_3^T N_3 C_{k+1}$. Along the same line in (5.27), we have

$$
\begin{aligned}
&\mathbb{E}\{x_{k+1}x_{k+1}^T\} \\
&\leq \mathbb{E}\{(1 + \eta_5)\tilde{x}_{k+1|k}\tilde{x}_{k+1|k}^T + (1 + \eta_5^{-1})\hat{x}_{k+1|k}\hat{x}_{k+1|k}^T\} \\
&= (1 + \eta_5)P_{k+1|k} + (1 + \eta_5^{-1})\hat{x}_{k+1|k}\hat{x}_{k+1|k}^T \\
&:= \Psi_{3,k+1}, \qquad (5.36)
\end{aligned}
$$

where η_5 is a positive scalar.

The others terms of (5.34) can be tackled as

$$
\begin{aligned}
&(K_{k+1} + \Delta K_{k+1})C_{k+1}\mathbb{E}\{x_{k+1}x_{k+1}^T\}C_{k+1}^T(K_{k+1} + \Delta K_{k+1})^T \\
&\leq K_{k+1}C_{k+1}(\Psi_{3,k+1}^{-1} - \gamma_{3,k+1}C_{k+1}^T N_3^T N_3 C_{k+1})^{-1}C_{k+1}^T K_{k+1}^T \\
&+ \gamma_{3,k+1}^{-1}M_3 M_3^T, \qquad (5.37) \\
&(K_{k+1} + \Delta K_{k+1})\mathbb{E}\{(\tilde{y}_{k+1} - y_{k+1})(\tilde{y}_{k+1} - y_{k+1})^T\}(K_{k+1} + \Delta K_{k+1})^T \\
&\leq \delta K_{k+1}(I - \gamma_{4,k+1}N_3^T N_3)^{-1}K_{k+1}^T + \delta\gamma_{4,k+1}^{-1}M_3 M_3^T, \qquad (5.38)
\end{aligned}
$$

$$
\begin{aligned}
&(K_{k+1} + \Delta K_{k+1})R_{k+1}(K_{k+1} + \Delta K_{k+1})^T \\
&\le K_{k+1}(R_{k+1}^{-1} - \gamma_{5,k+1}N_3^T N_3)^{-1}K_{k+1}^T + \gamma_{5,k+1}^{-1}M_3M_3^T.
\end{aligned} \tag{5.39}
$$

Substituting (5.35)–(5.39) into (5.34) leads to

$$
\begin{aligned}
P_{k+1|k+1} \le\ & \tau_4(I - \bar{\lambda}_{k+1}K_{k+1}C_{k+1})\Psi_{2,k+1}^{-1}(I - \bar{\lambda}_{k+1}K_{k+1}C_{k+1})^T \\
& + \tau_4\gamma_{2,k+1}^{-1}\bar{\lambda}_{k+1}^2 M_3M_3^T + (\tau_5\bar{\sigma}_{k+1} + \tau_5\kappa_2^2 + \eta_1^{-1}\kappa_2^2) \\
& \times [K_{k+1}C_{k+1}(\Psi_{3,k+1}^{-1} - \gamma_{3,k+1}C_{k+1}^T N_3^T N_3 C_{k+1})^{-1}C_{k+1}^T K_{k+1}^T \\
& + \gamma_{3,k+1}^{-1}M_3M_3^T] + \tau_6\delta K_{k+1}(I - \gamma_{4,k+1}N_3^T N_3)^{-1}K_{k+1}^T \\
& + \tau_6\delta\gamma_{4,k+1}^{-1}M_3M_3^T + \tau_7 K_{k+1}(R_{k+1}^{-1} - \gamma_{5,k+1}N_3^T N_3)^{-1}K_{k+1}^T \\
& + \tau_7\gamma_{5,k+1}^{-1}M_3M_3^T.
\end{aligned} \tag{5.40}
$$

Then, it is clear to obtain $P_{k+1|k+1} \le \Sigma_{k+1|k+1}$ by applying the mathematical induction method.

At last, in light of Lemma 2.3, taking the partial derivative of $\mathrm{tr}(\Sigma_{k+1|k+1})$ concerning on K_{k+1}, one gets

$$
\begin{aligned}
\frac{\partial \mathrm{tr}(\Sigma_{k+1|k+1})}{\partial K_{k+1}} =\ & -2\tau_4\bar{\lambda}_{k+1}(I - \bar{\lambda}_{k+1}K_{k+1}C_{k+1})\bar{\Psi}_{2,k+1}^{-1}C_{k+1}^T + 2K_{k+1} \\
& \times [(\tau_5\bar{\sigma}_{k+1} + \tau_5\kappa_2^2 + \eta_1^{-1}\kappa_2^2)C_{k+1}(\bar{\Psi}_{3,k+1}^{-1} - \gamma_{3,k+1}C_{k+1}^T \\
& \times N_3^T N_3 C_{k+1})^{-1}C_{k+1}^T + \tau_6\delta(I - \gamma_{4,k+1}N_3^T N_3)^{-1} \\
& + \tau_7(R_{k+1}^{-1} - \gamma_{5,k+1}N_3^T N_3)^{-1}].
\end{aligned}
$$

Letting the partial derivative be zero generates the result presented in (5.20), and the proof of this theorem is complete.

Remark 5.2 It is crucial to emphasize that the impacts from the RONs and missing measurements with UOPs on the filtering performance have been examined. Meanwhile, the related information has been reflected in the Theorem 5.3 and the concrete form of the optimized filter gain has also been exhibited.

5.2 Event-Triggered Filtering with Multiple Missing Measurements Under Uncertain Missing Probabilities

5.2.1 Problem Formulation

In this subsection, we consider the following class of time-varying nonlinear systems in the presence of multiple missing measurements:

$$
x_{k+1} = f(x_k) + g(x_k, \eta_k) + D_k\omega_k, \tag{5.41}
$$

$$y_k = \Xi_k C_k x_k + h(x_k, \zeta_k) + \nu_k, \tag{5.42}$$

where $x_k \in \mathbb{R}^n$ is the state, x_0 is the initial value with mean $\bar{x}_0$ and covariance P_0, $y_k \in \mathbb{R}^m$ is the measurement, η_k and ζ_k are zero-mean Gaussian noises, $\omega_k \in \mathbb{R}^p$ denotes the process noise with zero-mean and covariance $Q_k > 0$, $\nu_k \in \mathbb{R}^m$ is the measurement noise with zero-mean and covariance $R_k > 0$, C_k and D_k are given matrices of proper dimensions. $\Xi_k = \text{diag}\{\xi_{1,k}, \xi_{2,k}, \ldots, \xi_{m,k}\}$ with $\xi_{i,k}$ $(i = 1, 2, \ldots, m)$ denoting m independent random variables in k and i.

The random variables $\xi_{i,k} \in \mathbb{R}$ are employed to model the phenomena of the multiple missing measurements satisfying

$$\begin{aligned} &\text{Prob}\left\{\xi_{i,k} = 1\right\} = \bar{\xi}_i + \Delta\xi_i, \\ &\text{Prob}\left\{\xi_{i,k} = 0\right\} = 1 - (\bar{\xi}_i + \Delta\xi_i), \end{aligned} \tag{5.43}$$

where $\bar{\xi}_i + \Delta\xi_i \in [0, 1]$, $\bar{\xi}_i$ are pre-determined scalars, and $\Delta\xi_i$ $(i = 1, 2, \ldots, m)$ are used to describe the uncertain missing probabilities.

The functions $g(x_k, \eta_k)$ and $h(x_k, \zeta_k)$ stand for the stochastic nonlinearities with $g(0, \eta_k) = 0$, $h(0, \zeta_k) = 0$ and satisfy

$$\mathbb{E}\left\{\begin{bmatrix} g(x_k, \eta_k) \\ h(x_k, \zeta_k) \end{bmatrix} \middle| x_k \right\} = 0 \tag{5.44}$$

and the covariance presented by

$$\mathbb{E}\left\{\begin{bmatrix} g(x_k, \eta_k) \\ h(x_k, \zeta_k) \end{bmatrix}\begin{bmatrix} g(x_j, \eta_j) \\ h(x_j, \zeta_j) \end{bmatrix}^T \middle| x_k \right\} = 0, \quad k \neq j, \tag{5.45}$$

$$\mathbb{E}\left\{\begin{bmatrix} g(x_k, \eta_k) \\ h(x_k, \zeta_k) \end{bmatrix}\begin{bmatrix} g(x_k, \eta_k) \\ h(x_k, \zeta_k) \end{bmatrix}^T \middle| x_k \right\} = \sum_{i=1}^{s} \text{diag}\left\{\Pi_k^{1i} x_k^T \Gamma_k^i x_k, \Pi_k^{2i} x_k^T \Gamma_k^i x_k\right\} \tag{5.46}$$

with $s > 0$ denoting a given integer, Π_k^{1i}, Π_k^{2i} and Γ_k^i $(i = 1, 2, \ldots, s)$ are given matrices of suitable dimensions. Moreover, it is assumed that $\xi_{i,k}$, η_k, ζ_k, ω_k, ν_k and x_0 are mutually independent.

For the sake of reducing the network communication burden, the following event-based transmission mechanism is adopted:

$$(y_{k+j} - y_k)^T (y_{k+j} - y_k) > \sigma, \tag{5.47}$$

where $\sigma > 0$ is a given scalar. In other words, the current measurement y_{k+j} can be transmitted only when the inequality (5.47) holds. It can be obtained from (5.47) that the released measurement signal $\bar{y}_k$ is described as follows:

$$\bar{y}_k = y_{k_i}, \quad k \in \{k_i, k_i + 1, \ldots, k_{i+1} - 1\}, \tag{5.48}$$

where k_i is the triggering time with $k_0 = 0$.

Next, we construct the following filter:

$$\hat{x}_{k+1|k} = f(\hat{x}_{k|k}), \tag{5.49}$$
$$\hat{x}_{k+1|k+1} = \hat{x}_{k+1|k} + K_{k+1}(\bar{y}_{k+1} - \bar{\Xi}_{k+1}C_{k+1}\hat{x}_{k+1|k}), \tag{5.50}$$

where $\hat{x}_{k+1|k}$ is the one-step prediction, $\hat{x}_{k|k}$ is the estimation with $\hat{x}_{0|0} = \bar{x}_0$, $\bar{y}_{k+1}$ is the *actual* measurement, $\bar{\Xi}_{k+1} = \text{diag}\{\bar{\xi}_1, \bar{\xi}_2, \ldots, \bar{\xi}_m\}$, and K_{k+1} is the filter gain.

Denote the one-step prediction error as $\tilde{x}_{k+1|k} = x_{k+1} - \hat{x}_{k+1|k}$ and the filtering error as $\tilde{x}_{k+1|k+1} = x_{k+1} - \hat{x}_{k+1|k+1}$. As such, the one-step prediction error covariance is $P_{k+1|k} = \mathbb{E}\{\tilde{x}_{k+1|k}\tilde{x}^T_{k+1|k}\}$ and the FEC is $P_{k+1|k+1} = \mathbb{E}\{\tilde{x}_{k+1|k+1}\tilde{x}^T_{k+1|k+1}\}$. Subsequently, we aim to construct a time-varying filter of the form (5.49) and (5.50) such that, for multiple missing measurements, stochastic nonlinearities and ETM, an optimized upper bound for the FEC is derived, namely, there is a set of matrices $\Sigma_{k+1|k+1} > 0$ satisfying

$$P_{k+1|k+1} \leq \Sigma_{k+1|k+1}. \tag{5.51}$$

Furthermore, the trace of such an upper bound is optimized via appropriately constructing the filter gain K_{k+1}.

Remark 5.3 In (5.47), the ETM is adopted to schedule the data transmission. It is not difficult to observe that the scalar σ plays an important role on the triggering frequency and different σ can cause different data packet transmission rates. Furthermore, more events would be triggered and more updated measurement information could be delivered to the remote filter side when σ decreases. To be more specific, the event-based filtering method becomes the conventional clock-driven one when $\sigma = 0$. It is worthwhile to note that the triggered threshold σ strongly relies on the interaction effect between the limited network resource and whole performance requirements in practical engineering. For the purpose of obtaining a satisfactory trade-off scheme between the energy-saving and the filtering algorithm precision, the triggered threshold σ should be properly selected in the calculation process. Besides, the constraint condition (5.51) is introduced and the proposed upper bound is minimized through designing an appropriate filter gain. It is worth pointing out that, owing to the existences of multiple missing measurements and ETM, it is impossible to operate the real value of the $P_{k+1|k+1}$ and thus the optimized criterion (5.51) of $\Sigma_{k+1|k+1}$ is put forward. It will be indicated that the event-triggered threshold σ and the optimization of (5.51) would be explicitly analyzed.

5.2.2 Main Results

In this section, we are ready to establish a novel filtering algorithm which can handle the conducted filtering problem influenced by the multiple missing measurements with uncertain missing probabilities, stochastic nonlinearities and ETM. Firstly, the Taylor series expansion method is utilized to cope with the nonlinearity $f(x_k)$ and the linearization errors are estimated. Secondly, the concrete forms of the one-step prediction error covariance and FEC can be computed. Thirdly, the upper bound concerning on the FEC is solved and then optimized by constructing the filter gain.

Based on (5.41) and (5.49), the one-step prediction error is computed as

$$\tilde{x}_{k+1|k} = f(x_k) - f(\hat{x}_{k|k}) + g(x_k, \eta_k) + D_k \omega_k. \tag{5.52}$$

Adopting the Taylor series expansion method for $f(x_k)$ around $\hat{x}_{k|k}$, one has

$$f(x_k) = f(\hat{x}_{k|k}) + A_k \tilde{x}_{k|k} + o(|\tilde{x}_{k|k}^2|), \tag{5.53}$$

where $A_k = \frac{\partial f(x_k)}{\partial x_k}|_{x_k = \hat{x}_{k|k}}$ and $o(|\tilde{x}_{k|k}^2|)$ is the high-order term. It can be derived from Calafiore (2005), Hu et al. (2013), Xiong et al. (2010) via the approximated estimation that

$$o(|\tilde{x}_{k|k}^2|) = M_k \Delta_k L_k \tilde{x}_{k|k}, \tag{5.54}$$

where M_k is a scaling matrix, L_k is a known matrix that provides an extra freedom degree to adjust the filter, and Δ_k is an unknown matrix representing the linearization errors and satisfies

$$\Delta_k \Delta_k^T \leq I. \tag{5.55}$$

Then, according to (5.52)–(5.54), the one-step prediction error is rewritten by

$$\tilde{x}_{k+1|k} = (A_k + M_k \Delta_k L_k) \tilde{x}_{k|k} + g(x_k, \eta_k) + D_k \omega_k. \tag{5.56}$$

On the other hand, on the basis of (5.50), the filtering error is calculated as

$$\tilde{x}_{k+1|k+1} = \tilde{x}_{k+1|k} - K_{k+1}(\bar{y}_{k+1} - \bar{\Xi}_{k+1} C_{k+1} \hat{x}_{k+1|k}). \tag{5.57}$$

Adding the zero term $K_{k+1} y_{k+1} - K_{k+1} y_{k+1}$ to the right side of (5.57) yields

$$\begin{aligned} \tilde{x}_{k+1|k+1} = & (I - K_{k+1} \bar{\Xi}_{k+1} C_{k+1}) \tilde{x}_{k+1|k} - K_{k+1}(\bar{y}_{k+1} - y_{k+1}) \\ & - K_{k+1} \tilde{\Xi}_{k+1} C_{k+1} x_{k+1} - K_{k+1} \check{\Xi}_{k+1} C_{k+1} x_{k+1} \\ & - K_{k+1} h(x_{k+1}, \zeta_{k+1}) - K_{k+1} \nu_{k+1}, \end{aligned} \tag{5.58}$$

where

$$\tilde{\Xi}_{k+1} = \Xi_{k+1} - \hat{\Xi}_{k+1},$$
$$\hat{\Xi}_{k+1} = \text{diag}\{\bar{\xi}_1 + \Delta\xi_1, \bar{\xi}_2 + \Delta\xi_2, \ldots, \bar{\xi}_m + \Delta\xi_m\},$$
$$\check{\Xi}_{k+1} = \text{diag}\{\Delta\xi_1, \Delta\xi_2, \ldots, \Delta\xi_m\}.$$

Then, it is clear to see that $\mathbb{E}\{\tilde{\Xi}_{k+1}\} = 0$.

In the next step, we aim to derive the specific forms of the one-step prediction error covariance and FEC in light of (5.56) and (5.58), respectively. Firstly, the following theorem can be immediately received.

Theorem 5.4 *$P_{k+1|k}$ meets the following recursion:*

$$P_{k+1|k} = (A_k + M_k \Delta_k L_k) P_{k|k} (A_k + M_k \Delta_k L_k)^T + \sum_{i=1}^{s} \Pi_k^{1i} \text{tr}(\mathbb{E}\left\{x_k x_k^T\right\} \Gamma_k^i) + D_k Q_k D_k^T. \tag{5.59}$$

Proof By resorting to the definition of $P_{k+1|k}$ and utilizing (5.56), one obtains

$$P_{k+1|k} = (A_k + M_k \Delta_k L_k) P_{k|k} (A_k + M_k \Delta_k L_k)^T + D_k Q_k D_k^T + \sum_{i=1}^{s} \Pi_k^{1i} \text{tr}(\mathbb{E}\left\{x_k x_k^T\right\} \Gamma_k^i). \tag{5.60}$$

It is noteworthy that we have utilized the assumption that η_k and ω_k are mutually independent when deriving (5.60). Subsequently, the proof of this theorem is complete.

Next, according to (5.58) and the definition of the FEC, the recursion of the FEC is given as follows.

Theorem 5.5 *$P_{k+1|k+1}$ can be recursively computed*

$$\begin{aligned} P_{k+1|k+1} = {} & (I - K_{k+1}\bar{\Xi}_{k+1}C_{k+1}) P_{k+1|k} (I - K_{k+1}\bar{\Xi}_{k+1}C_{k+1})^T - \mathscr{A}_{k+1} - \mathscr{A}_{k+1}^T \\ & - \mathscr{B}_{k+1} - \mathscr{B}_{k+1}^T + K_{k+1}\mathbb{E}\{(\bar{y}_{k+1} - y_{k+1})(\bar{y}_{k+1} - y_{k+1})^T\} K_{k+1}^T \\ & + \mathscr{C}_{k+1} + \mathscr{C}_{k+1}^T + K_{k+1}\mathbb{E}\{\tilde{\Xi}_{k+1} C_{k+1} x_{k+1} x_{k+1}^T C_{k+1}^T \tilde{\Xi}_{k+1}^T\} K_{k+1}^T \\ & + K_{k+1}\check{\Xi}_{k+1} C_{k+1} \mathbb{E}\{x_{k+1} x_{k+1}^T\} C_{k+1}^T \check{\Xi}_{k+1}^T K_{k+1}^T \\ & + K_{k+1}\mathbb{E}\{h(x_{k+1}, \zeta_{k+1}) h^T(x_{k+1}, \zeta_{k+1})\} K_{k+1}^T \\ & + K_{k+1} R_{k+1} K_{k+1}^T, \end{aligned} \tag{5.61}$$

where

$$\begin{aligned}
\mathscr{A}_{k+1} &= (I - K_{k+1}\bar{\Xi}_{k+1}C_{k+1})\mathbb{E}\{\tilde{x}_{k+1|k}(\bar{y}_{k+1} - y_{k+1})^T\}K_{k+1}^T,\\
\mathscr{B}_{k+1} &= (I - K_{k+1}\bar{\Xi}_{k+1}C_{k+1})\mathbb{E}\{\tilde{x}_{k+1|k}x_{k+1}^T\}C_{k+1}^T\check{\Xi}_{k+1}^T K_{k+1}^T,\\
\mathscr{C}_{k+1} &= K_{k+1}\mathbb{E}\{(\bar{y}_{k+1} - y_{k+1})x_{k+1}^T\}C_{k+1}^T\check{\Xi}_{k+1}^T K_{k+1}^T.
\end{aligned} \tag{5.62}$$

Proof In light of the definition of $P_{k+1|k+1}$, we have

$$\begin{aligned}
&P_{k+1|k+1}\\
&= (I - K_{k+1}\bar{\Xi}_{k+1}C_{k+1})P_{k+1|k}(I - K_{k+1}\bar{\Xi}_{k+1}C_{k+1})^T\\
&- \mathscr{A}_{k+1} - \mathscr{A}_{k+1}^T - \mathcal{A}_{1,k+1} - \mathcal{A}_{1,k+1}^T - \mathscr{B}_{k+1} - \mathscr{B}_{k+1}^T\\
&- \mathcal{A}_{2,k+1} - \mathcal{A}_{2,k+1}^T - \mathcal{A}_{3,k+1} - \mathcal{A}_{3,k+1}^T\\
&+ K_{k+1}\mathbb{E}\{(\bar{y}_{k+1} - y_{k+1})(\bar{y}_{k+1} - y_{k+1})^T\}K_{k+1}^T + \mathcal{A}_{4,k+1}\\
&+ \mathcal{A}_{4,k+1}^T + \mathscr{C}_{k+1} + \mathscr{C}_{k+1}^T + \mathcal{A}_{5,k+1} + \mathcal{A}_{5,k+1}^T + \mathcal{A}_{6,k+1} + \mathcal{A}_{6,k+1}^T\\
&+ K_{k+1}\mathbb{E}\{\tilde{\Xi}_{k+1}C_{k+1}x_{k+1}x_{k+1}^T C_{k+1}^T\tilde{\Xi}_{k+1}^T\}K_{k+1}^T\\
&+ \mathcal{A}_{7,k+1} + \mathcal{A}_{7,k+1}^T + \mathcal{A}_{8,k+1} + \mathcal{A}_{8,k+1}^T + \mathcal{A}_{9,k+1} + \mathcal{A}_{9,k+1}^T\\
&+ K_{k+1}\check{\Xi}_{k+1}C_{k+1}\mathbb{E}\{x_{k+1}x_{k+1}^T\}C_{k+1}^T\check{\Xi}_{k+1}^T K_{k+1}^T\\
&+ \mathcal{A}_{10,k+1} + \mathcal{A}_{10,k+1}^T + \mathcal{A}_{11,k+1} + \mathcal{A}_{11,k+1}^T\\
&+ K_{k+1}\mathbb{E}\{h(x_{k+1},\zeta_{k+1})h^T(x_{k+1},\zeta_{k+1})\}K_{k+1}^T\\
&+ \mathcal{A}_{12,k+1} + \mathcal{A}_{12,k+1}^T + K_{k+1}\mathbb{E}\{v_{k+1}v_{k+1}^T\}K_{k+1}^T,
\end{aligned} \tag{5.63}$$

where

$$\begin{aligned}
\mathcal{A}_{1,k+1} &= (I - K_{k+1}\bar{\Xi}_{k+1}C_{k+1})\mathbb{E}\left\{\tilde{x}_{k+1|k}x_{k+1}^T C_{k+1}^T\tilde{\Xi}_{k+1}^T\right\}K_{k+1}^T,\\
\mathcal{A}_{2,k+1} &= (I - K_{k+1}\bar{\Xi}_{k+1}C_{k+1})\mathbb{E}\left\{\tilde{x}_{k+1|k}h^T(x_{k+1},\zeta_{k+1})\right\}K_{k+1}^T,\\
\mathcal{A}_{3,k+1} &= (I - K_{k+1}\bar{\Xi}_{k+1}C_{k+1})\mathbb{E}\left\{\tilde{x}_{k+1|k}v_{k+1}^T\right\}K_{k+1}^T,\\
\mathcal{A}_{4,k+1} &= K_{k+1}\mathbb{E}\{(\bar{y}_{k+1} - y_{k+1})x_{k+1}^T C_{k+1}^T\tilde{\Xi}_{k+1}^T\}K_{k+1}^T,\\
\mathcal{A}_{5,k+1} &= K_{k+1}\mathbb{E}\{(\bar{y}_{k+1} - y_{k+1})h^T(x_{k+1},\zeta_{k+1})\}K_{k+1}^T,\\
\mathcal{A}_{6,k+1} &= K_{k+1}\mathbb{E}\{(\bar{y}_{k+1} - y_{k+1})v_{k+1}^T\}K_{k+1}^T,\\
\mathcal{A}_{7,k+1} &= K_{k+1}\mathbb{E}\{\tilde{\Xi}_{k+1}C_{k+1}x_{k+1}x_{k+1}^T\}C_{k+1}^T\check{\Xi}_{k+1}^T K_{k+1}^T,\\
\mathcal{A}_{8,k+1} &= K_{k+1}\mathbb{E}\{\tilde{\Xi}_{k+1}C_{k+1}x_{k+1}h^T(x_{k+1},\zeta_{k+1})\}K_{k+1}^T,\\
\mathcal{A}_{9,k+1} &= K_{k+1}\mathbb{E}\{\tilde{\Xi}_{k+1}C_{k+1}x_{k+1}v_{k+1}^T\}K_{k+1}^T,\\
\mathcal{A}_{10,k+1} &= K_{k+1}\check{\Xi}_{k+1}C_{k+1}\mathbb{E}\{x_{k+1}h^T(x_{k+1},\zeta_{k+1})\}K_{k+1}^T,\\
\mathcal{A}_{11,k+1} &= K_{k+1}\check{\Xi}_{k+1}C_{k+1}\mathbb{E}\{x_{k+1}v_{k+1}^T\}K_{k+1}^T,\\
\mathcal{A}_{12,k+1} &= K_{k+1}\mathbb{E}\{h(x_{k+1},\zeta_{k+1})v_{k+1}^T\}K_{k+1}^T.
\end{aligned}$$

It is obvious to clarify that the terms $\mathcal{A}_{i,k+1}$ $(i = 1, 2, \ldots, 12)$ are all equal to zero matrices. After that, the recursion (5.61) can be derived directly, which finishes the proof of this theorem.

Remark 5.4 It is noticed that there exist some uncertain terms (e.g. Δ_k, $\mathscr{A}_{k+1}$, $\mathscr{B}_{k+1}$ and $\mathscr{C}_{k+1}$) in (5.61) owing to the existences of the linearization errors, ETM and uncertain missing probabilities. Consequently, the accurate FEC cannot be attained. For the purpose of designing a proper filter gain and achieving the satisfactory filtering performance, an alternative approach is adopted that an upper bound of the FEC is derived firstly and then optimized by suitably choosing the filter gain.

Next, we are in a position to search an upper bound of the FEC by properly dealing with the uncertain terms.

Theorem 5.6 *Let $\gamma_k > 0$ and $\varepsilon_i > 0$ $(i = 1, 2, \ldots, 5)$ represent scalars. If the following matrix difference equations:*

$$\begin{aligned}\Sigma_{k+1|k} &= A_k(\Sigma_{k|k}^{-1} - \gamma_k L_k^T L_k)^{-1} A_k^T + \gamma_k^{-1} M_k M_k^T + D_k Q_k D_k^T \\ &\quad + \sum_{i=1}^{s} \Pi_k^{1i} \mathrm{tr}\left(\left[(1+\varepsilon_1)\Sigma_{k|k} + (1+\varepsilon_1^{-1})\hat{x}_{k|k}\hat{x}_{k|k}^T\right]\Gamma_k^i\right),\end{aligned} \tag{5.64}$$

$$\begin{aligned}\Sigma_{k+1|k+1} &= (1+\varepsilon_2+\varepsilon_3)(I - K_{k+1}\bar{\Xi}_{k+1}C_{k+1})\Sigma_{k+1|k}(I - K_{k+1}\bar{\Xi}_{k+1}C_{k+1})^T \\ &\quad + K_{k+1}\Bigg[(1+\varepsilon_2^{-1}+\varepsilon_4)\sigma I + \frac{1}{4} I \circ (C_{k+1}\Psi_{k+1|k}C_{k+1}^T) \\ &\quad + (1+\varepsilon_3^{-1}+\varepsilon_4^{-1})\mathrm{tr}(C_{k+1}\Psi_{k+1|k}C_{k+1}^T)\rho^2 I \\ &\quad + \sum_{i=1}^{s} \Pi_{k+1}^{2i}\mathrm{tr}(\Psi_{k+1|k}\Gamma_{k+1}^i) + R_{k+1}\Bigg]K_{k+1}^T\end{aligned} \tag{5.65}$$

under the initial condition $\Sigma_{0|0} = P_{0|0} > 0$ have solutions $\Sigma_{k+1|k} > 0$ and $\Sigma_{k+1|k+1} > 0$ such that the following constrained condition:

$$\gamma_k^{-1} I - L_k \Sigma_{k|k} L_k^T > 0 \tag{5.66}$$

is satisfied where

$$\begin{aligned}\rho &= \max_i\{1 - \bar{\xi}_i, \bar{\xi}_i\}, \\ \Psi_{k+1|k} &= (1+\varepsilon_5)\Sigma_{k+1|k} + (1+\varepsilon_5^{-1})\hat{x}_{k+1|k}\hat{x}_{k+1|k}^T,\end{aligned} \tag{5.67}$$

then one obtains

$$P_{k+1|k+1} \leq \Sigma_{k+1|k+1}. \tag{5.68}$$

Proof The mathematical induction approach is adopted to testify this theorem.

Step 1: Let us prove $P_{k+1|k} \leq \Sigma_{k+1|k}$ under the assumption $P_{k|k} \leq \Sigma_{k|k}$. By resorting to Lemma 2.1, one has

$$\begin{aligned} P_{k+1|k} \leq & A_k(P_{k|k}^{-1} - \gamma_k L_k^T L_k)^{-1} A_k^T + \gamma_k^{-1} M_k M_k^T + D_k Q_k D_k^T \\ & + \sum_{i=1}^{s} \Pi_k^{1i} \mathrm{tr}(\mathbb{E}\left\{x_k x_k^T\right\} \Gamma_k^i). \end{aligned} \tag{5.69}$$

Next, based on Lemma 2.2, one gets

$$\mathbb{E}\left\{x_k x_k^T\right\} \leq (1+\varepsilon_1) P_{k|k} + (1+\varepsilon_1^{-1}) \hat{x}_{k|k} \hat{x}_{k|k}^T \tag{5.70}$$

with $\varepsilon_1 > 0$ denoting a scalar. Combining (5.69) and (5.70), we have

$$\begin{aligned} P_{k+1|k} \leq & A_k(\Sigma_{k|k}^{-1} - \gamma_k L_k^T L_k)^{-1} A_k^T + \gamma_k^{-1} M_k M_k^T + D_k Q_k D_k^T \\ & + \sum_{i=1}^{s} \Pi_k^{1i} \mathrm{tr}\left(\left[(1+\varepsilon_1)\Sigma_{k|k} + (1+\varepsilon_1^{-1}) \hat{x}_{k|k} \hat{x}_{k|k}^T\right] \Gamma_k^i\right) \\ := & \Sigma_{k+1|k}. \end{aligned} \tag{5.71}$$

Step 2: We are endeavored to verify that $P_{k+1|k+1} \leq \Sigma_{k+1|k+1}$. Using Lemma 2.2 again, we derive

$$\begin{aligned} -\mathscr{A}_{k+1} - \mathscr{A}_{k+1}^T \leq & \varepsilon_2 (I - K_{k+1}\bar{\Xi}_{k+1} C_{k+1}) P_{k+1|k} (I - K_{k+1}\bar{\Xi}_{k+1} C_{k+1})^T \\ & + \varepsilon_2^{-1} K_{k+1} \mathbb{E}\{(\bar{y}_{k+1} - y_{k+1})(\bar{y}_{k+1} - y_{k+1})^T\} K_{k+1}^T, \end{aligned} \tag{5.72}$$

$$\begin{aligned} -\mathscr{B}_{k+1} - \mathscr{B}_{k+1}^T \leq & \varepsilon_3 (I - K_{k+1}\bar{\Xi}_{k+1} C_{k+1}) P_{k+1|k} (I - K_{k+1}\bar{\Xi}_{k+1} C_{k+1})^T \\ & + \varepsilon_3^{-1} K_{k+1} \check{\Xi}_{k+1} C_{k+1} \mathbb{E}\{x_{k+1} x_{k+1}^T\} C_{k+1}^T \check{\Xi}_{k+1}^T K_{k+1}^T, \end{aligned} \tag{5.73}$$

$$\begin{aligned} \mathscr{C}_{k+1} + \mathscr{C}_{k+1}^T \leq & \varepsilon_4 K_{k+1} \mathbb{E}\{(\bar{y}_{k+1} - y_{k+1})(\bar{y}_{k+1} - y_{k+1})^T\} K_{k+1}^T \\ & + \varepsilon_4^{-1} K_{k+1} \check{\Xi}_{k+1} C_{k+1} \mathbb{E}\{x_{k+1} x_{k+1}^T\} C_{k+1}^T \check{\Xi}_{k+1}^T K_{k+1}^T, \end{aligned} \tag{5.74}$$

where $\varepsilon_i > 0$ $(i = 2, 3, 4)$ are scalars. Substituting (5.72)–(5.74) into (5.61) leads to

$$\begin{aligned} P_{k+1|k+1} \leq & (1+\varepsilon_2+\varepsilon_3)(I - K_{k+1}\bar{\Xi}_{k+1} C_{k+1}) P_{k+1|k} (I - K_{k+1}\bar{\Xi}_{k+1} C_{k+1})^T \\ & + (1+\varepsilon_2^{-1}+\varepsilon_4) K_{k+1} \mathbb{E}\{(\bar{y}_{k+1} - y_{k+1})(\bar{y}_{k+1} - y_{k+1})^T\} K_{k+1}^T \\ & + K_{k+1} \mathbb{E}\{\tilde{\Xi}_{k+1} C_{k+1} x_{k+1} x_{k+1}^T C_{k+1}^T \tilde{\Xi}_{k+1}^T\} K_{k+1}^T \\ & + (1+\varepsilon_3^{-1}+\varepsilon_4^{-1}) K_{k+1} \check{\Xi}_{k+1} C_{k+1} \mathbb{E}\{x_{k+1} x_{k+1}^T\} C_{k+1}^T \check{\Xi}_{k+1}^T K_{k+1}^T \\ & + K_{k+1} \mathbb{E}\{h(x_{k+1}, \zeta_{k+1}) h^T(x_{k+1}, \zeta_{k+1})\} K_{k+1}^T \\ & + K_{k+1} R_{k+1} K_{k+1}^T. \end{aligned} \tag{5.75}$$

Next, in terms of Lemma 2.4, the following term can be computed:

$$\begin{aligned}&\mathbb{E}\{\tilde{\Xi}_{k+1}C_{k+1}x_{k+1}x_{k+1}^{T}C_{k+1}^{T}\tilde{\Xi}_{k+1}^{T}\}\\&=\breve{\Xi}_{k+1}\circ\left(C_{k+1}\mathbb{E}\{x_{k+1}x_{k+1}^{T}\}C_{k+1}^{T}\right),\end{aligned}\tag{5.76}$$

where

$$\begin{aligned}\breve{\Xi}_{k+1}=\operatorname{diag}\big\{&(\bar{\xi}_1+\Delta\xi_1)\left[1-(\bar{\xi}_1+\Delta\xi_1)\right],(\bar{\xi}_2+\Delta\xi_2)\left[1-(\bar{\xi}_2+\Delta\xi_2)\right]\\&\ldots,(\bar{\xi}_m+\Delta\xi_m)\left[1-(\bar{\xi}_m+\Delta\xi_m)\right]\big\}.\end{aligned}\tag{5.77}$$

By means of the straightforward algebraic manipulations, it is easy to prove that $\breve{\Xi}_{k+1}\leq\frac{1}{4}I$. On the other hand, along the same line with (5.70), we get

$$\begin{aligned}&\mathbb{E}\left\{x_{k+1}x_{k+1}^{T}\right\}\\&\leq\mathbb{E}\left\{(1+\varepsilon_5)\tilde{x}_{k+1|k}\tilde{x}_{k+1|k}^{T}+(1+\varepsilon_5^{-1})\hat{x}_{k+1|k}\hat{x}_{k+1|k}^{T}\right\}\\&=(1+\varepsilon_5)P_{k+1|k}+(1+\varepsilon_5^{-1})\hat{x}_{k+1|k}\hat{x}_{k+1|k}^{T}\\&:=\bar{\Psi}_{k+1|k}.\end{aligned}\tag{5.78}$$

From (5.76) and (5.78), we have

$$\mathbb{E}\{\tilde{\Xi}_{k+1}C_{k+1}x_{k+1}x_{k+1}^{T}C_{k+1}^{T}\tilde{\Xi}_{k+1}^{T}\}\leq\frac{1}{4}I\circ\left(C_{k+1}\bar{\Psi}_{k+1|k}C_{k+1}^{T}\right).\tag{5.79}$$

Next, let us address the fourth term of the RHS of (5.75). We have

$$\begin{aligned}&(1+\varepsilon_3^{-1}+\varepsilon_4^{-1})K_{k+1}\breve{\Xi}_{k+1}C_{k+1}\mathbb{E}\{x_{k+1}x_{k+1}^{T}\}C_{k+1}^{T}\breve{\Xi}_{k+1}^{T}K_{k+1}^{T}\\&\leq(1+\varepsilon_3^{-1}+\varepsilon_4^{-1})\operatorname{tr}\left(C_{k+1}\bar{\Psi}_{k+1|k}C_{k+1}^{T}\right)K_{k+1}\breve{\Xi}_{k+1}\breve{\Xi}_{k+1}^{T}K_{k+1}^{T}\\&\leq(1+\varepsilon_3^{-1}+\varepsilon_4^{-1})\rho^2\operatorname{tr}\left(C_{k+1}\bar{\Psi}_{k+1|k}C_{k+1}^{T}\right)K_{k+1}K_{k+1}^{T},\end{aligned}\tag{5.80}$$

where ρ is defined as in (5.67). In view of (5.46), the fifth term of the RHS of (5.75) is computed as

$$\begin{aligned}&K_{k+1}\mathbb{E}\{h(x_{k+1},\zeta_{k+1})h^T(x_{k+1},\zeta_{k+1})\}K_{k+1}^{T}\\&=K_{k+1}\sum_{i=1}^{s}\Pi_{k+1}^{2i}\operatorname{tr}(\mathbb{E}\{x_{k+1}x_{k+1}^{T}\}\Gamma_{k+1}^{i})K_{k+1}^{T}\\&\leq K_{k+1}\sum_{i=1}^{s}\Pi_{k+1}^{2i}\operatorname{tr}(\bar{\Psi}_{k+1|k}\Gamma_{k+1}^{i})K_{k+1}^{T}.\end{aligned}\tag{5.81}$$

Together with (5.75) and (5.79)–(5.81), we have

$$\begin{aligned}&P_{k+1|k+1}\\&\leq(1+\varepsilon_2+\varepsilon_3)(I-K_{k+1}\bar{\Xi}_{k+1}C_{k+1})P_{k+1|k}(I-K_{k+1}\bar{\Xi}_{k+1}C_{k+1})^T\end{aligned}$$

$$
\begin{aligned}
&+ (1+\varepsilon_2^{-1}+\varepsilon_4)K_{k+1}\mathbb{E}\{(\bar{y}_{k+1}-y_{k+1})(\bar{y}_{k+1}-y_{k+1})^T\}K_{k+1}^T \\
&+ K_{k+1}\left[\frac{1}{4}I \circ \left(C_{k+1}\bar{\Psi}_{k+1|k}C_{k+1}^T\right)\right]K_{k+1}^T \\
&+ (1+\varepsilon_3^{-1}+\varepsilon_4^{-1})\rho^2\mathrm{tr}\left(C_{k+1}\bar{\Psi}_{k+1|k}C_{k+1}^T\right)K_{k+1}K_{k+1}^T \\
&+ K_{k+1}\sum_{i=1}^{s}\Pi_{k+1}^{2i}\mathrm{tr}(\bar{\Psi}_{k+1|k}\Gamma_{k+1}^i)K_{k+1}^T + K_{k+1}R_{k+1}K_{k+1}^T. \qquad (5.82)
\end{aligned}
$$

Noting the transmission condition (5.47), we obtain

$$
(\bar{y}_k - y_k)(\bar{y}_k - y_k)^T \le \sigma I. \qquad (5.83)
$$

Substituting (5.83) into (5.82) yields

$$
\begin{aligned}
&P_{k+1|k+1} \\
&\le (1+\varepsilon_2+\varepsilon_3)(I-K_{k+1}\bar{\Xi}_{k+1}C_{k+1})P_{k+1|k}(I-K_{k+1}\bar{\Xi}_{k+1}C_{k+1})^T \\
&+ (1+\varepsilon_2^{-1}+\varepsilon_4)\sigma K_{k+1}K_{k+1}^T + K_{k+1}\left[\frac{1}{4}I \circ \left(C_{k+1}\bar{\Psi}_{k+1|k}C_{k+1}^T\right)\right]K_{k+1}^T \\
&+ (1+\varepsilon_3^{-1}+\varepsilon_4^{-1})\rho^2\mathrm{tr}\left(C_{k+1}\bar{\Psi}_{k+1|k}C_{k+1}^T\right)K_{k+1}K_{k+1}^T \\
&+ K_{k+1}\sum_{i=1}^{s}\Pi_{k+1}^{2i}\mathrm{tr}(\bar{\Psi}_{k+1|k}\Gamma_{k+1}^i)K_{k+1}^T + K_{k+1}R_{k+1}K_{k+1}^T. \qquad (5.84)
\end{aligned}
$$

Then, compared with (5.65) and (5.84), it is easy to see that $P_{k+1|k+1} \le \Sigma_{k+1|k+1}$, which ends the proof of this theorem.

Finally, the objective is to design the filter gain and minimize the trace of this upper bound. Furthermore, the concrete expression form of the desired filter gain is proposed by solving two recursive matrix equations presented in Theorem 5.7.

Theorem 5.7 *If the filter gain is designed as follows:*

$$
\begin{aligned}
K_{k+1} = {} & (1+\varepsilon_2+\varepsilon_3)\Sigma_{k+1|k}C_{k+1}^T\bar{\Xi}_{k+1}^T\Big[(1+\varepsilon_2+\varepsilon_3)\bar{\Xi}_{k+1}C_{k+1}\Sigma_{k+1|k}C_{k+1}^T\bar{\Xi}_{k+1}^T \\
&+ (1+\varepsilon_2^{-1}+\varepsilon_4)\sigma I + \frac{1}{4}I \circ (C_{k+1}\Psi_{k+1|k}C_{k+1}^T) + (1+\varepsilon_3^{-1}+\varepsilon_4^{-1}) \\
&\times \mathrm{tr}(C_{k+1}\Psi_{k+1|k}C_{k+1}^T)\rho^2 I + \sum_{i=1}^{s}\Pi_{k+1}^{2i} \\
&\times \mathrm{tr}(\Psi_{k+1|k}\Gamma_{k+1}^i) + R_{k+1}\Big]^{-1}, \qquad (5.85)
\end{aligned}
$$

then $\mathrm{tr}(\Sigma_{k+1|k+1})$ *is minimized and* $\Sigma_{k+1|k+1}$ *can be given as the following form:*

$$
\begin{aligned}
\Sigma_{k+1|k+1} = \Sigma_{k+1|k} - \Sigma_{k+1|k}^{T} C_{k+1}^{T} \bar{\Xi}_{k+1}^{T} \Big[& (1+\varepsilon_2+\varepsilon_3)\bar{\Xi}_{k+1} C_{k+1} \Sigma_{k+1|k} C_{k+1}^{T} \bar{\Xi}_{k+1}^{T} \\
& + (1+\varepsilon_2^{-1}+\varepsilon_4)\sigma I + \frac{1}{4} I \circ (C_{k+1} \Psi_{k+1|k} C_{k+1}^{T}) \\
& + (1+\varepsilon_3^{-1}+\varepsilon_4^{-1}) \mathrm{tr}(C_{k+1} \Psi_{k+1|k} C_{k+1}^{T}) \rho^2 I \\
& + \sum_{i=1}^{s} \Pi_{k+1}^{2i} \mathrm{tr}(\Psi_{k+1|k} \Gamma_{k+1}^{i}) + R_{k+1} \Big]^{-1} \\
& \times \bar{\Xi}_{k+1} C_{k+1} \Sigma_{k+1|k}.
\end{aligned} \tag{5.86}
$$

Proof It is straightforward to obtain the proof of this theorem in view of Theorem 5.6. By resorting to Lemma 2.3, taking the partial derivative of $\mathrm{tr}(\Sigma_{k+1|k+1})$ regarding K_{k+1} and letting the result be zero, one obtains

$$
\begin{aligned}
& \frac{\partial \mathrm{tr}\left(\Sigma_{k+1|k+1}\right)}{\partial K_{k+1}} \\
& = -2(1+\varepsilon_2+\varepsilon_3)\left(I - K_{k+1}\bar{\Xi}_{k+1} C_{k+1}\right) \Sigma_{k+1|k} C_{k+1}^{T} \bar{\Xi}_{k+1}^{T} \\
& + 2K_{k+1} \Big[(1+\varepsilon_2^{-1}+\varepsilon_4)\sigma I + \frac{1}{4} I \circ (C_{k+1} \Psi_{k+1|k} C_{k+1}^{T}) \\
& + (1+\varepsilon_3^{-1}+\varepsilon_4^{-1}) \mathrm{tr}(C_{k+1} \Psi_{k+1|k} C_{k+1}^{T}) \rho^2 I \\
& + \sum_{i=1}^{s} \Pi_{k+1}^{2i} \mathrm{tr}(\Psi_{k+1|k} \Gamma_{k+1}^{i}) + R_{k+1} \Big] \\
& = 0.
\end{aligned} \tag{5.87}
$$

Then, it is easy to search for that the filter gain in (5.85) can be derived directly. Furthermore, the Eq. (5.86) also holds. As a result, the proof of this theorem is ended.

Remark 5.5 Until now, we have dealt with the recursive filtering problem for time-varying nonlinear systems involving ETM and multiple missing measurements under uncertain missing probabilities. Through using the Taylor series expansion method, the nonlinearity has been linearized by solving the Jacobian matrix and the high-order terms have been further estimated by means of the norm-bounded uncertainty. In what follows, a novel recursive filtering scheme has been presented, where an optimized upper bound concerning on the FEC has been solved and minimized via constructing the filter gain. Meanwhile, the effects from the considered phenomena (multiple missing measurements, ETM and stochastic nonlinearities) onto the filtering performance have been evaluated and the mentioned three factors have been clearly examined in the developed filtering strategy.

5.3 Three Illustrative Examples

In this section, some simulations are provided to validate the validity of the robust optimal event-based filtering methods.

Example 1 Consider the nonlinear stochastic systems (5.1) and (5.2) with following parameters:

$$
\begin{aligned}
A_k &= \begin{bmatrix} -0.3 \; 0.6 \\ -0.3 \; 0.6 \end{bmatrix}, \; B_k = \begin{bmatrix} 0.4 \; 0.64 \end{bmatrix}^T, \\
C_k &= \begin{bmatrix} -0.8 & 1 \\ 1 & 0.8 \end{bmatrix}, \; M_1 = M_2 = M_3 = \begin{bmatrix} 0.3 & 0 \\ 0 & 0.3 \end{bmatrix}, \\
N_1 = N_2 = N_3 &= \begin{bmatrix} 0.4 & 0 \\ 0 & 0.4 \end{bmatrix}, \; F_{1,k} = F_{2,k} = F_{3,k} = \begin{bmatrix} \sin(k) & 0 \\ 0 & \sin(k) \end{bmatrix}.
\end{aligned}
$$

In the simulation, let the initial values be $\hat{x}_{0|0} = \bar{x}_0 = \begin{bmatrix} 0.6 \; 0.6 \end{bmatrix}^T$ and $\Sigma_{0|0} = 10I_2$. Other coefficients are selected as

$$
\begin{aligned}
&\gamma_{4,k+1} = [1.2\lambda_{\max}(N_3 N_3^T) + 1]^{-1}, \\
&\gamma_{1,k} = [1.2\lambda_{\max}(N_2 \Sigma_{k|k} N_2^T) + 1]^{-1}, \\
&\gamma_{2,k+1} = [1.2\lambda_{\max}(N_3 C_{k+1} \Sigma_{k+1|k} C_{k+1}^T N_3^T) + 1]^{-1}, \\
&\gamma_{3,k+1} = [1.2\lambda_{\max}(N_3 C_{k+1} \bar{\Psi}_{3,k+1} C_{k+1}^T N_3^T) + 1]^{-1}, \\
&\gamma_{5,k+1} = [1.2\lambda_{\max}(N_3 R_{k+1} N_3^T) + 1]^{-1}, \\
&\varepsilon_1 = \varepsilon_2 = 0.8, \; \varepsilon_3 = 0.6, \; \varepsilon_4 = 0.5, \\
&\varepsilon_5 = 1, \; \varepsilon_6 = 0.5, \; \varepsilon_7 = 0.9, \; \eta_1 = 0.5, \\
&\eta_2 = 0.6, \; \eta_3 = 1, \; \eta_4 = 0.8, \; \eta_5 = 0.6, \\
&f(x_k) = \begin{bmatrix} 0.5\sin(x_{1,k})\cos(x_{2,k}) \; 0.5\cos(x_{2,k}) \end{bmatrix}^T, \; Q_k = 0.5, \\
&R_k = 0.4I, \; \bar{\alpha}_k = 0.75, \; \bar{\lambda}_k = 0.9, \; \kappa_1 = \kappa_2 = 0.05.
\end{aligned}
$$

To check the influence of the ETM, set $\delta = 0.01$. By using the Theorem 5.3, the upper bound matrix and filter gain are computed. Accordingly, the proposed time-varying filtering algorithm can be realized via (5.6)–(5.7). The curves of the state and its estimation are exhibited in Fig. 5.1, which reveals that the state can be well tracked. The log(MSE) of the state x_k and its upper bound are illustrated in Fig. 5.2. It is same as the analysis in Theorem 5.3 that the MSE keeps blow its upper bound.

Example 2 Let us consider the conducted time-varying nonlinear systems (5.41) and (5.42) with the following parameters:

$$
f(x_k) = \begin{bmatrix} 0.8x_{1,k} + x_{1,k}x_{2,k} \\ 1.05x_{2,k} - x_{1,k}x_{2,k} \end{bmatrix}, \; D_k = \begin{bmatrix} 0.1 \\ 0.08 + 0.4e^{-2k} \end{bmatrix},
$$

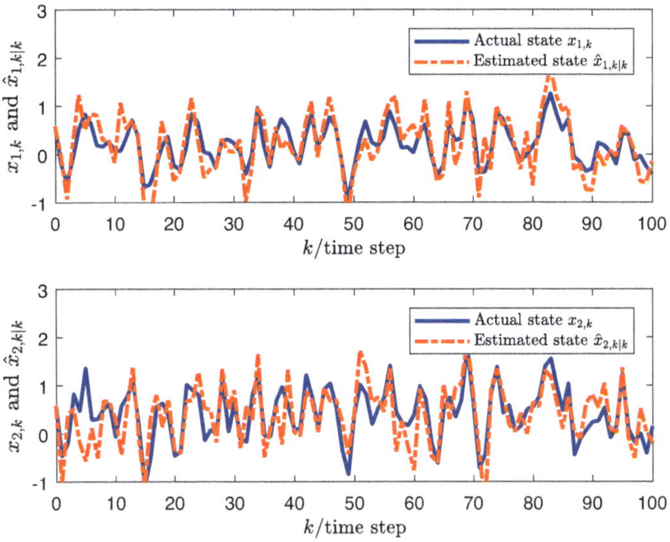

Fig. 5.1 The state trajectories of x_k and its estimation

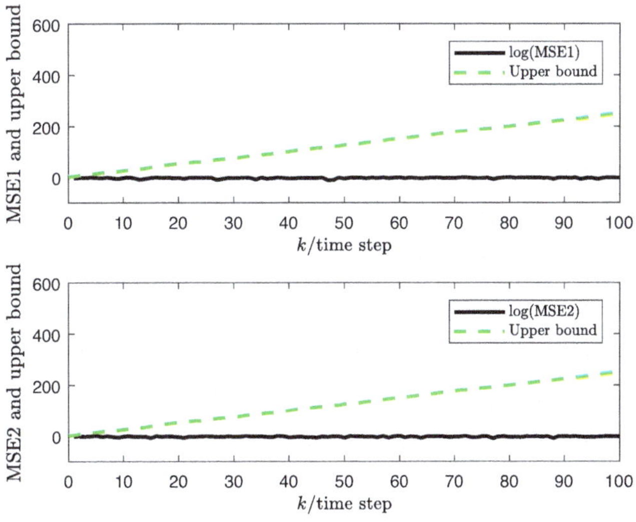

Fig. 5.2 log(MSE) and its upper bound

$$C_k = \begin{bmatrix} 1 & 0 \\ 0 & -1.2 + 0.03\sin(k+3) \end{bmatrix},$$

where $x_k = \begin{bmatrix} x_{1,k} & x_{2,k} \end{bmatrix}^T$ is the state vector with $x_{i,k}$ $(i = 1, 2)$ denoting the i-th element of the system state. The stochastic nonlinearities $g(x_k, \eta_k)$ and $h(x_k, \zeta_k)$ are modeled as

$$g(x_k, \eta_k) = \begin{bmatrix} 0.3 \\ 0.2 \end{bmatrix} [0.2\text{sign}\left(x_{1,k}\right) x_{1,k}\eta_{1,k} + 0.3\text{sign}\left(x_{2,k}\right) x_{2,k}\eta_{2,k}],$$

$$h(x_k, \zeta_k) = \begin{bmatrix} 0.2 \\ 0.4 \end{bmatrix} [0.2\text{sign}\left(x_{1,k}\right) x_{1,k}\zeta_{1,k} + 0.3\text{sign}\left(x_{2,k}\right) x_{1,k}\zeta_{2,k}],$$

where $\eta_{i,k}$ and $\zeta_{i,k}$ $(i = 1, 2)$ are the Gaussian white noises with zero-mean and unity covariances. It is obvious to see that the stochastic nonlinearities $g(x_k, \eta_k)$ and $h(x_k, \zeta_k)$ satisfy (5.44)–(5.46) with

$$\Pi_k^{11} = \begin{bmatrix} 0.09 & 0.06 \\ 0.06 & 0.04 \end{bmatrix}, \quad \Pi_k^{21} = \begin{bmatrix} 0.04 & 0.08 \\ 0.08 & 0.16 \end{bmatrix}, \quad \Gamma_k^1 = \begin{bmatrix} 0.04 & 0 \\ 0 & 0.09 \end{bmatrix}.$$

The initial values are selected as $x_0 = \begin{bmatrix} 1.2 & 1.2 \end{bmatrix}^T$ and $\Sigma_{0|0} = 15I$. The other parameters are selected as $\bar{\xi}_1 = 0.9$, $\bar{\xi}_2 = 0.85$, $M_k = \text{diag}\{0.3, 0.6\}$, $L_k = 0.2I$, $\gamma_k = [2\lambda_{\max}(L_k \Sigma_{k|k} L_k^T) + 0.5]^{-1}$, $\varepsilon_1 = 0.8$, $\varepsilon_2 = 0.4$, $\varepsilon_3 = 0.6$, $\varepsilon_4 = 0.2$, $\varepsilon_5 = 0.4$, $\sigma = 0.05$, $Q_k = 0.1$, and $R_k = 0.05I$.

By solving (5.64), (5.65) and (5.85), we can derive the filter gain and then the simulation results are indicated in Figs. 5.3, 5.4, 5.5, 5.6 and 5.7. Among the simulation results, Figs. 5.3 and 5.4 depict the trajectories of the real states $x_{i,k}$ and their estimations $\hat{x}_{i,k|k}$ $(i = 1, 2)$. From the Figs. 5.5 and 5.6, the log(MSE) for the states $x_{i,k}$ $(i = 1, 2)$ keep below its upper bounds, which also verify our results. The main reason is that a new scheme with compensation ability has been adopted by taking the available probability information into consideration and the proposed upper bound has been minimized by properly constructing the filter gain. In addition, with the aim to illustrate the influence from the triggered threshold σ onto the filtering performance, the values under different values of σ are shown in Fig. 5.7, which can be seen that the smaller triggered threshold σ is and then the smaller MSE is obtained, namely, when the triggered threshold σ becomes small, more data signals can be normally transmitted and thus the better filtering precision gets.

Example 3 Based on Lei et al. (2011), the online estimation/tracking problem of the position/velocity of a ballistic object is of vital significance. Furthermore, the dynamics of the ballistic object is nonlinear. Therefore, we deal with the event-based filtering issue when tracking a ballistic object through the network transmissions. For such a case, the multiple missing measurements might occur because of the unreliable communication links. Besides, the considered actual target motion may be contaminated with the stochastic nonlinearities owing to various reasons including the changes in the interconnections of subsystems or repairs of the components

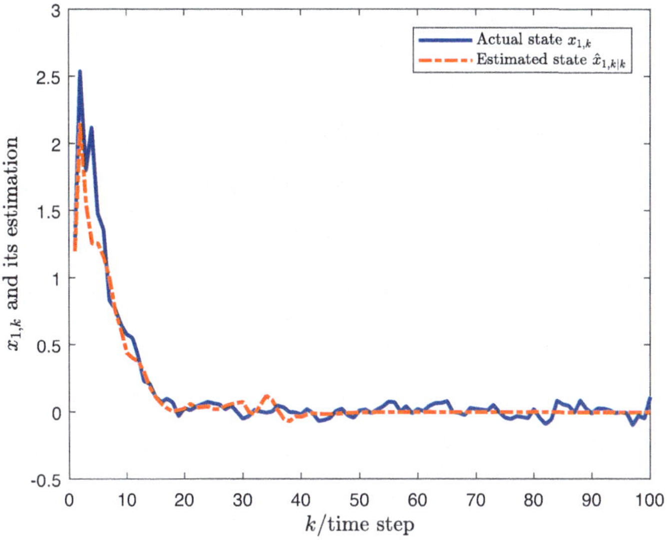

Fig. 5.3 State $x_{1,k}$ and its estimation $\hat{x}_{1,k|k}$

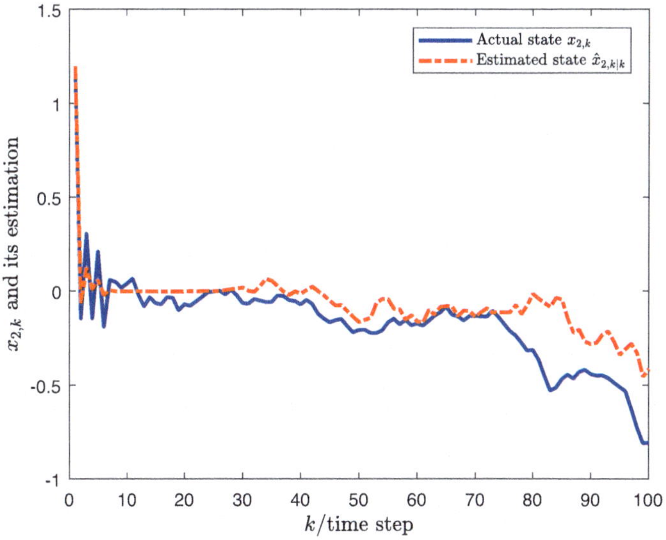

Fig. 5.4 State $x_{2,k}$ and its estimation $\hat{x}_{2,k|k}$

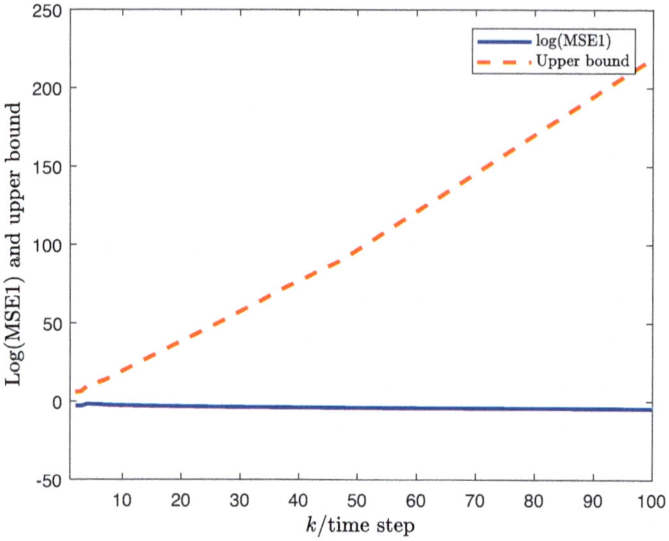

Fig. 5.5 log(MSE1) and its upper bound

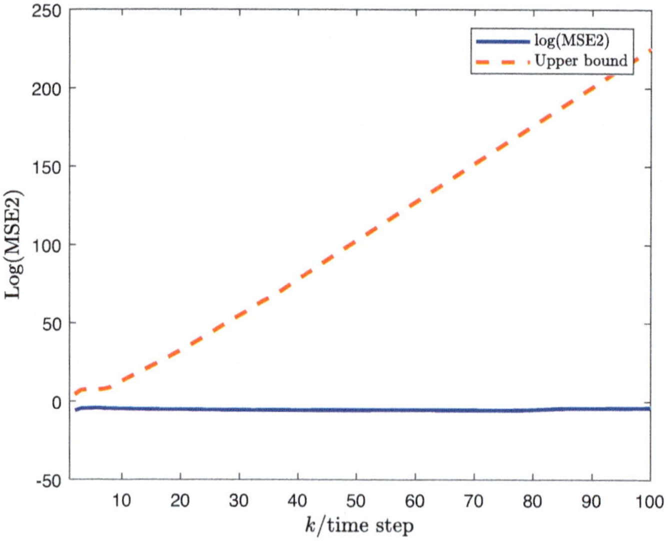

Fig. 5.6 log(MSE2) and its upper bound

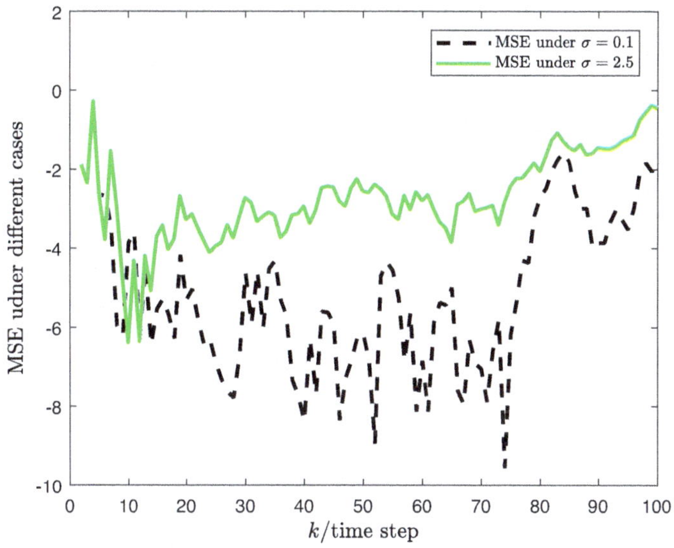

Fig. 5.7 MSE under different σ

and so on. Thus, the main aim is to construct the time-varying filtering method under the ETM. To be more specific, we are ready to examine the effects from the multiple missing measurements under uncertain missing probabilities and ETM onto the filtering performance. The conducted target motion is modeled by

$$\begin{aligned} x_{k+1} &= f(x_k) + g(x_k, \eta_k) + D_k \omega_k, \\ y_k &= \Xi_k C_k x_k + h(x_k, \zeta_k) + \nu_k, \end{aligned}$$

where

$$\begin{aligned} f(x_k) &= \Phi_k x_k + G(s(x_k) + H), \\ s(x_k) &= -\frac{g\rho(x_{2,k})}{2\beta}\sqrt{\dot{x}_{1,k}^2 + \dot{x}_{2,k}^2}\begin{bmatrix} \dot{x}_{1,k} \\ \dot{x}_{2,k} \end{bmatrix}, \\ \rho(x_{2,k}) &= \theta_1 \exp(-\theta_2 x_{2,k}), \\ \theta_1 &= \begin{cases} 1.227 & \text{if } x_{2,k} < 9144 \text{ m} \\ 1.754 & \text{if } x_{2,k} \geq 9144 \text{ m} \end{cases}, \\ \theta_2 &= \begin{cases} 1.093 \times 10^{-4} & \text{if } x_{2,k} < 9144 \text{ m} \\ 1.49 \times 10^{-4} & \text{if } x_{2,k} \geq 9144 \text{ m} \end{cases}, \end{aligned}$$

$$g(x_x, \eta_k) = \begin{bmatrix} 0.004 \\ 0.002 \\ 0.004 \\ 0.002 \end{bmatrix} [0.005x_{1,k}\eta_{1,k} + 0.005\dot{x}_{1,k}\eta_{2,k}$$
$$+ 0.0025x_{2,k}\eta_{3,k} + 0.005\dot{x}_{2,k}\eta_{4,k}],$$
$$h(x_k, \zeta_k) = \begin{bmatrix} 0.004 \\ 0.002 \end{bmatrix} [0.005x_{1,k}\zeta_{1,k} + 0.005\dot{x}_{1,k}\zeta_{2,k}$$
$$+ 0.0025x_{2,k}\zeta_{3,k} + 0.005\dot{x}_{2,k}\zeta_{4,k}],$$
$$\Phi_k = \begin{bmatrix} 1 & T & 0 & 0 \\ 0 & 1 & 0 & 0 \\ 0 & 0 & 1 & T \\ 0 & 0 & 0 & 1 \end{bmatrix}, \quad G = \begin{bmatrix} \frac{T^2}{2} & 0 \\ T & 0 \\ 0 & \frac{T^2}{2} \\ 0 & T \end{bmatrix},$$
$$C_k = \begin{bmatrix} 1 & 0 & 0 & 0 \\ 0 & 0 & 1 & 0 \end{bmatrix}, \quad H = \begin{bmatrix} 0 \\ -g \end{bmatrix},$$

where $x_k = \begin{bmatrix} x_{1,k} & \dot{x}_{1,k} & x_{2,k} & \dot{x}_{2,k} \end{bmatrix}^T$ is the system state, $x_{1,k}$ and $x_{2,k}$ are the target abscissa and the target ordinate, respectively. $\dot{x}_{1,k}$ and $\dot{x}_{2,k}$ stand for the velocity. T is the sampling period, g denotes the gravity acceleration, β represents the ballistic coefficient that relies on the object mass, shape, as well as cross-section area. $\rho(\cdot)$ is the air density. In addition, the zero-mean Gaussian white noises $\eta_{i,k} \in \mathbb{R}$, $\zeta_{i,k} \in \mathbb{R}$ $(i = 1, 2, 3, 4)$, $\omega_k \in \mathbb{R}^4$ and $\nu_k \in \mathbb{R}^2$ have covariances 1, 1, Q_k and $R_k = 100I_2$, respectively. Here,

$$Q_k = c \cdot \text{diag}\{q, q\}, \quad q = \begin{bmatrix} \frac{T^3}{3} & \frac{T^2}{2} \\ \frac{T^2}{2} & T \end{bmatrix}.$$

According to (5.44)–(5.46), we can obtain

$$\Pi_k^{11} = 10^{-4} \times \begin{bmatrix} 0.16 & 0.08 & 0.16 & 0.08 \\ 0.08 & 0.04 & 0.08 & 0.04 \\ 0.16 & 0.08 & 0.16 & 0.08 \\ 0.08 & 0.04 & 0.08 & 0.04 \end{bmatrix}, \quad \Pi_k^{21} = 10^{-4} \times \begin{bmatrix} 0.16 & 0.08 \\ 0.08 & 0.04 \end{bmatrix},$$
$$\Gamma_k^1 = 10^{-4} \times \text{diag}\{0.25,\ 0.25,\ 0.0625,\ 0.25\}.$$

The other parameters are set as

$$g = 9.81\,\text{m/s}^2, \ \beta = 4 \times 10^4\,\text{km/ms}^2, \ c = 0.1\,\text{m/s}^3, \ T = 1\,\text{s},$$
$$x_0 = 10^2 \times \begin{bmatrix} 30 & 4 & 90 & 3 \end{bmatrix}^T, \ M_k = \text{diag}\{9, 1.2, 9, 4\}, \ L_k = 0.01I_4,$$
$$\bar{\xi}_1 = 0.95, \ \bar{\xi}_2 = 0.98, \ \sigma = 0.04, \ \gamma_k = 0.04, \ \varepsilon_1 = 0.5,$$
$$\varepsilon_2 = 0.5, \ \varepsilon_3 = 0.5, \ \varepsilon_4 = 0.5, \ \varepsilon_5 = 0.5, \ \Sigma_{0|0} = 100I_4.$$

After that, the proposed event-based filtering issue can be addressed and the position/velocity of the ballistic object can be estimated well. In particular, according to the measurement information collected by the multiple sensors, it is obtained that the upper bound of the FEC can be provided in Theorem 5.6 and minimized by constructing the filter gain by resorting to Theorem 5.7. Next, the simulations can be shown in Figs. 5.8, 5.9, 5.10, 5.11, 5.12, 5.13, 5.14 and 5.15 by virtue of the developed filtering scheme under ETM.

For the simulation results, Fig. 5.8 plots the position and its related estimation of the target abscissa of the ballistic object by using the developed time-varying filtering scheme. In Fig. 5.9, the velocity and its corresponding estimation of the target abscissa of the ballistic object are depicted. Likewise, Figs. 5.10 and 5.11 are the position/velocity and their corresponding estimations of the target ordinate of the ballistic object. Besides, Figs. 5.12, 5.13, 5.14 and 5.15 depict the log(MSE) for the position/velocity components and their upper bounds. From the simulation results, it is concluded that the position and velocity components can be estimated, which further reveal the applicability and usefulness of main results.

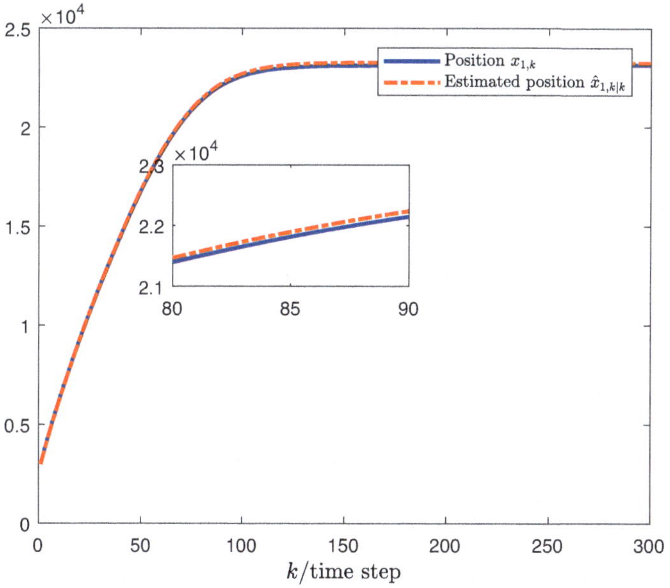

Fig. 5.8 Position $x_{1,k}$ and its estimation $\hat{x}_{1,k|k}$

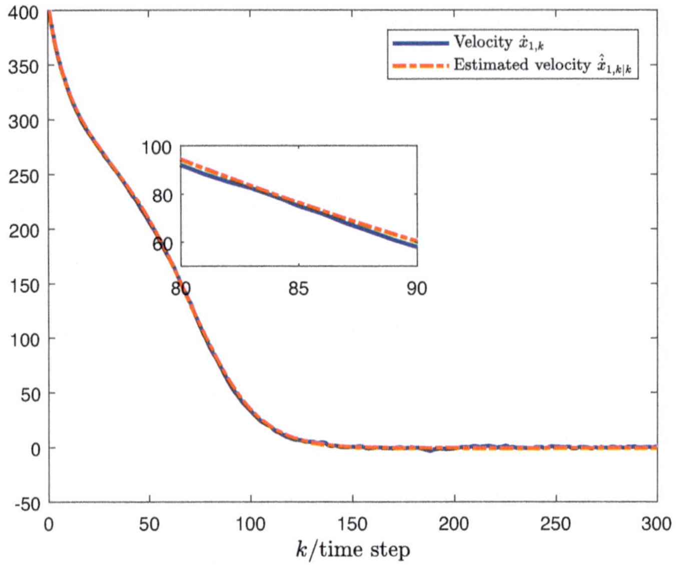

Fig. 5.9 Velocity $\dot{x}_{1,k}$ and its estimation $\hat{\dot{x}}_{1,k|k}$

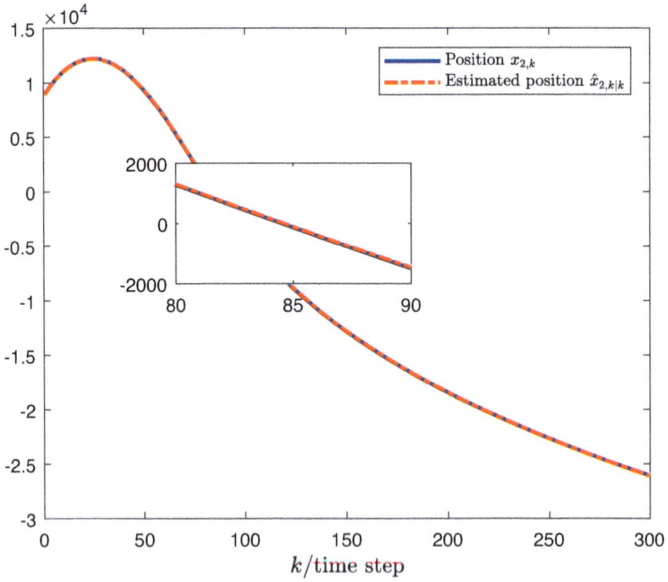

Fig. 5.10 Position $x_{2,k}$ and its estimation $\hat{x}_{2,k|k}$

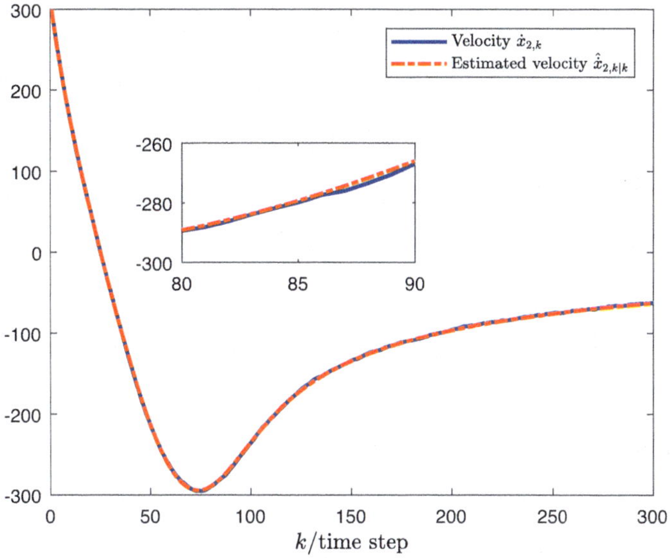

Fig. 5.11 Velocity $\dot{x}_{2,k}$ and its estimation $\hat{\dot{x}}_{2,k|k}$

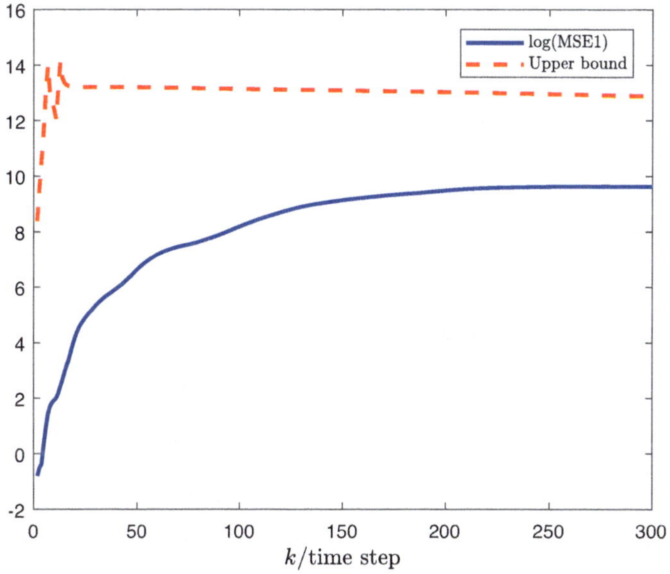

Fig. 5.12 log(MSE1) and its upper bound

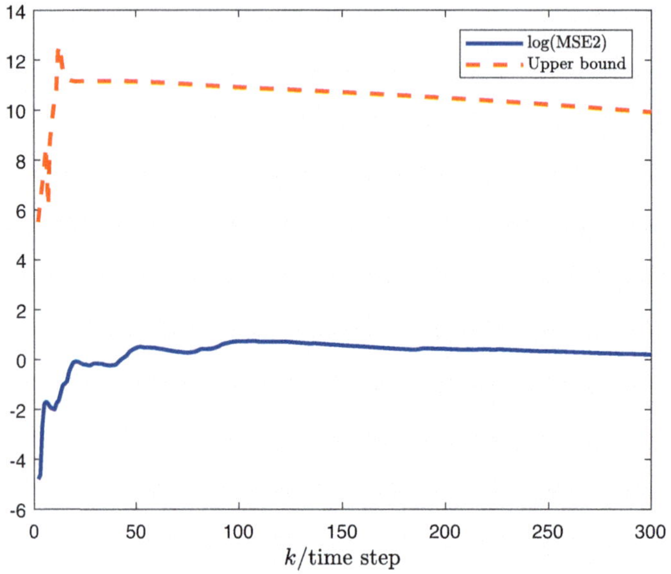

Fig. 5.13 log(MSE2) and its upper bound

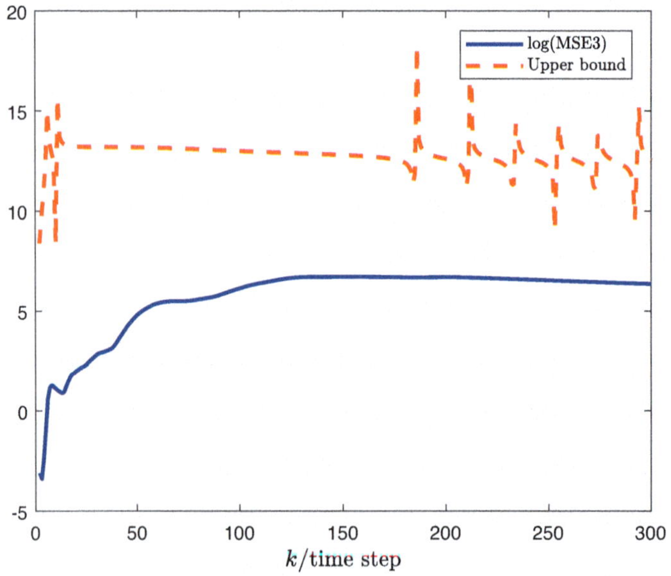

Fig. 5.14 log(MSE3) and its upper bound

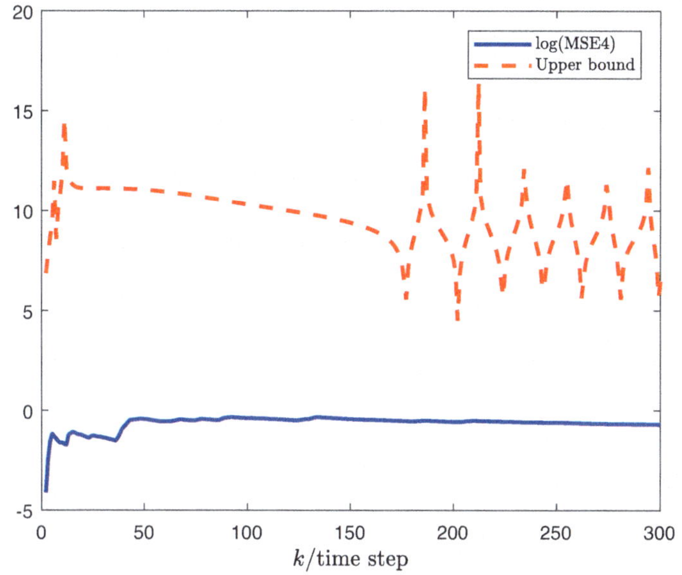

Fig. 5.15 log(MSE4) and its upper bound

5.4 Summary

In this chapter, the recursive filtering problems have been addressed for two classes of time-varying stochastic nonlinear systems under ETM and missing measurements. Firstly, an ETM has been utilized to arrange information transmission to save communication resources and enhance transmission efficiency. Through solving the Riccati-like difference equations, an upper bound of the FEC has been firstly calculated and subsequently minimized by constructing a desirable filter gain. Secondly, we have coped with the filtering problem for nonlinear stochastic systems involving ETM and multiple missing measurements. Specifically, the phenomenon of the missing measurements has been described by resorting to a set of random variables with uncertain occurrence probability. Besides, the ETM has also been used to decrease unnecessary data transmission and alleviate the network burden. A novel filtering compensation strategy has been established by taking the available probability information of the multiple missing measurements into consideration. In addition, the recursions of the one-step prediction error covariance and the FEC have been solved, and an upper bound concerning on the FEC has been received and optimized by appropriately constructing the filter gain. At last, some comparative simulations have been provided to show the validity and advantages of the presented event-triggered filtering approaches.

References

Calafiore G (2005) Reliable localization using set-valued nonlinear filters. IEEE Trans Syst Man Cybern Part A-Syst Hum 35(2):189–197
Hu J, Wang Z, Shen B, Gao H (2013) Quantized recursive filtering for a class of nonlinear systems with multiplicative noises and missing measurements. Int J Control 86(4):650–663
Lei M, van Wyk BJ, Qi Y (2011) Online estimation of the approximate posterior Cramer-Rao lower bound for discrete-time nonlinear filtering. IEEE Trans Aerosp Electron Syst 47(1):37–57
Xiong K, Wei C, Liu L (2010) Robust extended Kalman filtering for nonlinear systems with stochastic uncertainties. IEEE Trans Syst Man Cybern Part A-Syst Hum 40(2):399–405

Chapter 6
Joint Fault and State Estimation Against Randomly Occurring Deception Attacks

In this chapter, the fault estimation problem is handled for time-varying nonlinear stochastic systems subject to RONs and RODAs. Two mutually independent binary random variables are employed to characterize the RONs and RODAs, respectively. By using the augmentation approach, the co-design problem of the robust estimator and fault estimator is converted into the recursive estimator design problem. A new compensation strategy is designed such that, for both RONs and RODAs, an upper bound of the EEC can be derived and such an upper bound is minimized through selecting a desirable estimator gain. In addition, the estimator gain is shown via the solution to two Riccati-like matrix equations. It is indicated that the designed estimation scheme is of a recursive pattern that is suitable for online computations. Finally, a numerical simulation experiment is adopted to demonstrate the usefulness of the designed estimation scheme.

6.1 Problem Formulation

Consider the following class of time-varying nonlinear stochastic systems:

$$\breve{x}_{k+1} = \breve{A}_k\breve{x}_k + \alpha_k\breve{h}(\breve{x}_k) + \breve{F}_k f_k + \breve{B}_k\omega_k, \tag{6.1}$$

$$\breve{y}_k = \breve{C}_k\breve{x}_k + \nu_k, \tag{6.2}$$

where $\breve{x}_k \in \mathbb{R}^n$ is the system state with initial value $\breve{x}_0$, $\breve{y}_k \in \mathbb{R}^m$ is the measurement, $f_k \in \mathbb{R}^f$ is the additive fault, ω_k is the process noise with zero-mean and covariance $Q_k > 0$, ν_k is the measurement noise with zero-mean and covariance $R_k > 0$. $\breve{A}_k$, $\breve{F}_k$, $\breve{B}_k$ and $\breve{C}_k$ are pre-defined matrices of compatible dimensions.

The nonlinear function $\breve{h}(\breve{x}_k)$ is known with $\breve{h}(0) = 0$ and satisfies

$$\|\breve{h}(u) - \breve{h}(v)\| \leq \|H(u - v)\|, \ \ \forall\, u, v \in \mathbb{R}^n, \tag{6.3}$$

J. Hu et al., *Variance-Constrained Filtering for Stochastic Complex Systems*, Intelligent Control and Learning Systems 18,
https://doi.org/10.1007/978-981-96-2637-3_6

where H is a pre-determined matrix.

The binary random variable $\alpha_k \in \mathbb{R}$, which accounts for the phenomenon of RONs, takes the values on 0 or 1 with

$$\begin{aligned} \text{Prob}\{\alpha_k = 1\} &= \bar{\alpha}, \\ \text{Prob}\{\alpha_k = 0\} &= 1 - \bar{\alpha}, \end{aligned} \tag{6.4}$$

where $\bar{\alpha} \in [0, 1]$ is a given scalar.

In the sequel, we are interested in the co-design of state and fault estimator under the unreliable communication environments, where the deception attacks may occur randomly. The adopted randomly occurring attack model is given as follows:

$$\tilde{y}_k = \breve{y}_k + \gamma_k \bar{y}_k, \tag{6.5}$$

where $\tilde{y}_k$ is the received signal by the estimator, $\bar{y}_k$ represents the attack signal sent by adversaries. The deception attacks can be characterized by $\bar{y}_k = -\breve{y}_k + \eta_k$, where η_k is a non-zero energy-bounded signal satisfying $\|\eta_k\| \le \delta$ with δ being a known scalar. The random variable $\gamma_k \in \mathbb{R}$, which models the RODAs, satisfies the following Bernoulli distribution:

$$\begin{aligned} \text{Prob}\{\gamma_k = 1\} &= \bar{\gamma}, \\ \text{Prob}\{\gamma_k = 0\} &= 1 - \bar{\gamma} \end{aligned} \tag{6.6}$$

with $\bar{\gamma} \in [0, 1]$ being a pre-determined scalar. It is assumed that α_k, γ_k, ω_k, ν_k and $\breve{x}_0$ are mutually independent.

Letting $x_k = \begin{bmatrix} \breve{x}_k^T & f_k^T \end{bmatrix}^T$, we obtain

$$x_{k+1} = A_k x_k + \alpha_k h(x_k) + B_k \omega_k, \tag{6.7}$$

$$y_k = C_k x_k + \nu_k, \tag{6.8}$$

where

$$A_k = \begin{bmatrix} \breve{A}_k & \breve{F}_k \\ 0 & I \end{bmatrix}, \quad h(x_k) = \begin{bmatrix} \breve{h}(\breve{x}_k) \\ 0 \end{bmatrix}, \quad B_k = \begin{bmatrix} \breve{B}_k \\ 0 \end{bmatrix}, \quad C_k = \begin{bmatrix} \breve{C}_k & 0 \end{bmatrix}, \quad y_k = \breve{y}_k.$$

In this chapter, the following estimator is designed:

$$\hat{x}_{k+1|k} = A_k \hat{x}_{k|k} + \bar{\alpha} h(\hat{x}_{k|k}), \tag{6.9}$$

$$\hat{x}_{k+1|k+1} = \hat{x}_{k+1|k} + G_{k+1}[\tilde{y}_{k+1} - (1 - \bar{\gamma}) C_{k+1} \hat{x}_{k+1|k}], \tag{6.10}$$

where $\hat{x}_{k|k}$ is the state estimation, $\hat{x}_{k+1|k}$ denotes the one-step prediction, and G_{k+1} is the estimator gain.

Now, we are in a position to state the objective of this chapter.

(R1) Construct a time-varying estimator (6.9) and (6.10) such that an upper bound of the EEC $P_{k+1|k+1} = \mathbb{E}\{e_{k+1|k+1}e_{k+1|k+1}^T\}$ with $e_{k+1|k+1} = x_{k+1} - \hat{x}_{k+1|k+1}$ denoting the estimation error is ensured, that is to say, we aim to find a sequence of positive definite matrices $\Omega_{k+1|k+1}$ satisfying $P_{k+1|k+1} \leq \Omega_{k+1|k+1}$.

(R2) Deign the estimator gain via minimizing $\mathrm{tr}(\Omega_{k+1|k+1})$ at each time instant.

6.2 Design of Fault Estimation Algorithm

In this section, both the one-step prediction error covariance as well as the EEC are derived, and then the upper bound regarding the EEC is calculated by adopting the matrix analysis technique. Subsequently, according to the Riccati-like difference equation approach, the expression form of the estimator gain is given and the established upper bound regarding the EEC is optimized by employing the given estimator gain under the variance-constrained sense.

Firstly, we denote the one-step prediction error as $e_{k+1|k} = x_{k+1} - \hat{x}_{k+1|k}$. After that, according to (6.7) and (6.9), one has

$$e_{k+1|k} = A_k e_{k|k} + \bar{\alpha}[h(x_k) - h(\hat{x}_{k|k})] + \tilde{\alpha}_k h(x_k) + B_k \omega_k, \tag{6.11}$$

where $\tilde{\alpha}_k = \alpha_k - \bar{\alpha}$. Next, the estimation error $e_{k+1|k+1}$ is computed as

$$\begin{aligned} e_{k+1|k+1} = {} & [I - (1 - \bar{\gamma})G_{k+1}C_{k+1}]e_{k+1|k} + \tilde{\gamma}_{k+1}G_{k+1}C_{k+1}x_{k+1} \\ & -(1 - \gamma_{k+1})G_{k+1}\nu_{k+1} - \gamma_{k+1}G_{k+1}\eta_{k+1} \end{aligned} \tag{6.12}$$

with $\tilde{\gamma}_{k+1} = \gamma_{k+1} - \bar{\gamma}$.

Theorem 6.1 *The one-step prediction error covariance $P_{k+1|k} = \mathbb{E}\{e_{k+1|k}e_{k+1|k}^T\}$ is computed as*

$$\begin{aligned} P_{k+1|k} = {} & A_k P_{k|k} A_k^T + \mathscr{A}_k + \mathscr{A}_k^T + \bar{\alpha}^2 \mathbb{E}\{[h(x_k) - h(\hat{x}_{k|k})][h(x_k) - h(\hat{x}_{k|k})]^T\} \\ & + \bar{\alpha}(1 - \bar{\alpha})\mathbb{E}\{h(x_k)h^T(x_k)\} + B_k Q_k B_k^T, \end{aligned} \tag{6.13}$$

where

$$\mathscr{A}_k = \bar{\alpha}\mathbb{E}\{A_k e_{k|k}[h(x_k) - h(\hat{x}_{k|k})]^T\}. \tag{6.14}$$

Proof It follows from (6.4) and (6.11) that (6.13) holds directly. Thus, the proof of this theorem is omitted for brevity.

Theorem 6.2 *The EEC $P_{k+1|k+1}$ is computed as*

$$\begin{aligned}P_{k+1|k+1} =& [I-(1-\bar{\gamma})G_{k+1}C_{k+1}]P_{k+1|k}[I-(1-\bar{\gamma})G_{k+1}C_{k+1}]^T + \mathscr{B}_{k+1} + \mathscr{B}_{k+1}^T \\ &+\bar{\gamma}(1-\bar{\gamma})G_{k+1}C_{k+1}\mathbb{E}\{x_{k+1}x_{k+1}^T\}C_{k+1}^T G_{k+1}^T + \mathscr{C}_{k+1} + \mathscr{C}_{k+1}^T \\ &+(1-\bar{\gamma})G_{k+1}R_{k+1}G_{k+1}^T + \bar{\gamma}G_{k+1}\eta_{k+1}\eta_{k+1}^T G_{k+1}^T \end{aligned} \tag{6.15}$$

with

$$\begin{aligned}\mathscr{B}_{k+1} &= -\mathbb{E}\{\gamma_{k+1}[I-(1-\bar{\gamma})G_{k+1}C_{k+1}]e_{k+1|k}\eta_{k+1}^T G_{k+1}^T\}, \\ \mathscr{C}_{k+1} &= -\mathbb{E}\{\tilde{\gamma}_{k+1}\gamma_{k+1}G_{k+1}C_{k+1}x_{k+1}\eta_{k+1}^T G_{k+1}^T\}. \end{aligned} \tag{6.16}$$

Proof In terms of (6.12) and the definition of the EEC, it is derived that

$$\begin{aligned}P_{k+1|k+1} =& \mathbb{E}\{e_{k+1|k+1}e_{k+1|k+1}^T\} \\ =& [I-(1-\bar{\gamma})G_{k+1}C_{k+1}]P_{k+1|k}[I-(1-\bar{\gamma})G_{k+1}C_{k+1}]^T \\ &+\mathbb{E}\{\tilde{\gamma}_{k+1}[I-(1-\bar{\gamma})G_{k+1}C_{k+1}]e_{k+1|k}x_{k+1}^T C_{k+1}^T G_{k+1}^T\} \\ &+\mathbb{E}\{\tilde{\gamma}_{k+1}G_{k+1}C_{k+1}x_{k+1}e_{k+1|k}^T[I-(1-\bar{\gamma})G_{k+1}C_{k+1}]^T\} \\ &-\mathbb{E}\{(1-\gamma_{k+1})[I-(1-\bar{\gamma})G_{k+1}C_{k+1}]e_{k+1|k}v_{k+1}^T G_{k+1}^T\} \\ &-\mathbb{E}\{(1-\gamma_{k+1})G_{k+1}v_{k+1}e_{k+1|k}^T[I-(1-\bar{\gamma})G_{k+1}C_{k+1}]^T\} \\ &+\mathscr{B}_{k+1} + \mathscr{B}_{k+1}^T + \mathbb{E}\{\tilde{\gamma}_{k+1}^2 G_{k+1}C_{k+1}x_{k+1}x_{k+1}^T C_{k+1}^T G_{k+1}^T\} \\ &-\mathbb{E}\{\tilde{\gamma}_{k+1}(1-\gamma_{k+1})G_{k+1}C_{k+1}x_{k+1}v_{k+1}^T G_{k+1}^T\} \\ &-\mathbb{E}\{\tilde{\gamma}_{k+1}(1-\gamma_{k+1})G_{k+1}v_{k+1}x_{k+1}^T C_{k+1}^T G_{k+1}^T\} \\ &+\mathscr{C}_{k+1} + \mathscr{C}_{k+1}^T + (1-\bar{\gamma})G_{k+1}R_{k+1}G_{k+1}^T \\ &+\mathbb{E}\{(1-\gamma_{k+1})\gamma_{k+1}G_{k+1}v_{k+1}\eta_{k+1}^T G_{k+1}^T\} \\ &+\mathbb{E}\{(1-\gamma_{k+1})\gamma_{k+1}G_{k+1}\eta_{k+1}v_{k+1}^T G_{k+1}^T\} \\ &+\bar{\gamma}\mathbb{E}\{G_{k+1}\eta_{k+1}\eta_{k+1}^T G_{k+1}^T\} \\ =& [I-(1-\bar{\gamma})G_{k+1}C_{k+1}]P_{k+1|k}[I-(1-\bar{\gamma})G_{k+1}C_{k+1}]^T + \mathscr{B}_{k+1} + \mathscr{B}_{k+1}^T \\ &+\bar{\gamma}(1-\bar{\gamma})G_{k+1}C_{k+1}\mathbb{E}\{x_{k+1}x_{k+1}^T\}C_{k+1}^T G_{k+1}^T + \mathscr{C}_{k+1} + \mathscr{C}_{k+1}^T \\ &+(1-\bar{\gamma})G_{k+1}R_{k+1}G_{k+1}^T + \bar{\gamma}G_{k+1}\eta_{k+1}\eta_{k+1}^T G_{k+1}^T. \end{aligned} \tag{6.17}$$

When deriving (6.17), we have used the fact that γ_{k+1} and v_{k+1} are mutually independent. Then, the proof of this theorem is finished.

Remark 6.1 Until now, the recursions of the prediction error covariance $P_{k+1|k}$ and EEC $P_{k+1|k+1}$ have been derived according to the related definitions. However, considering the existence of the nonlinearity $h(\cdot)$ and some unknown terms in (6.13) and (6.15), it is impossible to calculate the exact expression forms associated with the prediction error covariance and EEC. With the objective of handling such an issue, we aim to derive the upper bounds of $P_{k+1|k}$ and $P_{k+1|k+1}$ respectively, and then construct the estimator gain by minimizing the trace of obtained upper bound regarding the EEC as an alternative way.

Theorem 6.3 *For positive constants* $\varepsilon_{k,1}$, $\varepsilon_{k,2}$, $\varepsilon_{k+1,3}$, $\varepsilon_{k+1,4}$ *and* $\varepsilon_{k+1,5}$, *the one-step prediction error covariance* $P_{k+1|k}$ *and EEC* $P_{k+1|k+1}$ *have the upper bounds* $\Omega_{k+1|k}$ *and* $\Omega_{k+1|k+1}$ *under the initial condition* $\Omega_{0|0} = P_{0|0}$ *respectively, where*

$$\begin{aligned}\Omega_{k+1|k} = &(1+\varepsilon_{k,1})A_k\Omega_{k|k}A_k^T + \left[\bar{\alpha}^2(1+\varepsilon_{k,1}^{-1}) + \bar{\alpha}(1-\bar{\alpha})(1+\varepsilon_{k,2})\right]\bar{H}\Omega_{k|k}\bar{H}^T \\ &+\bar{\alpha}(1-\bar{\alpha})(1+\varepsilon_{k,2}^{-1})\bar{H}\hat{x}_{k|k}\hat{x}_{k|k}^T\bar{H}^T + B_kQ_kB_k^T \end{aligned} \tag{6.18}$$

and

$$\begin{aligned}\Omega_{k+1|k+1} = &(1+\varepsilon_{k+1,3})[I-(1-\bar{\gamma})G_{k+1}C_{k+1}]\Omega_{k+1|k}[I-(1-\bar{\gamma})G_{k+1}C_{k+1}]^T \\ &+(1+\varepsilon_{k+1,4})\bar{\gamma}(1-\bar{\gamma})G_{k+1}C_{k+1}\bar{\Phi}_{k+1|k}C_{k+1}^TG_{k+1}^T \\ &+(1+\varepsilon_{k+1,3}^{-1}+\varepsilon_{k+1,4}^{-1})\bar{\gamma}\delta^2G_{k+1}G_{k+1}^T \\ &+(1-\bar{\gamma})G_{k+1}R_{k+1}G_{k+1}^T \end{aligned} \tag{6.19}$$

with

$$\begin{aligned}\bar{H} &= \mathrm{diag}\{H, 0\}, \\ \bar{\Phi}_{k+1|k} &= (1+\varepsilon_{k+1,5})\Omega_{k+1|k} + (1+\varepsilon_{k+1,5}^{-1})\hat{x}_{k+1|k}\hat{x}_{k+1|k}^T. \end{aligned} \tag{6.20}$$

Proof Employing Lemma 2.2, we have

$$\mathscr{A}_k + \mathscr{A}_k^T \le \varepsilon_{k,1}A_kP_{k|k}A_k^T + \varepsilon_{k,1}^{-1}\bar{\alpha}^2\mathbb{E}\{[h(x_k)-h(\hat{x}_{k|k})][h(x_k)-h(\hat{x}_{k|k})]^T\} \tag{6.21}$$

where $\varepsilon_{k,1} > 0$ is a constant. Subsequently, according to (6.3), one obtains

$$\mathbb{E}\{[h(x_k)-h(\hat{x}_{k|k})][h(x_k)-h(\hat{x}_{k|k})]^T\} \le \bar{H}P_{k|k}\bar{H}^T, \tag{6.22}$$

$$\mathbb{E}\{h(x_k)h^T(x_k)\} \le \bar{H}\mathbb{E}\{x_kx_k^T\}\bar{H}^T, \tag{6.23}$$

where $\bar{H} = \mathrm{diag}\{H, 0\}$. Again, in view of Lemma 2.2, we have

$$\begin{aligned}\mathbb{E}\{x_kx_k^T\} &= \mathbb{E}\{(e_{k|k}+\hat{x}_{k|k})(e_{k|k}+\hat{x}_{k|k})^T\} \\ &= \mathbb{E}\{e_{k|k}e_{k|k}^T + e_{k|k}\hat{x}_{k|k}^T + \hat{x}_{k|k}e_{k|k}^T + \hat{x}_{k|k}\hat{x}_{k|k}^T\} \\ &\le (1+\varepsilon_{k,2})P_{k|k} + (1+\varepsilon_{k,2}^{-1})\hat{x}_{k|k}\hat{x}_{k|k}^T, \end{aligned} \tag{6.24}$$

where $\varepsilon_{k,2} > 0$ is a scalar. Consequently, it follows from (6.21)–(6.24) that

$$\begin{aligned}P_{k+1|k} \le &(1+\varepsilon_{k,1})A_kP_{k|k}A_k^T + \left[\bar{\alpha}^2(1+\varepsilon_{k,1}^{-1}) + \bar{\alpha}(1-\bar{\alpha})(1+\varepsilon_{k,2})\right]\bar{H}P_{k|k}\bar{H}^T \\ &+\bar{\alpha}(1-\bar{\alpha})(1+\varepsilon_{k,2}^{-1})\bar{H}\hat{x}_{k|k}\hat{x}_{k|k}^T\bar{H}^T + B_kQ_kB_k^T. \end{aligned} \tag{6.25}$$

Next, we aim to handle the uncertain terms $\mathscr{B}_{k+1}$ and $\mathscr{C}_{k+1}$ of the RHS of (6.15). Similar to the derivation of (6.21), one has

$$\mathscr{B}_{k+1} + \mathscr{B}_{k+1}^T \leq \varepsilon_{k+1,3}[I - (1-\bar{\gamma})G_{k+1}C_{k+1}]P_{k+1|k}[I - (1-\bar{\gamma})G_{k+1}C_{k+1}]^T + \varepsilon_{k+1,3}^{-1}\bar{\gamma}G_{k+1}\eta_{k+1}\eta_{k+1}^T G_{k+1}^T \tag{6.26}$$

and

$$\mathscr{C}_{k+1} + \mathscr{C}_{k+1}^T \leq \varepsilon_{k+1,4}\bar{\gamma}(1-\bar{\gamma})G_{k+1}C_{k+1}\mathbb{E}\{x_{k+1}x_{k+1}^T\}C_{k+1}^T G_{k+1}^T + \varepsilon_{k+1,4}^{-1}\bar{\gamma}G_{k+1}\eta_{k+1}\eta_{k+1}^T G_{k+1}^T \tag{6.27}$$

with $\varepsilon_{k+1,3} > 0$ and $\varepsilon_{k+1,4} > 0$ being scalars. Substituting (6.26) and (6.27) into (6.15) yields

$$\begin{aligned} P_{k+1|k+1} \leq & (1+\varepsilon_{k+1,3})[I - (1-\bar{\gamma})G_{k+1}C_{k+1}]P_{k+1|k}[I - (1-\bar{\gamma})G_{k+1}C_{k+1}]^T \\ & + (1+\varepsilon_{k+1,4})\bar{\gamma}(1-\bar{\gamma})G_{k+1}C_{k+1}\mathbb{E}\{x_{k+1}x_{k+1}^T\}C_{k+1}^T G_{k+1}^T \\ & + (1-\bar{\gamma})G_{k+1}R_{k+1}G_{k+1}^T + (1+\varepsilon_{k+1,3}^{-1} + \varepsilon_{k+1,4}^{-1})\bar{\gamma} \\ & \times G_{k+1}\eta_{k+1}\eta_{k+1}^T G_{k+1}^T. \end{aligned} \tag{6.28}$$

Similarly, it follows from the derivation of (6.24) that

$$\begin{aligned} \mathbb{E}\{x_{k+1}x_{k+1}^T\} &= \mathbb{E}\{(e_{k+1|k} + \hat{x}_{k+1|k})(e_{k+1|k} + \hat{x}_{k+1|k})^T\} \\ &= \mathbb{E}\{e_{k+1|k}e_{k+1|k}^T + e_{k+1|k}\hat{x}_{k+1|k}^T + \hat{x}_{k+1|k}e_{k+1|k}^T + \hat{x}_{k+1|k}\hat{x}_{k+1|k}^T\} \\ &\leq (1+\varepsilon_{k+1,5})P_{k+1|k} + (1+\varepsilon_{k+1,5}^{-1})\hat{x}_{k+1|k}\hat{x}_{k+1|k}^T \end{aligned} \tag{6.29}$$

with $\varepsilon_{k+1,5} > 0$ being a scalar. Set

$$\Phi_{k+1|k} = (1+\varepsilon_{k+1,5})P_{k+1|k} + (1+\varepsilon_{k+1,5}^{-1})\hat{x}_{k+1|k}\hat{x}_{k+1|k}^T.$$

Based on (6.28) and (6.29), we have

$$\begin{aligned} P_{k+1|k+1} \leq & (1+\varepsilon_{k+1,3})[I - (1-\bar{\gamma})G_{k+1}C_{k+1}]P_{k+1|k}[I - (1-\bar{\gamma})G_{k+1}C_{k+1}]^T \\ & + (1+\varepsilon_{k+1,4})\bar{\gamma}(1-\bar{\gamma})G_{k+1}C_{k+1}\Phi_{k+1|k}C_{k+1}^T G_{k+1}^T \\ & + (1+\varepsilon_{k+1,3}^{-1} + \varepsilon_{k+1,4}^{-1})\bar{\gamma}\delta^2 G_{k+1}G_{k+1}^T \\ & + (1-\bar{\gamma})G_{k+1}R_{k+1}G_{k+1}^T. \end{aligned} \tag{6.30}$$

Subsequently, by employing the mathematical induction method, it is clear to obtain that

$$P_{k+1|k} \leq \Omega_{k+1|k}, \quad P_{k+1|k+1} \leq \Omega_{k+1|k+1}, \tag{6.31}$$

which finishes the proof of this theorem.

Finally, we are ready to design the specific form of the estimator gain and minimize this upper bound $\Omega_{k+1|k+1}$.

Theorem 6.4 *For positive constants* $\varepsilon_{k+1,3}$ *and* $\varepsilon_{k+1,4}$, *the estimator gain of (6.9)–(6.10) is given by*

$$G_{k+1} = (1+\varepsilon_{k+1,3})(1-\bar{\gamma})\Omega_{k+1|k}C_{k+1}^{T}M_{k+1}^{-1}, \tag{6.32}$$

where

$$\begin{aligned} M_{k+1} = &(1+\varepsilon_{k+1,3})(1-\bar{\gamma})^{2}C_{k+1}\Omega_{k+1|k}C_{k+1}^{T} + (1+\varepsilon_{k+1,4}) \\ &\times\bar{\gamma}(1-\bar{\gamma})C_{k+1}\bar{\Phi}_{k+1|k}C_{k+1}^{T} + (1-\bar{\gamma})R_{k+1} \\ &+(1+\varepsilon_{k+1,3}^{-1}+\varepsilon_{k+1,4}^{-1})\bar{\gamma}\delta^{2}I. \end{aligned} \tag{6.33}$$

With the estimator gain in (6.32), the obtained upper bound $\Omega_{k+1|k+1}$ *of the EEC* $P_{k+1|k+1}$ *in (6.19) can be minimized.*

Proof Rewrite the upper bound $\Omega_{k+1|k+1}$ in (6.19) as follows:

$$\begin{aligned} \Omega_{k+1|k+1} = &(1+\varepsilon_{k+1,3})\Omega_{k+1|k} - (1+\varepsilon_{k+1,3})(1-\bar{\gamma})\Omega_{k+1|k}C_{k+1}^{T}G_{k+1}^{T} \\ &-(1+\varepsilon_{k+1,3})(1-\bar{\gamma})G_{k+1}C_{k+1}\Omega_{k+1|k} + G_{k+1}M_{k+1}G_{k+1}^{T}, \end{aligned} \tag{6.34}$$

where M_{k+1} is defined as in (6.33). Noting $M_{k+1} > 0$ and completing the square with respect to G_{k+1} in (6.34), one has

$$\begin{aligned} \Omega_{k+1|k+1} = &(1+\varepsilon_{k+1,3})\Omega_{k+1|k} - (1+\varepsilon_{k+1,3})^{2}(1-\bar{\gamma})^{2}\Omega_{k+1|k}C_{k+1}^{T}M_{k+1}^{-1}C_{k+1}\Omega_{k+1|k} \\ &+[G_{k+1} - (1+\varepsilon_{k+1,3})(1-\bar{\gamma})\Omega_{k+1|k}C_{k+1}^{T}M_{k+1}^{-1}]M_{k+1} \\ &\times[G_{k+1} - (1+\varepsilon_{k+1,3})(1-\bar{\gamma})\Omega_{k+1|k}C_{k+1}^{T}M_{k+1}^{-1}]^{T}. \end{aligned} \tag{6.35}$$

Now, it is straightforward to see that $\Omega_{k+1|k+1}$ can be minimized by simply choosing

$$G_{k+1} = (1+\varepsilon_{k+1,3})(1-\bar{\gamma})\Omega_{k+1|k}C_{k+1}^{T}M_{k+1}^{-1}.$$

Hence, the proof of this theorem is ended.

Remark 6.2 In this chapter, we have discussed the co-design problem of robust estimator and fault estimator for time-varying systems impacted by RONs and RODAs. In particular, we have established a general model to better reflect the engineering practice and examined the influences from both RONs and RODAs onto the estimation performance. Accordingly, the available information of the RONs and RODAs (e.g. $\bar{\alpha}$, $\bar{\gamma}$ and δ) has been involved in the proposed estimation algorithm. Moreover, some scalars $\varepsilon_{k,1}$, $\varepsilon_{k,2}$, $\varepsilon_{k+1,3}$, $\varepsilon_{k+1,4}$ and $\varepsilon_{k+1,5}$ have been introduced to enhance the feasibility of the developed estimation scheme. During the implementation, the values of those scalars can be selected in prior and adjusted to strengthen the validity of the proposed estimation scheme.

6.3 An Illustrative Example

In this section, a simulation experiment is proposed to testify the correctness of the designed estimation method.

The corresponding system coefficients are given by

$$\breve{A}_k = \begin{bmatrix} 0.6 & 0.3 \\ 0.3 & 0.7 \end{bmatrix}, \quad \breve{F}_k = \begin{bmatrix} 1 \\ 0 \end{bmatrix}, \quad \breve{B}_k = \begin{bmatrix} 0.3 - 0.2\sin(4k) \\ 0.4 \end{bmatrix}, \quad \breve{C}_k = \begin{bmatrix} 1 & 0 \end{bmatrix},$$

and the state vector $\breve{x}_k = \begin{bmatrix} \breve{x}_{1,k} & \breve{x}_{2,k} \end{bmatrix}^T$ with $\breve{x}_{i,k}$ $(i = 1, 2)$ denoting the i-th element of the system state, the covariances of $\omega_k \in \mathbb{R}$ and $\nu_k \in \mathbb{R}$ are 0.08 and 0.06, respectively. The additive fault is selected as follows:

$$f_k = \begin{cases} 0.9, & \text{if } \ k \geq 20 \\ 0, & \text{otherwise.} \end{cases}$$

The nonlinear function $\breve{h}_k(\breve{x}_k)$ is set as

$$\breve{h}_k(\breve{x}_k) = \begin{bmatrix} 0.1\breve{x}_{1,k} - \tanh(0.2\breve{x}_{1,k}) \\ 0.1\breve{x}_{2,k} - \tanh(0.1\breve{x}_{2,k}) \end{bmatrix}$$

and $H = \mathrm{diag}\{0.45, 0.3\}$.

The initial conditions are $x_0 = \begin{bmatrix} 0.1 & 0.1 & 0 \end{bmatrix}^T$ and $\Omega_{0|0} = 5I_3$. Moreover, the other parameters are given by $\bar{\alpha} = 0.9$, $\bar{\gamma} = 0.05$, $\varepsilon_{k,1} = 0.6$, $\varepsilon_{k,2} = 0.5$, $\varepsilon_{k+1,3} = 0.3$, $\varepsilon_{k+1,4} = 0.2$, $\varepsilon_{k+1,5} = 0.6$, $\eta_k = 0.3\sin(2k)$ and $\delta = 0.1$. Via the Theorems 6.3–6.4, the estimator gain can be calculated recursively. Accordingly, the simulation results are given in Figs. 6.1, 6.2 and 6.3. Among the simulation results, it is seen that the unknown system states and the fault can be estimated effectively. This is due to the fact that we have proposed a new scheme when designing the time-varying estimator to compensate the influences from RONs and RODAs onto the estimation performance. On the other hand, for the sake of validating the effect from the deception attack onto the estimation performance, the curves of MSE under different $\bar{\gamma}$ are plotted in Fig. 6.4, which can be observed that the MSE becomes lower along with the smaller $\bar{\gamma}$, that is to say, when $\bar{\gamma}$ decreases, the more real measurement data can be obtained, and then the better estimation performance gets.

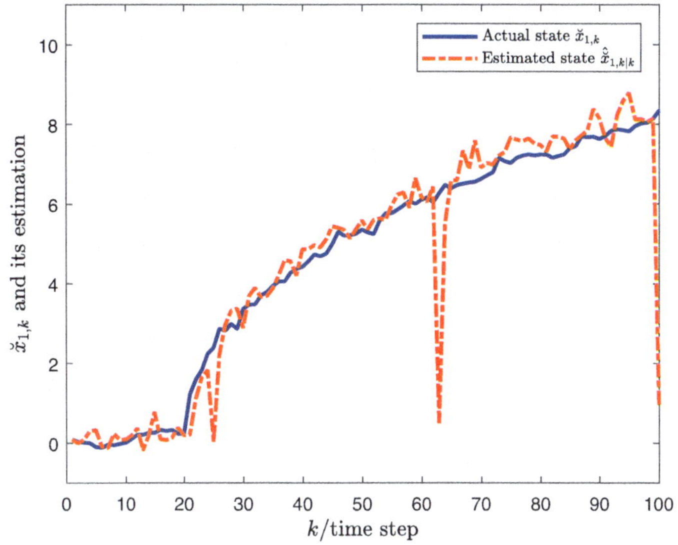

Fig. 6.1 The actual state $\breve{x}_{1,k}$ and its estimation $\hat{\breve{x}}_{1,k|k}$

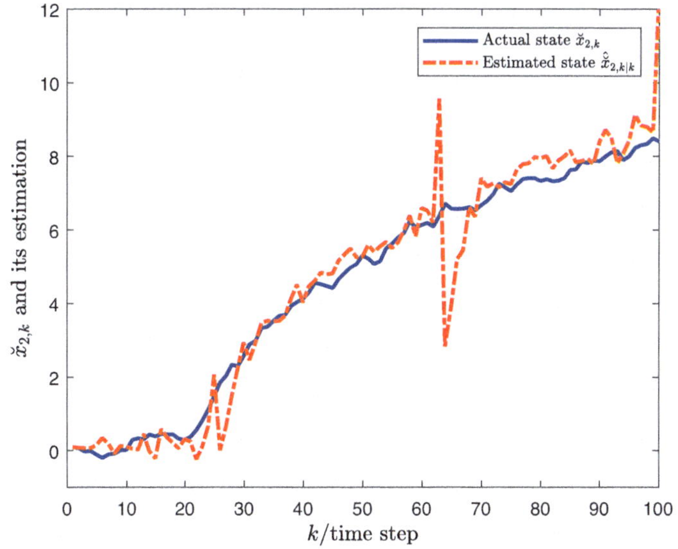

Fig. 6.2 The actual state $\breve{x}_{2,k}$ and its estimation $\hat{\breve{x}}_{2,k|k}$

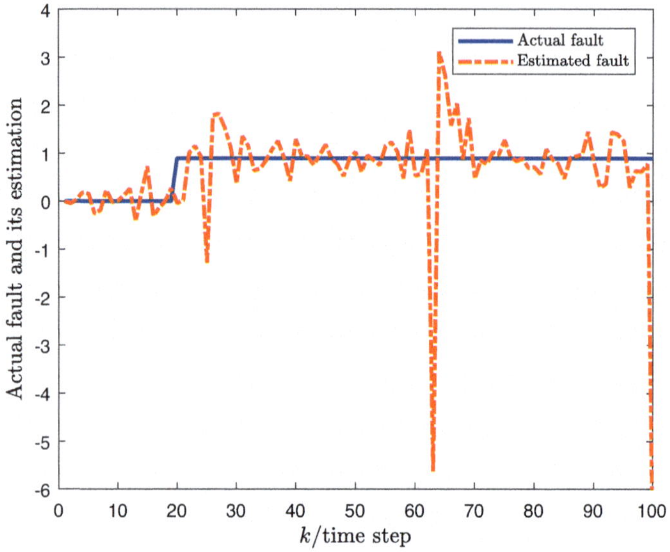

Fig. 6.3 The actual fault f_k and its estimation

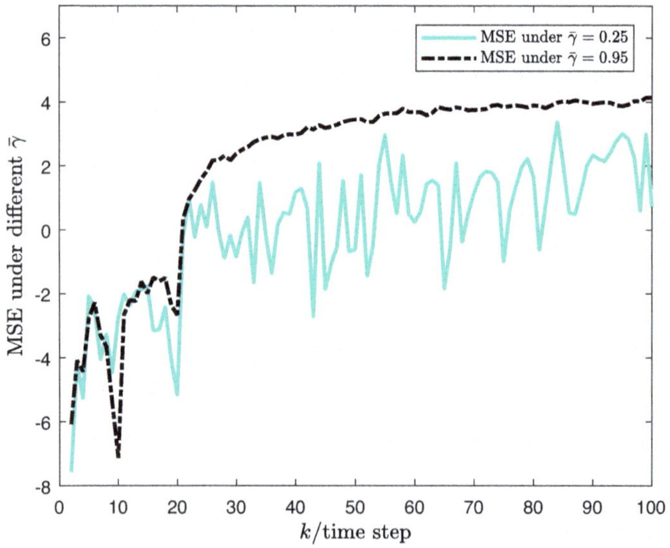

Fig. 6.4 MSE under different $\bar{\gamma}$

6.4 Summary

In this chapter, considerable effort has been made to investigate the co-design problem of the robust estimator and fault estimator for time-varying stochastic systems with RONs and RODAs. Two sets of Bernoulli random variables with given conditional probabilities have been employed to characterize the phenomena of RONs and RODAs, respectively. A novel compensation scheme has been proposed to examine the effects from RONs and RODAs onto the estimation performance. A locally optimal estimation scheme has been developed and the estimator gain has been designed on the basis of the solutions to two matrix equations. Finally, an illustrative example has been adopted to exhibit the validity of the presented fault estimation approach.

Chapter 7
Joint Fault and State Estimation with Packet Dropouts and Randomly Occurring Uncertainties

In this chapter, the fault estimation issue is addressed for time-varying networked systems influenced by ROUs and packet dropouts. Two sets of mutually independent random variables are employed to depict the ROUs and packet dropouts. The chief purpose is to develop a novel fault estimation scheme such that, for ROUs and packet dropouts, an optimized upper bound in relation to the EEC is obtained and the estimator gain with expression form is derived. A novel compensation strategy is proposed to enhance the estimation performance by resorting to the statistical characteristic concerning on the imperfect measurements. Furthermore, the monotonicity of the trace of this upper bound regarding the missing probability is examined. Finally, the effectiveness of the designed estimation scheme is illustrated via a simulation experiment.

7.1 Problem Formulation

Consider the following class of time-varying nonlinear systems:

$$\bar{x}_{k+1} = (\bar{A}_k + \alpha_k \Delta \bar{A}_k)\bar{x}_k + \bar{g}(\bar{x}_k, \xi_k) + \bar{F}_k f_k + \bar{B}_k \varpi_k, \tag{7.1}$$

$$\bar{y}_k = \Xi_k \bar{C}_k \bar{x}_k + \bar{h}(\bar{x}_k, \varsigma_k) + \upsilon_k, \tag{7.2}$$

where $\bar{x}_k \in \mathbb{R}^n$ denotes the system state, $\bar{x}_0$ is the initial state with mean $\breve{\bar{x}}_0$, $\bar{y}_k \in \mathbb{R}^m$ is the measurement signal, ξ_k and ς_k are zero-mean white noises, $f_k \in \mathbb{R}^f$ is the additive fault, $\Xi_k = \mathrm{diag}\{\beta_{1,k}, \beta_{2,k}, \ldots, \beta_{m,k}\}$ characterizes the random link failure case, ϖ_k is the zero-mean process noise with covariance Q_k, and υ_k is the zero-mean measurement noise with covariance $S_k > 0$. Here, $\bar{A}_k$, $\bar{B}_k$, $\bar{C}_k$ and $\bar{F}_k$ are given matrices of suitable dimensions.

J. Hu et al., *Variance-Constrained Filtering for Stochastic Complex Systems*, Intelligent Control and Learning Systems 18,
https://doi.org/10.1007/978-981-96-2637-3_7

The matrix $\Delta\bar{A}_k$ denotes the norm-bounded parameter uncertainty

$$\Delta\bar{A}_k = \bar{H}_k\mathscr{F}_k\bar{N}_k \tag{7.3}$$

with $\bar{H}_k$ and $\bar{N}_k$ being the known real-valued matrices, and $\mathscr{F}_k$ being an unknown matrix with $\mathscr{F}_k\mathscr{F}_k^T \leq I$.

The random variables $\alpha_k \in \mathbb{R}$ and $\beta_{i,k} \in \mathbb{R}$ $(i = 1, 2, \ldots, m)$ are introduced to account for the phenomena of ROUs and packet dropouts respectively and satisfy

$$\begin{aligned} \text{Prob}\{\alpha_k = 1\} = \bar{\alpha}_k,\ \text{Prob}\{\alpha_k = 0\} = 1 - \bar{\alpha}_k, \\ \text{Prob}\left\{\beta_{i,k} = 1\right\} = \bar{\beta}_{i,k},\ \text{Prob}\left\{\beta_{i,k} = 0\right\} = 1 - \bar{\beta}_{i,k}, \end{aligned} \tag{7.4}$$

where $\bar{\alpha}_k, \bar{\beta}_{i,k} \in [0, 1]$ are known constants.

The functions $\bar{g}(\bar{x}_k, \xi_k)$ and $\bar{h}(\bar{x}_k, \varsigma_k)$ denote the stochastic nonlinearities with $\bar{g}(0, \xi_k) = 0, \bar{h}(0, \varsigma_k)$ and possess the following first moment for all $\bar{x}_k$:

$$\mathbb{E}\left\{\begin{bmatrix} \bar{g}(\bar{x}_k, \xi_k) \\ \bar{h}(\bar{x}_k, \varsigma_k) \end{bmatrix} \middle| \bar{x}_k\right\} = 0 \tag{7.5}$$

and the covariances denoted by

$$\mathbb{E}\left\{\begin{bmatrix} \bar{g}(\bar{x}_k, \xi_k) \\ \bar{h}(\bar{x}_k, \varsigma_k) \end{bmatrix}\begin{bmatrix} \bar{g}(\bar{x}_j, \xi_j) \\ \bar{h}(\bar{x}_j, \varsigma_j) \end{bmatrix}^T \middle| \bar{x}_k\right\} = 0,\ \ k \neq j, \tag{7.6}$$

$$\mathbb{E}\left\{\begin{bmatrix} \bar{g}(\bar{x}_k, \xi_k) \\ \bar{h}(\bar{x}_k, \varsigma_k) \end{bmatrix}\begin{bmatrix} \bar{g}(\bar{x}_k, \xi_k) \\ \bar{h}(\bar{x}_k, \varsigma_k) \end{bmatrix}^T \middle| \bar{x}_k\right\} = \text{diag}\{\bar{\Gamma}_{1,k}\bar{x}_k^T\bar{\Omega}_{1,k}\bar{x}_k, \bar{\Gamma}_{2,k}\bar{x}_k^T\bar{\Omega}_{2,k}\bar{x}_k\}, \tag{7.7}$$

where $\bar{\Gamma}_{i,k}$ and $\bar{\Omega}_{i,k}$ $(i = 1, 2)$ are known matrices with proper dimensions. Throughout the chapter, we assume that α_k, $\beta_{i,k}$, ξ_k, ς_k, ϖ_k, υ_k and $\bar{x}_0$ are mutually independent.

Remark 7.1 In this chapter, the ROUs and packet dropouts are considered in a unified framework when designing the fault estimation algorithm, in which the random variables α_k and $\beta_{i,k}$ are utilized to describe them. In particular, the statistical information concerning on α_k and $\beta_{i,k}$ can be obtained through statistical tests.

Remark 7.2 As given in Yaz and Yaz (2012), Hu et al. (2012), the stochastic nonlinearities in (7.1) and (7.2) satisfying (7.5)–(7.7) could cover many nonlinear functions including: (1) the state-dependent multiplicative noises; (2) a stochastic sequence depending on the sign of a state nonlinear function; and (3) the random vectors depending on the norm of state. Particularly, the stochastic nonlinearity $\bar{h}(\bar{x}_k, \varsigma_k)$ in (7.2) occurs in the process of the signal transmission, which can be regarded as the internal nonlinear disturbance (e.g. state-dependent noise) and is independent on the sensor failures. Therefore, it is valuable to develop the current investigation.

To facilitate the subsequent derivation, set $x_k = \begin{bmatrix} \bar{x}_k^T & f_k^T \end{bmatrix}^T$. Then, the following compact form can be obtained:

$$x_{k+1} = A_k x_k + \alpha_k \Delta A_k x_k + g(x_k, \xi_k) + B_k \varpi_k, \tag{7.8}$$
$$y_k = \Xi_k C_k x_k + h(x_k, \varsigma_k) + \upsilon_k, \tag{7.9}$$

where

$$A_k = \begin{bmatrix} \bar{A}_k & \bar{F}_k \\ 0 & I \end{bmatrix}, \quad \Delta A_k = \begin{bmatrix} \Delta \bar{A}_k & 0 \\ 0 & 0 \end{bmatrix}, \quad g(x_k, \xi_k) = \begin{bmatrix} \bar{g}(\bar{x}_k, \xi_k) \\ 0 \end{bmatrix},$$
$$B_k = \begin{bmatrix} \bar{B}_k^T & 0^T \end{bmatrix}^T, \quad C_k = \begin{bmatrix} \bar{C}_k & 0 \end{bmatrix}, \quad h(x_k, \varsigma_k) = \bar{h}(\bar{x}_k, \varsigma_k). \tag{7.10}$$

According to (7.5)–(7.7), it is easy to obtain

$$\mathbb{E}\left\{ \begin{bmatrix} g(x_k, \xi_k) \\ h(x_k, \varsigma_k) \end{bmatrix} \middle| x_k \right\} = 0 \tag{7.11}$$

and

$$\mathbb{E}\left\{ \begin{bmatrix} g(x_k, \xi_k) \\ h(x_k, \varsigma_k) \end{bmatrix} \begin{bmatrix} g(x_j, \xi_j) \\ h(x_j, \varsigma_j) \end{bmatrix}^T \middle| x_k \right\} = 0, \quad k \neq j, \tag{7.12}$$
$$\mathbb{E}\left\{ \begin{bmatrix} g(x_k, \xi_k) \\ h(x_k, \varsigma_k) \end{bmatrix} \begin{bmatrix} g(x_k, \xi_k) \\ h(x_k, \varsigma_k) \end{bmatrix}^T \middle| x_k \right\} = \mathrm{diag}\{\Gamma_{1,k} x_k^T \Omega_{1,k} x_k, \bar{\Gamma}_{2,k} x_k^T \Omega_{2,k} x_k\} \tag{7.13}$$

with

$$\Gamma_{1,k} = \mathrm{diag}\{\bar{\Gamma}_{1,k}, 0\}, \quad \Omega_{1,k} = \mathrm{diag}\{\bar{\Omega}_{1,k}, 0\}, \quad \Omega_{2,k} = \mathrm{diag}\{\bar{\Omega}_{2,k}, 0\}.$$

In this chapter, we design the following time-varying estimator:

$$\hat{x}_{k+1|k} = A_k \hat{x}_{k|k}, \tag{7.14}$$
$$\hat{x}_{k+1|k+1} = \hat{x}_{k+1|k} + G_{k+1}(y_{k+1} - \bar{\Xi}_{k+1} C_{k+1} \hat{x}_{k+1|k}), \tag{7.15}$$

where $\hat{x}_{k|k}$ is the state estimate, $\hat{x}_{k+1|k}$ is the one-step prediction, G_{k+1} is the estimator gain, and $\bar{\Xi}_{k+1} = \mathrm{diag}\{\bar{\beta}_{1,k+1}, \bar{\beta}_{2,k+1}, \ldots, \bar{\beta}_{m,k+1}\}$.

The major purpose of this chapter is to construct a novel estimator (7.14) and (7.15) that can provide an algorithm for estimating the system states and fault signal jointly. Noting the relationship between the original system and augmented system, we can obtain the fault estimation $\hat{f}_k = \begin{bmatrix} 0 & I_f \end{bmatrix} \hat{x}_{k|k}$. In addition, the state estimation of the original system can be expressed as $\hat{\bar{x}}_k = \begin{bmatrix} I_n & 0 \end{bmatrix} \hat{x}_{k|k}$.

7.2 Design of Fault Estimator Gain

In this section, the explicit forms of the one-step prediction error covariance and EEC are firstly derived. In the subsequent derivations, an upper bound of the EEC is solved by the matrix analysis method and then the estimator gain is constructed by realizing the minimization of the trace of this upper bound.

Let the one-step prediction error be $e_{k+1|k} = x_{k+1} - \hat{x}_{k+1|k}$ and the estimation error be $e_{k+1|k+1} = x_{k+1} - \hat{x}_{k+1|k+1}$. It is clear that

$$
\begin{aligned}
e_{k+1|k} &= A_k e_{k|k} + \alpha_k \Delta A_k x_k + g(x_k, \xi_k) + B_k \varpi_k, &\quad (7.16)\\
e_{k+1|k+1} &= (I - G_{k+1}\bar{\Xi}_{k+1}C_{k+1})e_{k+1|k} - G_{k+1}\tilde{\Xi}_{k+1}C_{k+1}x_{k+1}\\
&\quad - G_{k+1}h(x_{k+1}, \varsigma_{k+1}) - G_{k+1}\upsilon_{k+1}, &\quad (7.17)
\end{aligned}
$$

where $\tilde{\Xi}_{k+1} = \Xi_{k+1} - \bar{\Xi}_{k+1}$.

In terms of (7.16) and (7.17), the one-step prediction error covariance and EEC are directly derived, which are exhibited in the following theorems.

Theorem 7.1 *The one-step prediction error covariance $P_{k+1|k}$ obeys the following recursion:*

$$
\begin{aligned}
P_{k+1|k} = {} & A_k P_{k|k} A_k^T + \bar{\alpha}_k A_k \mathbb{E}\{e_{k|k} x_k^T\} \Delta A_k^T + \bar{\alpha}_k \Delta A_k \mathbb{E}\{x_k e_{k|k}^T\} A_k^T\\
& + \bar{\alpha}_k \Delta A_k \mathbb{E}\{x_k x_k^T\} \Delta A_k^T + \mathbb{E}\{g(x_k, \xi_k) g^T(x_k, \xi_k)\} + B_k Q_k B_k^T, \quad (7.18)
\end{aligned}
$$

where $P_{k|k} = \mathbb{E}\{e_{k|k} e_{k|k}^T\}$ is the EEC.

Proof According to the definition of one-step prediction error covariance, we obtain

$$
\begin{aligned}
P_{k+1|k} &= \mathbb{E}\{e_{k+1|k} e_{k+1|k}^T\}\\
&= A_k P_{k|k} A_k^T + A_k \mathbb{E}\{\alpha_k e_{k|k} x_k^T\} \Delta A_k^T + \Delta A_k \mathbb{E}\{\alpha_k x_k e_{k|k}^T\} A_k^T\\
&\quad + A_k \mathbb{E}\{e_{k|k} g^T(x_k, \xi_k)\} + \mathbb{E}\{g(x_k, \xi_k) e_{k|k}^T\} A_k^T + A_k \mathbb{E}\{e_{k|k} \varpi_k^T\} B_k^T\\
&\quad + B_k \mathbb{E}\{\varpi_k e_{k|k}^T\} A_k^T + \Delta A_k \mathbb{E}\{\alpha_k^2 x_k x_k^T\} \Delta A_k^T + \Delta A_k \mathbb{E}\{\alpha_k x_k g^T(x_k, \xi_k)\}\\
&\quad + \mathbb{E}\{\alpha_k g(x_k, \xi_k) x_k^T\} \Delta A_k^T + \Delta A_k \mathbb{E}\{\alpha_k x_k \varpi_k^T\} B_k^T + B_k \mathbb{E}\{\alpha_k \varpi_k x_k^T\} \Delta A_k^T\\
&\quad + \mathbb{E}\{g(x_k, \xi_k) g^T(x_k, \xi_k)\} + \mathbb{E}\{g(x_k, \xi_k) \varpi_k^T\} B_k^T + B_k \mathbb{E}\{\varpi_k g^T(x_k, \xi_k)\}\\
&\quad + B_k Q_k B_k^T. \qquad (7.19)
\end{aligned}
$$

Note that α_k, ξ_k and ϖ_k are mutually independent, then it is not difficult to testify that (7.18) is true.

Theorem 7.2 *The EEC* $P_{k+1|k+1}$ *is calculated as*

$$\begin{aligned}P_{k+1|k+1} = &(I - G_{k+1}\bar{\Xi}_{k+1}C_{k+1})P_{k+1|k}(I - G_{k+1}\bar{\Xi}_{k+1}C_{k+1})^T \\ &+G_{k+1}\mathbb{E}\{\tilde{\Xi}_{k+1}C_{k+1}x_{k+1}x_{k+1}^T C_{k+1}^T \tilde{\Xi}_{k+1}^T\}G_{k+1}^T \\ &+G_{k+1}\mathbb{E}\{h(x_{k+1}, \varsigma_{k+1})h^T(x_{k+1}, \varsigma_{k+1})\}G_{k+1}^T \\ &+G_{k+1}S_{k+1}G_{k+1}^T. \end{aligned} \tag{7.20}$$

Proof It can be seen from the definition of the EEC that

$$\begin{aligned}P_{k+1|k+1} &= \mathbb{E}\{e_{k+1|k+1}e_{k+1|k+1}^T\} \\ &= (I - G_{k+1}\bar{\Xi}_{k+1}C_{k+1})P_{k+1|k}(I - G_{k+1}\bar{\Xi}_{k+1}C_{k+1})^T \\ &\quad -\mathcal{A}_{k+1} - \mathcal{A}_{k+1}^T - \mathcal{B}_{k+1} - \mathcal{B}_{k+1}^T - \mathcal{C}_{k+1} - \mathcal{C}_{k+1}^T \\ &\quad +G_{k+1}\mathbb{E}\{\tilde{\Xi}_{k+1}C_{k+1}x_{k+1}x_{k+1}^T C_{k+1}^T \tilde{\Xi}_{k+1}^T\}G_{k+1}^T + \mathcal{D}_{k+1} + \mathcal{D}_{k+1}^T \\ &\quad +\mathcal{E}_{k+1} + \mathcal{E}_{k+1}^T + G_{k+1}\mathbb{E}\{h(x_{k+1}, \varsigma_{k+1})h^T(x_{k+1}, \varsigma_{k+1})\}G_{k+1}^T \\ &\quad +\mathcal{F}_{k+1} + \mathcal{F}_{k+1}^T + G_{k+1}\mathbb{E}\{\upsilon_{k+1}\upsilon_{k+1}^T\}G_{k+1}^T, \end{aligned} \tag{7.21}$$

where

$$\begin{aligned}\mathcal{A}_{k+1} &= (I - G_{k+1}\bar{\Xi}_{k+1}C_{k+1})\mathbb{E}\{e_{k+1|k}x_{k+1}^T C_{k+1}^T \tilde{\Xi}_{k+1}^T\}G_{k+1}^T, \\ \mathcal{B}_{k+1} &= (I - G_{k+1}\bar{\Xi}_{k+1}C_{k+1})\mathbb{E}\{e_{k+1|k}h^T(x_{k+1}, \varsigma_{k+1})\}G_{k+1}^T, \\ \mathcal{C}_{k+1} &= (I - G_{k+1}\bar{\Xi}_{k+1}C_{k+1})\mathbb{E}\{e_{k+1|k}\upsilon_{k+1}^T\}G_{k+1}^T, \\ \mathcal{D}_{k+1} &= G_{k+1}\mathbb{E}\{\tilde{\Xi}_{k+1}C_{k+1}x_{k+1}h^T(x_{k+1}, \varsigma_{k+1})\}G_{k+1}^T, \\ \mathcal{E}_{k+1} &= G_{k+1}\mathbb{E}\{\tilde{\Xi}_{k+1}C_{k+1}x_{k+1}\upsilon_{k+1}^T\}G_{k+1}^T, \\ \mathcal{F}_{k+1} &= G_{k+1}\mathbb{E}\{h(x_{k+1}, \varsigma_{k+1})\upsilon_{k+1}^T\}G_{k+1}^T. \end{aligned}$$

Notice that ς_{k+1}, υ_{k+1} and $\beta_{i,k+1}$ $(i = 1, 2, \ldots, m)$ are mutually independent, together with (7.11), then it is obvious to find that $\mathcal{A}_{k+1}$, $\mathcal{B}_{k+1}$, $\mathcal{C}_{k+1}$, $\mathcal{D}_{k+1}$, $\mathcal{E}_{k+1}$ and $\mathcal{F}_{k+1}$ are equal to zero matrices. Thus, (7.20) can be calculated immediately.

It should be highlighted that there exist some uncertain terms in the related error covariances. To address such an issue, we will calculate a recursion of the upper bound of EEC in the following theorem.

Theorem 7.3 *Given* $\varepsilon_{1,k} > 0$, $\varepsilon_{2,k} > 0$ *and* $\varepsilon_{3,k+1} > 0$, *if the following recursive matrix equations under the initial condition* $\Omega_{0|0} = P_{0|0}$ *have positive definite solutions:*

$$\begin{aligned}\Omega_{k+1|k} = &(1 + \varepsilon_{1,k})A_k\Omega_{k|k}A_k^T + (1 + \varepsilon_{1,k}^{-1}\bar{\alpha}_k)\bar{\alpha}_k \\ &\times\text{tr}\big(N_k[(1 + \varepsilon_{2,k})\Omega_{k|k} + (1 + \varepsilon_{2,k}^{-1})\hat{x}_{k|k}\hat{x}_{k|k}^T]N_k^T\big)H_kH_k^T \\ &+\text{tr}\big([(1 + \varepsilon_{2,k})\Omega_{k|k} + (1 + \varepsilon_{2,k}^{-1})\hat{x}_{k|k}\hat{x}_{k|k}^T]\Omega_{1,k}\big)\Gamma_{1,k} \\ &+B_kQ_kB_k^T \end{aligned} \tag{7.22}$$

and

$$\begin{aligned}\Omega_{k+1|k+1} = &(I - G_{k+1}\bar{\Xi}_{k+1}C_{k+1})\Omega_{k+1|k}(I - G_{k+1}\bar{\Xi}_{k+1}C_{k+1})^T \\ &+G_{k+1}\big[\hat{\Xi}_{k+1} \circ (C_{k+1}\Lambda_{k+1}C_{k+1}^T)\big]G_{k+1}^T \\ &+\mathrm{tr}(\Lambda_{k+1}\Omega_{2,k+1})G_{k+1}\bar{\Gamma}_{2,k+1}G_{k+1}^T \\ &+G_{k+1}S_{k+1}G_{k+1}^T,\end{aligned} \tag{7.23}$$

where

$$\begin{aligned} H_k &= \big[\,\bar{H}_k^T\ 0^T\,\big]^T,\quad N_k = \big[\,\bar{N}_k\ 0\,\big], \\ \hat{\Xi}_{k+1} &= \mathrm{diag}\big\{\bar{\beta}_{1,k+1}(1-\bar{\beta}_{1,k+1}),\bar{\beta}_{2,k+1}(1-\bar{\beta}_{2,k+1}),\ldots,\bar{\beta}_{m,k+1}(1-\bar{\beta}_{m,k+1})\big\}, \\ \Lambda_{k+1} &= (1+\varepsilon_{3,k+1})\Omega_{k+1|k} + (1+\varepsilon_{3,k+1}^{-1})\hat{x}_{k+1|k}\hat{x}_{k+1|k}^T, \end{aligned} \tag{7.24}$$

then, one gets $P_{k+1|k+1} \le \Omega_{k+1|k+1}$.

Proof By resorting to the matrix analysis technique, we take two steps to derive the explicit forms of the upper bound of the EEC $P_{k+1|k+1}$.

Step 1: The unknown terms of the RHS of (7.18) will be tackled. In terms of Lemma 2.2, one has

$$\begin{aligned}&\bar{\alpha}_k A_k\mathbb{E}\{e_{k|k}x_k^T\}\Delta A_k^T + \bar{\alpha}_k\Delta A_k\mathbb{E}\{x_k e_{k|k}^T\}A_k^T \\ &\le \varepsilon_{1,k}A_kP_{k|k}A_k^T + \varepsilon_{1,k}^{-1}\bar{\alpha}_k^2\Delta A_k\mathbb{E}\{x_kx_k^T\}\Delta A_k^T\end{aligned} \tag{7.25}$$

with $\varepsilon_{1,k} > 0$ being a constant. It follows form (7.13) that

$$\begin{aligned}\mathbb{E}\{g(x_k,\xi_k)g^T(x_k,\xi_k)\} =&\mathbb{E}\{\Gamma_{1,k}x_k^T\Omega_{1,k}x_k\} \\ =&\mathrm{tr}\big(\mathbb{E}\{x_kx_k^T\}\Omega_{1,k}\big)\Gamma_{1,k}.\end{aligned} \tag{7.26}$$

Next, it follows from Lemma 2.2 again that

$$\begin{aligned}\mathbb{E}\{x_kx_k^T\} &= \mathbb{E}\{(e_{k|k}+\hat{x}_{k|k})(e_{k|k}+\hat{x}_{k|k})^T\} \\ &= \mathbb{E}\{e_{k|k}e_{k|k}^T + e_{k|k}\hat{x}_{k|k}^T + \hat{x}_{k|k}e_{k|k}^T + \hat{x}_{k|k}\hat{x}_{k|k}^T\} \\ &\le (1+\varepsilon_{2,k})P_{k|k} + (1+\varepsilon_{2,k}^{-1})\hat{x}_{k|k}\hat{x}_{k|k}^T,\end{aligned} \tag{7.27}$$

where $\varepsilon_{2,k} > 0$ is a scalar. Then, from the notation of matrix ΔA_k and (7.27), the term $\Delta A_k\mathbb{E}\{x_kx_k^T\}\Delta A_k^T$ can be calculated as

$$\begin{aligned}\Delta A_k\mathbb{E}\{x_kx_k^T\}\Delta A_k^T &\le H_k\mathscr{F}_kN_k\big[(1+\varepsilon_{2,k})P_{k|k} + (1+\varepsilon_{2,k}^{-1})\hat{x}_{k|k}\hat{x}_{k|k}^T\big]N_k^T\mathscr{F}_k^TH_k^T \\ &\le \mathrm{tr}\big(N_k\big[(1+\varepsilon_{2,k})P_{k|k} + (1+\varepsilon_{2,k}^{-1})\hat{x}_{k|k}\hat{x}_{k|k}^T\big]N_k^T\big)H_kH_k^T,\end{aligned} \tag{7.28}$$

where H_k and N_k are defined in (7.24). Substituting (7.25)–(7.28) into (7.18) leads to

$$
\begin{aligned}
P_{k+1|k} \leq\ & (1+\varepsilon_{1,k})A_k P_{k|k} A_k^T + (1+\varepsilon_{1,k}^{-1}\bar{\alpha}_k)\bar{\alpha}_k \\
& \times \mathrm{tr}\big(N_k[(1+\varepsilon_{2,k})P_{k|k} + (1+\varepsilon_{2,k}^{-1})\hat{x}_{k|k}\hat{x}_{k|k}^T]N_k^T\big)H_k H_k^T \\
& + \mathrm{tr}\big([(1+\varepsilon_{2,k})P_{k|k} + (1+\varepsilon_{2,k}^{-1})\hat{x}_{k|k}\hat{x}_{k|k}^T]\Omega_{1,k}\big)\Gamma_{1,k} \\
& + B_k Q_k B_k^T.
\end{aligned} \tag{7.29}
$$

Step 2: We devote to addressing the unknown terms of the RHS of (7.20). In view of Lemma 2.4, the term $\mathbb{E}\{\tilde{\Xi}_{k+1}C_{k+1}x_{k+1}x_{k+1}^T C_{k+1}^T \tilde{\Xi}_{k+1}^T\}$ can be calculated

$$
\mathbb{E}\{\tilde{\Xi}_{k+1}C_{k+1}x_{k+1}x_{k+1}^T C_{k+1}^T \tilde{\Xi}_{k+1}^T\} = \hat{\Xi}_{k+1} \circ \big(C_{k+1}\mathbb{E}\{x_{k+1}x_{k+1}^T\}C_{k+1}^T\big), \tag{7.30}
$$

where $\hat{\Xi}_{k+1}$ is defined as in (7.24). From (7.13), we have

$$
\mathbb{E}\{h(x_{k+1}, \varsigma_{k+1})h^T(x_{k+1}, \varsigma_{k+1})\} = \mathrm{tr}\big(\mathbb{E}\{x_{k+1}x_{k+1}^T\}\Omega_{2,k+1}\big)\bar{\Gamma}_{2,k+1}. \tag{7.31}
$$

Along the same line with the derivation of (7.27), we have

$$
\begin{aligned}
\mathbb{E}\{x_{k+1}x_{k+1}^T\} &= \mathbb{E}\{(e_{k+1|k} + \hat{x}_{k+1|k})(e_{k+1|k} + \hat{x}_{k+1|k})^T\} \\
&\leq (1+\varepsilon_{3,k+1})P_{k+1|k} + (1+\varepsilon_{3,k+1}^{-1})\hat{x}_{k+1|k}\hat{x}_{k+1|k}^T \\
&:= \bar{\Lambda}_{k+1},
\end{aligned} \tag{7.32}
$$

where $\varepsilon_{3,k+1} > 0$ is a scalar. Substituting (7.30)–(7.32) into (7.20) yields

$$
\begin{aligned}
P_{k+1|k+1} \leq\ & (I - G_{k+1}\bar{\Xi}_{k+1}C_{k+1})P_{k+1|k}(I - G_{k+1}\bar{\Xi}_{k+1}C_{k+1})^T \\
& + G_{k+1}\big[\hat{\Xi}_{k+1} \circ (C_{k+1}\bar{\Lambda}_{k+1}C_{k+1}^T)\big]G_{k+1}^T \\
& + \mathrm{tr}(\bar{\Lambda}_{k+1}\Omega_{2,k+1})G_{k+1}\bar{\Gamma}_{2,k+1}G_{k+1}^T \\
& + G_{k+1}S_{k+1}G_{k+1}^T.
\end{aligned} \tag{7.33}
$$

Together with (7.29), (7.33) and the mathematical induction method, it is obvious to prove that $P_{k+1|k+1} \leq \Omega_{k+1|k+1}$, which finishes the proof of this theorem.

Next, we aim to derive an optimal estimator gain and calculate the minimized upper bound of the EEC.

Theorem 7.4 *Assume that the estimator gain is determined by*

$$
G_{k+1} = \Omega_{k+1|k}C_{k+1}^T \bar{\Xi}_{k+1}^T \Phi_{k+1}^{-1} \tag{7.34}
$$

with

$$
\begin{aligned}
\Phi_{k+1} =\ & \bar{\Xi}_{k+1}C_{k+1}\Omega_{k+1|k}C_{k+1}^T\bar{\Xi}_{k+1}^T + \hat{\Xi}_{k+1} \circ (C_{k+1}\Lambda_{k+1}C_{k+1}^T) \\
& + \mathrm{tr}(\Lambda_{k+1}\Omega_{2,k+1})\bar{\Gamma}_{2,k+1} + S_{k+1}.
\end{aligned} \tag{7.35}
$$

Then, the minimal upper bound $\Omega_{k+1|k+1}$ is calculated as

$$\Omega_{k+1|k+1} = \Omega_{k+1|k} - \Omega_{k+1|k}^T C_{k+1}^T \bar{\Xi}_{k+1}^T \Phi_{k+1}^{-1} \bar{\Xi}_{k+1} C_{k+1} \Omega_{k+1|k}. \tag{7.36}$$

Proof By completing the square regarding G_{k+1} in (7.23), $\Omega_{k+1|k+1}$ is rewritten as:

$$\begin{aligned}\Omega_{k+1|k+1} = & \Omega_{k+1|k} - \Omega_{k+1|k}^T C_{k+1}^T \bar{\Xi}_{k+1}^T \Phi_{k+1}^{-1} \bar{\Xi}_{k+1} C_{k+1} \Omega_{k+1|k} \\ & + (G_{k+1} - \Omega_{k+1|k}^T C_{k+1}^T \bar{\Xi}_{k+1}^T \Phi_{k+1}^{-1}) \Phi_{k+1} \\ & \times (G_{k+1} - \Omega_{k+1|k}^T C_{k+1}^T \bar{\Xi}_{k+1}^T \Phi_{k+1}^{-1})^T. \end{aligned} \tag{7.37}$$

By setting $G_{k+1} = \Omega_{k+1|k} C_{k+1}^T \bar{\Xi}_{k+1}^T \Phi_{k+1}^{-1}$, it is easy to see that the upper bound $\Omega_{k+1|k+1}$ is minimized, which finishes the proof of this theorem.

Remark 7.3 Note that Lemma 2.2 has been utilized and some positive scalars (such as $\varepsilon_{1,k}$, $\varepsilon_{2,k}$ and $\varepsilon_{3,k+1}$) have been involved in the main results, which may bring certain conservatism to the estimation algorithm design. To overcome this difficulty, it is possible to develop a searching method for the scalars $\varepsilon_{1,k}$, $\varepsilon_{2,k}$ and $\varepsilon_{3,k+1}$ in the implementation process to decrease the conservatism of the proposed algorithm. It is crucial to note that the chief challenge of this chapter results from that, for all ROUs, stochastic nonlinearities and packet dropouts, the traditional estimation algorithms cannot maintain satisfactory performance. Therefore, a new fault estimation method is proposed by introducing a compensation scheme.

Based on Theorems 7.3 and 7.4, the recursive fault estimation (RFE) algorithm is given as follows:

RFE Algorithm:
Step 1 : Set $k = 0$, select $\varepsilon_{1,k}$, $\varepsilon_{2,k}$, $\varepsilon_{3,k+1}$ and $\Omega_{0\|0}$.
Step 2 : Calculate the matrix $\Omega_{k+1\|k}$ by (7.22).
Step 3 : Compute the matrices Λ_{k+1} and $\hat{\Xi}_{k+1}$ by (7.24).
Calculate Φ_{k+1} by (7.35), G_{k+1} by (7.34) and $\Omega_{k+1\|k+1}$ by (7.23).
Step 4 : Set $k = k + 1$, and go to *Step* 2.

7.3 Performance Analysis of Estimation Algorithm

In this section, we are dedicated to examining the effect from the packet dropouts onto the fault estimation performance. In other words, the relationship between the trace of the minimal upper bound and the occurrence probability of the packet dropouts will be revealed. Before proceeding, let $\bar{\beta}_{i,k} = \bar{\beta}_k$, i.e., all $\beta_{i,k}$ are with same statistical properties. Then, it is obvious to know that $\bar{\Xi}_k = \bar{\beta}_k I$ and $\hat{\Xi}_k = \bar{\beta}_k(1 - \bar{\beta}_k)I$.

Next, the relationship between the trace of the minimal upper bound and the missing measurements is adequately discussed.

Theorem 7.5 $\mathrm{tr}(\Omega_{k+1|k+1})$ *is non-decreasing when* $\bar{\beta}_{k+1}$ *decreases.*

Proof It can be derived from (7.36) that

$$\begin{aligned}
\Omega_{k+1|k+1} &= \Omega_{k+1|k} - \Omega^T_{k+1|k} C^T_{k+1} \bar{\Xi}^T_{k+1} \Phi^{-1}_{k+1} \bar{\Xi}_{k+1} C_{k+1} \Omega_{k+1|k} \\
&= \Omega_{k+1|k} - \bar{\beta}^2_{k+1} \Omega^T_{k+1|k} C^T_{k+1} \big[\bar{\beta}^2_{k+1} C_{k+1} \Omega_{k+1|k} C^T_{k+1} \\
&\quad + \bar{\beta}_{k+1}(1-\bar{\beta}_{k+1}) I \circ (C_{k+1} \Lambda_{k+1} C^T_{k+1}) \\
&\quad + \mathrm{tr}(\Lambda_{k+1} \Omega_{2,k+1}) \bar{\Gamma}_{2,k+1} + S_{k+1}\big]^{-1} C_{k+1} \Omega_{k+1|k}.
\end{aligned} \tag{7.38}$$

Subsequently, based on Lemma 2.3, taking the partial derivative of $\mathrm{tr}(\Omega_{k+1|k+1})$ in relation to $\bar{\beta}_{k+1}$, one has

$$\begin{aligned}
&\frac{\partial \mathrm{tr}(\Omega_{k+1|k+1})}{\partial \bar{\beta}_{k+1}} \\
&= \frac{\partial \mathrm{tr}}{\partial \bar{\beta}_{k+1}} \bigg(\Omega_{k+1|k} - \bar{\beta}^2_{k+1} \Omega^T_{k+1|k} C^T_{k+1} \big[\bar{\beta}^2_{k+1} C_{k+1} \Omega_{k+1|k} C^T_{k+1} \\
&\quad + \bar{\beta}_{k+1}(1-\bar{\beta}_{k+1}) I \circ (C_{k+1} \Lambda_{k+1} C^T_{k+1}) + \mathrm{tr}(\Lambda_{k+1} \Omega_{2,k+1}) \bar{\Gamma}_{2,k+1} \\
&\quad + S_{k+1}\big]^{-1} C_{k+1} \Omega_{k+1|k} \bigg) \\
&= \mathrm{tr}\bigg(-2\bar{\beta}_{k+1} \Omega^T_{k+1|k} C^T_{k+1} \big[\bar{\beta}^2_{k+1} C_{k+1} \Omega_{k+1|k} C^T_{k+1} + \bar{\beta}_{k+1}(1-\bar{\beta}_{k+1}) I \\
&\quad \circ (C_{k+1} \Lambda_{k+1} C^T_{k+1}) + \mathrm{tr}(\Lambda_{k+1} \Omega_{2,k+1}) \bar{\Gamma}_{2,k+1} + S_{k+1}\big]^{-1} C_{k+1} \Omega_{k+1|k} \\
&\quad + \bar{\beta}^2_{k+1} \Omega^T_{k+1|k} C^T_{k+1} \big[\bar{\beta}^2_{k+1} C_{k+1} \Omega_{k+1|k} C^T_{k+1} \\
&\quad + \bar{\beta}_{k+1}(1-\bar{\beta}_{k+1}) I \circ (C_{k+1} \Lambda_{k+1} C^T_{k+1}) \\
&\quad + \mathrm{tr}(\Lambda_{k+1} \Omega_{2,k+1}) \bar{\Gamma}_{2,k+1} + S_{k+1}\big]^{-1} \big[2\bar{\beta}_{k+1} C_{k+1} \Omega_{k+1|k} C^T_{k+1} \\
&\quad + (1-2\bar{\beta}_{k+1}) I \circ (C_{k+1} \Lambda_{k+1} C^T_{k+1})\big] \big[\bar{\beta}^2_{k+1} C_{k+1} \Omega_{k+1|k} C^T_{k+1} \\
&\quad + \bar{\beta}_{k+1}(1-\bar{\beta}_{k+1}) I \circ (C_{k+1} \Lambda_{k+1} C^T_{k+1}) + \mathrm{tr}(\Lambda_{k+1} \Omega_{2,k+1}) \\
&\quad \times \bar{\Gamma}_{2,k+1} + S_{k+1}\big]^{-1} C_{k+1} \Omega_{k+1|k} \bigg) \\
&\le \mathrm{tr}\bigg(-2\bar{\beta}_{k+1} \Omega^T_{k+1|k} C^T_{k+1} \big[\bar{\beta}^2_{k+1} C_{k+1} \Omega_{k+1|k} C^T_{k+1} + \bar{\beta}_{k+1}(1-\bar{\beta}_{k+1}) I \\
&\quad \circ (C_{k+1} \Lambda_{k+1} C^T_{k+1}) + \mathrm{tr}(\Lambda_{k+1} \Omega_{2,k+1}) \bar{\Gamma}_{2,k+1} + S_{k+1}\big]^{-1} \\
&\quad \times C_{k+1} \Omega_{k+1|k} + \Omega^T_{k+1|k} C^T_{k+1} \big[\bar{\beta}^2_{k+1} C_{k+1} \Omega_{k+1|k} C^T_{k+1} + \bar{\beta}_{k+1}(1-\bar{\beta}_{k+1}) I \\
&\quad \circ (C_{k+1} \Lambda_{k+1} C^T_{k+1}) + \mathrm{tr}(\Lambda_{k+1} \Omega_{2,k+1}) \bar{\Gamma}_{2,k+1} + S_{k+1}\big]^{-1} \\
&\quad \times \Big[2\bar{\beta}_{k+1} \big[\bar{\beta}^2_{k+1} C_{k+1} \Omega_{k+1|k} C^T_{k+1} + \bar{\beta}_{k+1}(1-\bar{\beta}_{k+1}) I \\
&\quad \circ (C_{k+1} \Lambda_{k+1} C^T_{k+1}) + \mathrm{tr}(\Lambda_{k+1} \Omega_{2,k+1}) \bar{\Gamma}_{2,k+1} + S_{k+1}\big]
\end{aligned}$$

$$
\begin{aligned}
&-\bar{\beta}_{k+1}^2 I \circ (C_{k+1}\Lambda_{k+1}C_{k+1}^T)\Big]\Big[\bar{\beta}_{k+1}^2 C_{k+1}\Omega_{k+1|k}C_{k+1}^T \\
&+\bar{\beta}_{k+1}(1-\bar{\beta}_{k+1})I \circ (C_{k+1}\Lambda_{k+1}C_{k+1}^T) \\
&+\mathrm{tr}(\Lambda_{k+1}\Omega_{2,k+1})\bar{\Gamma}_{2,k+1} + S_{k+1}\Big]^{-1} C_{k+1}\Omega_{k+1|k}\Big) \\
=\ &\mathrm{tr}\Big(-\bar{\beta}_{k+1}^2 \Omega_{k+1|k}^T C_{k+1}^T \Big[\bar{\beta}_{k+1}^2 C_{k+1}\Omega_{k+1|k}C_{k+1}^T \\
&+\bar{\beta}_{k+1}(1-\bar{\beta}_{k+1})I \circ (C_{k+1}\Lambda_{k+1}C_{k+1}^T) \\
&+\mathrm{tr}(\Lambda_{k+1}\Omega_{2,k+1})\bar{\Gamma}_{2,k+1} + S_{k+1}\Big]^{-1}\Big[I \circ (C_{k+1}\Lambda_{k+1}C_{k+1}^T)\Big] \\
&\times\Big[\bar{\beta}_{k+1}^2 C_{k+1}\Omega_{k+1|k}C_{k+1}^T + \bar{\beta}_{k+1}(1-\bar{\beta}_{k+1})I \circ (C_{k+1}\Lambda_{k+1}C_{k+1}^T) \\
&+\mathrm{tr}(\Lambda_{k+1}\Omega_{2,k+1})\bar{\Gamma}_{2,k+1} + S_{k+1}\Big]^{-1} C_{k+1}\Omega_{k+1|k}\Big). \qquad (7.39)
\end{aligned}
$$

Now, it is obvious to see that

$$
\frac{\partial \mathrm{tr}(\Omega_{k+1|k+1})}{\partial \bar{\beta}_{k+1}} \leq 0. \qquad (7.40)
$$

The above inequality implies that $\mathrm{tr}(\Omega_{k+1|k+1})$ is non-decreasing when $\bar{\beta}_{k+1}$ decreases, which ends the proof of this theorem.

Remark 7.4 As exhibited in the mentioned theorem, the monotonicity relationship between the trace of the minimal upper bound and the missing probability is shown with rigorous theoretical analysis. Specifically, when the missing probability $\bar{\beta}_{k+1}$ increases, then the estimation performance of the designed algorithm can be improved due to the fact that more measurement information can be utilized.

7.4 An Illustrative Example

In this section, some simulations are proposed to certify the usefulness of the developed fault estimation strategy.

The parameters of the time-varying systems (7.1) and (7.2) are set as

$$
\bar{A}_k = \begin{bmatrix} 0.75 & 0.2 \\ 0.2 & 0.85 \end{bmatrix}, \ \bar{F}_k = \begin{bmatrix} 1 \\ 0 \end{bmatrix}, \ \bar{B}_k = \begin{bmatrix} 0.5 \\ 0.3 - 0.1\sin(5k) \end{bmatrix}, \ \bar{C}_k = \begin{bmatrix} 0.9 & 0.9 \end{bmatrix},
$$
$$
\bar{H}_k = \begin{bmatrix} 0.05 & 0.1 \end{bmatrix}^T, \ \bar{N}_k = \begin{bmatrix} 0.24 & 0.08 \end{bmatrix}, \ \mathscr{F}_k = \sin(4k),
$$

and the state $\bar{x}_k = \begin{bmatrix} \bar{x}_{1,k} & \bar{x}_{2,k} \end{bmatrix}^T$ with $\bar{x}_{i,k}$ $(i = 1, 2)$ being the i-th element of the system state, the covariances of $\varpi_k \in \mathbb{R}$ and $\upsilon_k \in \mathbb{R}$ are set as 0.1 and 0.1, respectively. The additive fault is set as

$$f_k = \begin{cases} 0.8, & \text{if } k \geq 10 \\ 0, & \text{otherwise.} \end{cases}$$

The stochastic nonlinearities $\bar{g}(\bar{x}_k, \xi_k)$ and $\bar{h}(\bar{x}_k, \varsigma_k)$ are depicted as

$$\begin{aligned} \bar{g}(\bar{x}_k, \xi_k) &= \begin{bmatrix} 0.3 \\ 0.5 \end{bmatrix} [0.2\text{sign}\left(\bar{x}_{1,k}\right) \bar{x}_{1,k}\xi_{1,k} + 0.3\text{sign}\left(\bar{x}_{2,k}\right) \bar{x}_{2,k}\xi_{2,k}], \\ \bar{h}(\bar{x}_k, \varsigma_k) &= 0.4\left[0.2\text{sign}\left(\bar{x}_{1,k}\right) \bar{x}_{1,k}\varsigma_{1,k} + 0.3\text{sign}\left(\bar{x}_{2,k}\right) \bar{x}_{2,k}\varsigma_{2,k}\right], \end{aligned}$$

where $\xi_{i,k}$ and $\varsigma_{i,k}$ $(i = 1, 2)$ represent the zero-mean Gaussian white noises with unity covariances. It is seen that the stochastic nonlinearities $\bar{g}(\bar{x}_k, \xi_k)$ and $\bar{h}(\bar{x}_k, \varsigma_k)$ meet

$$\begin{aligned} &\mathbb{E}\left\{\begin{bmatrix} \bar{g}(\bar{x}_k, \xi_k) \\ \bar{h}(\bar{x}_k, \varsigma_k) \end{bmatrix} \middle| \bar{x}_k \right\} = 0, \\ &\mathbb{E}\left\{\begin{bmatrix} \bar{g}(\bar{x}_k, \xi_k) \\ \bar{h}(\bar{x}_k, \varsigma_k) \end{bmatrix} \begin{bmatrix} \bar{g}(\bar{x}_k, \xi_k) \\ \bar{h}(\bar{x}_k, \varsigma_k) \end{bmatrix}^T \middle| \bar{x}_k \right\} = \text{diag}\{\bar{\Gamma}_{1,k}\bar{x}_k^T \bar{\Omega}_{1,k}\bar{x}_k, \bar{\Gamma}_{2,k}\bar{x}_k^T \bar{\Omega}_{2,k}\bar{x}_k\}, \end{aligned}$$

where

$$\bar{\Gamma}_{1,k} = \begin{bmatrix} 0.09 & 0.15 \\ 0.15 & 0.25 \end{bmatrix}, \quad \bar{\Gamma}_{2,k} = 0.16, \quad \bar{\Omega}_{1,k} = \bar{\Omega}_{2,k} = \begin{bmatrix} 0.04 & 0 \\ 0 & 0.09 \end{bmatrix}.$$

The initial conditions are chosen as $x_0 = \begin{bmatrix} 1 & 1 & 0 \end{bmatrix}^T$ and $\Omega_{0|0} = 5I_3$. Other parameters are selected as $\bar{\alpha}_k = 0.8$, $\bar{\beta}_{i,k} = 0.85$ (Case I), $\bar{\beta}_{i,k} = 0.4$ (Case II), $\varepsilon_{1,k} = 0.5$, $\varepsilon_{2,k} = 0.5$, and $\varepsilon_{3,k+1} = 0.5$. Here, MSE$i$ is the mean-square error for the state estimation, i.e., MSE$i = (1/K) \sum_{s=1}^{K} \left(x_{i,k}^{(s)} - \hat{x}_{i,k|k}^{(s)}\right)^2$ with $K = 200$ being the number of simulation test. $x_{i,k}^{(s)}$ and $\hat{x}_{i,k|k}^{(s)}$ denote the values of $x_{i,k}$ and $\hat{x}_{i,k|k}$ in the s-th simulation test. From Figs. 7.1, 7.2 and 7.3, it can be observed that the fault signal and unknown system states are estimated well.

To illustrate the result given in Theorem 7.5, some comparisons under different probabilities of packet dropouts are considered, i.e.,

$$\bar{\beta}_{i,k} = 0.85 \quad (\text{Case I}) \quad \text{and} \quad \bar{\beta}_{i,k} = 0.4 \quad (\text{Case II}).$$

The specific experiment results are shown in Figs. 7.4, 7.5 and 7.6. Obviously, the estimation performance of the developed algorithm is improved for Case I because more measurement information has been employed when designing the estimation algorithm, which further testifies the result presented in Theorem 7.5.

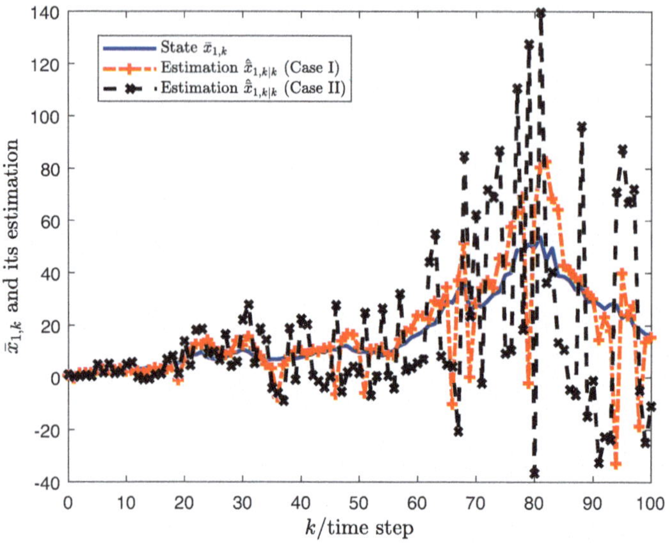

Fig. 7.1 State $\bar{x}_{1,k}$ and its estimation $\hat{\bar{x}}_{1,k|k}$

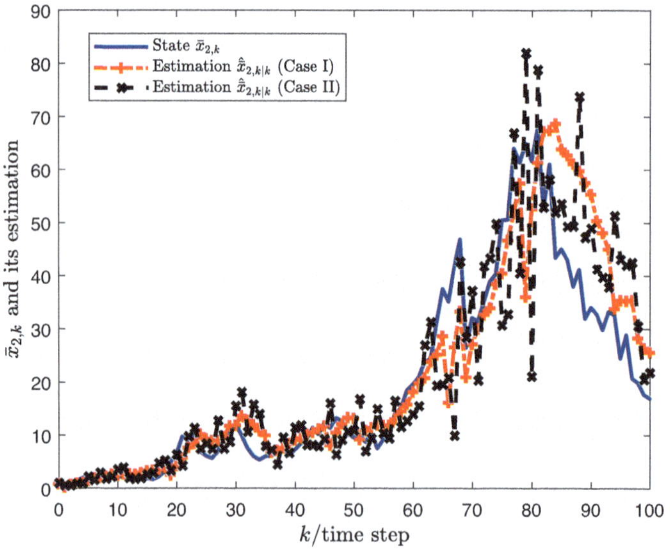

Fig. 7.2 State $\bar{x}_{2,k}$ and its estimation $\hat{\bar{x}}_{2,k|k}$

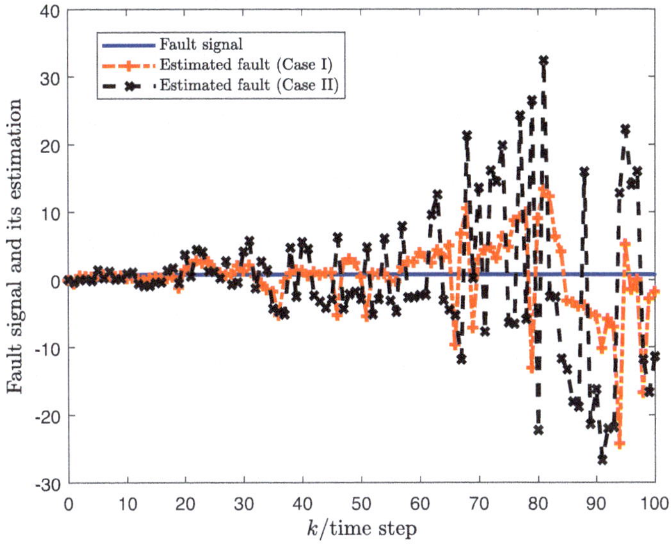

Fig. 7.3 Fault signal f_k and its estimation

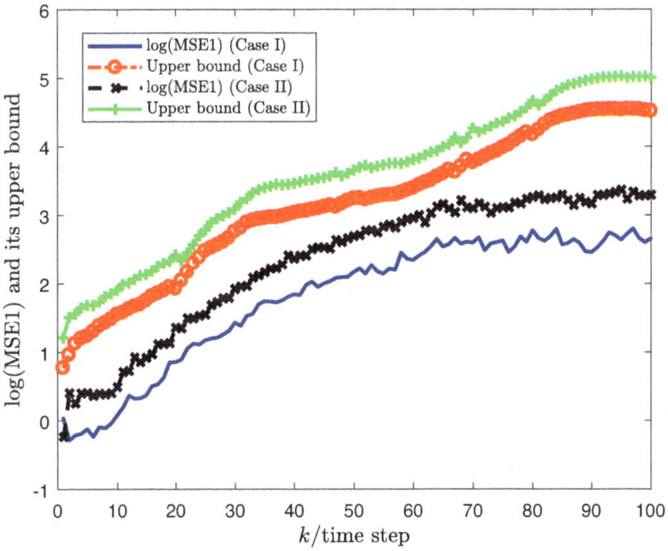

Fig. 7.4 log(MSE1) and its upper bound

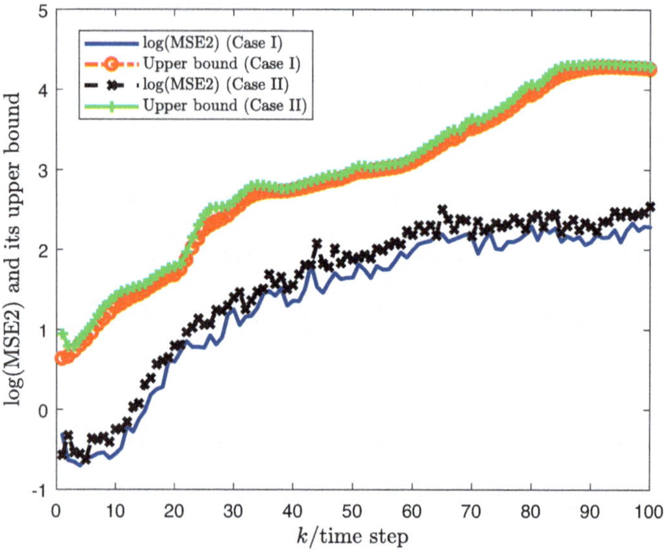

Fig. 7.5 log(MSE2) and its upper bound

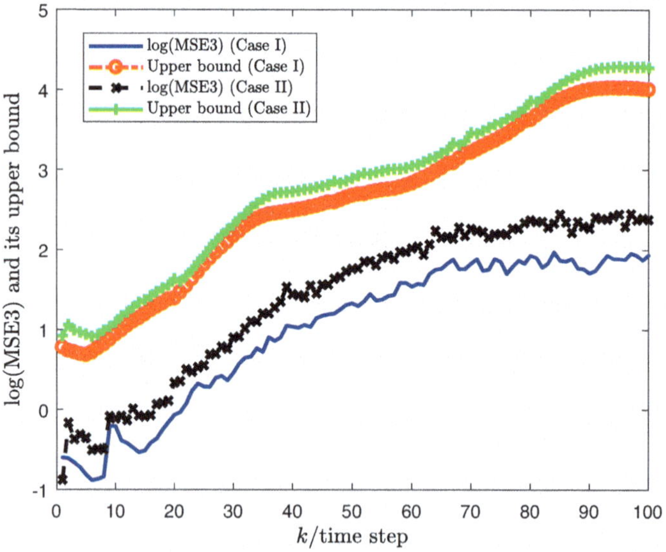

Fig. 7.6 log(MSE3) and its upper bound

7.5 Summary

In this chapter, we have tackled the fault estimation issue for time-varying networked systems incorporating with the ROUs, stochastic nonlinearities and packet dropouts. A set of mutually independent random variables has been applied to model the packet dropouts and all measurements have been authorized to possess individual missing probability in the given model. A novel fault estimation method has been presented such that an optimized upper bound of the EEC has been ensured and the estimator gain has been derived. Moreover, the relationship between this minimal upper bound and the missing probability has been examined from a theoretical viewpoint. Finally, the simulation experiments have been conducted to validate the usefulness of the designed estimation scheme.

References

Hu J, Wang Z, Gao H, Stergioulas LK (2012) Probability-guaranteed H_∞ finite-horizon filtering for a class of nonlinear time-varying systems with sensor saturations. Syst Control Lett 61(4):477–484

Yaz E, Yaz Y (2001) State estimation of uncertain nonlinear stochastic systems with general criteria. Appl Math Lett 14(5):605–610

Chapter 8
Joint Fault and State Estimation with Randomly Occurring Faults and Sensor Saturations

In this chapter, the fault estimation issue is handled for time-varying nonlinear stochastic systems with ROFs and sensor saturations. A set of Bernoulli random variables is employed to model the ROFs and then the signum function is adopted to model the phenomenon of sensor saturations. Next, we are in a position to design a locally optimal estimator providing the joint estimation of the system states and fault signals such that the EEC possesses an optimal upper bound that can be minimized by appropriately constructing the estimator gain. It is observed that the established recursive estimation algorithm is suitable for online calculations. Furthermore, a sufficient condition is proposed to clarify the boundedness of the estimation error dynamics. Finally, the illustrative simulation experiments are used to exhibit the validity and applicability of the presented joint estimation strategy.

8.1 Problem Formulation

In this chapter, we consider the following class of time-varying nonlinear systems:

$$\breve{x}_{k+1} = (\breve{A}_k + \Delta\breve{A}_k)\breve{x}_k + \breve{g}(\breve{x}_k) + \alpha_k \breve{F}_k f_k + \breve{B}_k \omega_k, \tag{8.1}$$

$$\breve{y}_k = \sigma(\breve{C}_k \breve{x}_k) + \nu_k, \tag{8.2}$$

where $\breve{x}_k \in \mathbb{R}^n$ stands for the system state, $\breve{x}_0$ represents the initial value with mean $\bar{\breve{x}}_0$, $\breve{y}_k \in \mathbb{R}^m$ denotes the measurement, $f_k \in \mathbb{R}^f$ is the fault signal, ω_k is the process noise with zero-mean and covariance $Q_k > 0$, and ν_k is the measurement noise with zero-mean and covariance $R_k > 0$. $\breve{A}_k$, $\breve{F}_k$, $\breve{B}_k$ and $\breve{C}_k$ are known matrices of suitable dimensions. $\Delta\breve{A}_k$ is an uncertain matrix with

$$\Delta\breve{A}_k = \breve{M}_k \mathscr{F}_k \breve{N}_k, \quad \mathscr{F}_k \mathscr{F}_k^T \leq I, \tag{8.3}$$

J. Hu et al., *Variance-Constrained Filtering for Stochastic Complex Systems*, Intelligent Control and Learning Systems 18,
https://doi.org/10.1007/978-981-96-2637-3_8

where $\mathscr{F}_k$ denotes the uncertainty, $\breve{M}_k$ and $\breve{N}_k$ are given matrices of proper dimensions. The nonlinear function $\breve{g}(\breve{x}_k)$ with $\breve{g}(0) = 0$ satisfies

$$\|\breve{g}(a) - \breve{g}(b)\| \leq \|G(a - b)\|, \quad \forall\, a, b \in \mathbb{R}^n \tag{8.4}$$

with G being a given matrix.

The fault f_k is modeled as

$$f_{k+1} = A_{f,k} f_k, \tag{8.5}$$

where $A_{f,k}$ represents a given matrix. In order to characterize the ROFs, the binary random variable $\alpha_k \in \mathbb{R}$ is introduced, which takes the values of 0 or 1 with

$$\text{Prob}\{\alpha_k = 1\} = \bar{\alpha}, \ \text{Prob}\{\alpha_k = 0\} = 1 - \bar{\alpha}, \tag{8.6}$$

where $\bar{\alpha} \in [0, 1]$ is a pre-defined scalar.

The saturation function $\sigma : \mathbb{R}^m \to \mathbb{R}^m$ is given as

$$\sigma(u) = \left[\sigma_1(u_1)\ \sigma_2(u_2)\ \cdots\ \sigma_m(u_m)\right]^T, \tag{8.7}$$

where $u = [u_1,\ u_2, \ldots, u_m]^T$, $\sigma_i(u_i) = \text{sign}(u_i)\min\{\varrho_i, |u_i|\}$ $(i = 1, 2, \ldots, m)$, $\text{sign}(\cdot)$ is the signum function and ϱ_i represents the saturation level.

Remark 8.1 The assumption (8.4) reflects the Lipschitz nonlinearities which have been widely investigated in a great deal of literature. It is worth emphasizing that, based on the same assumption condition, the issues of fault detection and estimation have been widely investigated for time-varying systems with the Lipschitz nonlinearities, see Boulkroune et al. (2013), Koenig et al. (2016) for more details.

Remark 8.2 In this chapter, the fault estimation issue is investigated for time-varying nonlinear systems with sensor saturated measurements, in which the additive fault is also handled here. Particularly, the faults given in (8.5) could cover the constant faults as special case. Moreover, a linear fault model with time-varying $A_{f,k}$ is introduced to describe the time-varying characteristic of the conducted system model. The primary reason lies in that certain priori knowledge of the fault can be obtained according to the engineering reality after the fault detection. In fact, this type of fault is usually common in reality Klatt and Engell (1998). Further, a binary variable α_k is introduced to model the behavior of the ROFs in (8.1). Specifically, the fault does not occur when $\alpha_k = 0$, and the fault occurs when $\alpha_k = 1$. The main purpose of this chapter is to develop a joint state and fault estimation strategy. For instance, when the fault randomly occurs, the sudden change would be detected via adopting the fault detection scheme, and then the proposed fault estimation algorithm would be executed. In such a case, the initial implementation time of the developed estimation method is equal to the sudden change time. As a result, the sudden change can be viewed as the appearance of the fault.

Throughout this chapter, it assumes that α_k, ω_k, v_k and $\breve{x}_0$ are mutually independent. Setting $x_k = \begin{bmatrix} \breve{x}_k^T & f_k^T \end{bmatrix}^T$, we have

$$x_{k+1} = [\bar{A}_k + \bar{A}_{1,k} + (\alpha_k - \bar{\alpha})\bar{A}_{2,k}]x_k + g(x_k) + B_k\omega_k, \tag{8.8}$$

$$y_k = \sigma(C_k x_k) + v_k, \tag{8.9}$$

where

$$\bar{A}_k = \begin{bmatrix} \breve{A}_k & \bar{\alpha}\breve{F}_k \\ 0 & A_{f,k} \end{bmatrix}, \quad \bar{A}_{1,k} = \begin{bmatrix} \Delta\breve{A}_k & 0 \\ 0 & 0 \end{bmatrix}, \quad \bar{A}_{2,k} = \begin{bmatrix} 0 & \breve{F}_k \\ 0 & 0 \end{bmatrix},$$

$$g(x_k) = \begin{bmatrix} \breve{g}(\breve{x}_k) \\ 0 \end{bmatrix}, \quad B_k = \begin{bmatrix} \breve{B}_k \\ 0 \end{bmatrix}, \quad C_k = \begin{bmatrix} \breve{C}_k & 0 \end{bmatrix}, \quad y_k = \breve{y}_k. \tag{8.10}$$

In the sequel, we construct the following estimator:

$$\hat{x}_{k+1|k} = \bar{A}_k\hat{x}_{k|k} + g(\hat{x}_{k|k}), \tag{8.11}$$

$$\hat{x}_{k+1|k+1} = \hat{x}_{k+1|k} + G_{k+1}(y_{k+1} - C_{k+1}\hat{x}_{k+1|k}), \tag{8.12}$$

where $\hat{x}_{k|k}$ is the state estimation with $\hat{x}_{0|0} = \begin{bmatrix} \bar{\breve{x}}_0^T & 0^T \end{bmatrix}^T$, $\hat{x}_{k+1|k}$ is the one-step prediction, and G_{k+1} is the estimator gain matrix.

Let the estimation error and its covariance be

$$e_{k+1|k+1} = x_{k+1} - \hat{x}_{k+1|k+1},$$

$$P_{k+1|k+1} = \mathbb{E}\{e_{k+1|k+1}e_{k+1|k+1}^T\}.$$

Now, we summarize the main research purpose, where the ROFs and sensor saturations are considered.

(R1) We are ready to construct an estimator (8.11) and (8.12) such that there is an upper bound of the EEC $P_{k+1|k+1}$, that is to say, we are searching a set of positive definite matrices $\Omega_{k+1|k+1}$ meeting

$$P_{k+1|k+1} \le \Omega_{k+1|k+1}. \tag{8.13}$$

Besides, such an upper bound $\Omega_{k+1|k+1}$ is optimized via suitably selecting the estimator gain G_{k+1}.

(R2) A sufficient condition is given to illustrate the exponential boundedness with respect to the estimation error in the mean-square sense from the mathematical analysis viewpoint.

8.2 Main Results

In this section, the explicit forms of the one-step prediction error covariance and the EEC are firstly given. Then, the upper bounds of them are attained by adopting the stochastic analysis approach. In what follows, the estimator gain is designed through solving two matrix difference equations. Finally, a sufficient condition is proposed to verify the exponential boundedness of the estimation error dynamics in mean-square sense.

8.2.1 Design of the Fault Estimator Gain

Let the one-step prediction error be $e_{k+1|k} = x_{k+1} - \hat{x}_{k+1|k}$. Next, it can be deduced from (8.8) and (8.11) that

$$e_{k+1|k} = \bar{A}_k e_{k|k} + [\bar{A}_{1,k} + (\alpha_k - \bar{\alpha})\bar{A}_{2,k}]x_k + g(x_k) - g(\hat{x}_{k|k}) + B_k \omega_k \tag{8.14}$$

with $\bar{A}_k$, $\bar{A}_{1,k}$ and $\bar{A}_{2,k}$ being given in (8.10). Further, the estimation error can be computed from (8.9) and (8.12) that

$$e_{k+1|k+1} = e_{k+1|k} - G_{k+1}[\sigma(C_{k+1}x_{k+1}) - C_{k+1}\hat{x}_{k+1|k} + \nu_{k+1}]. \tag{8.15}$$

In terms of (8.14) and (8.15), the recursive forms of the one-step prediction error covariance and EEC are given based on related definitions.

Lemma 8.1 *The one-step prediction error covariance $P_{k+1|k}$ has the following recursion:*

$$\begin{aligned} P_{k+1|k} = & \bar{A}_k P_{k|k} \bar{A}_k^T + \bar{A}_k \mathbb{E}\{e_{k|k} x_k^T\} \bar{A}_{1,k}^T + \bar{A}_{1,k} \mathbb{E}\{x_k e_{k|k}^T\} \bar{A}_k^T \\ & + \mathbb{E}\{\mathcal{A}_k + \mathcal{A}_k^T\} + \bar{A}_{1,k} \mathbb{E}\{x_k x_k^T\} \bar{A}_{1,k}^T + \mathbb{E}\{\mathcal{B}_k + \mathcal{B}_k^T\} \\ & + \bar{\alpha}(1-\bar{\alpha}) \bar{A}_{2,k} \mathbb{E}\{x_k x_k^T\} \bar{A}_{2,k}^T + \mathbb{E}\{[g(x_k) - g(\hat{x}_{k|k})] \\ & \times [g(x_k) - g(\hat{x}_{k|k})]^T\} + B_k Q_k B_k^T, \end{aligned} \tag{8.16}$$

where

$$\begin{aligned} \mathcal{A}_k &= \bar{A}_k e_{k|k} [g(x_k) - g(\hat{x}_{k|k})]^T, \\ \mathcal{B}_k &= \bar{A}_{1,k} x_k [g(x_k) - g(\hat{x}_{k|k})]^T. \end{aligned} \tag{8.17}$$

Proof By resorting to the definition of the one-step prediction error covariance, one gets

$$\begin{aligned}
P_{k+1|k} &= \mathbb{E}\{e_{k+1|k}e_{k+1|k}^T\} \\
&= \bar{A}_k P_{k|k}\bar{A}_k^T + \bar{A}_k\mathbb{E}\{e_{k|k}x_k^T\}\bar{A}_{1,k}^T + \bar{A}_{1,k}\mathbb{E}\{x_k e_{k|k}^T\}\bar{A}_k^T \\
&\quad + \mathcal{P}_{1,k} + \mathcal{P}_{1,k}^T + \bar{A}_k\mathbb{E}\{e_{k|k}[g(x_k) - g(\hat{x}_{k|k})]^T\} \\
&\quad + \mathbb{E}\{[g(x_k) - g(\hat{x}_{k|k})]e_{k|k}^T\}\bar{A}_k^T + \mathcal{P}_{2,k} + \mathcal{P}_{2,k}^T \\
&\quad + \bar{A}_{1,k}\mathbb{E}\{x_k x_k^T\}\bar{A}_{1,k}^T + \bar{A}_{1,k}\mathbb{E}\{x_k[g(x_k) - g(\hat{x}_{k|k})]^T\} \\
&\quad + \mathbb{E}\{[g(x_k) - g(\hat{x}_{k|k})]x_k^T\}\bar{A}_{1,k}^T + \mathcal{P}_{3,k} + \mathcal{P}_{3,k}^T \\
&\quad + \mathcal{P}_{4,k} + \mathcal{P}_{4,k}^T + \bar{\alpha}(1-\bar{\alpha})\bar{A}_{2,k}\mathbb{E}\{x_k x_k^T\}\bar{A}_{2,k}^T \\
&\quad + \mathcal{P}_{5,k} + \mathcal{P}_{5,k}^T + \mathcal{P}_{6,k} + \mathcal{P}_{6,k}^T \\
&\quad + \mathbb{E}\{[g(x_k) - g(\hat{x}_{k|k})][g(x_k) - g(\hat{x}_{k|k})]^T\} \\
&\quad + \mathcal{P}_{7,k} + \mathcal{P}_{7,k}^T + B_k Q_k B_k^T,
\end{aligned}$$

where

$$\begin{aligned}
\mathcal{P}_{1,k} &= \mathbb{E}\{(\alpha_k - \bar{\alpha})\bar{A}_k e_{k|k}x_k^T\bar{A}_{2,k}^T\}, \\
\mathcal{P}_{2,k} &= \mathbb{E}\{\bar{A}_k e_{k|k}\omega_k^T B_k^T\}, \\
\mathcal{P}_{3,k} &= \mathbb{E}\{(\alpha_k - \bar{\alpha})\bar{A}_{1,k}x_k x_k^T\bar{A}_{2,k}^T\}, \\
\mathcal{P}_{4,k} &= \mathbb{E}\{\bar{A}_{1,k}x_k\omega_k^T B_k^T\}, \\
\mathcal{P}_{5,k} &= \mathbb{E}\{(\alpha_k - \bar{\alpha})\bar{A}_{2,k}x_k[g(x_k) - g(\hat{x}_{k|k})]^T\}, \\
\mathcal{P}_{6,k} &= \mathbb{E}\{(\alpha_k - \bar{\alpha})\bar{A}_{2,k}x_k\omega_k^T B_k^T\}, \\
\mathcal{P}_{7,k} &= \mathbb{E}\{[g(x_k) - g(\hat{x}_{k|k})]\omega_k^T B_k^T\}.
\end{aligned}$$

It is straightforward to find that $\mathcal{P}_{i,k}$ $(i = 1, 2, \ldots, 7)$ are equal to zero and then (8.16) holds.

Lemma 8.2 *The EEC* $P_{k+1|k+1}$ *is obtained as*

$$\begin{aligned}
P_{k+1|k+1} &= (I - G_{k+1}C_{k+1})P_{k+1|k}(I - G_{k+1}C_{k+1})^T \\
&\quad + \mathbb{E}\{\mathcal{C}_{k+1} + \mathcal{C}_{k+1}^T\} + \mathbb{E}\{\mathcal{D}_{k+1} + \mathcal{D}_{k+1}^T\} \\
&\quad + G_{k+1}C_{k+1}\mathbb{E}\{x_{k+1}x_{k+1}^T\}C_{k+1}^T G_{k+1}^T \\
&\quad + G_{k+1}\mathbb{E}\{\sigma(C_{k+1}x_{k+1})\sigma^T(C_{k+1}x_{k+1})\}G_{k+1}^T \\
&\quad + \mathbb{E}\{\mathcal{E}_{k+1} + \mathcal{E}_{k+1}^T\} + G_{k+1}R_{k+1}G_{k+1}^T,
\end{aligned} \tag{8.18}$$

where

$$\begin{aligned}
\mathcal{C}_{k+1} &= (I - G_{k+1}C_{k+1})e_{k+1|k}x_{k+1}^T C_{k+1}^T G_{k+1}^T, \\
\mathcal{D}_{k+1} &= -(I - G_{k+1}C_{k+1})e_{k+1|k}\sigma^T(C_{k+1}x_{k+1})G_{k+1}^T, \\
\mathcal{E}_{k+1} &= -G_{k+1}C_{k+1}x_{k+1}\sigma^T(C_{k+1}x_{k+1})G_{k+1}^T.
\end{aligned} \tag{8.19}$$

Proof Adding the zero term

$$G_{k+1}C_{k+1}x_{k+1} - G_{k+1}C_{k+1}x_{k+1}$$

to the RHS of (8.15) yields

$$\begin{aligned} e_{k+1|k+1} = &(I - G_{k+1}C_{k+1})e_{k+1|k} + G_{k+1}C_{k+1}x_{k+1} \\ &-G_{k+1}\sigma(C_{k+1}x_{k+1}) - G_{k+1}v_{k+1}. \end{aligned} \tag{8.20}$$

It can be obtained from (8.20) that

$$\begin{aligned} P_{k+1|k+1} &= \mathbb{E}\{e_{k+1|k+1}e^T_{k+1|k+1}\} \\ &= (I - G_{k+1}C_{k+1})P_{k+1|k}(I - G_{k+1}C_{k+1})^T \\ &\quad +(I - G_{k+1}C_{k+1})\mathbb{E}\{e_{k+1|k}x^T_{k+1}\}C^T_{k+1}G^T_{k+1} \\ &\quad +G_{k+1}C_{k+1}\mathbb{E}\{x_{k+1}e^T_{k+1|k}\}(I - G_{k+1}C_{k+1})^T \\ &\quad -(I - G_{k+1}C_{k+1})\mathbb{E}\{e_{k+1|k}\sigma^T(C_{k+1}x_{k+1})\}G^T_{k+1} \\ &\quad -G_{k+1}\mathbb{E}\{\sigma(C_{k+1}x_{k+1})e^T_{k+1|k}\}(I - G_{k+1}C_{k+1})^T \\ &\quad -\mathcal{Q}_{1,k+1} - \mathcal{Q}^T_{1,k+1} + G_{k+1}C_{k+1}\mathbb{E}\{x_{k+1}x^T_{k+1}\}C^T_{k+1}G^T_{k+1} \\ &\quad -G_{k+1}C_{k+1}\mathbb{E}\{x_{k+1}\sigma^T(C_{k+1}x_{k+1})\}G^T_{k+1} \\ &\quad -G_{k+1}\mathbb{E}\{\sigma(C_{k+1}x_{k+1})x^T_{k+1}\}C^T_{k+1}G^T_{k+1} - \mathcal{Q}_{2,k+1} \\ &\quad -\mathcal{Q}^T_{2,k+1} + G_{k+1}\mathbb{E}\{\sigma(C_{k+1}x_{k+1})\sigma^T(C_{k+1}x_{k+1})\}G^T_{k+1} \\ &\quad +\mathcal{Q}_{3,k+1} + \mathcal{Q}^T_{3,k+1} + G_{k+1}R_{k+1}G^T_{k+1}, \end{aligned}$$

where

$$\begin{aligned} \mathcal{Q}_{1,k+1} &= \mathbb{E}\{(I - G_{k+1}C_{k+1})e_{k+1|k}v^T_{k+1}G^T_{k+1}\}, \\ \mathcal{Q}_{2,k+1} &= \mathbb{E}\{G_{k+1}C_{k+1}x_{k+1}v^T_{k+1}G^T_{k+1}\}, \\ \mathcal{Q}_{3,k+1} &= \mathbb{E}\{G_{k+1}\sigma(C_{k+1}x_{k+1})v^T_{k+1}G^T_{k+1}\}. \end{aligned}$$

In what follows, it is not difficult to observe that $\mathcal{Q}_{i,k+1} = 0$ $(i = 1, 2, 3)$. As a result, it is summarized that (8.18) holds, which ends the proof of this lemma.

Remark 8.3 It is worth emphasizing that some unknown terms are appeared in Lemmas 8.1 and 8.2 on account of the parameter uncertainties, nonlinearities as well as sensor saturations. Thus, it is difficult to obtain an accurate EEC. For the sake of tackling this issue, in the next step, we are in a position to search for an upper bound of the EEC and minimize this upper bound by suitably constructing the estimator gain via the recursive method.

Theorem 8.1 *Let* $\gamma_{k,i} > 0 (i = 1, 2, 3, 4)$ *and* $\gamma_{k+1,j} > 0 (j = 5, 6, 7, 8)$ *be scalars. Under the initial condition* $\Omega_{0|0} = P_{0|0}$, *assume that the following two difference equations have solutions* $\Omega_{k+1|k}$ *and* $\Omega_{k+1|k+1}$*:*

$$\Omega_{k+1|k} = (1+\gamma_{k,1}+\gamma_{k,2})\bar{A}_k\Omega_{k|k}\bar{A}_k^T + (1+\gamma_{k,1}^{-1}+\gamma_{k,3})\mathrm{tr}\left(\bar{N}_k\Theta_k\bar{N}_k^T\right)\bar{M}_k\bar{M}_k^T$$
$$+\bar{\alpha}(1-\bar{\alpha})\bar{A}_{2,k}\Theta_k\bar{A}_{2,k}^T + (1+\gamma_{k,2}^{-1}+\gamma_{k,3}^{-1})\bar{G}\Omega_{k|k}\bar{G}^T + B_kQ_kB_k^T, \quad (8.21)$$

and

$$\Omega_{k+1|k+1} = (1+\gamma_{k+1,5}+\gamma_{k+1,6})(I-G_{k+1}C_{k+1})\Omega_{k+1|k}(I-G_{k+1}C_{k+1})^T$$
$$+(1+\gamma_{k+1,5}^{-1}+\gamma_{k+1,7})G_{k+1}C_{k+1}\Omega_{k+1}C_{k+1}^TG_{k+1}^T$$
$$+G_{k+1}[\bar{\varrho}(1+\gamma_{k+1,6}^{-1}+\gamma_{k+1,7}^{-1})I + R_{k+1}]G_{k+1}^T, \quad (8.22)$$

where

$$\bar{\varrho} = \sum_{i=1}^{m}\varrho_i^2, \quad \bar{M}_k^T = \left[\,\breve{M}_k^T\ 0\,\right],$$
$$\bar{N}_k = \left[\,\breve{N}_k\ 0\,\right], \quad \bar{G} = \mathrm{diag}\{G, 0\},$$
$$\Theta_k = (1+\gamma_{k,4})\Omega_{k|k} + (1+\gamma_{k,4}^{-1})\hat{x}_{k|k}\hat{x}_{k|k}^T,$$
$$\Omega_{k+1} = (1+\gamma_{k+1,8})\Omega_{k+1|k} + (1+\gamma_{k+1,8}^{-1})\hat{x}_{k+1|k}\hat{x}_{k+1|k}^T. \quad (8.23)$$

Then, one has

$$P_{k+1|k+1} \le \Omega_{k+1|k+1}. \quad (8.24)$$

Furthermore, such an upper bound $\Omega_{k+1|k+1}$ can be minimized by employing the following estimator gain:

$$G_{k+1} = (1+\gamma_{k+1,5}+\gamma_{k+1,6})\Omega_{k+1|k}C_{k+1}^T\mathcal{K}_{k+1}^{-1}, \quad (8.25)$$

where

$$\mathcal{K}_{k+1} = (1+\gamma_{k+1,5}+\gamma_{k+1,6})C_{k+1}\Omega_{k+1|k}C_{k+1}^T + (1+\gamma_{k+1,5}^{-1}+\gamma_{k+1,7})$$
$$\times C_{k+1}\Omega_{k+1}C_{k+1}^T + \bar{\varrho}(1+\gamma_{k+1,6}^{-1}+\gamma_{k+1,7}^{-1})I + R_{k+1}. \quad (8.26)$$

Proof Firstly, we aim to cope with some unknown terms of the RHS of (8.16) and then present the upper bound of the one-step prediction error covariance. By virtue of the Lemma 2.2, it can be obtained that

$$\bar{A}_k\mathbb{E}\{e_{k|k}x_k^T\}\bar{A}_{1,k}^T + \bar{A}_{1,k}\mathbb{E}\{x_ke_{k|k}^T\}\bar{A}_k^T$$
$$\le \gamma_{k,1}\bar{A}_kP_{k|k}\bar{A}_k^T + \gamma_{k,1}^{-1}\bar{A}_{1,k}\mathbb{E}\{x_kx_k^T\}\bar{A}_{1,k}^T, \quad (8.27)$$

where $\gamma_{k,1} > 0$ is a scalar. Based on Lemma 2.2, one has

$$\mathbb{E}\{\mathcal{A}_k + \mathcal{A}_k^T\}$$
$$\le \gamma_{k,2}\bar{A}_kP_{k|k}\bar{A}_k^T + \gamma_{k,2}^{-1}\mathbb{E}\{[g(x_k)-g(\hat{x}_{k|k})][g(x_k)-g(\hat{x}_{k|k})]^T\}, \quad (8.28)$$

$$\begin{aligned}
&\mathbb{E}\{\mathcal{B}_k + \mathcal{B}_k^T\} \\
&\leq \gamma_{k,3}\bar{A}_{1,k}\mathbb{E}\{x_k x_k^T\}\bar{A}_{1,k}^T + \gamma_{k,3}^{-1}\mathbb{E}\{[g(x_k) - g(\hat{x}_{k|k})][g(x_k) - g(\hat{x}_{k|k})]^T\},
\end{aligned} \quad (8.29)$$

where $\gamma_{k,2} > 0$ and $\gamma_{k,3} > 0$ are scalars. Substituting (8.27)–(8.29) into (8.16) yields

$$\begin{aligned}
P_{k+1|k} \leq\ & (1 + \gamma_{k,1} + \gamma_{k,2})\bar{A}_k P_{k|k}\bar{A}_k^T + (1 + \gamma_{k,1}^{-1} + \gamma_{k,3})\bar{A}_{1,k}\mathbb{E}\{x_k x_k^T\}\bar{A}_{1,k}^T \\
& + (1 + \gamma_{k,2}^{-1} + \gamma_{k,3}^{-1})\mathbb{E}\{[g(x_k) - g(\hat{x}_{k|k})][g(x_k) - g(\hat{x}_{k|k})]^T\} \\
& + \bar{\alpha}(1 - \bar{\alpha})\bar{A}_{2,k}\mathbb{E}\{x_k x_k^T\}\bar{A}_{2,k}^T + B_k Q_k B_k^T.
\end{aligned} \quad (8.30)$$

Subsequently, from Lemma 2.2, one obtains

$$\begin{aligned}
&\mathbb{E}\{x_k x_k^T\} \\
&= \mathbb{E}\{(e_{k|k} + \hat{x}_{k|k})(e_{k|k} + \hat{x}_{k|k})^T\} \\
&= \mathbb{E}\{e_{k|k}e_{k|k}^T + e_{k|k}\hat{x}_{k|k}^T + \hat{x}_{k|k}e_{k|k}^T + \hat{x}_{k|k}\hat{x}_{k|k}^T\} \\
&\leq (1 + \gamma_{k,4})P_{k|k} + (1 + \gamma_{k,4}^{-1})\hat{x}_{k|k}\hat{x}_{k|k}^T
\end{aligned} \quad (8.31)$$

with $\gamma_{k,4} > 0$ denoting a scalar. Then, it can be deduced from (8.3) and (8.31) that

$$\begin{aligned}
&\bar{A}_{1,k}\mathbb{E}\{x_k x_k^T\}\bar{A}_{1,k}^T \\
&\leq \bar{M}_k \mathscr{F}_k \bar{N}_k \left[(1 + \gamma_{k,4})P_{k|k} + (1 + \gamma_{k,4}^{-1})\hat{x}_{k|k}\hat{x}_{k|k}^T\right] \bar{N}_k^T \mathscr{F}_k^T \bar{M}_k^T \\
&\leq \mathrm{tr}\left(\bar{N}_k \left[(1 + \gamma_{k,4})P_{k|k} + (1 + \gamma_{k,4}^{-1})\hat{x}_{k|k}\hat{x}_{k|k}^T\right] \bar{N}_k^T\right) \bar{M}_k \bar{M}_k^T,
\end{aligned} \quad (8.32)$$

where $\bar{M}_k$ and $\bar{N}_k$ are given in (8.23). According to (8.4), one has

$$\begin{aligned}
&\mathbb{E}\{[g(x_k) - g(\hat{x}_{k|k})][g(x_k) - g(\hat{x}_{k|k})]^T\} \\
&\leq \mathbb{E}\{\bar{G}e_{k|k}e_{k|k}^T\bar{G}^T\} \\
&= \bar{G}P_{k|k}\bar{G}^T,
\end{aligned} \quad (8.33)$$

where $\bar{G}$ is defined as in (8.23). Thus, according to (8.30)–(8.33), we have

$$\begin{aligned}
P_{k+1|k} \leq\ & (1 + \gamma_{k,1} + \gamma_{k,2})\bar{A}_k P_{k|k}\bar{A}_k^T + (1 + \gamma_{k,1}^{-1} + \gamma_{k,3}) \\
& \times \mathrm{tr}\left(\bar{N}_k \left[(1 + \gamma_{k,4})P_{k|k} + (1 + \gamma_{k,4}^{-1})\hat{x}_{k|k}\hat{x}_{k|k}^T\right] \bar{N}_k^T\right) \bar{M}_k \bar{M}_k^T \\
& + \bar{\alpha}(1 - \bar{\alpha})\bar{A}_{2,k}\left[(1 + \gamma_{k,4})P_{k|k} + (1 + \gamma_{k,4}^{-1})\hat{x}_{k|k}\hat{x}_{k|k}^T\right]\bar{A}_{2,k}^T \\
& + (1 + \gamma_{k,2}^{-1} + \gamma_{k,3}^{-1})\bar{G}P_{k|k}\bar{G}^T + B_k Q_k B_k^T.
\end{aligned} \quad (8.34)$$

Next, adopting Lemma 2.2 again, one gets

$$\begin{aligned}
&\mathbb{E}\{\mathcal{C}_{k+1} + \mathcal{C}_{k+1}^T\} \\
&\leq \gamma_{k+1,5}(I - G_{k+1}C_{k+1})P_{k+1|k}(I - G_{k+1}C_{k+1})^T
\end{aligned}$$

$$
\begin{aligned}
&+\gamma_{k+1,5}^{-1}G_{k+1}C_{k+1}\mathbb{E}\{x_{k+1}x_{k+1}^T\}C_{k+1}^TG_{k+1}^T, \quad (8.35)\\
&\mathbb{E}\{\mathcal{D}_{k+1}+\mathcal{D}_{k+1}^T\}\\
&\leq \gamma_{k+1,6}(I-G_{k+1}C_{k+1})P_{k+1|k}(I-G_{k+1}C_{k+1})^T\\
&+\gamma_{k+1,6}^{-1}G_{k+1}\mathbb{E}\{\sigma(C_{k+1}x_{k+1})\sigma^T(C_{k+1}x_{k+1})\}G_{k+1}^T, \quad (8.36)\\
&\mathbb{E}\{\mathcal{E}_{k+1}+\mathcal{E}_{k+1}^T\}\\
&\leq \gamma_{k+1,7}G_{k+1}C_{k+1}\mathbb{E}\{x_{k+1}x_{k+1}^T\}C_{k+1}^TG_{k+1}^T\\
&+\gamma_{k+1,7}^{-1}G_{k+1}\mathbb{E}\{\sigma(C_{k+1}x_{k+1})\sigma^T(C_{k+1}x_{k+1})\}G_{k+1}^T, \quad (8.37)
\end{aligned}
$$

where $\gamma_{k+1,i}>0$ $(i=5,6,7)$ are scalars. Employing the addressing method as in (8.31), we have

$$
\mathbb{E}\{x_{k+1}x_{k+1}^T\}\leq(1+\gamma_{k+1,8})P_{k+1|k}+(1+\gamma_{k+1,8}^{-1})\hat{x}_{k+1|k}\hat{x}_{k+1|k}^T \quad (8.38)
$$

with $\gamma_{k+1,8}>0$ representing a scalar. Furthermore, it can be attained from (8.7) that

$$
\mathbb{E}\{\sigma(C_{k+1}x_{k+1})\sigma^T(C_{k+1}x_{k+1})\}\leq\bar{\varrho}I, \quad (8.39)
$$

where $\bar{\varrho}$ is defined as in (8.23). Substituting (8.35)–(8.39) into (8.18) results in

$$
\begin{aligned}
P_{k+1|k+1}\leq&(1+\gamma_{k+1,5}+\gamma_{k+1,6})(I-G_{k+1}C_{k+1})P_{k+1|k}(I-G_{k+1}C_{k+1})^T\\
&+(1+\gamma_{k+1,5}^{-1}+\gamma_{k+1,7})G_{k+1}C_{k+1}\bar{\Omega}_{k+1}C_{k+1}^TG_{k+1}^T\\
&+G_{k+1}[\bar{\varrho}(1+\gamma_{k+1,6}^{-1}+\gamma_{k+1,7}^{-1})I+R_{k+1}]G_{k+1}^T, \quad (8.40)
\end{aligned}
$$

where

$$
\bar{\Omega}_{k+1}=(1+\gamma_{k+1,8})P_{k+1|k}+(1+\gamma_{k+1,8}^{-1})\hat{x}_{k+1|k}\hat{x}_{k+1|k}^T.
$$

Then, in terms of the mathematical induction method, it is obvious to see that

$$
P_{k+1|k+1}\leq\Omega_{k+1|k+1}. \quad (8.41)
$$

Lastly, we aim to design the optimal estimator gain that can minimize this upper bound $\Omega_{k+1|k+1}$. Based on Lemma 2.3, taking the partial derivative of the trace of (8.22) concerning on G_{k+1} and letting the result be zero, we arrive at

$$
\begin{aligned}
\frac{\partial\operatorname{tr}(\Omega_{k+1|k+1})}{\partial G_{k+1}}=&-2(1+\gamma_{k+1,5}+\gamma_{k+1,6})(I-G_{k+1}C_{k+1})\Omega_{k+1|k}C_{k+1}^T\\
&+2G_{k+1}\Big[(1+\gamma_{k+1,5}^{-1}+\gamma_{k+1,7})C_{k+1}\Omega_{k+1}C_{k+1}^T\\
&+\bar{\varrho}(1+\gamma_{k+1,6}^{-1}+\gamma_{k+1,7}^{-1})I+R_{k+1}\Big]\\
=&\ 0. \quad (8.42)
\end{aligned}
$$

In light of (8.42) and the tedious algebraic manipulations, the estimator gain can be determined as in (8.25). Accordingly, the proof of this theorem is complete.

Up to now, we have obtained an upper bound regarding the EEC and optimized this upper bound via selecting an appropriate estimator gain. Through solving a set of difference equations, the estimator gain has also been derived. It is noteworthy to see that the designed estimation algorithm can better estimate the system states and faults simultaneously and the developed recursive estimation scheme is suitable for online applications.

Remark 8.4 It is seen that $\breve{x}_k \in \mathbb{R}^n$ and $f_k \in \mathbb{R}^f$, thus the computational complexity of the proposed estimation algorithm can be described as $O((n+f)^3)$, which exhibits a close relationship with the variable dimension. It is clearly found that the computational burden is primarily resulted from the basic mathematical computation. Notably, many handling methods can be utilized to improve the computational efficiency of mathematical calculations.

8.2.2 Performance Analysis

According to Definition 2.1, a sufficient condition is proposed in the following theorem to testify the boundedness of the estimation error dynamics.

Theorem 8.2 *Assume that there are positive real scalars $\bar{a}$, $\underline{b}$, $\bar{b}$, $\underline{c}$, $\bar{c}$, $\bar{f}$, $\bar{m}$, $\bar{n}$, $\bar{g}$, $\underline{q}_1$, $\bar{q}_1$, $\bar{q}_2$, $\bar{\vartheta}_1$, $\bar{\vartheta}_2$ and $\bar{\vartheta}_3$ such that the following conditions:*

$$
\begin{aligned}
&\|\bar{A}_k\| \le \bar{a}, \quad \underline{b}I \le \breve{B}_k \breve{B}_k^T \le \bar{b}I, \quad \underline{c} \le \|\breve{C}_k\| \le \bar{c},\\
&\|\breve{F}_k\| \le \bar{f}, \quad \|\breve{M}_k\| \le \bar{m}, \quad \|\breve{N}_k\| \le \bar{n}, \quad \|G\| \le \bar{g},\\
&\underline{q}_1 I \le Q_k \le \bar{q}_1 I, \quad R_k \le \bar{q}_2 I, \quad \mathrm{tr}(\Theta_k) \le \bar{\vartheta}_1,\\
&\mathrm{tr}(\Omega_{k|k}) \le \bar{\vartheta}_2, \quad \mathrm{tr}(\Omega_{k+1}) \le \bar{\vartheta}_3,\\
&\rho = \bar{a}^2 \left(1 + \frac{\bar{c}^2}{\underline{c}^2}\right)^2 < 1
\end{aligned}
\tag{8.43}
$$

hold, then the estimation error is exponentially bounded in the mean-square sense.

Proof Substituting (8.14) into (8.20) leads to

$$
e_{k+1|k+1} = \Lambda_{k+1} \bar{A}_k e_{k|k} + r_{k+1} + s_{k+1}, \tag{8.44}
$$

where

$$
\begin{aligned}
\Lambda_{k+1} &= I - G_{k+1}C_{k+1},\\
r_{k+1} &= \Lambda_{k+1}\bar{A}_{1,k}x_k + \Lambda_{k+1}[g(x_k) - g(\hat{x}_{k|k})] + G_{k+1}[C_{k+1}x_{k+1} - \sigma(C_{k+1}x_{k+1})],\\
s_{k+1} &= (\alpha_k - \bar{\alpha})\Lambda_{k+1}\bar{A}_{2,k}x_k + \Lambda_{k+1}B_k\omega_k - G_{k+1}\nu_{k+1}.
\end{aligned} \tag{8.45}
$$

From (8.25) and (8.26), one has

$$
\begin{aligned}
&\|G_{k+1}\|\\
&= \|(1 + \gamma_{k+1,5} + \gamma_{k+1,6})\Omega_{k+1|k}C_{k+1}^T\mathcal{K}_{k+1}^{-1}\|\\
&\le \frac{\bar{c}}{\underline{c}^2} := \bar{k}.
\end{aligned} \tag{8.46}
$$

Similarly, one has

$$
\|\Lambda_{k+1}\| = \|I - G_{k+1}C_{k+1}\| \le 1 + \frac{\bar{c}^2}{\underline{c}^2} := \bar{\lambda}. \tag{8.47}
$$

Based on Lemma 2.2 and (8.43), one obtains

$$
\begin{aligned}
\mathbb{E}\{r_{k+1}^T r_{k+1}\} \le\ & (1 + \varsigma_1 + \varsigma_2)\mathbb{E}\{x_k^T\bar{A}_{1,k}^T\Lambda_{k+1}^T\Lambda_{k+1}\bar{A}_{1,k}x_k\} + (1 + \varsigma_1^{-1} + \varsigma_3)\\
& \times\mathbb{E}\{[g(x_k) - g(\hat{x}_{k|k})]^T\Lambda_{k+1}^T\Lambda_{k+1}[g(x_k) - g(\hat{x}_{k|k})]\}\\
& + (1 + \varsigma_2^{-1} + \varsigma_3^{-1})\mathbb{E}\{[C_{k+1}x_{k+1} - \sigma(C_{k+1}x_{k+1})]^T G_{k+1}^T G_{k+1}\\
& \times[C_{k+1}x_{k+1} - \sigma(C_{k+1}x_{k+1})]\}\\
\le\ & (1 + \varsigma_1 + \varsigma_2)\mathrm{tr}\big(\mathbb{E}\{x_k x_k^T\}\bar{N}_k^T\mathscr{F}_k^T\bar{M}_k^T\Lambda_{k+1}^T\Lambda_{k+1}\bar{M}_k\mathscr{F}_k\bar{N}_k\big)\\
& + (1 + \varsigma_1^{-1} + \varsigma_3)\mathrm{tr}(\bar{G}\Omega_{k|k}\bar{G}^T\Lambda_{k+1}^T\Lambda_{k+1}) + (1 + \varsigma_2^{-1} + \varsigma_3^{-1})\\
& \times\Big[(1 + \varsigma_4)\mathrm{tr}(\mathbb{E}\{x_{k+1}x_{k+1}^T\}C_{k+1}^T G_{k+1}^T G_{k+1}C_{k+1}) + (1 + \varsigma_4^{-1})\\
& \times\mathrm{tr}\big(\mathbb{E}\{\sigma(C_{k+1}x_{k+1})\sigma^T(C_{k+1}x_{k+1})\}G_{k+1}^T G_{k+1}\big)\Big]\\
\le\ & (1 + \varsigma_1 + \varsigma_2)\bar{m}^2\bar{\lambda}^2\bar{n}^2\bar{\vartheta}_1 + (1 + \varsigma_1^{-1} + \varsigma_3)\bar{\lambda}^2\bar{g}^2\bar{\vartheta}_2\\
& + (1 + \varsigma_2^{-1} + \varsigma_3^{-1})\big[(1 + \varsigma_4)\bar{c}^2\bar{k}^2\bar{\vartheta}_3 + (1 + \varsigma_4^{-1})\bar{k}^2\bar{\varrho}m\big]\\
:=\ & \bar{r}
\end{aligned} \tag{8.48}
$$

and

$$
\begin{aligned}
\mathbb{E}\{s_{k+1}^T s_{k+1}\} &= \bar{\alpha}(1 - \bar{\alpha})\mathbb{E}\{x_k^T\bar{A}_{2,k}^T\Lambda_{k+1}^T\Lambda_{k+1}\bar{A}_{2,k}x_k\}\\
&\quad + \mathbb{E}\{\omega_k^T B_k^T\Lambda_{k+1}^T\Lambda_{k+1}B_k\omega_k\}\\
&\quad + \mathbb{E}\{\nu_{k+1}^T G_{k+1}^T G_{k+1}\nu_{k+1}\}\\
&\le \bar{\alpha}(1 - \bar{\alpha})\bar{\lambda}^2\bar{f}^2\bar{\vartheta}_1 + \bar{q}_1 n\bar{b}\bar{\lambda}^2 + \bar{q}_2 m\bar{k}^2\\
&:= \bar{s},
\end{aligned} \tag{8.49}
$$

where $\varsigma_i > 0$ $(i = 1, 2, 3, 4)$ are scalars.

Next, consider the following iterative matrix equation of Ψ_k:

$$\Psi_{k+1} = \Lambda_{k+1}\bar{A}_k\Psi_k\bar{A}_k^T\Lambda_{k+1}^T + B_kQ_kB_k^T + \varepsilon I, \tag{8.50}$$

where $\Psi_0 = B_0Q_0B_0^T + \varepsilon I$ and $\varepsilon > 0$ is a scalar. In terms of the above recursive matrix equation, it can be found that

$$\begin{aligned}\|\Psi_{k+1}\| &\leq \|\Lambda_{k+1}\|^2\|\bar{A}_k\|^2\|\Psi_k\| + \|B_kQ_kB_k^T\| + \|\varepsilon I\| \\ &\leq \rho\|\Psi_k\| + \bar{b}\bar{q}_1 + \varepsilon,\end{aligned} \tag{8.51}$$

where ρ is given in (8.43). Besides, the following inequality is obtained directly:

$$\|\Psi_k\| \leq \rho^k\|\Psi_0\| + (\bar{b}\bar{q}_1 + \varepsilon)\sum_{i=0}^{k-1}\rho^i. \tag{8.52}$$

It follows from $\rho < 1$ that

$$\|\Psi_k\| < \|\Psi_0\| + (\bar{b}\bar{q}_1 + \varepsilon)\sum_{i=0}^{\infty}\rho^i = \|\Psi_0\| + \frac{\bar{b}\bar{q}_1 + \varepsilon}{1 - \rho}. \tag{8.53}$$

Furthermore, it can be derived that

$$\Psi_k \geq \varepsilon I. \tag{8.54}$$

According to (8.53) and (8.54), there are two scalars $\bar{\psi} > 0$ and $\underline{\psi} > 0$ such that $\underline{\psi} I \leq \Psi_k \leq \bar{\psi} I$ holds for all $k \geq 0$.

Let $V_k(e_{k|k}) = e_{k|k}^T\Psi_k^{-1}e_{k|k}$. For a scalar $\delta > 0$, it can be obtained from Lemma 2.2 and (8.44) that

$$\begin{aligned}&\mathbb{E}\{V_{k+1}(e_{k+1|k+1})|e_{k|k}\} - (1+\delta)V_k(e_{k|k}) \\ &= \mathbb{E}\Big\{ \left(\Lambda_{k+1}\bar{A}_ke_{k|k} + r_{k+1} + s_{k+1}\right)^T \Psi_{k+1}^{-1}\left(\Lambda_{k+1}\bar{A}_ke_{k|k} + r_{k+1} + s_{k+1}\right)\Big\} \\ &\quad -(1+\delta)e_{k|k}^T\Psi_k^{-1}e_{k|k} \\ &= \mathbb{E}\left\{e_{k|k}^T[\bar{A}_k^T\Lambda_{k+1}^T\Psi_{k+1}^{-1}\Lambda_{k+1}\bar{A}_k - (1+\delta)\Psi_k^{-1}]e_{k|k}\right\} \\ &\quad +2\mathbb{E}\left\{e_{k|k}^T\bar{A}_k^T\Lambda_{k+1}^T\Psi_{k+1}^{-1}r_{k+1}\right\} \\ &\quad +\mathbb{E}\left\{r_{k+1}^T\Psi_{k+1}^{-1}r_{k+1}\right\} + \mathbb{E}\left\{s_{k+1}^T\Psi_{k+1}^{-1}s_{k+1}\right\} \\ &\leq (1+\delta)\mathbb{E}\left\{e_{k|k}^T\left(\bar{A}_k^T\Lambda_{k+1}^T\Psi_{k+1}^{-1}\Lambda_{k+1}\bar{A}_k - \Psi_k^{-1}\right)e_{k|k}\right\} \\ &\quad +(1+\delta^{-1})\mathbb{E}\left\{r_{k+1}^T\Psi_{k+1}^{-1}r_{k+1}\right\} + \mathbb{E}\left\{s_{k+1}^T\Psi_{k+1}^{-1}s_{k+1}\right\}.\end{aligned} \tag{8.55}$$

In view of the definition of matrix Ψ_{k+1} and employing the matrix inversion lemma, one has

$$\begin{aligned}
&\bar{A}_k^T \Lambda_{k+1}^T \Psi_{k+1}^{-1} \Lambda_{k+1} \bar{A}_k - \Psi_k^{-1} \\
&= \bar{A}_k^T \Lambda_{k+1}^T \left(\Lambda_{k+1} \bar{A}_k \Psi_k \bar{A}_k^T \Lambda_{k+1}^T + B_k Q_k B_k^T + \varepsilon I\right)^{-1} \Lambda_{k+1} \bar{A}_k - \Psi_k^{-1} \\
&= -\left[\Psi_k + \Psi_k \bar{A}_k^T \Lambda_{k+1}^T \left(B_k Q_k B_k^T + \varepsilon I\right)^{-1} \Lambda_{k+1} \bar{A}_k \Psi_k\right]^{-1} \\
&= -\left[I + \bar{A}_k^T \Lambda_{k+1}^T \left(B_k Q_k B_k^T + \varepsilon I\right)^{-1} \Lambda_{k+1} \bar{A}_k \Psi_k\right]^{-1} \Psi_k^{-1} \\
&\leq -\left(1 + \frac{\bar{a}^2 \bar{\lambda}^2 \bar{\psi}}{\underline{q}_1 \underline{b}}\right)^{-1} \Psi_k^{-1}.
\end{aligned} \tag{8.56}$$

Substituting (8.56) into (8.55) implies

$$\begin{aligned}
&\mathbb{E}\{V_{k+1}(e_{k+1|k+1})|e_{k|k}\} - (1+\delta) V_k(e_{k|k}) \\
&\leq -(1+\delta)\left(1 + \frac{\bar{a}^2 \bar{\lambda}^2 \bar{\psi}}{\underline{q}_1 \underline{b}}\right)^{-1} V_k(e_{k|k}) + \kappa
\end{aligned} \tag{8.57}$$

with $\kappa = (1+\delta^{-1})\frac{\bar{r}}{\underline{\psi}} + \frac{\bar{s}}{\underline{\psi}}$. Then, it is seen from (8.57) that

$$\mathbb{E}\{V_{k+1}(e_{k+1|k+1})|e_{k|k}\} \leq \chi V_k(e_{k|k}) + \kappa, \tag{8.58}$$

where $\chi = (1+\delta)\left[1 - \left(1 + \frac{\bar{a}^2 \bar{\lambda}^2 \bar{\psi}}{\underline{q}_1 \underline{b}}\right)^{-1}\right]$. It is obvious to find that $\chi \in (0, 1)$ for $\delta > 0$. In addition, we get

$$\begin{aligned}
\mathbb{E}\{\|e_{k+1|k+1}\|^2\} &\leq \frac{\bar{\psi}}{\underline{\psi}} \mathbb{E}\{\|e_{0|0}\|^2\} \chi^{k+1} + \kappa \bar{\psi} \sum_{i=0}^{k} \chi^i \\
&\leq \frac{\bar{\psi}}{\underline{\psi}} \mathbb{E}\{\|e_{0|0}\|^2\} \chi^{k+1} + \kappa \bar{\psi} \sum_{i=0}^{\infty} \chi^i \\
&= \frac{\bar{\psi}}{\underline{\psi}} \mathbb{E}\{\|e_{0|0}\|^2\} \chi^{k+1} + \frac{\kappa \bar{\psi}}{1-\chi}.
\end{aligned} \tag{8.59}$$

By virtue of Definition 2.1, it can be summarized that $e_{k|k}$ is exponentially bounded in mean-square sense.

On the other hand, it is noted that the estimation is biased due mainly to the considered nonlinearity, sensor saturations and ROFs. Since

$$\begin{aligned}
&\mathbb{E}\{(e_{k+1|k+1} - \mathbb{E}\{e_{k+1|k+1}\})^T (e_{k+1|k+1} - \mathbb{E}\{e_{k+1|k+1}\})\} \\
&= \mathbb{E}\{\|e_{k+1|k+1}\|^2\} - \|\mathbb{E}\{e_{k+1|k+1}\}\|^2 \\
&\geq 0,
\end{aligned}$$

one has

$$\begin{aligned}\|\mathbb{E}\{e_{k+1|k+1}\}\| &\le \sqrt{\mathbb{E}\{\|e_{k+1|k+1}\|^2\}}\\ &\le \sqrt{\frac{\bar{\psi}}{\underline{\psi}}\mathbb{E}\{\|e_{0|0}\|^2\}\chi^{k+1}+\frac{\kappa\bar{\psi}}{1-\chi}}.\end{aligned}$$

The proof of this theorem is ended now.

Remark 8.5 So far, the fault estimation issue has been addressed for time-varying uncertain nonlinear systems with ROFs and sensor saturations. In the introduced time-varying stochastic system model, there are mainly the following four factors that complicate the estimation algorithm, i.e., parameter uncertainty, nonlinearity, ROFs and sensor saturations. The merits of the developed fault estimation algorithm can be summarized as: 1) the constructed estimator is fairly simple and of a recursive manner, thus the proposed estimation approach is easy to implement; and 2) the influences from the parameter uncertainty, nonlinearity, ROFs and sensor saturations are all examined in the proposed estimation algorithm.

8.3 Three Illustrative Examples

In this section, three examples are applied to show the correctness and applicability of the designed joint state and fault estimation algorithm.

Example 8.1 Consider a nonlinear system (8.1) and (8.2) with the following parameters:

$$\begin{aligned}\breve{A}_k &= \begin{bmatrix} 0.35 & -0.2+0.1\sin(6k) \\ 0.26 & -0.08 \end{bmatrix}, \quad \breve{F}_k = \begin{bmatrix} 0.05 \\ 0.02 \end{bmatrix},\\ \breve{B}_k &= \begin{bmatrix} 0.2-0.1\sin(5k) \\ 0.3 \end{bmatrix}, \quad \breve{M}_k = \begin{bmatrix} 0.1 \\ 0.3 \end{bmatrix},\\ \breve{C}_k &= \begin{bmatrix} 0.95 & 0.8 \end{bmatrix}, \quad \breve{N}_k = \begin{bmatrix} 0.2 & 0 \end{bmatrix}, \quad \mathscr{F}_k = \sin(8k).\end{aligned}$$

The state vector is denoted by $\breve{x}_k = \begin{bmatrix} \breve{x}_{1,k} & \breve{x}_{2,k} \end{bmatrix}^T$ with $\breve{x}_{i,k}$ $(i=1,2)$ being the i-th element of the system state, the covariances of $\omega_k \in \mathbb{R}$ and $v_k \in \mathbb{R}$ are 0.03 and 0.06, respectively. The saturation level is set to be 0.3. The nonlinear function $\breve{g}(\breve{x}_k)$ is given as

$$\breve{g}(\breve{x}_k) = \begin{bmatrix} 0.2\breve{x}_{1,k} - \tanh(0.1\breve{x}_{1,k}) \\ 0.15\breve{x}_{2,k} - \tanh(0.15\breve{x}_{2,k}) \end{bmatrix}$$

and $G = \mathrm{diag}\{0.3, 0.3\}$.

For simulation purpose, the fault is set as

$$f_k = \begin{cases} 3, & \text{if} \quad k \geq 40 \\ 0, & \text{otherwise} \end{cases}$$

and the matrix $A_{f,k}$ is described by $A_{f,k} = 1$. Moreover, the other parameters are chosen as $\bar{\alpha} = 0.95$, $\gamma_{k,i} = 0.2$ and $\gamma_{k+1,j} = 0.3$ $(i = 1, 2, 3, 4;\ j = 5, 6, 7, 8)$. According to Theorem 8.1, we can have the estimator gain recursively and then the designed estimator in (8.11) and (8.12) can be implemented. The simulations are given in Figs. 8.1, 8.2, 8.3 and 8.4. Among them, the trajectories of the real states $\breve{x}_{i,k}$ and their estimations $\hat{\breve{x}}_{i,k}$ $(i = 1, 2)$ are depicted in Figs. 8.1 and 8.2, the actual fault and its estimation are depicted in Figs. 8.3, and 8.4 plots the measurement signals without and with saturations in the case of no measurement noises. Among the simulation results, it is observed that the system states and fault can be estimated effectively which further confirms the usefulness of the proposed estimation algorithm. This is mainly due to the fact that special efforts have been made to compensate the impacts from the ROFs and sensor saturations onto the estimation performance when designing the estimator.

Example 8.2 Consider a continuous stirred tank reactor (CSTR) shown in Fig. 8.5, where A and B are the educt and the desired product, respectively. Moreover, C_{A0} is the low concentration of educt A, C_A and C_B represent the concentrations of the educt A and the desired product B within the reactor, ϑ stands for the reactor temperature and ϑ_k is the coolant temperature. The balance equations of the CSTR are given below Zhang et al. (2014):

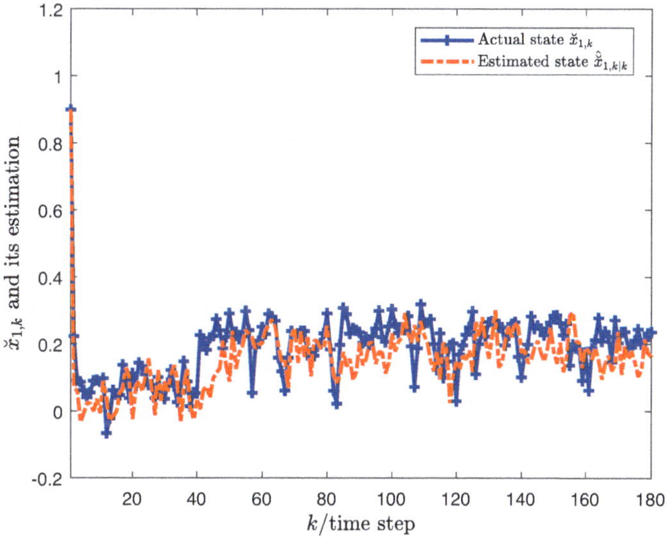

Fig. 8.1 The actual state $\breve{x}_{1,k}$ and its estimation $\hat{\breve{x}}_{1,k|k}$

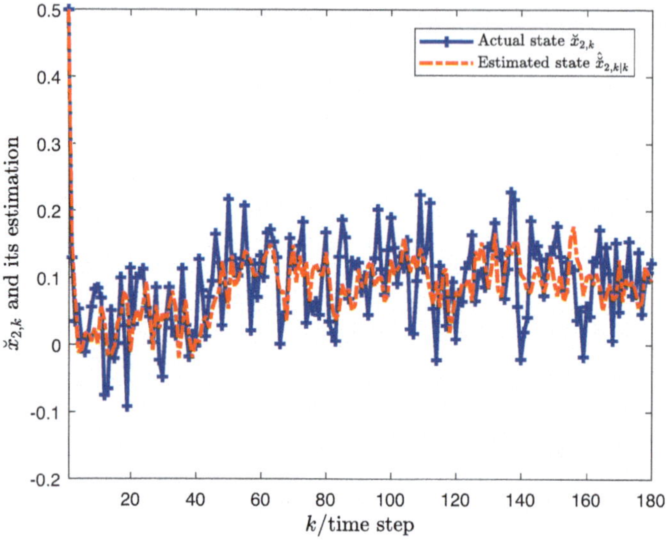

Fig. 8.2 The actual state $\breve{x}_{2,k}$ and its estimation $\hat{\breve{x}}_{2,k|k}$

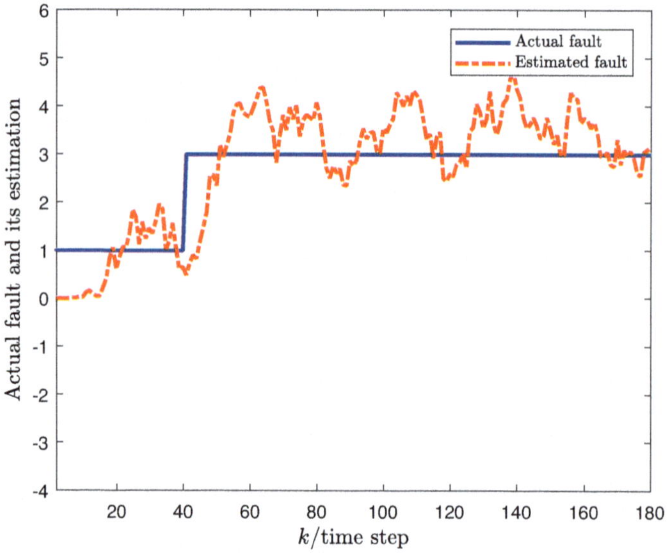

Fig. 8.3 The actual fault f_k and its estimation

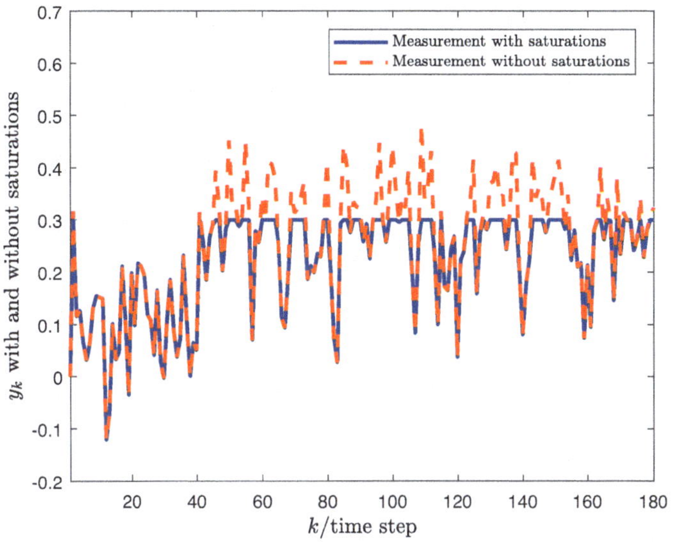

Fig. 8.4 The measurements y_k with/without saturations

Fig. 8.5 The continuous stirred tank reactor

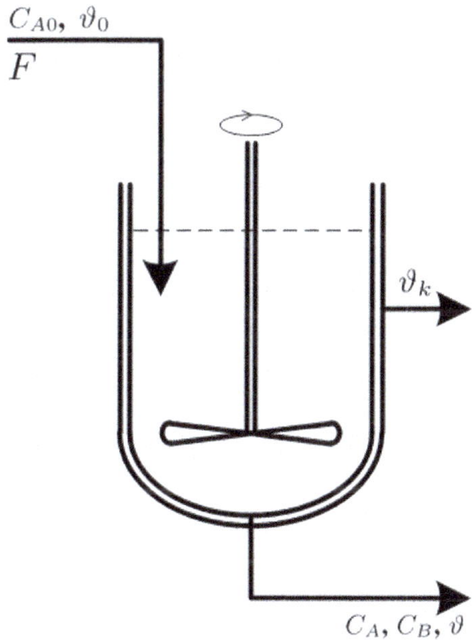

$$
\begin{cases}
\frac{\mathrm{d}C_A}{\mathrm{d}t} = \frac{F}{V}(C_{A0} - C_A) - k_1 C_A - k_3 C_A^2 \\
\frac{\mathrm{d}C_B}{\mathrm{d}t} = -\frac{F}{V}C_B + k_1 C_A - k_2 C_B \\
\frac{\mathrm{d}\vartheta}{\mathrm{d}t} = \frac{F}{V}(\vartheta_0 - \vartheta) + \frac{k_\omega A_R}{\varrho C_P V}(\vartheta_k - \vartheta) \\
\qquad - \frac{k_1 C_A \Delta H_R^{AB} + k_2 C_B \Delta H_R^{BC} + k_3 C_A^2 \Delta H_R^{AD}}{\varrho C_P}
\end{cases}
\tag{8.60}
$$

where F is the normalized process stream inflow, V is the volume flow, ϱ is the density, C_P is the heat capacity, and the other parameters can be found in Zhang et al. (2014).

Along the similar lines in Zhang et al. (2014), Klatt and Engell (1998), by defining $\breve{x}_k = \left[C_A^T \; C_B^T \; \vartheta^T \right]^T$, the system (8.60) can be reformulated by the discrete-time system described as in (8.1)–(8.2) with

$$
\breve{A}_k = \begin{bmatrix} 0.3872 & 0.0222 & 0.01823 \\ 0.2444 & 0.3897 & 0.0007102 \\ -0.06849 & 0.9711 & 0.4008 \end{bmatrix},
$$

$$
\breve{B}_k = \begin{bmatrix} 0.009496 \\ 0.6474 \\ 0.6779 \end{bmatrix}, \quad \breve{C}_k = \begin{bmatrix} 0 & 0 & 1 \end{bmatrix}.
$$

When modeling the CSTR system, the sudden changes of the low concentration of the educt could be described by a fault input which is modeled by an additive fault. Moreover, there exist the modeling errors (parameter uncertainties), nonlinearities and sensor saturations. Our purpose is to design a fault estimation scheme for time-varying systems with parameter uncertainties, nonlinearities and sensor saturations. Other parameters are given as follows:

$$
\breve{F}_k = \begin{bmatrix} 0.01 \\ 0 \\ 0 \end{bmatrix}, \quad \breve{M}_k = \begin{bmatrix} 0.1 \\ 0.3 \\ 0.2 \end{bmatrix},
$$

$$
\begin{aligned}
\breve{N}_k &= \begin{bmatrix} 0.1 & 0 & 0.2 \end{bmatrix}, \quad \mathscr{F}_k = \sin(3k), \\
A_{f,k} &= 2\sin(k), \quad \bar{\alpha} = 0.95, \quad Q_k = 0.05, \quad R_k = 0.02, \\
\gamma_{k,1} &= \gamma_{k,2} = 0.3, \quad \gamma_{k,3} = \gamma_{k,4} = 0.5, \\
\gamma_{k+1,5} &= \gamma_{k+1,6} = \gamma_{k+1,7} = \gamma_{k+1,8} = 0.3.
\end{aligned}
$$

Moreover, the saturation level is selected to be 0.2.

Let the nonlinear function $\breve{g}(\breve{x}_k)$ be

$$\breve{g}(\breve{x}_k) = \begin{bmatrix} 0.1\breve{x}_{1,k} - \tanh(0.2\breve{x}_{1,k}) \\ 0.15\breve{x}_{2,k} - \tanh(0.15\breve{x}_{2,k}) \\ 0.2\breve{x}_{3,k} - \tanh(0.1\breve{x}_{3,k}) \end{bmatrix}$$

with $G = \mathrm{diag}\{0.3, 0.3, 0.3\}$.

By using the Theorem 8.1, the simulations can be shown as in Figs. 8.6, 8.7, 8.8, 8.9 and 8.10. From the Fig. 8.9, it can be found that the additive fault can be well estimated through the proposed estimation algorithm, which further reveals the applicability of the established fault estimation method.

Example 8.3 Following Farina et al. (2002), Lei et al. (2011), the joint recursive robust state/fault estimation issue is addressed for a ballistic object tracking system. When tracking a ballistic object, the measurements are received by a radar system with various sensors communicating over a communication network. The sensor saturations might appear because of the physical limitations. In addition, the system might encounter the fault due to the unpredictable variations of the network media, and the ROFs may stem from the issues of the aging and electromagnetic interference. Furthermore, the norm-bounded uncertainties are used to reflect the imperfections of the mathematical model. Our main purpose is, therefore, to establish a fault estimation scheme for time-varying nonlinear systems with parameter uncertainties and sensor saturations. Therefore, we aim to design a recursive estimator such that, for parameter uncertainties, ROFs and sensor saturations, this upper bound can be optimized by

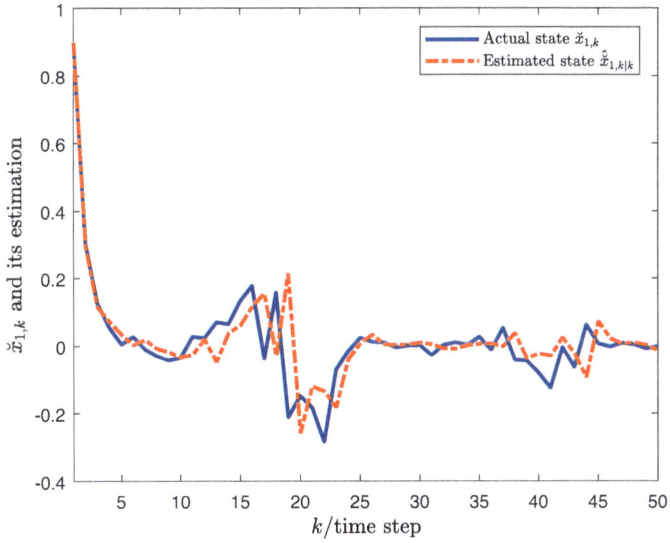

Fig. 8.6 The actual state $\breve{x}_{1,k}$ and its estimation $\hat{\breve{x}}_{1,k|k}$

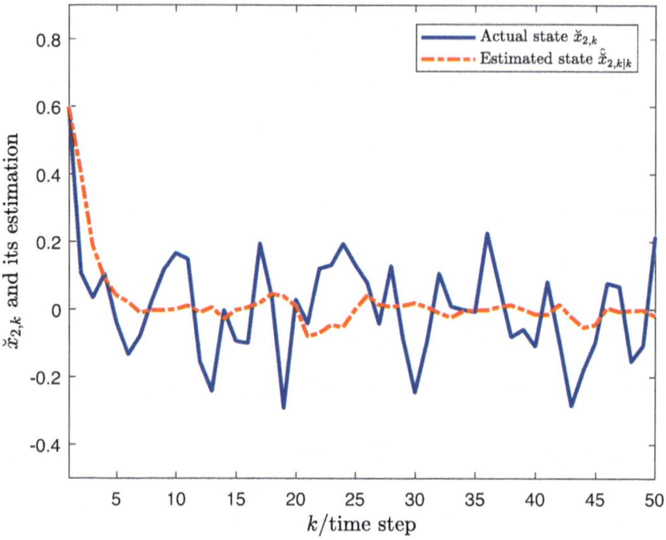

Fig. 8.7 The actual state $\breve{x}_{2,k}$ and its estimation $\hat{\breve{x}}_{2,k|k}$

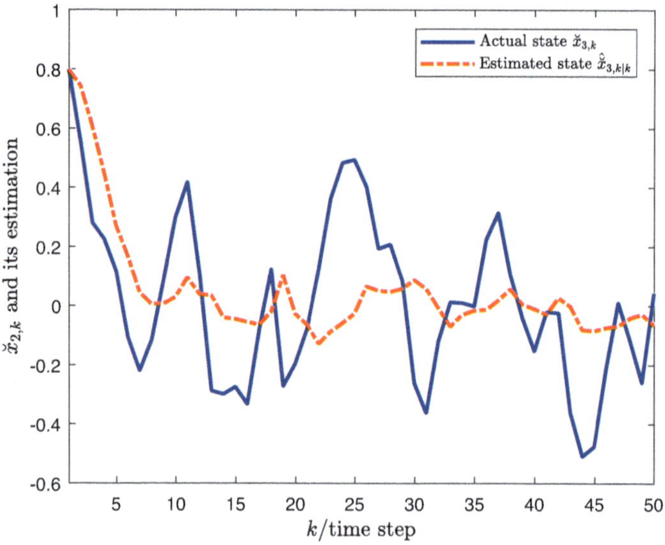

Fig. 8.8 The actual state $\breve{x}_{3,k}$ and its estimation $\hat{\breve{x}}_{3,k|k}$

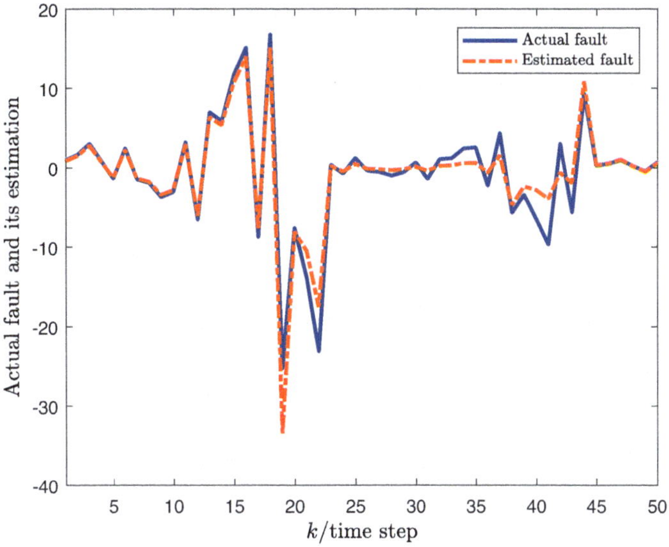

Fig. 8.9 The actual fault f_k and its estimation

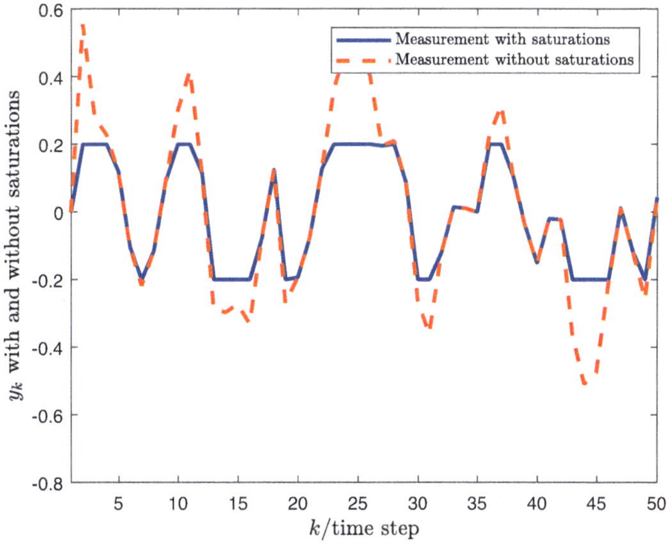

Fig. 8.10 The measurements y_k with/without saturations

appropriately choosing the estimator gain. Particularly, the target abscissa, target ordinate and the fault are all estimated jointly. The proposed ballistic object tracking systems are modeled by

$$\begin{cases} \breve{x}_{k+1} = (\breve{A}_k + \Delta\breve{A}_k)\breve{x}_k + \breve{g}(\breve{x}_k) + \alpha_k \breve{F}_k f_k + \omega_k \\ \breve{y}_k = \sigma(\breve{C}_k \breve{x}_k) + \nu_k \end{cases}$$

with

$$\begin{aligned}
&\breve{g}(\breve{x}_k) = G(h(\breve{x}_k) + H), \ \rho(\breve{x}_{2,k}) = \theta_1 \exp(-\theta_2 \breve{x}_{2,k}), \\
&h(\breve{x}_k) = -\frac{g\rho(\breve{x}_{2,k})}{2\beta}\sqrt{\dot{\breve{x}}_{1,k}^2 + \dot{\breve{x}}_{2,k}^2}\begin{bmatrix} \dot{\breve{x}}_{1,k} \\ \dot{\breve{x}}_{2,k} \end{bmatrix}, \\
&\breve{A}_k = \begin{bmatrix} 1 & T & 0 & 0 \\ 0 & 1 & 0 & 0 \\ 0 & 0 & 1 & T \\ 0 & 0 & 0 & 1 \end{bmatrix}, \ G = \begin{bmatrix} \frac{T^2}{2} & 0 \\ T & 0 \\ 0 & \frac{T^2}{2} \\ 0 & T \end{bmatrix}, \ \breve{F}_k = \begin{bmatrix} 0.01 \\ 0 \\ 0.01 \\ 0 \end{bmatrix}, \ \breve{M}_k = \begin{bmatrix} 0.1 \\ 0.3 \\ 0.2 \\ 0.1 \end{bmatrix}, \\
&\breve{C}_k = \begin{bmatrix} 1 & 0 & 0 & 0 \\ 0 & 0 & 1 & 0 \end{bmatrix}, \ H = \begin{bmatrix} 0 \\ -g \end{bmatrix}, \ \breve{N}_k = \begin{bmatrix} 0.1 & 0 & 0.2 & 0.1 \end{bmatrix}, \ \mathscr{F}_k = \sin(3k),
\end{aligned}$$

where $\breve{x}_k = \begin{bmatrix} \breve{x}_{1,k} & \dot{\breve{x}}_{1,k} & \breve{x}_{2,k} & \dot{\breve{x}}_{2,k} \end{bmatrix}^T$, $\breve{x}_{1,k}$ is the target abscissa, $\breve{x}_{2,k}$ is the target ordinate, T denotes the sampling period, g represents the gravity acceleration, β is the ballistic parameter that relies on the object mass, the shape and the cross-section area. The function $\rho(\cdot)$ is the air density, typically being an exponentially attenuation function of the object height ($\theta_1 = 1.227$, $\theta_2 = 1.093 \times 10^{-4}$ for the object height $\breve{x}_{2,k} <$ 9144m, and $\theta_1 = 1.754$, $\theta_2 = 1.49 \times 10^{-4}$ for the object height $\breve{x}_{2,k} \geq 9144$m). The noise covariances are $Q_k = 10I$ and $R_k = 10I$. Here,

$$Q_k = \text{cdiag}\{\bar{Q}, \bar{Q}\}, \ \ \bar{Q} = \begin{bmatrix} \frac{T^3}{3} & \frac{T^2}{2} \\ \frac{T^2}{2} & T \end{bmatrix}.$$

Other parameters are set as

$$\begin{aligned}
&g = 9.81\text{m/s}^2, \ \beta = 4 \times 10^4\text{kg/ms}^2, \ c = 0.1\text{m}^2/\text{s}^3, \\
&T = 1\text{s}, \ \bar{\breve{x}}_0 = 10^2 \times \begin{bmatrix} 300 & 4 & 90 & 3 \end{bmatrix}^T, \ A_{f,k} = 2\sin(k), \\
&\bar{\alpha} = 0.95, \ \gamma_{k,1} = \gamma_{k,2} = 0.3, \ \gamma_{k,3} = \gamma_{k,4} = 0.5, \\
&\gamma_{k+1,5} = \gamma_{k+1,6} = \gamma_{k+1,7} = \gamma_{k+1,8} = 0.3.
\end{aligned}$$

Moreover, the saturation level is selected as $10^4 \times \begin{bmatrix} 5 & 5 \end{bmatrix}^T$.

By resorting to Theorem 8.1, the simulation results are given in Figs. 8.11, 8.12 and 8.13. Among the results, the target abscissa, target ordinate and their estimations are plotted in Figs. 8.11 and 8.12 respectively. From the Fig. 8.13, it is observed that

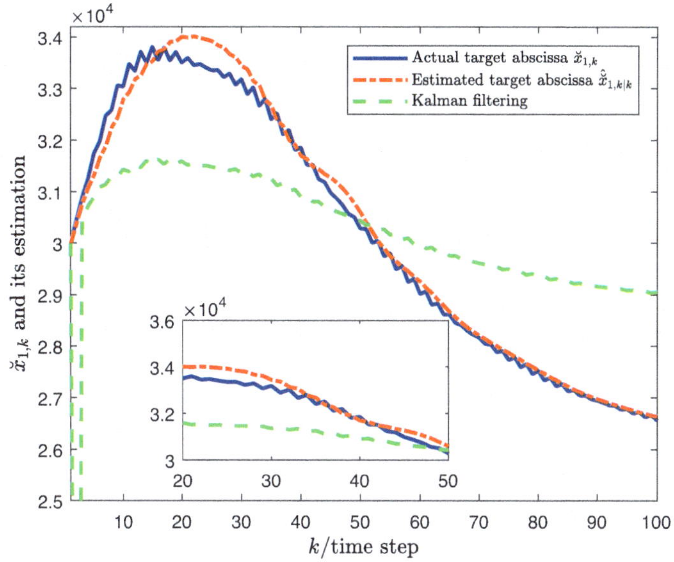

Fig. 8.11 The actual target abscissa and its estimation

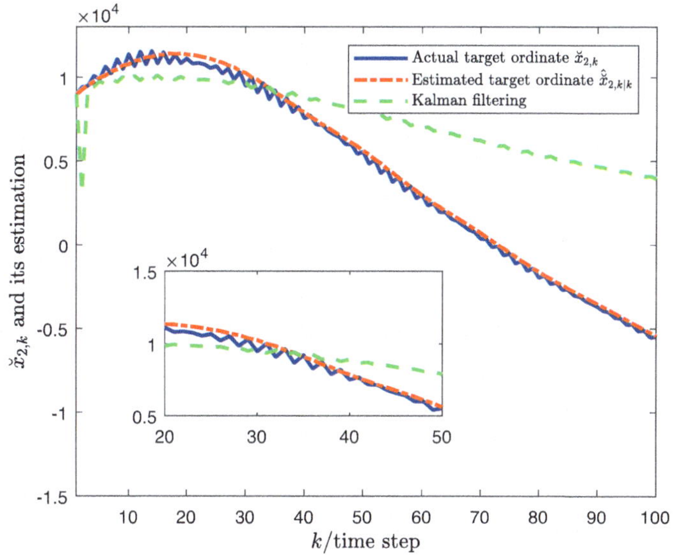

Fig. 8.12 The actual target ordinate and its estimation

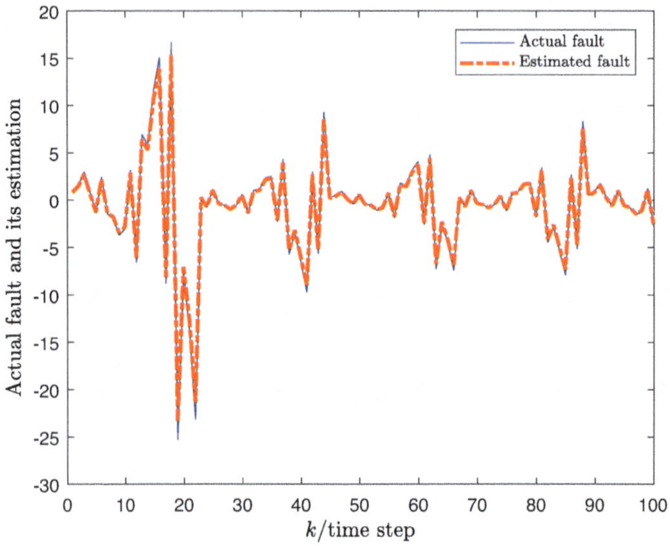

Fig. 8.13 The actual fault and its estimation

the established estimation scheme can estimate the additive fault, target abscissa and target ordinate well.

8.4 Summary

In this chapter, we have studied the joint fault and state estimation issue for time-varying nonlinear stochastic systems subjected to ROFs and sensor saturations. A set of Bernoulli distributed random variables with known conditional probability has been utilized to describe the ROFs. A new compensation strategy has been presented to alleviate the affects from ROFs and sensor saturations onto the estimation performance. On the basis of the available information, a novel estimator has been designed in which an upper bound of the EEC has been solved and minimized through constructing a proper estimator gain. Besides, a sufficient condition has been adopted to prove the exponentially boundedness of the estimation error dynamics in the mean-square sense. Finally, the feasibility of the established estimation method has been testified via some simulation examples.

References

Boulkroune B, Halabi S, Zemouche A (2013) H_{-}/H_{∞} fault detection filter for a class of nonlinear descriptor systems. Int J Control 86(2):253–262

Farina A, Ristic B, Benvenuti D (2002) Tracking a ballistic target: comparison of several nonlinear filters. IEEE Trans Aerosp Electron Syst 38(3):854–867

Klatt KU, Engell S (1998) Gain-scheduling trajectory control of a continuous stirred tank reactor. Comput Chem Eng 22(4–5):491–502

Koenig D, Marx B, Varrier S (2016) Filtering and fault estimation of descriptor switched systems. Automatica 63:116–121

Lei M, van Wyk BJ, Qi Y (2011) Online estimation of the approximate posterior cramer-rao lower bound for discrete-time nonlinear filtering. IEEE Trans Aerosp Electron Syst 47(1):37–57

Zhang W, Dong H, Guo G, Yu L (2014) Distributed sampled-data H_{∞} filtering for sensor networks with nonuniform sampling periods. IEEE Trans Ind Inf 10(2):871–881

Chapter 9
State Estimation for Complex Networks with Missing Measurements

In this chapter, the state estimation issue is handled for time-varying coupled stochastic complex networks involving missing measurements in the variance-constrained framework. A set of Bernoulli distributed random variables is introduced to describe the missing measurements. The Taylor series expansion method is utilized to address the nonlinear function and the linearization errors are estimated. The objective of this chapter is to construct a desirable state estimator such that, for both the missing measurements and random disturbances, an upper bound on the EEC can be ensured and the estimator gain is derived through optimizing the obtained upper bound. It will be shown that the developed recursive state estimation strategy is applicable for online calculations. Lastly, a simulation experiment is utilized to validate the usefulness of the presented state estimation method.

9.1 Problem Formulation

Consider the following time-varying stochastic complex networks with N coupled nodes:

$$x_{i,k+1} = f(x_{i,k}) + \sum_{j=1}^{N} \omega_{ij} \Gamma x_{j,k} + B_{i,k} \varpi_{i,k}, \tag{9.1}$$

$$y_{i,k} = \lambda_{i,k} C_{i,k} x_{i,k} + \nu_{i,k}, \quad i = 1, 2, \ldots, N, \tag{9.2}$$

where $x_{i,k} \in \mathbb{R}^n$ is the state, its initial state value $x_{i,0}$ is with the mean $\bar{x}_{i,0}$, the nonlinear function $f(\cdot)$ has continuous and bounded second order partial derivatives, $y_{i,k} \in \mathbb{R}^m$ is the measurement, $\varpi_{i,k}$ is the process noise with zero-mean and covariance $Q_{i,k} > 0$, and $\nu_{i,k}$ is the measurement noise with zero-mean and

J. Hu et al., *Variance-Constrained Filtering for Stochastic Complex Systems*, Intelligent Control and Learning Systems 18,
https://doi.org/10.1007/978-981-96-2637-3_9

covariance $R_{i,k} > 0$. To describe the probabilistic missing phenomena, the random variable $\lambda_{i,k} \in \mathbb{R}$ is introduced, which possesses the probability mass function $p_{i,k}(s)$ on the interval $[\alpha_{i,k}, \beta_{i,k}]$ $(0 \le \alpha_{i,k} \le \beta_{i,k} \le 1)$ with mathematical expectation $\mu_{i,k}$ and variance $\sigma_{i,k}^2$. $B_{i,k}$ and $C_{i,k}$ are known matrices of proper dimensions. $\Gamma = \text{diag}\{\dot{\gamma}_1, \dot{\gamma}_2, \ldots, \dot{\gamma}_n\}$ denotes a matrix linking the j-th state variable if $\dot{\gamma}_j \neq 0$, and $W = [\omega_{ij}] \in \mathbb{R}^{N \times N}$ is the coupling configuration matrix of the network with $\omega_{ij} \ge 0$ $(i \neq j)$ but not all zero.

Remark 9.1 In the considered time-varying complex networks (9.1) and (9.2), the coupled cases of the system states are characterized by the linking matrix Γ and the configuration matrix W. It should be noticed that such a state-connected model possesses its distinguished features in analyzing the dynamical behaviors of complex networks. However, in many practical engineering applications, the subsystems are often coupled through their measurements and, in such a case, the measurement-connected models can better preserve the large-scale nature of the networks (e.g. the sparsity).

In this chapter, the following state estimator is constructed:

$$\hat{x}_{i,k+1|k} = f(\hat{x}_{i,k|k}) + \sum_{j=1}^{N} \omega_{ij} \Gamma \hat{x}_{j,k|k}, \tag{9.3}$$

$$\hat{x}_{i,k+1|k+1} = \hat{x}_{i,k+1|k} + G_{i,k+1}(y_{i,k+1} - \mu_{i,k+1} C_{i,k+1} \hat{x}_{i,k+1|k}), \tag{9.4}$$

where $\hat{x}_{i,k|k}$ $(i = 1, 2, \ldots, N)$ is the state estimate, $\hat{x}_{i,k+1|k}$ is the one-step prediction at time k, $y_{i,k+1}$ is the *actual* measurement from the i-th node of the complex networks, $\mathbb{E}\{\lambda_{i,k+1}\} = \mu_{i,k+1}$, and $G_{i,k+1}$ is the estimator gain.

For node i, we define the one-step prediction error as $e_{i,k+1|k} = x_{i,k+1} - \hat{x}_{i,k+1|k}$ and the estimation error as $e_{i,k+1|k+1} = x_{i,k+1} - \hat{x}_{i,k+1|k+1}$. Then, one has

$$e_{i,k+1|k} = f(x_{i,k}) - f(\hat{x}_{i,k|k}) + \sum_{j=1}^{N} \omega_{ij} \Gamma (x_{j,k} - \hat{x}_{j,k|k}) + B_{i,k} \varpi_{i,k}. \tag{9.5}$$

By using the Taylor series expansion method around $\hat{x}_{i,k|k}$, one has

$$f(x_{i,k}) = f(\hat{x}_{i,k|k}) + A_{i,k} e_{i,k|k} + o(|e_{i,k|k}^2|), \tag{9.6}$$

where

$$A_{i,k} = \frac{\partial f(x_{i,k})}{\partial x_{i,k}} \Big|_{x_{i,k} = \hat{x}_{i,k|k}},$$

and $o(|e_{i,k|k}^2|)$ represents the high-order term of the Taylor series expansion. As in Calafiore (2005), Xiong (2010), the high-order term is estimated by the following form:

$$o(|e_{i,k|k}^2|) = L_{i,k}\aleph_{i,k}e_{i,k|k}, \tag{9.7}$$

where $L_{i,k}$ denotes scaling matrix and $\aleph_{i,k}$ is unknown matrix representing the linearization errors of the dynamical model satisfying $\aleph_{i,k}\aleph_{i,k}^T \leq I$. Then, based on (9.5)–(9.7), we have

$$e_{i,k+1|k} = (A_{i,k} + L_{i,k}\aleph_{i,k})e_{i,k|k} + \sum_{j=1}^{N}\omega_{ij}\Gamma(x_{j,k} - \hat{x}_{j,k|k}) + B_{i,k}\varpi_{i,k}. \tag{9.8}$$

On the other hand, it follows from (9.2) and (9.4) that

$$\begin{aligned} e_{i,k+1|k+1} &= e_{i,k+1|k} - G_{i,k+1}(y_{i,k+1} - \mu_{i,k+1}C_{i,k+1}\hat{x}_{i,k+1|k}) \\ &= (I - \mu_{i,k+1}G_{i,k+1}C_{i,k+1})e_{i,k+1|k} - (\lambda_{i,k+1} - \mu_{i,k+1}) \\ &\quad \times G_{i,k+1}C_{i,k+1}x_{i,k+1} - G_{i,k+1}\nu_{i,k+1}. \end{aligned} \tag{9.9}$$

For the purpose of simplicity, set

$$\begin{aligned} x_k &= \left[x_{1,k}^T \; x_{2,k}^T \; \cdots \; x_{N,k}^T\right]^T, \\ \varpi_k &= \left[\varpi_{1,k}^T \; \varpi_{2,k}^T \; \cdots \; \varpi_{N,k}^T\right]^T, \\ \nu_k &= \left[\nu_{1,k}^T \; \nu_{2,k}^T \; \cdots \; \nu_{N,k}^T\right]^T, \\ \hat{x}_{k+1|k} &= \left[\hat{x}_{1,k+1|k}^T \; \hat{x}_{2,k+1|k}^T \; \cdots \; \hat{x}_{N,k+1|k}^T\right]^T, \\ \hat{x}_{k+1|k+1} &= \left[\hat{x}_{1,k+1|k+1}^T \; \hat{x}_{2,k+1|k+1}^T \; \cdots \; \hat{x}_{N,k+1|k+1}^T\right]^T, \\ e_{k+1|k} &= \left[e_{1,k+1|k}^T \; e_{2,k+1|k}^T \; \cdots \; e_{N,k+1|k}^T\right]^T, \\ e_{k+1|k+1} &= \left[e_{1,k+1|k+1}^T \; e_{2,k+1|k+1}^T \; \cdots \; e_{N,k+1|k+1}^T\right]^T, \\ A_k &= \text{diag}\{A_{1,k}, A_{2,k}, \ldots, A_{N,k}\}, \\ L_k &= \text{diag}\{L_{1,k}, L_{2,k}, \ldots, L_{N,k}\}, \\ \aleph_k &= \text{diag}\{\aleph_{1,k}, \aleph_{2,k}, \ldots, \aleph_{N,k}\}, \\ G_k &= \text{diag}\{G_{1,k}, G_{2,k}, \ldots, G_{N,k}\}, \\ B_k &= \text{diag}\{B_{1,k}, B_{2,k}, \ldots, B_{N,k}\}, \\ C_k &= \text{diag}\{C_{1,k}, C_{2,k}, \ldots, C_{N,k}\}, \\ \mathbb{I} &= \text{diag}\{I, I, \ldots, I\}, \\ \bar{\Lambda}_k &= \text{diag}\{\mu_{1,k}I, \mu_{2,k}I, \ldots, \mu_{N,k}I\}, \\ \Lambda_k &= \text{diag}\{\lambda_{1,k}I, \lambda_{2,k}I, \ldots, \lambda_{N,k}I\}. \end{aligned}$$

Subsequently, by resorting to the Kronecker product, we have

$$
\begin{aligned}
e_{k+1|k} &= (A_k + L_k \aleph_k) e_{k|k} + (W \otimes \Gamma) e_{k|k} + B_k \varpi_k, &(9.10)\\
e_{k+1|k+1} &= (\mathbb{I} - G_{k+1} \bar{\Lambda}_{k+1} C_{k+1}) e_{k+1|k} - G_{k+1} (\Lambda_{k+1} - \bar{\Lambda}_{k+1}) \\
&\quad \times C_{k+1} x_{k+1} - G_{k+1} \nu_{k+1}. &(9.11)
\end{aligned}
$$

Now, in terms of the definition of the estimation error, we define the EEC as $P_{k+1|k+1} = \mathbb{E}\{e_{k+1|k+1} e_{k+1|k+1}^T\}$. The objective includes twofold. Firstly, we are in a position to construct an estimator of form (9.3) and (9.4) such that an upper bound concerning on the EEC is provided, that is, we are searching for a set of matrices $\Omega_{k+1|k+1} > 0$ satisfying

$$
P_{k+1|k+1} \leq \Omega_{k+1|k+1}. \tag{9.12}
$$

Secondly, the obtained upper bound $\Omega_{k+1|k+1}$ is minimized through suitably selecting the estimator gain.

9.2 State Estimation for Complex Networks

In this section, according to (9.10)–(9.11), the one-step prediction error covariance and the EEC are computed. In what follows, an upper bound regarding the EEC is solved and the estimator gain is constructed to minimize this upper bound.

Theorem 9.1 *The one-step prediction error covariance* $P_{k+1|k} = \mathbb{E}\{e_{k+1|k} e_{k+1|k}^T\}$ *is computed as*

$$
P_{k+1|k} = (A_k + W \otimes \Gamma + L_k \aleph_k) P_{k|k} (A_k + W \otimes \Gamma + L_k \aleph_k)^T + B_k Q_k B_k^T, \tag{9.13}
$$

where

$$
Q_k = \mathrm{diag}\{Q_{1,k}, Q_{2,k}, \ldots, Q_{N,k}\}.
$$

Proof Notice that (9.13) follows from (9.10) directly, thus the proof is omitted here for brevity.

Theorem 9.2 *The EEC* $P_{k+1|k+1}$ *is computed as*

$$
\begin{aligned}
P_{k+1|k+1} &= \left(\mathbb{I} - G_{k+1} \bar{\Lambda}_{k+1} C_{k+1}\right) P_{k+1|k} \left(\mathbb{I} - G_{k+1} \bar{\Lambda}_{k+1} C_{k+1}\right)^T \\
&\quad + G_{k+1} \left[\check{\Lambda}_{k+1} \circ \left(C_{k+1} \mathbb{E}\left\{x_{k+1} x_{k+1}^T\right\} C_{k+1}^T\right)\right] G_{k+1}^T \\
&\quad + G_{k+1} R_{k+1} G_{k+1}^T
\end{aligned} \tag{9.14}
$$

with

$$\begin{aligned}\check{\Lambda}_{k+1} &= \text{diag}\left\{\sigma_{1,k+1}^2 I, \sigma_{2,k+1}^2 I, \ldots, \sigma_{N,k+1}^2 I\right\}, \\ R_{k+1} &= \text{diag}\{R_{1,k+1}, R_{2,k+1}, \ldots, R_{N,k+1}\}. \end{aligned} \tag{9.15}$$

Proof In light of (9.11), we obtain

$$\begin{aligned}P_{k+1|k+1} =& \left(\mathbb{I} - G_{k+1}\bar{\Lambda}_{k+1}C_{k+1}\right) P_{k+1|k} \left(\mathbb{I} - G_{k+1}\bar{\Lambda}_{k+1}C_{k+1}\right)^T \\ &+ G_{k+1}\mathbb{E}\left\{\left(\Lambda_{k+1} - \bar{\Lambda}_{k+1}\right) C_{k+1}x_{k+1}x_{k+1}^T C_{k+1}^T \left(\Lambda_{k+1} - \bar{\Lambda}_{k+1}\right)\right\} G_{k+1}^T \\ &+ G_{k+1}R_{k+1}G_{k+1}^T. \end{aligned} \tag{9.16}$$

Subsequently, according to Lemma 2.4 and the characteristic of the conditional expectation, one has

$$\begin{aligned}&\mathbb{E}\left\{\left(\Lambda_{k+1} - \bar{\Lambda}_{k+1}\right) C_{k+1}x_{k+1}x_{k+1}^T C_{k+1}^T \left(\Lambda_{k+1} - \bar{\Lambda}_{k+1}\right)\right\} \\ &= \mathbb{E}\left\{\mathbb{E}\left\{\left(\Lambda_{k+1} - \bar{\Lambda}_{k+1}\right) C_{k+1}x_{k+1}x_{k+1}^T C_{k+1}^T \left(\Lambda_{k+1} - \bar{\Lambda}_{k+1}\right)\right\} |\Lambda_{k+1}\right\} \\ &= \mathbb{E}\left\{\left(\Lambda_{k+1} - \bar{\Lambda}_{k+1}\right) C_{k+1}\mathbb{E}\left\{x_{k+1}x_{k+1}^T\right\} C_{k+1}^T \left(\Lambda_{k+1} - \bar{\Lambda}_{k+1}\right)\right\} \\ &= \check{\Lambda}_{k+1} \circ \left(C_{k+1}\mathbb{E}\left\{x_{k+1}x_{k+1}^T\right\} C_{k+1}^T\right), \end{aligned} \tag{9.17}$$

where $\check{\Lambda}_{k+1}$ is denoted in (9.15). Then, it is observed from (9.16) and (9.17) that (9.14) holds, which completes the proof of this theorem.

Remark 9.2 Now, we have derived the explicit forms of the one-step prediction error covariance and the EEC. Nevertheless, there exist some unknown terms, such as $\aleph_k$ and $\mathbb{E}\left\{x_{k+1}x_{k+1}^T\right\}$ in (9.13) and (9.14). Therefore, this brings some difficulties for the calculation of the EEC. In order to solve such an issue, an alternative method is to calculate the upper bound of the EEC and then minimize this upper bound through choosing an appropriate estimator gain.

Theorem 9.3 *Let $\varepsilon > 0$ and $\gamma_k > 0$ be scalars. If the following two matrix equations:*

$$\begin{aligned}\Omega_{k+1|k} =& (A_k + W \otimes \Gamma)\left(\Omega_{k|k}^{-1} - \gamma_k I\right)^{-1} (A_k + W \otimes \Gamma)^T \\ &+ \gamma_k^{-1} L_k L_k^T + B_k Q_k B_k^T \end{aligned} \tag{9.18}$$

and

$$\begin{aligned}\Omega_{k+1|k+1} =& \left(\mathbb{I} - G_{k+1}\bar{\Lambda}_{k+1}C_{k+1}\right) \Omega_{k+1|k} \left(\mathbb{I} - G_{k+1}\bar{\Lambda}_{k+1}C_{k+1}\right)^T \\ &+ G_{k+1}\left[\check{\Lambda}_{k+1} \circ \left(C_{k+1}\Omega_{k+1}C_{k+1}^T\right)\right] G_{k+1}^T \\ &+ G_{k+1}R_{k+1}G_{k+1}^T \end{aligned} \tag{9.19}$$

with the initial condition $P_{0|0} \leq \Omega_{0|0}$ *have solutions* $\Omega_{k+1|k} > 0$ *and* $\Omega_{k+1|k+1} > 0$ *such that the inequality* $\Omega_{k|k} < \gamma_k^{-1} I$ *holds for all* $k \geq 0$*, then with the estimator parameter* $G_{k+1} = \text{diag}\{G_{1,k+1}, G_{2,k+1}, \ldots, G_{N,k+1}\}$ *and*

$$G_{i,k+1} = \Phi_{1,i} \Omega_{k+1|k} C_{k+1}^T \bar{\Lambda}_{k+1}^T S_{k+1}^{-1} \Phi_{2,i}^T, \tag{9.20}$$

the matrix $\Omega_{k+1|k+1}$ *is an upper bound of* $P_{k+1|k+1}$*, i.e.,*

$$P_{k+1|k+1} \leq \Omega_{k+1|k+1}, \tag{9.21}$$

where

$$\begin{aligned}
\Phi_{1,i} &= \big[\underbrace{0\ 0\ \cdots\ 0}_{i-1}\ I_{n\times n}\ \underbrace{0\ 0\ \cdots\ 0}_{N-i}\big],\\
\Phi_{2,i} &= \big[\underbrace{0\ 0\ \cdots\ 0}_{i-1}\ I_{m\times m}\ \underbrace{0\ 0\ \cdots\ 0}_{N-i}\big],\\
\Omega_{k+1} &= (1+\varepsilon)\,\Omega_{k+1|k} + \left(1+\varepsilon^{-1}\right)\hat{x}_{k+1|k}\hat{x}_{k+1|k}^T,\\
S_{k+1} &= \bar{\Lambda}_{k+1} C_{k+1} \Omega_{k+1|k} C_{k+1}^T \bar{\Lambda}_{k+1}^T + \check{\Lambda}_{k+1}\\
&\quad \circ \left(C_{k+1}\Omega_{k+1} C_{k+1}^T\right) + R_{k+1}.
\end{aligned} \tag{9.22}$$

Moreover, it can be shown that $\text{tr}(\Omega_{k+1|k+1})$ *is minimized by employing the estimator gain as in (9.20).*

Proof We adopt the mathematical induction method to prove this theorem. Noting the initial condition, we obtain $P_{0|0} \leq \Omega_{0|0}$. Assuming $P_{k|k} \leq \Omega_{k|k}$, the aim is to testify that $P_{k+1|k+1} \leq \Omega_{k+1|k+1}$ holds.

In terms of Lemma 2.1, one has

$$\begin{aligned}
P_{k+1|k} &\leq (A_k + W \otimes \Gamma)\left(P_{k|k}^{-1} - \gamma_k I\right)^{-1}(A_k + W \otimes \Gamma)^T\\
&\quad + \gamma_k^{-1} L_k L_k^T + B_k Q_k B_k^T\\
&\leq (A_k + W \otimes \Gamma)\left(\Omega_{k|k}^{-1} - \gamma_k I\right)^{-1}(A_k + W \otimes \Gamma)^T\\
&\quad + \gamma_k^{-1} L_k L_k^T + B_k Q_k B_k^T\\
&:= \Omega_{k+1|k},
\end{aligned} \tag{9.23}$$

where $\gamma_k > 0$ is a scalar. Then, it is concluded that

$$P_{k+1|k} \leq \Omega_{k+1|k}.$$

Next, let us address the second term of the RHS of (9.14). Employing Lemma 2.2, one has

$$e_{k+1|k}\hat{x}_{k+1|k}^T + \hat{x}_{k+1|k} e_{k+1|k}^T \leq \varepsilon e_{k+1|k} e_{k+1|k}^T + \varepsilon^{-1}\hat{x}_{k+1|k}\hat{x}_{k+1|k}^T.$$

Then, one has

$$\mathbb{E}\left\{x_{k+1}x_{k+1}^T\right\} \leq (1+\varepsilon)\, P_{k+1|k} + \left(1+\varepsilon^{-1}\right)\hat{x}_{k+1|k}\hat{x}_{k+1|k}^T, \tag{9.24}$$

where $\varepsilon > 0$ is a constant. Combining (9.14) with (9.24), one obtains

$$\begin{aligned} P_{k+1|k+1} \leq\ & \left(\mathbb{I} - G_{k+1}\bar{\Lambda}_{k+1}C_{k+1}\right) P_{k+1|k} \left(\mathbb{I} - G_{k+1}\bar{\Lambda}_{k+1}C_{k+1}\right)^T \\ & + G_{k+1}\left[\check{\Lambda}_{k+1} \circ \left(C_{k+1}\Psi_{k+1}C_{k+1}^T\right)\right] G_{k+1}^T \\ & + G_{k+1}R_{k+1}G_{k+1}^T, \end{aligned} \tag{9.25}$$

where

$$\Psi_{k+1} = (1+\varepsilon)\, P_{k+1|k} + \left(1+\varepsilon^{-1}\right)\hat{x}_{k+1|k}\hat{x}_{k+1|k}^T.$$

Thus, it is clear to obtain that

$$P_{k+1|k+1} \leq \Omega_{k+1|k+1}.$$

Next, we aim to calculate the estimator gain $G_{i,k+1}$ in (9.20). It is obvious that $\Omega_{k+1|k+1}$ in (9.19) can be computed as

$$\begin{aligned} \Omega_{k+1|k+1} =\ & \Omega_{k+1|k} - G_{k+1}\bar{\Lambda}_{k+1}C_{k+1}\Omega_{k+1|k} - \Omega_{k+1|k}C_{k+1}^T\bar{\Lambda}_{k+1}^T G_{k+1}^T \\ & + G_{k+1}S_{k+1}G_{k+1}^T, \end{aligned} \tag{9.26}$$

where S_{k+1} is denoted in (9.22). Substituting $G_{k+1} = \sum_{i=1}^N \left(\Phi_{1,i}^T G_{i,k+1}\Phi_{2,i}\right)$ into (9.26), one has

$$\begin{aligned} \operatorname{tr}\left(\Omega_{k+1|k+1}\right) = \operatorname{tr}\Bigg(& \Omega_{k+1|k} - \sum_{i=1}^N \left(\Phi_{1,i}^T G_{i,k+1}\Phi_{2,i}\right)\bar{\Lambda}_{k+1}C_{k+1}\Omega_{k+1|k} \\ & - \Omega_{k+1|k}C_{k+1}^T\bar{\Lambda}_{k+1}^T \sum_{i=1}^N \left(\Phi_{1,i}^T G_{i,k+1}\Phi_{2,i}\right)^T \\ & + \sum_{i=1}^N \left(\Phi_{1,i}^T G_{i,k+1}\Phi_{2,i}\right) S_{k+1} \left(\Phi_{1,i}^T G_{i,k+1}\Phi_{2,i}\right)^T \Bigg) \end{aligned} \tag{9.27}$$

with $\Phi_{1,i}$ and $\Phi_{2,i}$ denoting in (9.22). It is noted that, when deriving (9.27), we have utilized the following fact:

$$\operatorname{tr}\left(\left(\Phi_{1,i}^T G_{i,k+1}\Phi_{2,i}\right) S_{k+1} \left(\Phi_{1,i}^T G_{i,k+1}\Phi_{2,i}\right)^T\right) = 0, \quad \text{if } i \neq j.$$

By virtue of Lemma 2.3, taking the partial derivative of $\operatorname{tr}(\Omega_{k+1|k+1})$ regarding $G_{i,k+1}$ and letting the derivative be zero, we arrive at

$$
\begin{aligned}
\frac{\partial \operatorname{tr}\left(\Omega_{k+1|k+1}\right)}{\partial G_{i,k+1}} &= -2\Phi_{1,i}\Omega_{k+1|k}C_{k+1}^{T}\bar{\Lambda}_{k+1}^{T}\Phi_{2,i}^{T} \\
&\quad +2\Phi_{1,i}\Phi_{1,i}^{T}G_{i,k+1}\Phi_{2,i}S_{k+1}\Phi_{2,i}^{T} \\
&= 0. \qquad (9.28)
\end{aligned}
$$

Note that $S_{k+1} > 0$ is a diagonal matrix and $\Phi_{1,i}\Phi_{1,i}^{T} = I_{n\times n}$. Through the algebraic manipulations, it follows from (9.28) that

$$
\begin{aligned}
G_{i,k+1} &= \Phi_{1,i}\Omega_{k+1|k}C_{k+1}^{T}\bar{\Lambda}_{k+1}^{T}\Phi_{2,i}^{T}\left(\Phi_{2,i}S_{k+1}\Phi_{2,i}^{T}\right)^{-1} \\
&= \Phi_{1,i}\Omega_{k+1|k}C_{k+1}^{T}\bar{\Lambda}_{k+1}^{T}S_{k+1}^{-1}\Phi_{2,i}^{T}. \qquad (9.29)
\end{aligned}
$$

So far, the proof of this theorem is finished.

Remark 9.3 A great deal of effort has been dedicated to addressing the state estimation issues for complex networks. Nevertheless, it is worthwhile to note that most of state estimation algorithms can only be utilized for the *time-invariant* complex networks regarding the asymptotical behavior of the estimation error model. It is worth mentioning that the estimation method proposed in this chapter can be applied for the *time-varying* complex networks through evaluating the transient behavior of the EEC over a certain time horizon. In the process of estimation algorithm design, the scalars ε and γ_k have been introduced to help strengthen the feasibility and solvability of the proposed estimation scheme. It is highlighted that the parameter γ_k affects the solvability of the matrix difference equations (9.18) and (9.19). During the implementation, γ_k could be provided a prior and adjusted to ensure the inequality $\Omega_{k|k} < \gamma_k^{-1}I$ to improve the solvability and feasibility of the developed estimation scheme.

9.3 An Illustrative Example

In this section, an illustrative simulation experiment is shown to verify the validity and correctness of the designed estimation strategy.

Consider a time-varying stochastic complex network (9.1) and (9.2) with four nodes. The coupling configuration matrix is set as $W = [\omega_{ij}]_{4\times 4}$ with

$$
\omega_{ij} = \begin{cases} -0.06, & i = j \\ 0.02, & i \neq j \end{cases}
$$

and the inner-coupling matrix is set as $\Gamma = I_2$. The noise covariances are set as $Q_{1,k} = 0.01$, $Q_{2,k} = 0.02$, $Q_{3,k} = 0.02$ and $Q_{4,k} = 0.03$, $v_{i,k}$ are zero-mean Gaussian white noises with covariances $R_{1,k} = 0.02$, $R_{2,k} = 0.02$, $R_{3,k} = 0.01$ and $R_{4,k} = 0.01$. The matrices $B_{i,k}$ and $C_{i,k}$ are chosen as

$$B_{1,k} = \begin{bmatrix} 0.1 - 0.2\sin(2k) \\ 0.1 \end{bmatrix}, \quad B_{2,k} = \begin{bmatrix} -0.08 \\ 0.18 \end{bmatrix},$$
$$B_{3,k} = \begin{bmatrix} 0.04 \\ 0.05 \end{bmatrix}, \quad B_{4,k} = \begin{bmatrix} 0.07 \\ -0.02 \end{bmatrix},$$
$$C_{1,k} = \begin{bmatrix} 1 & 0.3 \end{bmatrix}, \quad C_{2,k} = \begin{bmatrix} 1 & 0.8 \end{bmatrix},$$
$$C_{3,k} = \begin{bmatrix} 1 & 1.5 \end{bmatrix}, \quad C_{4,k} = \begin{bmatrix} 0.9 & 1.2 \end{bmatrix}.$$

The nonlinear function $f(x_{i,k})$ is selected as

$$f(x_{i,k}) = \begin{bmatrix} 0.7x_{i,k}^1 + \sin(x_{i,k}^1 x_{i,k}^2) \\ 1.48x_{i,k}^2 - \sin(x_{i,k}^1 x_{i,k}^2) \end{bmatrix},$$

where $x_{i,k} = \begin{bmatrix} x_{i,k}^1 & x_{i,k}^2 \end{bmatrix}^T$ is the system state. $x_{i,k}^1$ and $x_{i,k}^2$ are the first and second components of $x_{i,k}$, respectively. Assume that the probability mass functions for $\lambda_{i,k}$ ($i = 1, 2, 3, 4$) are given as

$$p_{i,k}(s) = \begin{cases} 0.1, & s = 0 \\ 0.2, & s = 0.5 \\ 0.7, & s = 1. \end{cases}$$

Consequently, the expectation and variance of $\lambda_{i,k}$ are computed as $\mu_{i,k} = 0.8$ and $\sigma_{i,k}^2 = 0.11$. Other parameters are chosen as

$$\begin{aligned} \bar{x}_{1,0} = \bar{x}_{2,0} = \bar{x}_{3,0} = \bar{x}_{4,0} &= \begin{bmatrix} 2 & -0.3 \end{bmatrix}^T, \\ \Omega_{1,0|0} = \Omega_{2,0|0} = \Omega_{3,0|0} = \Omega_{4,0|0} &= 10I_2, \\ \varepsilon = 0.6, \ \gamma_k = 0.03, \ L_{1,k} &= \text{diag}\{0.01, 0.02\}, \\ L_{2,k} = \text{diag}\{0.01, 0.04\}, \ L_{3,k} &= \text{diag}\{0.02, 0.04\}, \\ L_{4,k} &= \text{diag}\{0.03, 0.06\}. \end{aligned}$$

The simulations are exhibited in Figs. 9.1, 9.2, 9.3, 9.4, 9.5, 9.6, 9.7, 9.8, 9.9, 9.10, 9.11 and 9.12, where the trajectories of the actual states $x_{i,k}^j$ and their estimates $\hat{x}_{i,k|k}^j$ ($j = 1, 2$) are displayed, which indicate that the developed state estimation scheme has a satisfactory tracking performance. In addition, in order to clarify the impact from the missing probability onto the estimation performance, some comparative simulation results are provided in Figs. 9.9, 9.10, 9.11 and 9.12. Specifically, the values of missing probabilities are set as

Case I:

$$p_{i,k}(s) = \begin{cases} 0.1, & s = 0 \\ 0.1, & s = 0.5 \\ 0.8, & s = 1 \end{cases}$$

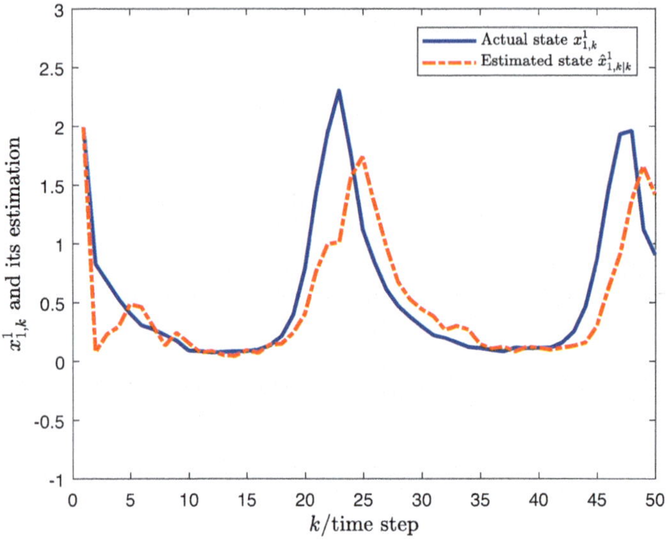

Fig. 9.1 Actual state $x^1_{1,k}$ and its estimation $\hat{x}^1_{1,k|k}$

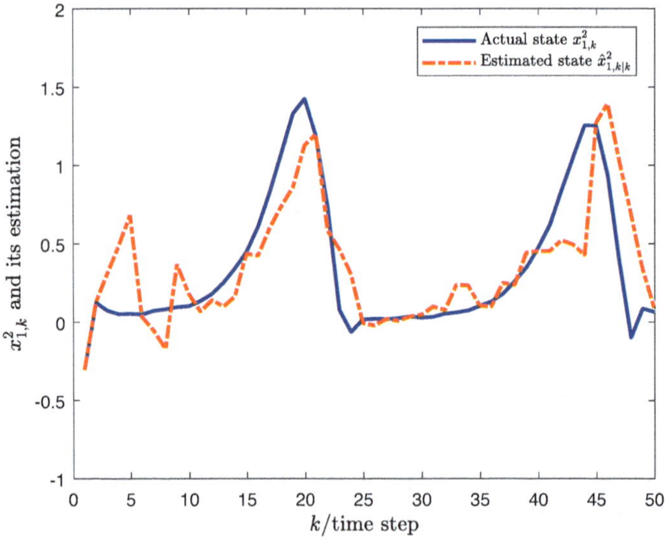

Fig. 9.2 Actual state $x^2_{1,k}$ and its estimation $\hat{x}^2_{1,k|k}$

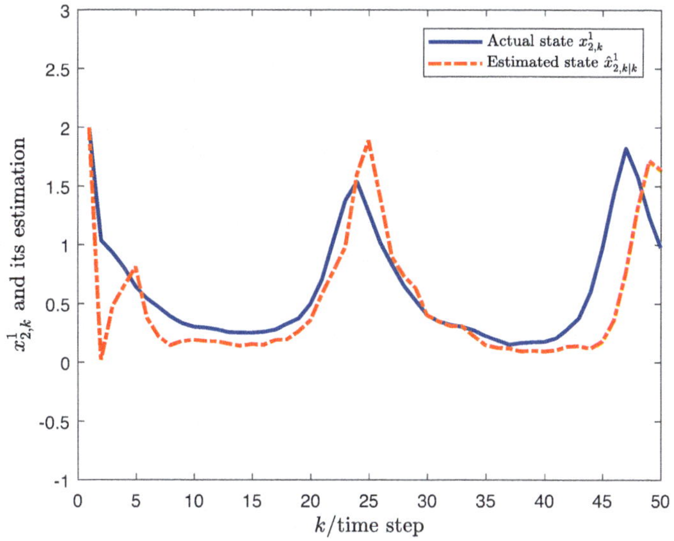

Fig. 9.3 Actual state $x^1_{2,k}$ and its estimation $\hat{x}^1_{2,k|k}$

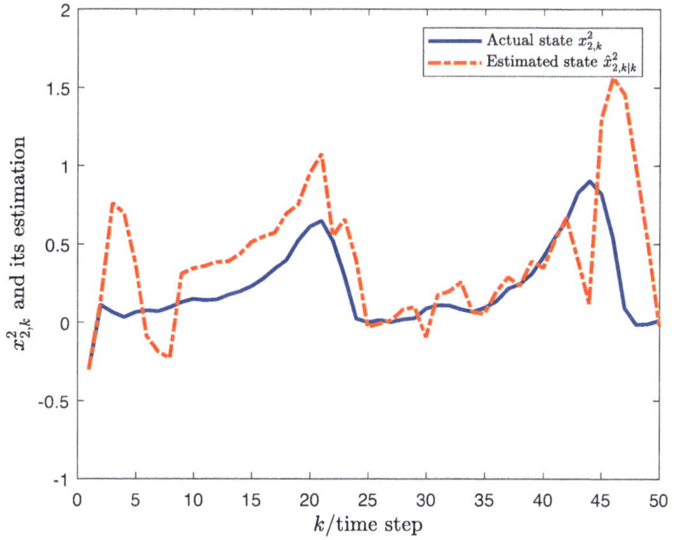

Fig. 9.4 Actual state $x^2_{2,k}$ and its estimation $\hat{x}^2_{2,k|k}$

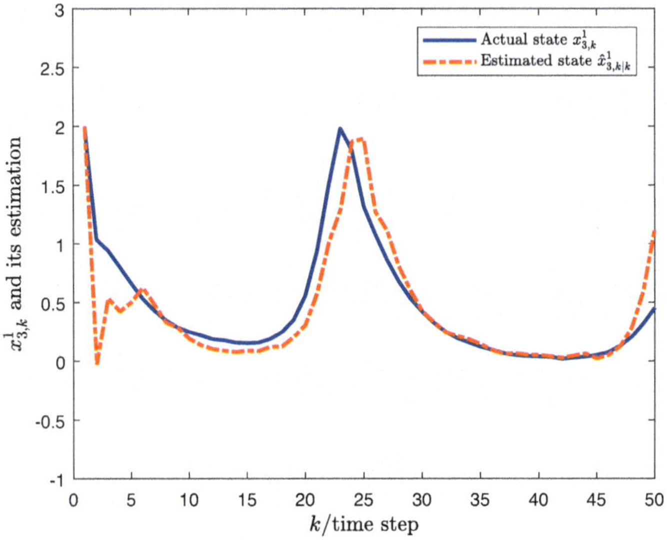

Fig. 9.5 Actual state $x^1_{3,k}$ and its estimation $\hat{x}^1_{3,k|k}$

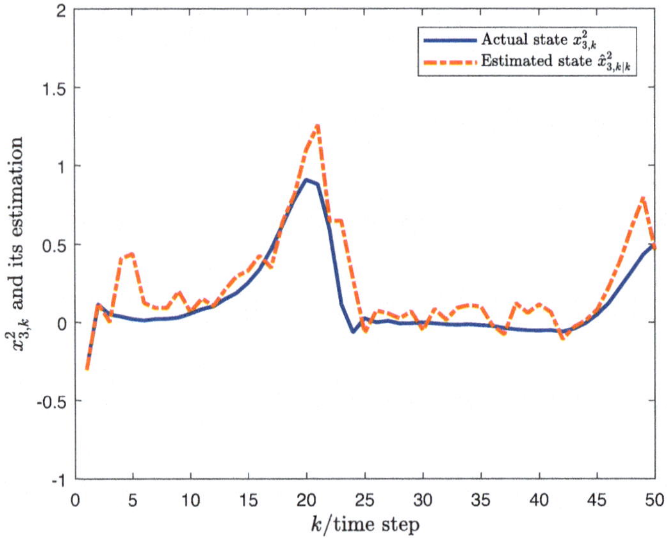

Fig. 9.6 Actual state $x^2_{3,k}$ and its estimation $\hat{x}^2_{3,k|k}$

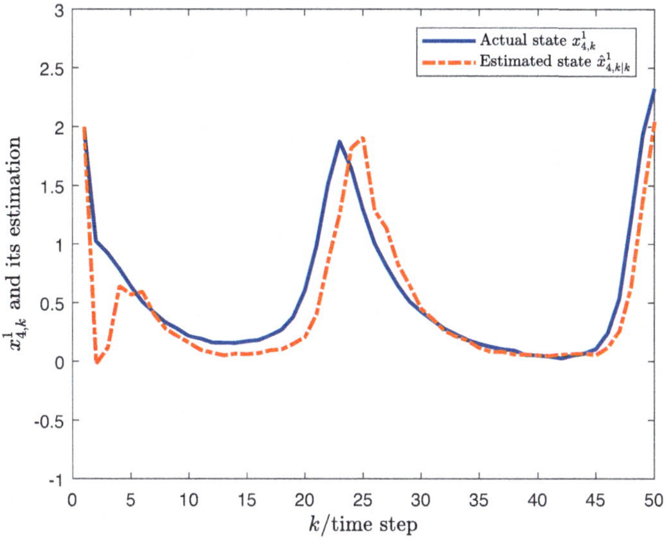

Fig. 9.7 Actual state $x^1_{4,k}$ and its estimation $\hat{x}^1_{4,k|k}$

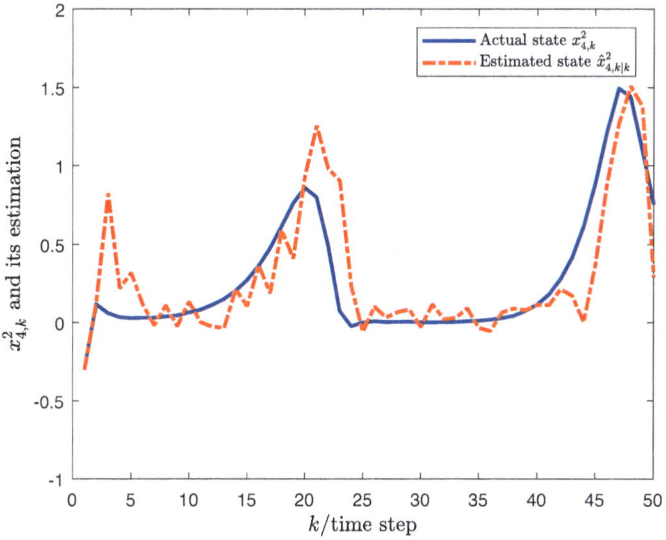

Fig. 9.8 Actual state $x^2_{4,k}$ and its estimation $\hat{x}^2_{4,k|k}$

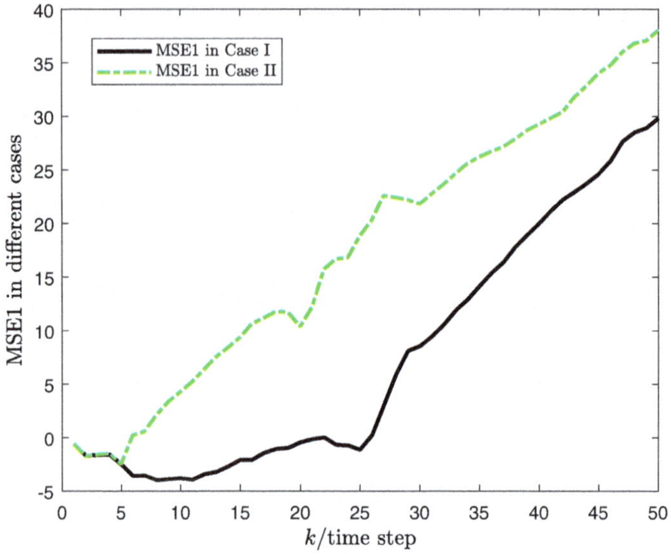

Fig. 9.9 MSE1 under Case I and Case II

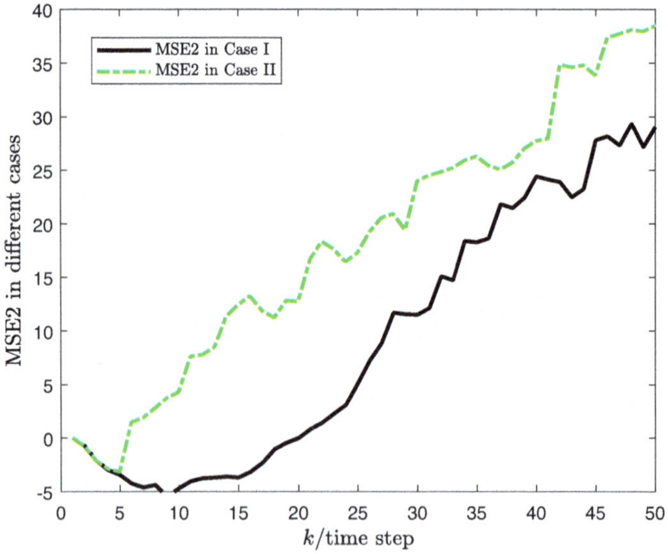

Fig. 9.10 MSE2 under Case I and Case II

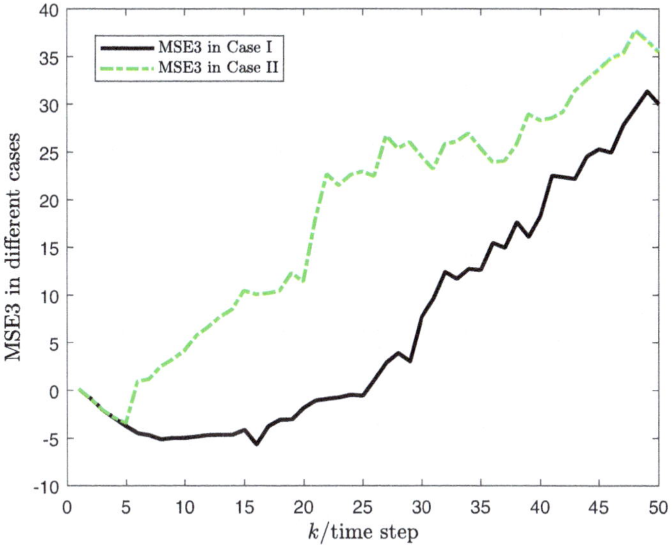

Fig. 9.11 MSE3 under Case I and Case II

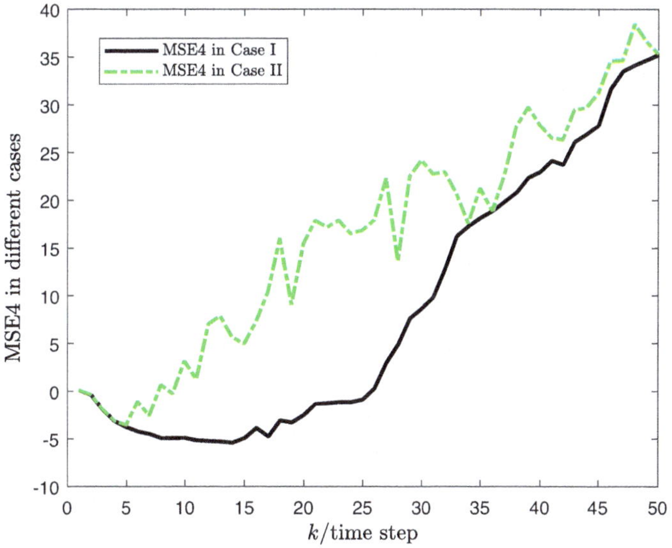

Fig. 9.12 MSE4 under Case I and Case II

Case II:

$$p_{i,k}(s) = \begin{cases} 0.8, & s = 0 \\ 0.1, & s = 0.5 \\ 0.1, & s = 1 \end{cases}$$

It can be obviously seen from these figures that the MSE becomes smaller along with the smaller missing probability, which is accordant to the theoretical results.

9.4 Summary

In this chapter, the recursive state estimation strategy has been designed for time-varying stochastic complex networks with missing measurements and random disturbances. An array of Bernoulli distributed random variables has been employed to describe the phenomenon of the missing measurements. On the basis of the available measurement, a novel state estimator has been designed, where the missing probability has been introduced to compensate the influence from the missing measurements onto the estimation performance. Accordingly, the covariance performance of the estimation error has been considered when deriving the estimator parameters for the addressed time-varying complex networks. It has been shown that the desired estimator parameters can be derived by solving the difference matrix equations. Finally, a simulation has been provided to illustrate the validity of the estimation method presented in this chapter.

References

Calafiore G (2005) Reliable localization using set-valued nonlinear filters. IEEE Trans Syst Man Cybern Part A-Syst Hum 35(2):189–197

Xiong K, Wei C, Liu L (2010) Robust extended Kalman filtering for nonlinear systems with stochastic uncertainties. IEEE Trans Syst Man Cybern Part A-Syst Hum 40(2):399–405

Chapter 10
Quantized State Estimation for Complex Networks with Uncertain Inner Coupling

In this chapter, we aim to provide a novel recursive state estimation algorithm for stochastic complex networks in the presence of uncertain inner coupling and measurement quantization mechanism. The coupling strengths are varying over certain intervals and the measurements are quantized before being delivered to the remote estimator side. The main attention of this chapter is to design a state estimation scheme so as to guarantee a locally minimized upper bound regarding the EEC. In addition, a sufficient condition is given to guarantee the exponential boundedness of the estimation error under the mean-square sense. At last, the comparative simulations are adopted to validate the usefulness of the designed state estimation approach.

10.1 Problem Formulation

Consider the following class of time-varying stochastic complex networks with N coupled nodes:

$$x_{i,k+1} = A_{i,k}x_{i,k} + \sum_{j=1}^{N} \omega_{ij}\Gamma x_{j,k} + B_{i,k}\varpi_{i,k}, \tag{10.1}$$

$$y_{i,k} = C_{i,k}x_{i,k} + \nu_{i,k}, \tag{10.2}$$

where $x_{i,k} \in \mathbb{R}^n$ is the state with the initial state $x_{i,0}$ $(i = 1, 2, \ldots, N)$, $y_{i,k} \in \mathbb{R}^m$ is the measurement signal, $\varpi_{i,k}$ and $\nu_{i,k}$ denote zero-mean noises with covariances $Q_{i,k} > 0$ and $R_{i,k} > 0$, respectively. $A_{i,k}$, $B_{i,k}$ and $C_{i,k}$ denote given matrices of proper dimensions. $W = [\omega_{ij}]_{N\times N}$ is the coupling configuration matrix with $\omega_{ij} \geq 0$ $(i \neq j)$.

J. Hu et al., *Variance-Constrained Filtering for Stochastic Complex Systems*, Intelligent Control and Learning Systems 18,
https://doi.org/10.1007/978-981-96-2637-3_10

In (10.1), $\Gamma = \text{diag}\{\gamma_1, \gamma_2, \ldots, \gamma_n\}$ is the inner coupling matrix with $\gamma_l \neq 0$ being the linking. Here, the coupling strength γ_l $(l = 1, 2, \ldots, n)$ is unknown but belongs to a given interval $[\underline{\gamma}_l, \bar{\gamma}_l]$, where $\underline{\gamma}_l$ and $\bar{\gamma}_l$ denote known scalars with $\underline{\gamma}_l < \bar{\gamma}_l$. Denoting

$$\bar{\Gamma} = \text{diag}\left\{\frac{\underline{\gamma}_1 + \bar{\gamma}_1}{2}, \frac{\underline{\gamma}_2 + \bar{\gamma}_2}{2}, \ldots, \frac{\underline{\gamma}_n + \bar{\gamma}_n}{2}\right\},$$

$$\underline{\Gamma} = \text{diag}\left\{\frac{\bar{\gamma}_1 - \underline{\gamma}_1}{2}, \frac{\bar{\gamma}_2 - \underline{\gamma}_2}{2}, \ldots, \frac{\bar{\gamma}_n - \underline{\gamma}_n}{2}\right\},$$

then the matrix Γ is rewritten as $\Gamma = \bar{\Gamma} + \tilde{\Gamma}$ with $\tilde{\Gamma} \in \left[-\underline{\Gamma}, \underline{\Gamma}\right]$. In what follows, Γ can be modeled as

$$\Gamma = \bar{\Gamma} + \mathscr{F}\underline{\Gamma}, \quad \mathscr{F}\mathscr{F}^T = \mathscr{F}^T\mathscr{F} \leq I \tag{10.3}$$

with $\mathscr{F} = \tilde{\Gamma}\underline{\Gamma}^{-1}$. Moreover, it is assumed that $\varpi_{i,k}$, $\nu_{i,k}$ and $x_{i,0}$ are mutually independent.

Remark 10.1 The considered system model (10.1) and (10.2) is fairly comprehensive. To be more specific, the first term of (10.1) could denote its evolution behavior regarding the i-th node, the second term reflects the influences resulted from the neighboring nodes, and the third term denotes the process noise. Furthermore, the measurements are modeled by (10.2). Generally, the proposed complex networks model can be employed to describe various practical systems.

In the process of signal transmission, the signal quantization might appear owing to the constrained network communication capacity. Here, the quantization process is modeled by

$$q(y_{i,k}) = \left[q_1(y_{i,k}^{(1)})\ q_2(y_{i,k}^{(2)}) \cdots q_m(y_{i,k}^{(m)})\right]^T,$$

where $y_{i,k}^{(1)}, y_{i,k}^{(2)}, \ldots, y_{i,k}^{(m)}$ are the components of $y_{i,k}$.

The set of quantization levels of each $q_j(\cdot)$ is described as

$$\mathcal{U}_j = \left\{\pm u_l^{(j)},\ u_l^{(j)} = (\chi^{(j)})^l u_0^{(j)},\ l = 0, \pm 1, \pm 2, \ldots\right\} \cup \{0\},\ u_0^{(j)} > 0,$$

where $\chi^{(j)} \in (0, 1)$ $(j = 1, 2, \ldots, m)$ denote the quantization densities, and $u_l^{(j)}$ is the quantization level. Then, the following logarithmic quantizer is used:

$$q_j(y_{i,k}^{(j)}) = \begin{cases} u_l^{(j)}, & \frac{1}{1+\delta_j}u_l^{(j)} < y_{i,k}^{(j)} \leq \frac{1}{1-\delta_j}u_l^{(j)} \\ 0, & y_{i,k}^{(j)} = 0 \\ -q_j(-y_{i,k}^{(j)}), & y_{i,k}^{(j)} < 0 \end{cases} \tag{10.4}$$

where $\delta_j = \frac{1-\chi^{(j)}}{1+\chi^{(j)}}$. From the mentioned definition, we obtain

$$q_j(y_{i,k}^{(j)}) = \left(1 + \Delta_k^{(j)}\right) y_{i,k}^{(j)}$$

with $|\Delta_k^{(j)}| \leq \delta_j$.

Setting $\Delta_k = \text{diag}\left\{\Delta_k^{(1)}, \Delta_k^{(2)}, \ldots, \Delta_k^{(m)}\right\}$, the measurements received by the state estimator are modeled as

$$\tilde{y}_{i,k} = q(y_{i,k}) = (I + \Delta_k)C_{i,k}x_{i,k} + (I + \Delta_k)v_{i,k}. \tag{10.5}$$

Denoting $\Lambda = \text{diag}\{\delta_1, \delta_2, \ldots, \delta_m\}$ and $\mathfrak{F}_k = \Delta_k \Lambda^{-1}$, one has $\Delta_k = \mathfrak{F}_k \Lambda$ with $\mathfrak{F}_k$ satisfying $\mathfrak{F}_k \mathfrak{F}_k^T = \mathfrak{F}_k^T \mathfrak{F}_k \leq I$ (Fig. 10.1).

The following time-varying state estimator is constructed:

$$\hat{x}_{i,k+1|k} = A_{i,k}\hat{x}_{i,k|k} + \sum_{j=1}^{N} \omega_{ij} \bar{\Gamma} \hat{x}_{j,k|k}, \tag{10.6}$$

$$\hat{x}_{i,k+1|k+1} = \hat{x}_{i,k+1|k} + G_{i,k+1}(\tilde{y}_{i,k+1} - C_{i,k+1}\hat{x}_{i,k+1|k}), \tag{10.7}$$

where $\hat{x}_{i,k+1|k}$ $(i = 1, 2, \ldots, N)$ is the one-step prediction, $\hat{x}_{i,k|k}$ is the state estimate, $\tilde{y}_{i,k+1}$ is the *actual* measurement, and $G_{i,k+1}$ is the desired estimator gain.

Let $e_{i,k+1|k} = x_{i,k+1} - \hat{x}_{i,k+1|k}$ be the one-step prediction error and $e_{i,k+1|k+1} = x_{i,k+1} - \hat{x}_{i,k+1|k+1}$ represent the estimation error. It can be obtained from (10.1) and (10.6) that

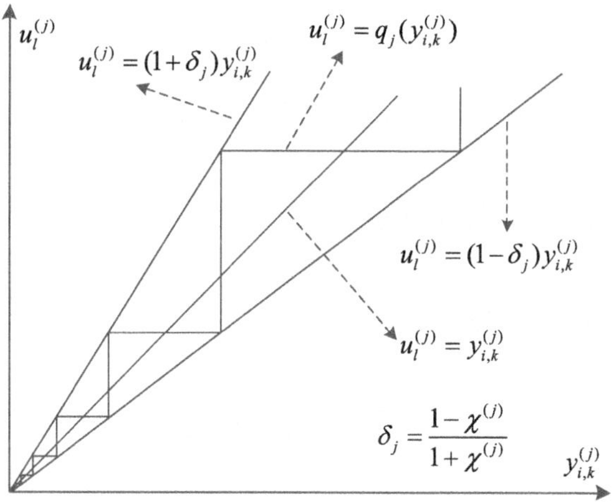

Fig. 10.1 The logarithmic quantizer

$$e_{i,k+1|k} = A_{i,k}e_{i,k|k} + \sum_{j=1}^{N} \omega_{ij}\bar{\Gamma}(x_{j,k} - \hat{x}_{j,k|k}) + \sum_{j=1}^{N} \omega_{ij}\tilde{\Gamma}x_{j,k} + B_{i,k}\varpi_{i,k}. \quad (10.8)$$

Together with (10.5) and (10.7), we get

$$\begin{aligned} e_{i,k+1|k+1} = &(I - G_{i,k+1}C_{i,k+1})e_{i,k+1|k} - G_{i,k+1}\Delta_{k+1}C_{i,k+1}x_{i,k+1} \\ &-G_{i,k+1}(I + \Delta_{k+1})\nu_{i,k+1}. \end{aligned} \quad (10.9)$$

Before proceeding, we denote

$$\begin{aligned} \mathbb{I} &= \text{diag}\{\underbrace{I, I, \ldots, I}_{N}\}, \\ \tilde{\Delta}_k &= \text{diag}\{\underbrace{\Delta_k, \Delta_k, \ldots, \Delta_k}_{N}\}, \\ A_k &= \text{diag}\{A_{1,k}, A_{2,k}, \ldots, A_{N,k}\}, \\ G_k &= \text{diag}\{G_{1,k}, G_{2,k}, \ldots, G_{N,k}\}, \\ B_k &= \text{diag}\{B_{1,k}, B_{2,k}, \ldots, B_{N,k}\}, \\ C_k &= \text{diag}\{C_{1,k}, C_{2,k}, \ldots, C_{N,k}\}, \\ x_k &= \left[x_{1,k}^T \; x_{2,k}^T \; \cdots \; x_{N,k}^T\right]^T, \\ \varpi_k &= \left[\varpi_{1,k}^T \; \varpi_{2,k}^T \; \cdots \; \varpi_{N,k}^T\right]^T, \\ \nu_k &= \left[\nu_{1,k}^T \; \nu_{2,k}^T \; \cdots \; \nu_{N,k}^T\right]^T, \\ \hat{x}_{k+1|k} &= \left[\hat{x}_{1,k+1|k}^T \; \hat{x}_{2,k+1|k}^T \; \cdots \; \hat{x}_{N,k+1|k}^T\right]^T, \\ \hat{x}_{k+1|k+1} &= \left[\hat{x}_{1,k+1|k+1}^T \; \hat{x}_{2,k+1|k+1}^T \; \cdots \; \hat{x}_{N,k+1|k+1}^T\right]^T, \\ e_{k+1|k} &= \left[e_{1,k+1|k}^T \; e_{2,k+1|k}^T \; \cdots \; e_{N,k+1|k}^T\right]^T, \\ e_{k+1|k+1} &= \left[e_{1,k+1|k+1}^T \; e_{2,k+1|k+1}^T \; \cdots \; e_{N,k+1|k+1}^T\right]^T. \end{aligned}$$

According to the corresponding definitions and the Kronecker product, the one-step prediction error $e_{k+1|k}$ and the estimation error $e_{k+1|k+1}$ can be rewritten as

$$e_{k+1|k} = (A_k + W \otimes \bar{\Gamma})e_{k|k} + (W \otimes \tilde{\Gamma})x_k + B_k\varpi_k, \quad (10.10)$$

$$e_{k+1|k+1} = \Pi_{k+1}e_{k+1|k} - G_{k+1}\tilde{\Delta}_{k+1}C_{k+1}x_{k+1} - G_{k+1}(\mathbb{I} + \tilde{\Delta}_{k+1})\nu_{k+1}, \quad (10.11)$$

where $\Pi_{k+1} = \mathbb{I} - G_{k+1}C_{k+1}$.

Now, let $P_{k+1|k+1} = \mathbb{E}\{e_{k+1|k+1}e_{k+1|k+1}^T\}$ represent the EEC. Then, the purposes of the chapter are summarized as follows.

(R1) We endeavor to design a new estimator (10.6)–(10.7) such that an upper bound regarding the EEC $\Omega_{k+1|k+1}$ is ensured, i.e.,

$$P_{k+1|k+1} \leq \Omega_{k+1|k+1}. \quad (10.12)$$

Furthermore, the trace of this upper bound is optimized by constructing an estimator gain.

(R2) A sufficient condition with mathematical proof is presented to guarantee the exponential boundedness of the estimation error dynamics in the sense of mean-square.

10.2 Main Results

According to the definitions of the one-step prediction error covariance and the EEC, the following essential lemmas are given.

Lemma 10.1 *The one-step prediction error covariance $P_{k+1|k}$ is given as*

$$\begin{aligned}P_{k+1|k} &= (A_k + W\otimes\bar{\Gamma})P_{k|k}(A_k + W\otimes\bar{\Gamma})^T + (A_k + W\otimes\bar{\Gamma})\mathbb{E}\{e_{k|k}x_k^T\}(W\otimes\tilde{\Gamma})^T\\&\quad +(W\otimes\tilde{\Gamma})\mathbb{E}\{x_k e_{k|k}^T\}(A_k + W\otimes\bar{\Gamma})^T + (W\otimes\tilde{\Gamma})\mathbb{E}\{x_k x_k^T\}(W\otimes\tilde{\Gamma})^T\\&\quad +B_k Q_k B_k^T,\end{aligned}\tag{10.13}$$

where

$$Q_k = \mathrm{diag}\{Q_{1,k}, Q_{2,k}, \ldots, Q_{N,k}\}.\tag{10.14}$$

Proof By resorting to the definition of $P_{k+1|k}$ and (10.10), we get

$$\begin{aligned}P_{k+1|k} &= \mathbb{E}\{e_{k+1|k}e_{k+1|k}^T\}\\&= (A_k + W\otimes\bar{\Gamma})P_{k|k}(A_k + W\otimes\bar{\Gamma})^T + (A_k + W\otimes\bar{\Gamma})\mathbb{E}\{e_{k|k}x_k^T\}(W\otimes\tilde{\Gamma})^T\\&\quad +(W\otimes\tilde{\Gamma})\mathbb{E}\{x_k e_{k|k}^T\}(A_k + W\otimes\bar{\Gamma})^T + (W\otimes\tilde{\Gamma})\mathbb{E}\{x_k x_k^T\}(W\otimes\tilde{\Gamma})^T\\&\quad +B_k Q_k B_k^T,\end{aligned}\tag{10.15}$$

where Q_k is given in (10.14). Noting $\mathbb{E}\{\varpi_k\} = 0$, (10.15) can be derived, which finishes the proof of this theorem.

Lemma 10.2 *The EEC $P_{k+1|k+1}$ is calculated as*

$$\begin{aligned}P_{k+1|k+1} &= \Pi_{k+1}P_{k+1|k}\Pi_{k+1}^T - \Pi_{k+1}\mathbb{E}\{e_{k+1|k}x_{k+1}^T\}C_{k+1}^T\tilde{\Delta}_{k+1}^T G_{k+1}^T\\&\quad -G_{k+1}\tilde{\Delta}_{k+1}C_{k+1}\mathbb{E}\{x_{k+1}e_{k+1|k}^T\}\Pi_{k+1}^T\\&\quad +G_{k+1}\tilde{\Delta}_{k+1}C_{k+1}\mathbb{E}\{x_{k+1}x_{k+1}^T\}C_{k+1}^T\tilde{\Delta}_{k+1}^T G_{k+1}^T\\&\quad +G_{k+1}(\mathbb{I} + \tilde{\Delta}_{k+1})R_{k+1}(\mathbb{I} + \tilde{\Delta}_{k+1})^T G_{k+1}^T,\end{aligned}\tag{10.16}$$

where

$$R_{k+1} = \mathrm{diag}\{R_{1,k+1}, R_{2,k+1}, \ldots, R_{N,k+1}\}.\tag{10.17}$$

Proof In terms of (10.11), one has

$$
\begin{aligned}
&P_{k+1|k+1} \\
&= \mathbb{E}\{e_{k+1|k+1}e_{k+1|k+1}^T\} \\
&= \mathbb{E}\{[\Pi_{k+1}e_{k+1|k} - G_{k+1}\tilde{\Delta}_{k+1}C_{k+1}x_{k+1} - G_{k+1}(\mathbb{I} + \tilde{\Delta}_{k+1})\nu_{k+1}] \\
&\quad \times [\Pi_{k+1}e_{k+1|k} - G_{k+1}\tilde{\Delta}_{k+1}C_{k+1}x_{k+1} - G_{k+1}(\mathbb{I} + \tilde{\Delta}_{k+1})\nu_{k+1}]^T\} \\
&= \Pi_{k+1}P_{k+1|k}\Pi_{k+1}^T - \Pi_{k+1}\mathbb{E}\{e_{k+1|k}x_{k+1}^T\}C_{k+1}^T\tilde{\Delta}_{k+1}^T G_{k+1}^T \\
&\quad -G_{k+1}\tilde{\Delta}_{k+1}C_{k+1}\mathbb{E}\{x_{k+1}e_{k+1|k}^T\}\Pi_{k+1}^T \\
&\quad +G_{k+1}\tilde{\Delta}_{k+1}C_{k+1}\mathbb{E}\{x_{k+1}x_{k+1}^T\}C_{k+1}^T\tilde{\Delta}_{k+1}^T G_{k+1}^T \\
&\quad +G_{k+1}(\mathbb{I} + \tilde{\Delta}_{k+1})\mathbb{E}\{\nu_{k+1}\nu_{k+1}^T\}(\mathbb{I} + \tilde{\Delta}_{k+1})^T G_{k+1}^T \\
&\quad -\mathcal{A}_{k+1} - \mathcal{A}_{k+1}^T + \mathcal{B}_{k+1} + \mathcal{B}_{k+1}^T,
\end{aligned} \tag{10.18}
$$

where

$$
\begin{aligned}
\mathcal{A}_{k+1} &= \Pi_{k+1}\mathbb{E}\{e_{k+1|k}\nu_{k+1}^T\}(\mathbb{I} + \tilde{\Delta}_{k+1})^T G_{k+1}^T, \\
\mathcal{B}_{k+1} &= G_{k+1}\tilde{\Delta}_{k+1}C_{k+1}\mathbb{E}\{x_{k+1}\nu_{k+1}^T\}(\mathbb{I} + \tilde{\Delta}_{k+1})^T G_{k+1}^T.
\end{aligned}
$$

Next, it is not difficult to see that $\mathcal{A}_{k+1} = 0$ and $\mathcal{B}_{k+1} = 0$. Then, it is concluded that (10.16) holds, which completes the proof of this theorem.

It is worth pointing out that the proposed EEC includes some unknown terms, thus it is not easy to derive the estimator gain on the premise of minimizing the trace of the EEC. Next, an alternative approach is utilized, where an upper bound of the resulting EEC is obtained via employing the matrix theory, and then the trace of the solved upper bound can be minimized by resorting to the desirable estimator gain.

Theorem 10.1 *For known scalars $\varepsilon_{l,k} > 0$ $(l = 1, 2, 3, 4)$ and $\varepsilon_{k+1} > 0$, if the following difference equations:*

$$
\begin{aligned}
\Omega_{k+1|k} = &(1 + \varepsilon_{1,k})(A_k + W \otimes \bar{\Gamma})\Omega_{k|k}(A_k + W \otimes \bar{\Gamma})^T \\
&+(1 + \varepsilon_{1,k}^{-1})\mathrm{tr}\Big(\check{\Gamma}[(1 + \varepsilon_{2,k})\Omega_{k|k} + (1 + \varepsilon_{2,k}^{-1})\hat{x}_{k|k}\hat{x}_{k|k}^T]\check{\Gamma}^T\Big)\check{W}\check{W}^T \\
&+B_k Q_k B_k^T
\end{aligned} \tag{10.19}
$$

and

$$
\begin{aligned}
\Omega_{k+1|k+1} = &(1 + \varepsilon_{3,k+1})\Pi_{k+1}\Omega_{k+1|k}\Pi_{k+1}^T \\
&+G_{k+1}\Big\{(1 + \varepsilon_{3,k+1}^{-1})\mathrm{tr}\big(\tilde{\Lambda}C_{k+1}[(1 + \varepsilon_{4,k+1})\Omega_{k+1|k} \\
&+(1 + \varepsilon_{4,k+1}^{-1})\hat{x}_{k+1|k}\hat{x}_{k+1|k}^T]C_{k+1}^T\tilde{\Lambda}^T\big)\mathbb{I} \\
&+(R_{k+1}^{-1} - \varepsilon_{k+1}\tilde{\Lambda}^T\tilde{\Lambda})^{-1} + \varepsilon_{k+1}^{-1}\mathbb{I}\Big\}G_{k+1}^T
\end{aligned} \tag{10.20}
$$

under $\Omega_{0|0} = P_{0|0}$ *and the following inequality constraint:*

$$\varepsilon_{k+1}^{-1}\mathbb{I} - \tilde{\Lambda} R_{k+1} \tilde{\Lambda}^T > 0 \tag{10.21}$$

have solutions $\Omega_{k+1|k} > 0$ *and* $\Omega_{k+1|k+1} > 0$*, then it can be obtained that*

$$P_{k+1|k+1} \leq \Omega_{k+1|k+1}, \tag{10.22}$$

where

$$\begin{aligned}
\check{W} &= W \otimes I, \\
\tilde{\Lambda} &= \mathrm{diag}\{\underbrace{\Lambda, \Lambda, \ldots, \Lambda}_{N}\}, \\
\check{\Gamma} &= \mathrm{diag}\{\underbrace{\underline{\Gamma}, \underline{\Gamma}, \ldots, \underline{\Gamma}}_{N}\}.
\end{aligned} \tag{10.23}$$

Furthermore, if the estimator gain is designed as

$$G_{k+1} = \mathrm{diag}\{G_{1,k+1}, G_{2,k+1}, \ldots, G_{N,k+1}\} \tag{10.24}$$

with

$$\begin{aligned}
G_{i,k+1} &= (1 + \varepsilon_{3,k+1})\Psi_{1,i}\Omega_{k+1|k}C_{k+1}^T\Psi_{2,i}^T(\Psi_{2,i}\Xi_{k+1}\Psi_{2,i}^T)^{-1}, \\
\Xi_{k+1} &= (1 + \varepsilon_{3,k+1})C_{k+1}\Omega_{k+1|k}C_{k+1}^T + (R_{k+1}^{-1} - \varepsilon_{k+1}\tilde{\Lambda}^T\tilde{\Lambda})^{-1} + \varepsilon_{k+1}^{-1}\mathbb{I} \\
&\quad + (1 + \varepsilon_{3,k+1}^{-1})\mathrm{tr}\big(\tilde{\Lambda}C_{k+1}[(1 + \varepsilon_{4,k+1})\Omega_{k+1|k} \\
&\quad + (1 + \varepsilon_{4,k+1}^{-1})\hat{x}_{k+1|k}\hat{x}_{k+1|k}^T]C_{k+1}^T\tilde{\Lambda}^T\big)\mathbb{I}, \\
\Psi_{1,i} &= \big[\underbrace{0\ 0\ \cdots\ 0}_{i-1}\ \ I_{n\times n}\ \ \underbrace{0\ 0\ \cdots\ 0}_{N-i}\big], \\
\Psi_{2,i} &= \big[\underbrace{0\ 0\ \cdots\ 0}_{i-1}\ \ I_{m\times m}\ \ \underbrace{0\ 0\ \cdots\ 0}_{N-i}\big],
\end{aligned} \tag{10.25}$$

then $\mathrm{tr}(\Omega_{k+1|k+1})$ *can be minimized.*

Proof According to Lemma 2.2, one obtains

$$\begin{aligned}
&(A_k + W \otimes \bar{\Gamma})\mathbb{E}\{e_{k|k}x_k^T\}(W \otimes \tilde{\Gamma})^T + (W \otimes \tilde{\Gamma})\mathbb{E}\{x_k e_{k|k}^T\}(A_k + W \otimes \bar{\Gamma})^T \\
&\leq \varepsilon_{1,k}(A_k + W \otimes \bar{\Gamma})P_{k|k}(A_k + W \otimes \bar{\Gamma})^T \\
&\quad + \varepsilon_{1,k}^{-1}(W \otimes \tilde{\Gamma})\mathbb{E}\{x_k x_k^T\}(W \otimes \tilde{\Gamma})^T,
\end{aligned} \tag{10.26}$$

where $\varepsilon_{1,k} > 0$ is a constant. By virtue of Lemma 2.2, we obtain

$$\begin{aligned}\mathbb{E}\{x_k x_k^T\} &= \mathbb{E}\{(e_{k|k} + \hat{x}_{k|k})(e_{k|k} + \hat{x}_{k|k})^T\} \\ &= \mathbb{E}\{e_{k|k}e_{k|k}^T + e_{k|k}\hat{x}_{k|k}^T + \hat{x}_{k|k}e_{k|k}^T + \hat{x}_{k|k}\hat{x}_{k|k}^T\} \\ &\leq (1 + \varepsilon_{2,k})P_{k|k} + (1 + \varepsilon_{2,k}^{-1})\hat{x}_{k|k}\hat{x}_{k|k}^T \\ &:= \bar{\mathscr{S}}_k,\end{aligned} \tag{10.27}$$

where $\varepsilon_{2,k} > 0$ is a scalar. In light of (10.13), (10.26) and (10.27), we have

$$\begin{aligned}P_{k+1|k} \leq\ & (1 + \varepsilon_{1,k})(A_k + W \otimes \bar{\Gamma})P_{k|k}(A_k + W \otimes \bar{\Gamma})^T \\ & + (1 + \varepsilon_{1,k}^{-1})(W \otimes \tilde{\Gamma})\bar{\mathscr{S}}_k(W \otimes \tilde{\Gamma})^T + B_k Q_k B_k^T.\end{aligned} \tag{10.28}$$

Next, noting $W \otimes \tilde{\Gamma} = \check{W}\check{\mathscr{F}}\check{\Gamma}$, with $\check{W}$ and $\check{\Gamma}$ being given in (10.23), and $\check{\mathscr{F}} = \text{diag}\{\underbrace{\mathscr{F}, \mathscr{F}, \ldots, \mathscr{F}}_{N}\}$, we have

$$\begin{aligned}(1 + \varepsilon_{1,k}^{-1})(W \otimes \tilde{\Gamma})\bar{\mathscr{S}}_k(W \otimes \tilde{\Gamma})^T &= (1 + \varepsilon_{1,k}^{-1})\check{W}\check{\mathscr{F}}\check{\Gamma}\bar{\mathscr{S}}_k\check{\Gamma}^T\check{\mathscr{F}}^T\check{W}^T \\ &\leq (1 + \varepsilon_{1,k}^{-1})\text{tr}(\check{\Gamma}\bar{\mathscr{S}}_k\check{\Gamma}^T)\check{W}\check{W}^T.\end{aligned} \tag{10.29}$$

Therefore, from (10.28) and (10.29), one arrives at

$$\begin{aligned}P_{k+1|k} \leq\ & (1 + \varepsilon_{1,k})(A_k + W \otimes \bar{\Gamma})P_{k|k}(A_k + W \otimes \bar{\Gamma})^T \\ & + (1 + \varepsilon_{1,k}^{-1})\text{tr}(\check{\Gamma}\bar{\mathscr{S}}_k\check{\Gamma}^T)\check{W}\check{W}^T + B_k Q_k B_k^T.\end{aligned} \tag{10.30}$$

Secondly, we aim to tackle the unknown terms of the RHS of (10.16). With the help of Lemma 2.2 again, it is derived that

$$\begin{aligned}& -\Pi_{k+1}\mathbb{E}\{e_{k+1|k}x_{k+1}^T\}C_{k+1}^T\tilde{\Delta}_{k+1}^T G_{k+1}^T \\ & -G_{k+1}\tilde{\Delta}_{k+1}C_{k+1}\mathbb{E}\{x_{k+1}e_{k+1|k}^T\}\Pi_{k+1}^T \\ \leq\ & \varepsilon_{3,k+1}\Pi_{k+1}P_{k+1|k}\Pi_{k+1}^T + \varepsilon_{3,k+1}^{-1}G_{k+1}\tilde{\Delta}_{k+1} \\ & \times C_{k+1}\mathbb{E}\{x_{k+1}x_{k+1}^T\}C_{k+1}^T\tilde{\Delta}_{k+1}^T G_{k+1}^T\end{aligned} \tag{10.31}$$

with $\varepsilon_{3,k+1} > 0$ being a scalar. Substituting (10.31) into (10.16) leads to

$$\begin{aligned}P_{k+1|k+1} \leq\ & (1 + \varepsilon_{3,k+1})\Pi_{k+1}P_{k+1|k}\Pi_{k+1}^T + (1 + \varepsilon_{3,k+1}^{-1})G_{k+1}\tilde{\Delta}_{k+1}C_{k+1}\mathbb{E}\{x_{k+1}x_{k+1}^T\} \\ & \times C_{k+1}^T\tilde{\Delta}_{k+1}^T G_{k+1}^T + G_{k+1}(\mathbb{I} + \tilde{\Delta}_{k+1})R_{k+1}(\mathbb{I} + \tilde{\Delta}_{k+1})^T G_{k+1}^T.\end{aligned} \tag{10.32}$$

Along the same line with (10.27), one has

$$\begin{aligned}& \mathbb{E}\{x_{k+1}x_{k+1}^T\} \\ \leq\ & (1 + \varepsilon_{4,k+1})P_{k+1|k} + (1 + \varepsilon_{4,k+1}^{-1})\hat{x}_{k+1|k}\hat{x}_{k+1|k}^T \\ :=\ & \bar{\mathscr{T}}_{k+1}\end{aligned} \tag{10.33}$$

with $\varepsilon_{4,k+1} > 0$ denoting a scalar. After that, it can be derived from (10.33) and $\tilde{\Delta}_{k+1} = \tilde{\mathfrak{F}}_{k+1}\tilde{\Lambda}$ that

$$\begin{aligned}
& G_{k+1}\tilde{\Delta}_{k+1}C_{k+1}\mathbb{E}\{x_{k+1}x_{k+1}^T\}C_{k+1}^T\tilde{\Delta}_{k+1}^T G_{k+1}^T \\
&= G_{k+1}\tilde{\mathfrak{F}}_{k+1}\tilde{\Lambda}C_{k+1}\mathbb{E}\{x_{k+1}x_{k+1}^T\}C_{k+1}^T\tilde{\Lambda}^T\tilde{\mathfrak{F}}_{k+1}^T G_{k+1}^T \\
&\le G_{k+1}\tilde{\mathfrak{F}}_{k+1}\tilde{\Lambda}C_{k+1}\bar{\mathscr{T}}_{k+1}C_{k+1}^T\tilde{\Lambda}^T\tilde{\mathfrak{F}}_{k+1}^T G_{k+1}^T \\
&\le \mathrm{tr}(\tilde{\Lambda}C_{k+1}\bar{\mathscr{T}}_{k+1}C_{k+1}^T\tilde{\Lambda}^T)G_{k+1}G_{k+1}^T,
\end{aligned} \tag{10.34}$$

where $\tilde{\mathfrak{F}}_{k+1} = \mathrm{diag}\{\underbrace{\mathfrak{F}_{k+1}, \mathfrak{F}_{k+1}, \ldots, \mathfrak{F}_{k+1}}_{N}\}$ and $\tilde{\Lambda}$ is given in (10.23). Subsequently, combining Lemma 2.1 with (10.21), one has

$$\begin{aligned}
& G_{k+1}(\mathbb{I} + \tilde{\Delta}_{k+1})R_{k+1}(\mathbb{I} + \tilde{\Delta}_{k+1})^T G_{k+1}^T \\
&= G_{k+1}(\mathbb{I} + \tilde{\mathfrak{F}}_{k+1}\tilde{\Lambda})R_{k+1}(\mathbb{I} + \tilde{\mathfrak{F}}_{k+1}\tilde{\Lambda})^T G_{k+1}^T \\
&\le G_{k+1}[(R_{k+1}^{-1} - \varepsilon_{k+1}\tilde{\Lambda}^T\tilde{\Lambda})^{-1} + \varepsilon_{k+1}^{-1}\mathbb{I}]G_{k+1}^T.
\end{aligned} \tag{10.35}$$

Substituting (10.34) and (10.35) into (10.32) yields

$$\begin{aligned}
P_{k+1|k+1} \le\ & (1 + \varepsilon_{3,k+1})\Pi_{k+1}P_{k+1|k}\Pi_{k+1}^T \\
& + (1 + \varepsilon_{3,k+1}^{-1})\mathrm{tr}(\tilde{\Lambda}C_{k+1}\bar{\mathscr{T}}_{k+1}C_{k+1}^T\tilde{\Lambda}^T)G_{k+1}G_{k+1}^T \\
& + G_{k+1}\big[(R_{k+1}^{-1} - \varepsilon_{k+1}\tilde{\Lambda}^T\tilde{\Lambda})^{-1} + \varepsilon_{k+1}^{-1}\mathbb{I}\big]G_{k+1}^T.
\end{aligned} \tag{10.36}$$

Thirdly, the aim is to exhibit that $P_{k+1|k+1} \le \Omega_{k+1|k+1}$ in light of the mathematical induction method. In order to testify this assertion, note the initial condition $P_{0|0} = \Omega_{0|0}$ and assume $P_{k|k} \le \Omega_{k|k}$. Then, it can be obtained from (10.19), (10.30), $\mathscr{S}_k$ in (10.27) and $P_{k|k} \le \Omega_{k|k}$ that

$$P_{k+1|k} \le \Omega_{k+1|k}.$$

Based on (10.20), (10.36), $\bar{\mathscr{T}}_{k+1}$ in (10.33) and $P_{k+1|k} \le \Omega_{k+1|k}$, we have

$$P_{k+1|k+1} \le \Omega_{k+1|k+1}. \tag{10.37}$$

Finally, the objective is to minimize the trace of the presented upper bound $\Omega_{k+1|k+1}$ by selecting the estimator gain $G_{i,k+1}$. Rewriting (10.20), one has

$$\begin{aligned}
& \Omega_{k+1|k+1} \\
&= (1 + \varepsilon_{3,k+1})\Omega_{k+1|k} - (1 + \varepsilon_{3,k+1})\Omega_{k+1|k}C_{k+1}^T G_{k+1}^T \\
&\quad - (1 + \varepsilon_{3,k+1})G_{k+1}C_{k+1}\Omega_{k+1|k} + G_{k+1}\Xi_{k+1}G_{k+1}^T
\end{aligned} \tag{10.38}$$

with Ξ_{k+1} being defined as in (10.25). Consider that $G_{k+1} = \sum_{i=1}^{N}(\Psi_{1,i}^T G_{i,k+1}\Psi_{2,i})$, where $\Psi_{1,i}$ and $\Psi_{2,i}$ are defined as in (10.25). It is easily seen that $\Psi_{1,j}\Psi_{1,i}^T = 0$ for $i \neq j$. Therefore, it is clear to check that

$$\begin{aligned}
&\operatorname{tr}\big((\Psi_{1,i}^T G_{i,k+1}\Psi_{2,i})\Xi_{k+1}(\Psi_{1,j}^T G_{j,k+1}\Psi_{2,j})^T\big) \\
&= \operatorname{tr}\big(\Psi_{1,i}^T G_{i,k+1}\Psi_{2,i}\Xi_{k+1}\Psi_{2,j}^T G_{j,k+1}^T \Psi_{1,j}\big) \\
&= \operatorname{tr}\big(\Psi_{1,j}\Psi_{1,i}^T G_{i,k+1}\Psi_{2,i}\Xi_{k+1}\Psi_{2,j}^T G_{j,k+1}^T\big) \\
&= 0
\end{aligned} \tag{10.39}$$

for $i \neq j$. Therefore, $\operatorname{tr}(\Omega_{k+1|k+1})$ is computed as

$$\begin{aligned}
&\operatorname{tr}(\Omega_{k+1|k+1}) \\
&= \operatorname{tr}\Bigg((1+\varepsilon_{3,k+1})\Omega_{k+1|k} - (1+\varepsilon_{3,k+1})\Omega_{k+1|k}C_{k+1}^T\sum_{i=1}^{N}(\Psi_{1,i}^T G_{i,k+1}\Psi_{2,i})^T \\
&\quad -(1+\varepsilon_{3,k+1})\sum_{i=1}^{N}(\Psi_{1,i}^T G_{i,k+1}\Psi_{2,i})C_{k+1}\Omega_{k+1|k} \\
&\quad +\sum_{i=1}^{N}(\Psi_{1,i}^T G_{i,k+1}\Psi_{2,i})\Xi_{k+1}\sum_{i=1}^{N}(\Psi_{1,i}^T G_{i,k+1}\Psi_{2,i})^T\Bigg) \\
&= \operatorname{tr}\Bigg((1+\varepsilon_{3,k+1})\Omega_{k+1|k} - (1+\varepsilon_{3,k+1})\Omega_{k+1|k}C_{k+1}^T\sum_{i=1}^{N}(\Psi_{1,i}^T G_{i,k+1}\Psi_{2,i})^T \\
&\quad -(1+\varepsilon_{3,k+1})\sum_{i=1}^{N}(\Psi_{1,i}^T G_{i,k+1}\Psi_{2,i})C_{k+1}\Omega_{k+1|k} \\
&\quad +\sum_{i=1}^{N}(\Psi_{1,i}^T G_{i,k+1}\Psi_{2,i})\Xi_{k+1}(\Psi_{1,i}^T G_{i,k+1}\Psi_{2,i})^T\Bigg).
\end{aligned} \tag{10.40}$$

It should be noted that, during the process of deriving the second equation of (10.40), the equation (10.39) is employed.

In what follows, take the partial derivative of $\operatorname{tr}(\Omega_{k+1|k+1})$ regarding the estimator gain $G_{i,k+1}$ and let the derivative denote zero. Besides, by means of Lemma 2.3, we have

$$\begin{aligned}
&\frac{\partial \operatorname{tr}(\Omega_{k+1|k+1})}{\partial G_{i,k+1}} \\
&= -2(1+\varepsilon_{3,k+1})\Psi_{1,i}\Omega_{k+1|k}C_{k+1}^T\Psi_{2,i}^T + 2\Psi_{1,i}\Psi_{1,i}^T G_{i,k+1}\Psi_{2,i}\Xi_{k+1}\Psi_{2,i}^T \\
&= 0.
\end{aligned} \tag{10.41}$$

Notice that $\Psi_{1,i}\Psi_{1,i}^T = I_{n\times n}$, then the optimized estimator gain is calculated as

$$G_{i,k+1} = (1+\varepsilon_{3,k+1})\Psi_{1,i}\Omega_{k+1|k}C_{k+1}^T\Psi_{2,i}^T(\Psi_{2,i}\Xi_{k+1}\Psi_{2,i}^T)^{-1}, \tag{10.42}$$

which finishes the proof of this theorem.

Remark 10.2 It is essential to note that the solutions to (10.19) and (10.20) are related with the values of some positive scalars $\varepsilon_{1,k}$, $\varepsilon_{2,k}$, $\varepsilon_{3,k+1}$, $\varepsilon_{4,k+1}$ and ε_{k+1}. As a result, it is crucial to select appropriate values of them to complete the estimation algorithm design. The common addressing methods include, but are not limited to, the genetic algorithm and gradient descent method. Particularly, in the implementation process, ε_{k+1} can be properly selected to guarantee the feasibility of (10.21).

10.3 Boundedness Analysis

In this section, a sufficient condition is provided to ensure the exponential mean-square boundedness of the estimation error dynamics.

Based on Definition 2.1 and Theorem 10.1, the following theorem has a sufficient condition to prove the exponential mean-square boundedness of the obtained estimation error.

Theorem 10.2 *If there are positive scalars $\bar{a}$, $\underline{b}$, $\bar{b}$, $\underline{c}$, $\bar{c}$, $\bar{w}$, $\underline{q}$, $\bar{q}$, $\bar{\tau}$, $\bar{\xi}_1$ and $\bar{\xi}_2$ satisfying the following conditions:*

$$\begin{aligned}
&\|A_{i,k}\| \le \bar{a}, \quad \underline{b}I \le B_{i,k}B_{i,k}^T \le \bar{b}I, \quad \underline{c} \le \|C_{i,k}\| \le \bar{c}, \\
&\|\check{W}\| \le \bar{w}, \quad \underline{q}I \le Q_k \le \bar{q}I, \quad R_k \le \bar{\tau}I, \\
&\mathrm{tr}(\Xi_{k|k}) \le \bar{\xi}_1, \quad \mathrm{tr}(\Xi_{k+1|k}) \le \bar{\xi}_2, \\
&\varrho = 2\mathbf{a}\left(1+\frac{\bar{c}^2}{\underline{c}^2}\right)^2 < 1,
\end{aligned} \tag{10.43}$$

where

$$\begin{aligned}
\mathbf{a} &= \bar{a}^2 + \left[\bar{w}\max_l\left(\frac{\underline{\gamma}_l+\bar{\gamma}_l}{2}\right)\right]^2, \\
\Xi_{k|k} &= (1+\varepsilon_{2,k})\Omega_{k|k} + (1+\varepsilon_{2,k}^{-1})\hat{x}_{k|k}\hat{x}_{k|k}^T, \\
\Xi_{k+1|k} &= (1+\varepsilon_{4,k+1})\Omega_{k+1|k} + (1+\varepsilon_{4,k+1}^{-1})\hat{x}_{k+1|k}\hat{x}_{k+1|k}^T,
\end{aligned} \tag{10.44}$$

then it can be concluded that the estimation error is exponentially mean-square bounded.

Proof Representing $\mathscr{A}_k = A_k + W \otimes \bar{\Gamma}$ and noting $\Pi_{k+1} = \mathbb{I} - G_{k+1}C_{k+1}$, combining (10.10) and (10.11), one obtains

$$e_{k+1|k+1} = \Pi_{k+1}\mathscr{A}_k e_{k|k} + r_{k+1} + s_{k+1}, \tag{10.45}$$

where

$$\begin{aligned} r_{k+1} &= \Pi_{k+1}(W \otimes \tilde{\Gamma})x_k - G_{k+1}\tilde{\Delta}_{k+1}C_{k+1}x_{k+1}, \\ s_{k+1} &= \Pi_{k+1}B_k\varpi_k - G_{k+1}(\mathbb{I} + \tilde{\Delta}_{k+1})v_{k+1}. \end{aligned} \tag{10.46}$$

On the basis of (10.24) and (10.25), one gets

$$\begin{aligned} \|G_{k+1}\| &\le \max_i \|G_{i,k+1}\| \\ &< \max_i \|(1 + \varepsilon_{3,k+1})\Psi_{1,i}\Omega_{k+1|k}C_{k+1}^T\Psi_{2,i}^T \\ &\quad \times[(1 + \varepsilon_{3,k+1})\Psi_{2,i}C_{k+1}\Omega_{k+1|k}C_{k+1}^T\Psi_{2,i}^T]^{-1}\| \\ &< \frac{\bar{c}}{\underline{c}^2} := \bar{\kappa} \end{aligned} \tag{10.47}$$

with $\underline{c}$ and $\bar{c}$ being given in (10.43). In what follows, it is clear to find that

$$\|\Pi_{k+1}\| = \|\mathbb{I} - G_{k+1}C_{k+1}\| \le 1 + \frac{\bar{c}^2}{\underline{c}^2} := \bar{\pi}. \tag{10.48}$$

Then, it can be obtained from Lemma 2.2 and the properties of trace that

$$\begin{aligned} &\mathbb{E}\{r_{k+1}^T r_{k+1}\} \\ &\le (1+\eta)\mathbb{E}\{x_k^T(W \otimes \tilde{\Gamma})^T\Pi_{k+1}^T\Pi_{k+1}(W \otimes \tilde{\Gamma})x_k\} \\ &\quad +(1+\eta^{-1})\mathbb{E}\{x_{k+1}^T C_{k+1}^T\tilde{\Delta}_{k+1}^T G_{k+1}^T G_{k+1}\tilde{\Delta}_{k+1}C_{k+1}x_{k+1}\} \\ &\le (1+\eta)\mathrm{tr}\big(\mathbb{E}\{x_k x_k^T\}(W \otimes \tilde{\Gamma})^T\Pi_{k+1}^T\Pi_{k+1}(W \otimes \tilde{\Gamma})\big) \\ &\quad +(1+\eta^{-1})\mathrm{tr}\big(\mathbb{E}\{x_{k+1}x_{k+1}^T\}C_{k+1}^T\tilde{\Delta}_{k+1}^T G_{k+1}^T G_{k+1}\tilde{\Delta}_{k+1}C_{k+1}\big), \end{aligned}$$

where $\eta > 0$ is a scalar.

Next, in light of (10.43), (10.46)–(10.48) and $W \otimes \tilde{\Gamma} = \check{W}\check{\mathscr{F}}\check{\Gamma}$, one gets

$$\begin{aligned} &\mathbb{E}\{r_{k+1}^T r_{k+1}\} \\ &\le (1+\eta)\mathrm{tr}\big(\Xi_{k|k}\check{\Gamma}^T\check{\mathscr{F}}^T\check{W}^T\Pi_{k+1}^T\Pi_{k+1}\check{W}\check{\mathscr{F}}\check{\Gamma}\big) \\ &\quad +(1+\eta^{-1})\mathrm{tr}\big(\Xi_{k+1|k}C_{k+1}^T\tilde{\Lambda}^T\tilde{\mathfrak{F}}_{k+1}^T G_{k+1}^T G_{k+1}\tilde{\mathfrak{F}}_{k+1}\tilde{\Lambda}C_{k+1}\big) \\ &\le (1+\eta)\bar{\pi}^2\bar{w}^2\|\underline{\Gamma}\|^2\bar{\xi}_1 + (1+\eta^{-1})\bar{\kappa}^2\|\tilde{\Lambda}\|^2\bar{c}^2\bar{\xi}_2 \\ &:= \bar{r}. \end{aligned} \tag{10.49}$$

Likewise, one obtains

$$\begin{aligned}
&\mathbb{E}\{s_{k+1}^T s_{k+1}\} \\
&= \mathbb{E}\{\varpi_k^T B_k^T \Pi_{k+1}^T \Pi_{k+1} B_k \varpi_k\} \\
&\quad + \mathbb{E}\{\nu_{k+1}^T (\mathbb{I} + \tilde{\Delta}_{k+1})^T G_{k+1}^T G_{k+1} (\mathbb{I} + \tilde{\Delta}_{k+1}) \nu_{k+1}\} \\
&\leq \bar{b}\bar{q}\bar{\pi}^2 + 2\bar{\kappa}^2 (1 + \|\tilde{\Lambda}\|^2)\bar{\tau} \\
&:= \bar{s}.
\end{aligned} \tag{10.50}$$

Notice the following matrix equation about Φ_k:

$$\Phi_{k+1} = \Pi_{k+1} \mathscr{A}_k \Phi_k \mathscr{A}_k^T \Pi_{k+1}^T + B_k Q_k B_k^T + \delta_0 I, \tag{10.51}$$

where $\Phi_0 = B_0 Q_0 B_0^T + \delta_0 I$ with $\delta_0 > 0$ denoting a constant. Then, in light of the above iterative matrix equation, it is clear to observe that

$$\begin{aligned}
\|\Phi_{k+1}\| &\leq \|\Pi_{k+1}\|^2 \|\mathscr{A}_k\|^2 \|\Phi_k\| + \|B_k Q_k B_k^T\| + \|\delta_0 I\| \\
&\leq \varrho \|\Phi_k\| + \bar{b}\bar{q} + \delta_0
\end{aligned} \tag{10.52}$$

with ϱ being defined in (10.43). Furthermore, we have the following inequality:

$$\|\Phi_k\| \leq \varrho^k \|\Phi_0\| + (\bar{b}\bar{q} + \delta_0) \sum_{i=0}^{k-1} \varrho^i. \tag{10.53}$$

Noting $\varrho < 1$, we have

$$\begin{aligned}
\|\Phi_k\| &< \|\Phi_0\| + (\bar{b}\bar{q} + \delta_0) \sum_{i=0}^{\infty} \varrho^i \\
&= \|\Phi_0\| + \frac{\bar{b}\bar{q} + \delta_0}{1 - \varrho}.
\end{aligned} \tag{10.54}$$

On the other hand, one has

$$\Phi_k \geq \delta_0 I. \tag{10.55}$$

By resorting to (10.54) and (10.55), there are two scalars $\bar{\phi} > 0$ and $\underline{\phi} > 0$ such that $\underline{\phi} I \leq \Phi_k \leq \bar{\phi} I$ holds for all $k \geq 0$.

In the sequel, by virtue of the matrix inversion lemma, one has

$$
\begin{aligned}
&\mathscr{A}_k^T \Pi_{k+1}^T \Phi_{k+1}^{-1} \Pi_{k+1} \mathscr{A}_k - \Phi_k^{-1} \\
=& \mathscr{A}_k^T \Pi_{k+1}^T \left(\Pi_{k+1} \mathscr{A}_k \Phi_k \mathscr{A}_k^T \Pi_{k+1}^T + B_k Q_k B_k^T + \varepsilon I\right)^{-1} \Pi_{k+1} \mathscr{A}_k - \Phi_k^{-1} \\
=& -\left[\Phi_k + \Phi_k \mathscr{A}_k^T \Pi_{k+1}^T \left(B_k Q_k B_k^T + \varepsilon I\right)^{-1} \Pi_{k+1} \mathscr{A}_k \Phi_k\right]^{-1} \\
=& -\left[\mathbb{I} + \mathscr{A}_k^T \Pi_{k+1}^T \left(B_k Q_k B_k^T + \varepsilon I\right)^{-1} \Pi_{k+1} \mathscr{A}_k \Phi_k\right]^{-1} \Phi_k^{-1} \\
\leq& -\bar{\alpha} \Phi_k^{-1},
\end{aligned} \tag{10.56}
$$

where $\bar{\alpha} = \left(1 + \frac{2\mathbf{a}\bar{\pi}^2\bar{\phi}}{q\underline{b}}\right)^{-1}$ and $\mathbf{a}$ is defined in (10.44). Notice that $\bar{\alpha} \in (0, 1)$, then there is a scalar $\varepsilon > 0$ satisfying $\vartheta = (1+\varepsilon)(1-\bar{\alpha}) < 1$. Letting $V_k(e_{k|k}) = e_{k|k}^T \Phi_k^{-1} e_{k|k}$, we have

$$
\begin{aligned}
&\mathbb{E}\{V_{k+1}(e_{k+1|k+1})|e_{k|k}\} - (1+\varepsilon)V_k(e_{k|k}) \\
=& \mathbb{E}\Big\{ \left(\Pi_{k+1}\mathscr{A}_k e_{k|k} + r_{k+1} + s_{k+1}\right)^T \Phi_{k+1}^{-1} \left(\Pi_{k+1}\mathscr{A}_k e_{k|k} + r_{k+1} + s_{k+1}\right) \Big\} \\
&-(1+\varepsilon) e_{k|k}^T \Phi_k^{-1} e_{k|k} \\
=& \mathbb{E}\left\{e_{k|k}^T [\mathscr{A}_k^T \Pi_{k+1}^T \Phi_{k+1}^{-1} \Pi_{k+1} \mathscr{A}_k - (1+\varepsilon)\Phi_k^{-1}] e_{k|k}\right\} \\
&+2\mathbb{E}\left\{e_{k|k}^T \mathscr{A}_k^T \Pi_{k+1}^T \Phi_{k+1}^{-1} r_{k+1}\right\} + \mathbb{E}\left\{r_{k+1}^T \Phi_{k+1}^{-1} r_{k+1}\right\} \\
&+\mathbb{E}\left\{s_{k+1}^T \Phi_{k+1}^{-1} s_{k+1}\right\}.
\end{aligned}
$$

Then, by virtue of the Lemma 2.2 again, we have the following inequality:

$$
\begin{aligned}
&\mathbb{E}\{V_{k+1}(e_{k+1|k+1})|e_{k|k}\} - (1+\varepsilon)V_k(e_{k|k}) \\
\leq& (1+\varepsilon)\mathbb{E}\left\{e_{k|k}^T \left(\mathscr{A}_k^T \Pi_{k+1}^T \Phi_{k+1}^{-1} \Pi_{k+1} \mathscr{A}_k - \Phi_k^{-1}\right) e_{k|k}\right\} \\
&+(1+\varepsilon^{-1})\mathbb{E}\left\{r_{k+1}^T \Phi_{k+1}^{-1} r_{k+1}\right\} \\
&+\mathbb{E}\left\{s_{k+1}^T \Phi_{k+1}^{-1} s_{k+1}\right\} \\
\leq& -\bar{\alpha}(1+\varepsilon)V_k(e_{k|k}) + \varsigma,
\end{aligned} \tag{10.57}
$$

where $\varsigma = (1+\varepsilon^{-1})\frac{\bar{r}^2}{\underline{\phi}} + \frac{\bar{s}^2}{\underline{\phi}}$. Then, it follows from (10.57) that

$$
\mathbb{E}\{V_{k+1}(e_{k+1|k+1})|e_{k|k}\} \leq \vartheta V_k(e_{k|k}) + \varsigma, \tag{10.58}
$$

where $\vartheta = (1+\varepsilon)(1-\bar{\alpha})$.

Finally, we have

$$\begin{aligned}\mathbb{E}\{\|e_{k+1|k+1}\|^2\} &\le \frac{\bar{\phi}}{\underline{\phi}}\mathbb{E}\{\|e_{0|0}\|^2\}\vartheta^{k+1} + \varsigma\bar{\phi}\sum_{i=0}^{k}\vartheta^i \\ &\le \frac{\bar{\phi}}{\underline{\phi}}\mathbb{E}\{\|e_{0|0}\|^2\}\vartheta^{k+1} + \varsigma\bar{\phi}\sum_{i=0}^{\infty}\vartheta^i \\ &= \frac{\bar{\phi}}{\underline{\phi}}\mathbb{E}\{\|e_{0|0}\|^2\}\vartheta^{k+1} + \frac{\varsigma\bar{\phi}}{1-\vartheta}. \end{aligned} \quad (10.59)$$

As such, in terms of Definition 2.1, it can be found that $e_{k|k}$ is exponentially mean-square bounded, which completes the proof of this theorem.

Remark 10.3 Based on the main results given in Theorem 10.2, it is obvious to see that $\mathbb{E}\{\|e_{k+1|k+1}\|\} \le \sqrt{\frac{\bar{\phi}}{\underline{\phi}}\mathbb{E}\{\|e_{0|0}\|^2\}\vartheta^{k+1} + \frac{\varsigma\bar{\phi}}{1-\vartheta}}$, which indicates that the upper bound on the estimation error relies on the inner coupling strength, quantization coefficients and various noise covariances. Specifically, it is found that, the bigger the inner coupling error $\tilde{\Gamma}$ is, the bigger the parameter $\bar{r}$ obtains, the bigger the parameter ς is, and then the bigger upper bound associated with the estimation error gets. In addition, the smaller the quantization density $\chi^{(j)}$ is, the bigger the parameters $\bar{s}$ and $\bar{r}$ are, the bigger the parameter ς obtains, thus the bigger upper bound regarding the estimation error reaches.

Remark 10.4 It is worthwhile to emphasize that some effective synchronization approaches have been shown in Huang et al. (2015, 2018) within the context of *time-invariant* dynamical networks under communication constraints. Different from the existing approaches, a great deal of effort has been dedicated to providing an estimation scheme for *time-varying* complex networks. To be more specific, the distinctive features of our main results lie in that: (1) the uncertain inner coupling and the quantization errors are firstly addressed for stochastic complex networks; (2) the comprehensive analyses are developed on the premise of ensuring two important performance indices; and (3) a recursive estimation algorithm is proposed that relates to the time-varying property of the networks, thus providing an attractive characteristic for online implementations.

Remark 10.5 It is pointed out that the time-varying complex networks with stochastic noises are taken into consideration. In order to estimate the system states, the state estimation approach is presented that can address the influences from the Gaussian noises, uncertain inner coupling as well as quantization errors. Besides, the newly designed recursive estimation scheme can offer satisfactory estimation performance in spite of the worst situation. Specifically, the proposed estimation scheme has the following distinguished merits: (1) the estimator structure is fairly simple especially in the time-varying situation; (2) the state estimations coming from neighboring nodes and its innovation information are both utilized for the estimator design; (3)

the effects generated by the uncertain inner coupling, quantization errors as well as different noises are examined in the developed estimation scheme; and (4) a novel recursive estimation method is proposed that is suitable for online calculations.

10.4 An illustrative Example

In this section, a numerical simulation experiment is used to demonstrate the effectiveness of the developed state estimation scheme. The localization of multiple mobile robots is used to demonstrate the usefulness of the designed state estimation algorithm.

Referencing Li et al. (2018), the localization problem regarding multiple mobile robots is addressed in this chapter. The state is composed of the position and the orientation of the robot. Moreover, the dynamics behavior of each mobile robot is related to other robots and thus the coupling terms inevitably exist. When tracking a robot, the signal quantization might occur on account of the bandwidth limitations of the network channels. For real-time tracking, the system coefficients would be time-varying and there exist different noises. Consequently, the main purpose is to establish a novel recursive estimation scheme with tracking capability such that a locally optimal upper bound regarding the EEC is proposed.

Stimulated by this background, the stochastic complex networks as in (10.1)–(10.2) with three nodes are considered, and the coupling configuration matrix is modeled as

$$W = \begin{bmatrix} 0.3 & 0.2 & 0.2 \\ 0.2 & 0.3 & 0.2 \\ 0.2 & 0.2 & 0.3 \end{bmatrix}.$$

The corresponding parameters are set as

$$\begin{aligned}
A_{1,k} &= \begin{bmatrix} 0.55 + 0.1\cos(2k) & 0.35 \\ -0.6 & 0.75 - 0.1\sin(k) \end{bmatrix}, \\
A_{2,k} &= \begin{bmatrix} 0.45 + 0.1\sin(2k) & -0.45 \\ 0.75 & 0.7 \end{bmatrix}, \\
A_{3,k} &= \begin{bmatrix} 0.85 & 0.75 \\ -0.35 & 0.35 - 0.1\sin(6k) \end{bmatrix}, \\
B_{1,k} &= \begin{bmatrix} 0.6 - 0.1\sin(5k) \\ -0.6 \end{bmatrix}, \\
B_{2,k} &= \begin{bmatrix} 0.4 \\ 0.3 - 0.12\cos(3k) \end{bmatrix}, \quad B_{3,k} = \begin{bmatrix} 0.3 \\ 0.5 \end{bmatrix}, \\
C_{1,k} &= \begin{bmatrix} 1.96 & 1.3 + 0.1\sin(2k) \end{bmatrix}, \\
C_{2,k} &= \begin{bmatrix} 3 + \cos(k)\sin(3k) & 1.5 \end{bmatrix}, \\
C_{3,k} &= \begin{bmatrix} 1.9 + 2\cos(k) & 3.6 + 1.5\sin(3k) \end{bmatrix}.
\end{aligned}$$

Here, $\Gamma = \text{diag}\{\gamma_1, \gamma_2\}$ is the inner coupling matrix, where $\gamma_i \in [0.3, 0.5]$ $(i = 1, 2)$ denotes the unknown coupling strength. After that, it is obvious to obtain that $\bar{\Gamma} = \text{diag}\{0.4, 0.4\}$ and $\underline{\Gamma} = \text{diag}\{0.1, 0.1\}$.

Among the simulations, the other corresponding parameters are chosen as $x_{1,0} = \begin{bmatrix} 1.8 & 1.2 \end{bmatrix}^T, x_{2,0} = \begin{bmatrix} 1.2 & -1.6 \end{bmatrix}^T, x_{3,0} = \begin{bmatrix} 1.6 & 0.8 \end{bmatrix}^T, \hat{x}_{1,0|0} = \begin{bmatrix} 1.5 & -1.2 \end{bmatrix}^T, \hat{x}_{2,0|0} = \begin{bmatrix} 1 & -2.3 \end{bmatrix}^T$, $\hat{x}_{3,0|0} = \begin{bmatrix} -0.5 & 1.5 \end{bmatrix}^T$, $\Omega_{1,0|0} = 30I_2$, $\Omega_{2,0|0} = 20I_2$, $\Omega_{3,0|0} = 25I_2$, $\Omega_{0|0} = \text{diag}\{\Omega_{1,0|0}, \Omega_{2,0|0}, \Omega_{3,0|0}\}$, $\varepsilon_{1,k} = 0.5$, $\varepsilon_{2,k} = 0.5$, $\varepsilon_{3,k+1} = 0.1$, $\varepsilon_{4,k+1} = 1$, $\varepsilon_{k+1} = 1/(1.2\lambda_{\max}(\tilde{\Lambda} R_{k+1} \tilde{\Lambda}) + 0.5)$, $Q_{1,k} = Q_{3,k} = 0.1$, $Q_{2,k} = 0.2$, $R_{1,k} = R_{2,k} = 0.2$ and $R_{3,k} = 0.1$.

In order to reflect the superiority of the established algorithm, the parameters of the logarithmic quantizer are set in two cases, i.e., Case I: $u_0^{(1)} = 0.5$, $\chi^{(1)} = 0.8$, Case II: $u_0^{(1)} = 0.5$, $\chi^{(1)} = 0.3$. Therefore, the newly established variance-constrained state estimation approach can be realized and the related simulation results are depicted in Figs. 10.2, 10.3, 10.4, 10.5, 10.6, 10.7, 10.8, 10.9, 10.10 and 10.11. Here, the real states of all nodes and their estimations in two cases are depicted in Figs. 10.2, 10.3 and 10.4. In Figs. 10.5, 10.6 and 10.7, the ideal measurements of all nodes and the quantization signals are given. It can be observed from Figs. 10.2, 10.3, 10.4, 10.5, 10.6 and 10.7 that the estimation results under Case I are better, which means that the established variance-constrained estimation algorithm performs well when estimating the state of proposed complex networks.

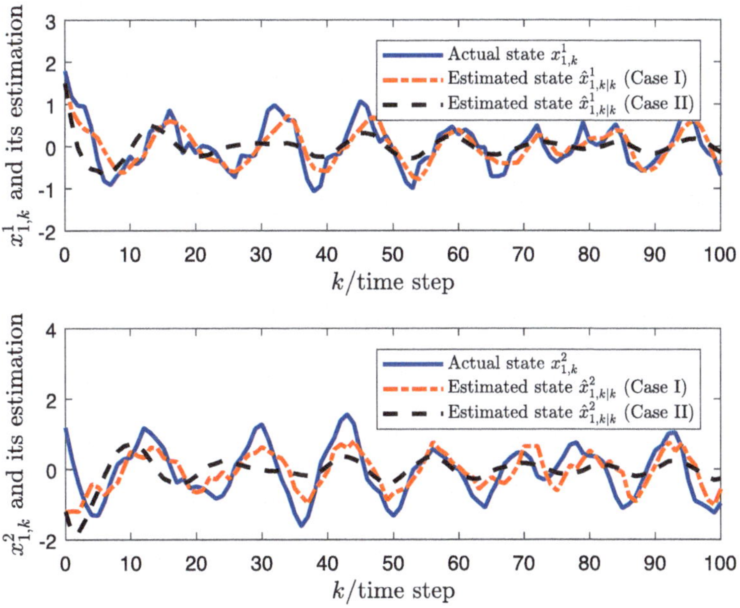

Fig. 10.2 Actual state $x_{1,k}$ and estimates under Cases I–II

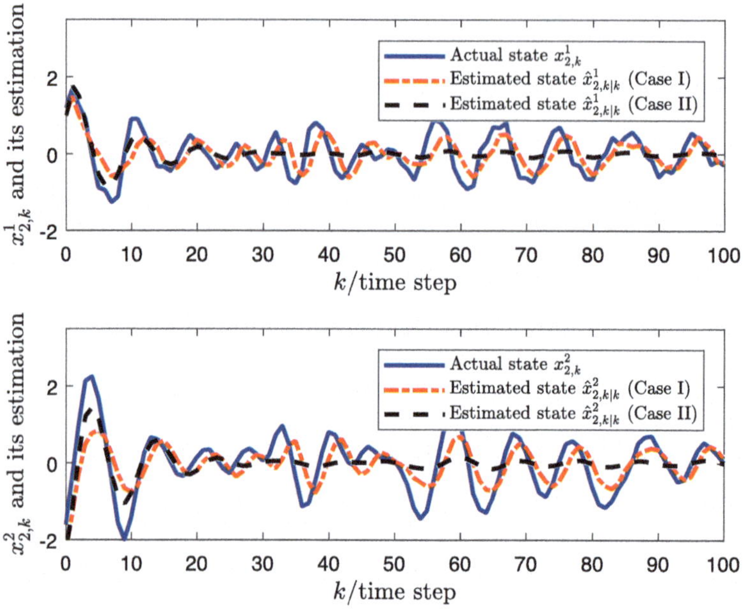

Fig. 10.3 Actual state $x_{2,k}$ and estimates under Cases I–II

On the other hand, the corresponding simulations have been shown in Figs. 10.8, 10.9 and 10.10, in which the *average estimation error* of each node has been given. It can be observed from Figs. 10.8, 10.9 and 10.10 that the smaller final bound of the estimation error is obtained if the bigger quantization density $\chi^{(1)}$ is selected (i.e., Case I). In other words, the estimation accuracy under Case I is better in the comparison with the one under Case II.

In order to comparison, $\mathrm{tr}(P_{k+1|k+1})$ and $\mathrm{tr}(\Omega_{k+1|k+1})$ under Cases I–II are explicitly plotted in Fig. 10.11. Based on the simulations, the larger the quantization density $\chi^{(1)}$ leads to the better the estimation performance. To sum up, this chapter has handled the estimation problem for stochastic complex networks subject to the uncertain inner coupling and the quantization errors under variance-constrained framework, where the proposed state estimation method has performed well.

10.5 Summary

In this chapter, the state estimation issue has been addressed for stochastic complex networks involving uncertain inner coupling and quantization effects. A novel state estimator has been designed on the basis of the quantized measurements, in which the coupling configuration information of the networks has been sufficiently considered. By resorting to the stochastic analysis method, the upper bound regarding the EEC

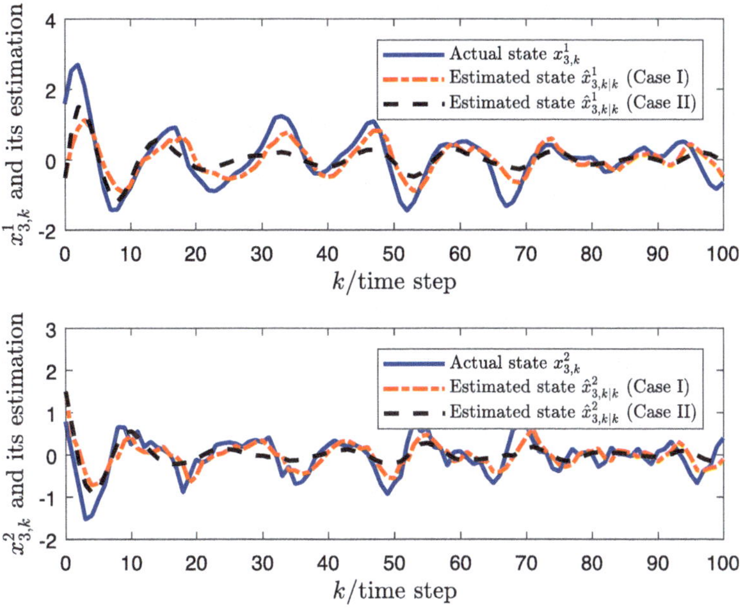

Fig. 10.4 Actual state $x_{3,k}$ and estimates under Cases I–II

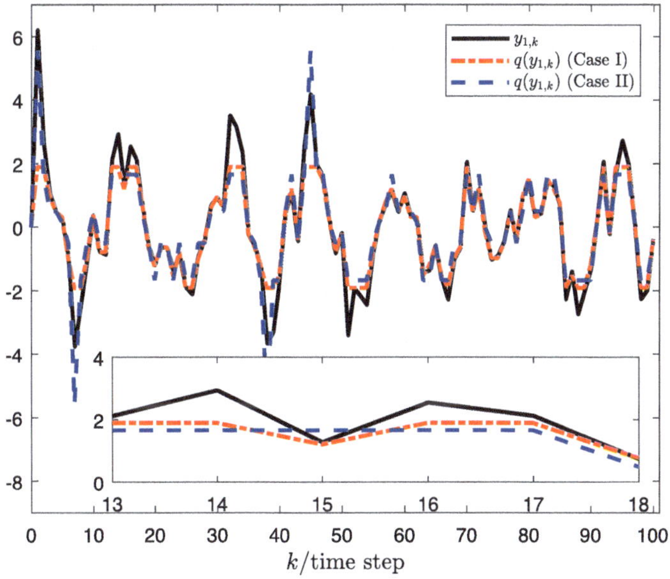

Fig. 10.5 The measurement output and quantization signal of $y_{1,k}$

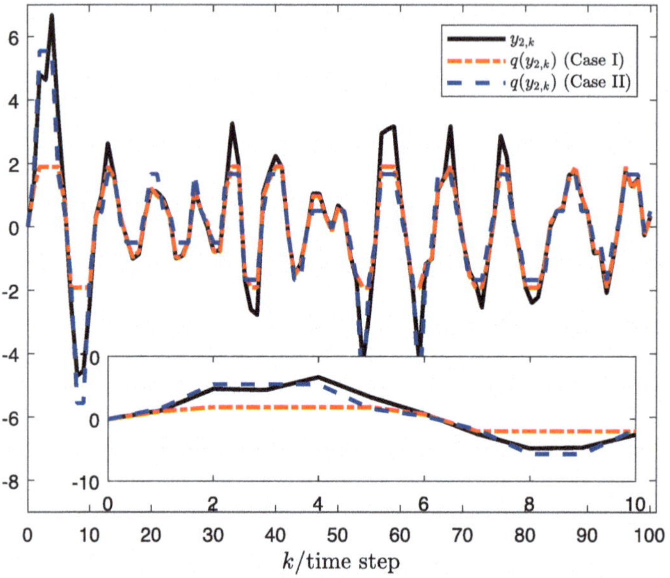

Fig. 10.6 The measurement output and quantization signal of $y_{2,k}$

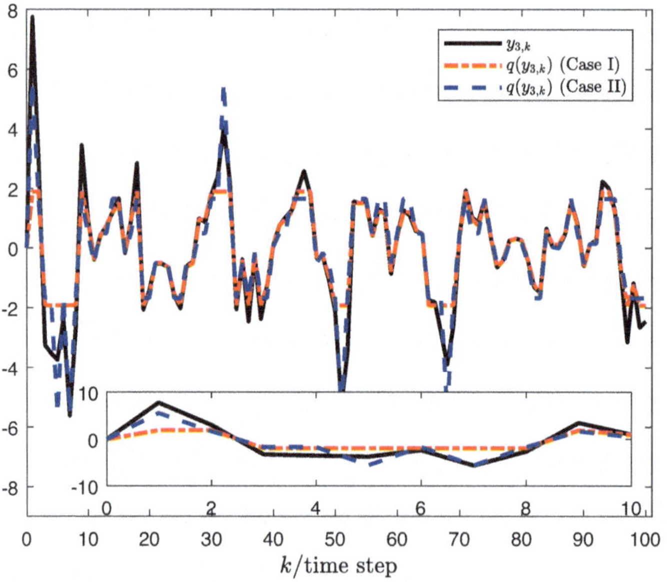

Fig. 10.7 The measurement output and quantization signal of $y_{3,k}$

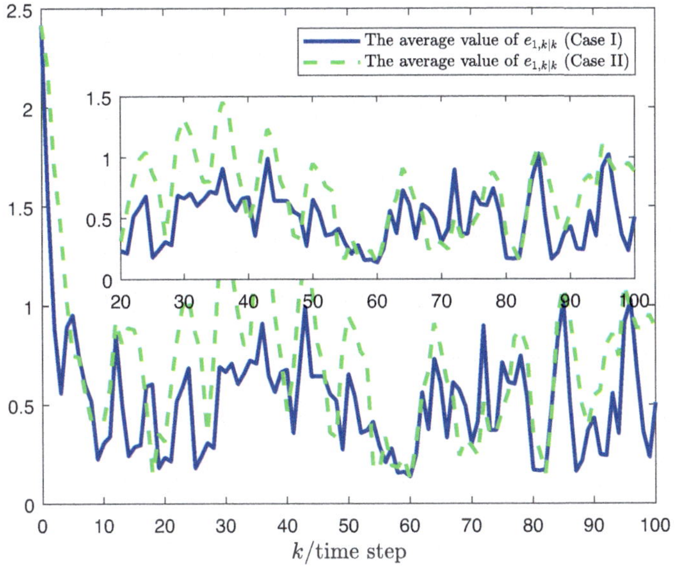

Fig. 10.8 The average value of $e_{1,k|k}$

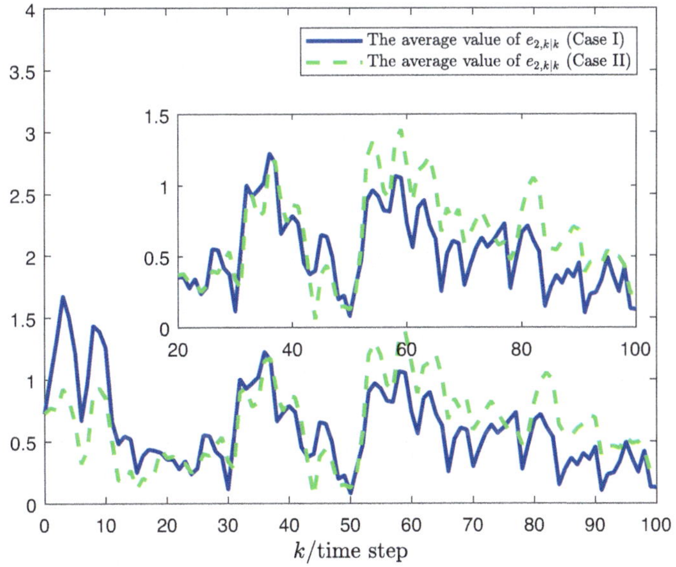

Fig. 10.9 The average value of $e_{2,k|k}$

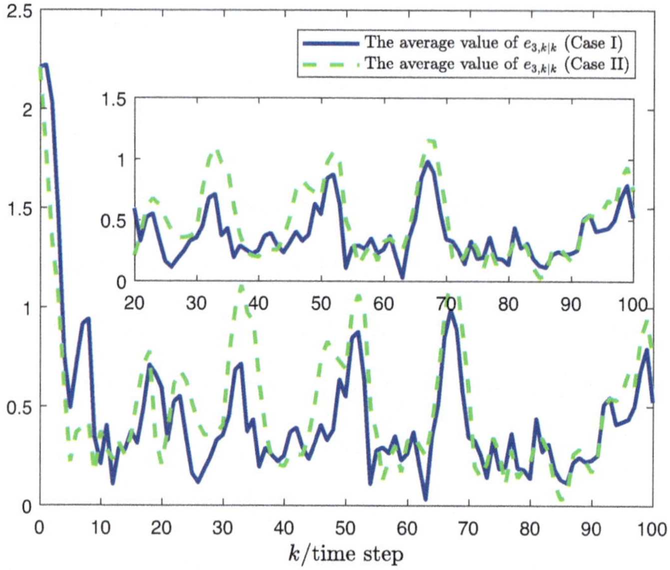

Fig. 10.10 The average value of $e_{3,k|k}$

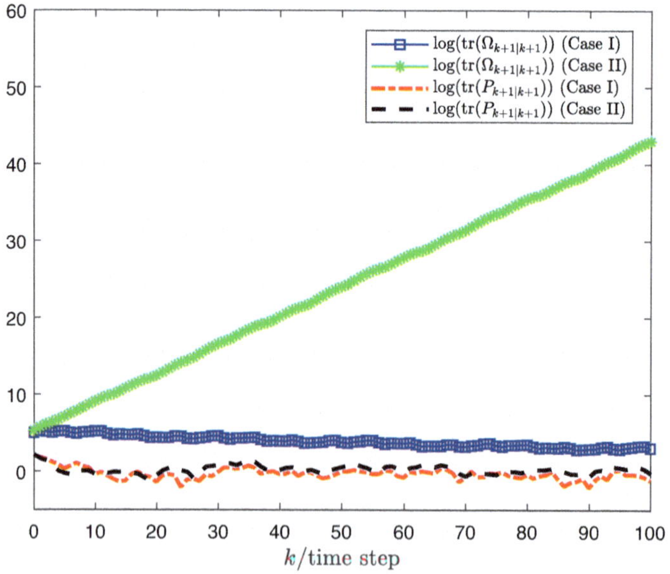

Fig. 10.11 The traces of $P_{k+1|k+1}$ and their upper bound $\Omega_{k+1|k+1}$

has been presented and then this upper bound has been minimized through properly selecting the estimator gain. Moreover, a sufficient condition with rigorous theoretical proof has been proposed to clarify the boundedness problem of the estimation error dynamics. Lastly, a simulation experiment with comparisons has been provided to exhibit the effectiveness of the designed state estimation scheme.

References

Huang C, Ho DWC, Lu J, Kurths J (2015) Pinning synchronization in T-S fuzzy complex networks with partial and discrete-time couplings. IEEE Trans Fuzzy Syst 23(4):1274–1285

Huang C, Wang W, Cao J, Lu J (2018) Synchronization-based passivity of partially coupled neural networks with event-triggered communication. Neurocomputing 319:134–143

Li W, Jia Y, Du J (2018) Variance-constrained state estimation for nonlinearly coupled complex networks. IEEE Trans Cybern 48(2):818–824

Chapter 11
Event-Based State Estimation with Missing Measurements Under Uncertain Occurrence Probabilities

In this chapter, we address the state estimation problem for a class of time-varying complex networks subject to ETM and stochastic coupling with UOPs. Firstly, two sets of Bernoulli random variables are used to model the stochastic coupling and missing measurements, where the UOPs are clearly described here. For the sake of decreasing transmission frequency, the ETM is adopted to arrange signal transmission. Next, a robust state estimation approach is established for considered complex networks. In addition, the monotonicity is shown between the trace of proposed upper bound on the EEC and the deterministic missing occurrence probability. At last, a simulation and comparisons are given to validate the usefulness of the developed robust state estimation method.

11.1 Problem Formulation

Consider the following time-varying stochastic coupling networks including N coupled nodes:

$$x_{i,k+1} = A_{i,k}x_{i,k} + A^d_{i,k}x_{i,k-d} + \gamma_{i,k}\sum_{j=1}^{N} w_{ij}\Gamma x_{j,k} + B_{i,k}\varpi_{i,k}, \tag{11.1}$$

$$y_{i,k} = \lambda_{i,k}C_{i,k}x_{i,k} + \nu_{i,k}, \quad i = 1, 2, \ldots, N, \tag{11.2}$$

$$x_{i,k} = \varphi_{i,k}, \quad k = -d, -d+1, \ldots, 0, \tag{11.3}$$

where $x_{i,k}$ and $y_{i,k}$ stand for the state and measurement, Γ stands for a given linking matrix, and $W = [w_{ij}]_{N\times N}$ with $w_{ij} \geq 0\,(i \neq j)$ represents a coupling configuration matrix, $\varpi_{i,k}$ and $\nu_{i,k}$ represent zero-mean process noise and measurement noise with covariances $Q_{i,k}$ and $R_{i,k} > 0$, respectively. The time delay d is a positive scalar.

J. Hu et al., *Variance-Constrained Filtering for Stochastic Complex Systems*, Intelligent Control and Learning Systems 18,
https://doi.org/10.1007/978-981-96-2637-3_11

$A_{i,k}$, $A^d_{i,k}$, $B_{i,k}$ and $C_{i,k}$ refer to known real value matrices of proper dimensions. $\varphi_{i,k}$ ($k = -d, -d+1, \ldots, 0$) denote known initial conditions.

The binary random variables $\gamma_{i,k}$ and $\lambda_{i,k}$ having UOPs satisfy

$$\text{Prob}\{\gamma_{i,k} = 1\} = \bar{\gamma}_{i,k} + \Delta\gamma_{i,k}, \tag{11.4}$$

$$\text{Prob}\{\lambda_{i,k} = 1\} = \bar{\lambda}_{i,k} + \Delta\lambda_{i,k}, \tag{11.5}$$

where $\bar{\gamma}_{i,k}$ and $\bar{\lambda}_{i,k}$ denote the nominal mathematical expectations, $|\Delta\gamma_{i,k}| \le \rho_{i,k}$ and $|\Delta\lambda_{i,k}| \le \bar{\rho}_{i,k}$ with $\rho_{i,k}$ and $\bar{\rho}_{i,k}$ denoting given constants. Furthermore, it assumes that the variables $\gamma_{i,k}$, $\varpi_{i,k}$, $\lambda_{i,k}$ and $\nu_{i,k}$ are mutually independent. Moreover, define the variance of $\gamma_{i,k}$ as $\Omega_{i,k}$ and the variance of $\lambda_{i,k}$ as $\bar{\sigma}_{i,k}$.

To decrease unnecessary data transmission, the ETM is applied to decide whether the measurement $y_{i,k}$ is delivered to the remote estimator side. Specifically, the event triggering instants sequence is set as $0 \le k_0^i \le k_1^i \le \cdots \le k_l^i \le \cdots$ and the released data signal is denoted by $\bar{y}_{i,k} = y_{i,k_l^i}$ for $k \in [k_l^i, k_{l+1}^i)$. The event generator function $g(y_{i,k}, \bar{y}_{i,k}, \delta_i)$ is defined as

$$g(y_{i,k}, \bar{y}_{i,k}, \delta_i) = (y_{i,k} - \bar{y}_{i,k})^T (y_{i,k} - \bar{y}_{i,k}) - \delta_i \tag{11.6}$$

with $\delta_i > 0$ being an adjustable threshold. Then, the triggering sequence is decided by

$$k_{l+1}^i = \inf\{k \in \mathbb{N}^+ | k > k_l^i,\ g(y_{i,k}, \bar{y}_{i,k}, \delta_i) > 0\}. \tag{11.7}$$

Remark 11.1 When the conventional time-triggered strategy becomes a concern, the measurements are transmitted in a fixed time interval and thus excessive network resources are occupied. For the sake of decreasing resource consumption and improving communication efficiency, the ETM is employed in the network channel to arrange data transmission. Specifically, the ETM executes only if the specific condition is met, which indeed decreases energy consumption and provides the desirable estimation performance.

In this chapter, we design the following state estimator:

$$\hat{x}_{i,k+1|k} = A_{i,k}\hat{x}_{i,k|k} + A^d_{i,k}\hat{x}_{i,k-d|k-d} + \bar{\gamma}_{i,k}\sum_{j=1}^{N} w_{ij}\Gamma\hat{x}_{j,k|k}, \tag{11.8}$$

$$\hat{x}_{i,k+1|k+1} = \hat{x}_{i,k+1|k} + G_{i,k+1}(\bar{y}_{i,k+1} - \bar{\lambda}_{i,k+1}C_{i,k+1}\hat{x}_{i,k+1|k}), \tag{11.9}$$

where $\hat{x}_{i,k+1|k}$ is the one-step prediction, $\hat{x}_{i,k+1|k+1}$ denotes the state estimation, and $G_{i,k+1}$ is the estimator gain.

To facilitate the subsequent development of the main results, denote the one-step prediction error as $e_{i,k+1|k} = x_{i,k+1} - \hat{x}_{i,k+1|k}$ and the estimation error as $e_{i,k+1|k+1} = x_{i,k+1} - \hat{x}_{i,k+1|k+1}$. Thus, their related covariances are defined as

$$P_{i,k+1|k} = \mathbb{E}\{e_{i,k+1|k}e^T_{i,k+1|k}\},\ P_{i,k+1|k+1} = \mathbb{E}\{e_{i,k+1|k+1}e^T_{i,k+1|k+1}\}.$$

The chief objective is to provide an efficient state estimation method with performance evaluation criterion. In particular, a time-varying state estimator (11.8) and (11.9) is designed such that

(R1) there is a sequence of matrices $\Omega_{i,k+1|k+1} > 0$ meeting $P_{i,k+1|k+1} \leq \Omega_{i,k+1|k+1}$;

(R2) the estimator gain matrix $G_{i,k+1}$ is derived via minimizing $\text{tr}(\Omega_{i,k+1|k+1})$;

(R3) the relationship between the trace of proposed upper bound matrix and the missing probability is discussed.

11.2 Design of Estimation Algorithm

In this section, a local optimal upper bound regarding the EEC is attained and the expression form of the estimator gain is provided for helping the implementation.

Adding $G_{i,k+1}y_{i,k+1} - G_{i,k+1}y_{i,k+1}$ to $e_{i,k+1|k+1}$, one has

$$\begin{aligned} e_{i,k+1|k} &= A_{i,k}e_{i,k|k} + A^d_{i,k}e_{i,k-d|k-d} + (\Delta\gamma_{i,k} + \tilde{\gamma}_{i,k})\sum_{j=1}^{N} w_{ij}\Gamma x_{j,k} \\ &\quad + \bar{\gamma}_{i,k}\sum_{j=1}^{N} w_{ij}\Gamma e_{j,k|k} + B_{i,k}\varpi_{i,k}, \end{aligned} \tag{11.10}$$

$$\begin{aligned} e_{i,k+1|k+1} &= (I - \bar{\lambda}_{i,k+1}G_{i,k+1}C_{i,k+1})e_{i,k+1|k} - G_{i,k+1}(\bar{y}_{i,k+1} - y_{i,k+1}) \\ &\quad - G_{i,k+1}\nu_{i,k+1} - \tilde{\lambda}_{i,k+1}G_{i,k+1}C_{i,k+1}x_{i,k+1} \\ &\quad - \Delta\lambda_{i,k+1}G_{i,k+1}C_{i,k+1}x_{i,k+1}, \end{aligned} \tag{11.11}$$

where $\tilde{\gamma}_{i,k} = \gamma_{i,k} - \bar{\gamma}_{i,k} - \Delta\gamma_{i,k}$, $\tilde{\lambda}_{i,k+1} = \lambda_{i,k+1} - \bar{\lambda}_{i,k+1} - \Delta\lambda_{i,k+1}$. It is immediate that $\mathbb{E}\{\tilde{\gamma}_{i,k}\} = 0$ and $\mathbb{E}\{\tilde{\lambda}_{i,k+1}\} = 0$.

Now, the specific forms of the one-step prediction error covariance and the EEC are exhibited, and the proofs are omitted here for brevity.

Lemma 11.1 *The one-step prediction error covariance $P_{i,k+1|k}$ satisfies the following form:*

$$\begin{aligned} P_{i,k+1|k} &= A_{i,k}P_{i,k|k}A^T_{i,k} + A^d_{i,k}P_{i,k-d|k-d}(A^d_{i,k})^T + (\Omega_{i,k} + \Delta\gamma^2_{i,k})\sum_{j=1}^{N}\sum_{l=1}^{N} w_{ij}w_{il}\Gamma \\ &\quad \times\mathbb{E}\{x_{j,k}x^T_{l,k}\}\Gamma^T + \bar{\gamma}^2_{i,k}\sum_{j=1}^{N}\sum_{l=1}^{N} w_{ij}w_{il}\Gamma\mathbb{E}\{e_{j,k|k}e^T_{l,k|k}\}\Gamma^T \\ &\quad + \sum_{l=1}^{6}(\mathscr{A}_l + \mathscr{A}_l^T) + B_{i,k}Q_{i,k}B^T_{i,k}, \end{aligned} \tag{11.12}$$

where

$$\begin{aligned}
\mathscr{A}_1 &= A_{i,k}\mathbb{E}\{e_{i,k|k}e^T_{i,k-d|k-d}\}(A^d_{i,k})^T,\\
\mathscr{A}_2 &= A_{i,k}\mathbb{E}\{\Delta\gamma_{i,k}e_{i,k|k}\sum_{j=1}^{N}w_{ij}x^T_{j,k}\}\Gamma^T,\\
\mathscr{A}_3 &= A_{i,k}\mathbb{E}\{\bar{\gamma}_{i,k}\sum_{j=1}^{N}w_{ij}e_{i,k|k}e^T_{j,k|k}\}\Gamma^T,\\
\mathscr{A}_4 &= A^d_{i,k}\mathbb{E}\{\Delta\gamma_{i,k}e_{i,k-d|k-d}\sum_{j=1}^{N}w_{ij}x^T_{j,k}\}\Gamma^T,\\
\mathscr{A}_5 &= A^d_{i,k}\mathbb{E}\{\bar{\gamma}_{i,k}e_{i,k-d|k-d}\sum_{j=1}^{N}w_{ij}e^T_{j,k|k}\}\Gamma^T,\\
\mathscr{A}_6 &= \Gamma\mathbb{E}\{\Delta\gamma_{i,k}\bar{\gamma}_{i,k}\sum_{j=1}^{N}\sum_{l=1}^{N}w_{ij}w_{il}x_{j,k}e^T_{l,k|k}\}\Gamma^T.
\end{aligned} \tag{11.13}$$

Lemma 11.2 *The EEC* $P_{i,k+1|k+1}$ *is determined as*

$$\begin{aligned}
P_{i,k+1|k+1} = {} & (I-\bar{\lambda}_{i,k+1}G_{i,k+1}C_{i,k+1})P_{i,k+1|k}(I-\bar{\lambda}_{i,k+1}G_{i,k+1}C_{i,k+1})^T\\
&+G_{i,k+1}\mathbb{E}\{(\bar{y}_{i,k+1}-y_{i,k+1})(\bar{y}_{i,k+1}-y_{i,k+1})^T\}G^T_{i,k+1}\\
&+\bar{\sigma}_{i,k+1}G_{i,k+1}C_{i,k+1}\mathbb{E}\{x_{i,k+1}x^T_{i,k+1}\}C^T_{i,k+1}G^T_{i,k+1}-\mathscr{B}_1-\mathscr{B}_1^T\\
&+\Delta\lambda^2_{i,k+1}G_{i,k+1}C_{i,k+1}\mathbb{E}\{x_{i,k+1}x^T_{i,k+1}\}C^T_{i,k+1}G^T_{i,k+1}-\mathscr{B}_2-\mathscr{B}_2^T\\
&+G_{i,k+1}R_{i,k+1}G^T_{i,k+1}+\mathscr{B}_3+\mathscr{B}_3^T+\mathscr{B}_4+\mathscr{B}_4^T+\mathscr{B}_5+\mathscr{B}_5^T,
\end{aligned} \tag{11.14}$$

where

$$\begin{aligned}
\mathscr{B}_1 &= (I-\bar{\lambda}_{i,k+1}G_{i,k+1}C_{i,k+1})\mathbb{E}\{e_{i,k+1|k}(\bar{y}_{i,k+1}-y_{i,k+1})^T\}G^T_{i,k+1},\\
\mathscr{B}_2 &= \Delta\lambda_{i,k+1}(I-\bar{\lambda}_{i,k+1}G_{i,k+1}C_{i,k+1})\mathbb{E}\{e_{i,k+1|k}x^T_{i,k+1}\}C^T_{i,k+1}G^T_{i,k+1},\\
\mathscr{B}_3 &= G_{i,k+1}\mathbb{E}\{\tilde{\lambda}_{i,k+1}(\bar{y}_{i,k+1}-y_{i,k+1})x^T_{i,k+1}\}C^T_{i,k+1}G^T_{i,k+1},\\
\mathscr{B}_4 &= \Delta\lambda_{i,k+1}G_{i,k+1}\mathbb{E}\{(\bar{y}_{i,k+1}-y_{i,k+1})x^T_{i,k+1}\}C^T_{i,k+1}G^T_{i,k+1},\\
\mathscr{B}_5 &= G_{i,k+1}\mathbb{E}\{(\bar{y}_{i,k+1}-y_{i,k+1})v^T_{i,k+1}\}G^T_{i,k+1}.
\end{aligned} \tag{11.15}$$

Remark 11.2 Up to now, the detailed forms of the one-step prediction error covariance and the EEC have been provided in Lemmas 11.1 and 11.2. However, there exist some unknown terms (e.g. $\Delta\lambda_{i,k}$, $\Omega_{i,k}$, $\bar{\sigma}_{i,k+1}$ and $\mathbb{E}\{x_{i,k+1}x^T_{i,k+1}\}$), then the accurate EEC cannot be derived directly, which brings certain challenges to the estimation scheme design.

Next, an optimal upper bound on the EEC is derived and the estimator gain is given by solving the matrix equations.

Theorem 11.1 *Letting* $\varepsilon_i > 0$ $(i = 1, 2, \ldots, 7)$ *and* $\eta_j > 0$ $(j = 1, 2, \ldots, 6)$ *be scalars, if the following matrix equations:*

$$\begin{aligned}
\Omega_{i,k+1|k} = {} & \tau_1 A_{i,k}\Omega_{i,k|k}A_{i,k}^T + \tau_2 A_{i,k}^d\Omega_{i,k-d|k-d}(A_{i,k}^d)^T + \tau_5 w_i \sum_{j=1}^N w_{ij}\Gamma\Omega_{j,k|k}\Gamma^T \\
& + \tau_4(1+\varepsilon_7^{-1})w_i\sum_{j=1}^N w_{ij}\Gamma\hat{x}_{j,k|k}\hat{x}_{j,k|k}^T\Gamma^T + B_{i,k}Q_{i,k}B_{i,k}^T,
\end{aligned} \tag{11.16}$$

$$\begin{aligned}
\Omega_{i,k+1|k+1} = {} & \tau_6(I - \bar{\lambda}_{i,k+1}G_{i,k+1}C_{i,k+1})\Omega_{i,k+1|k}(I - \bar{\lambda}_{i,k+1}G_{i,k+1}C_{i,k+1})^T \\
& + \tau_7\delta_i G_{i,k+1}G_{i,k+1}^T + \left[(1+\eta_3^{-1})\check{\bar{\rho}}_{i,k+1} + \tau_8\bar{\rho}_{i,k+1}^2\right]G_{i,k+1}C_{i,k+1} \\
& \times\bar{\Psi}_{i,k+1|k}C_{i,k+1}^T G_{i,k+1}^T + (1+\eta_5^{-1})G_{i,k+1}R_{i,k+1}G_{i,k+1}^T
\end{aligned} \tag{11.17}$$

under the initial condition $\Omega_{i,0|0} \geq P_{i,0|0}$ *have solutions* $\Omega_{i,k+1|k} > 0$ *and* $\Omega_{i,k+1|k+1} > 0$*, then one has*

$$P_{i,k+1|k+1} \leq \Omega_{i,k+1|k+1},$$

where

$$\begin{aligned}
\tau_1 &= 1 + \varepsilon_1 + \varepsilon_2 + \varepsilon_3, \quad \tau_2 = 1 + \varepsilon_1^{-1} + \varepsilon_4 + \varepsilon_5, \\
\tau_3 &= 1 + \varepsilon_3^{-1} + \varepsilon_5^{-1} + \varepsilon_6^{-1}, \quad \tau_4 = (1 + \varepsilon_2^{-1} + \varepsilon_4^{-1} + \varepsilon_6)\rho_{i,k}^2 + \check{\rho}_{i,k}, \\
\tau_5 &= \tau_3\bar{\gamma}_{i,k}^2 + \tau_4(1+\varepsilon_7), \quad \tau_6 = 1 + \eta_1 + \eta_2, \\
\tau_7 &= 1 + \eta_1^{-1} + \eta_3 + \eta_4 + \eta_5, \quad \tau_8 = 1 + \eta_2^{-1} + \eta_4^{-1}, \\
w_i &= \sum_{j=1}^N w_{ij}, \quad \check{\rho}_{i,k} = \min\left\{\frac{1}{4}, (\bar{\gamma}_{i,k} + \rho_{i,k})(1 - \bar{\gamma}_{i,k} + \rho_{i,k})\right\}, \\
\check{\bar{\rho}}_{i,k+1} &= \min\left\{\frac{1}{4}, (\bar{\lambda}_{i,k+1} + \bar{\rho}_{i,k+1})(1 - \bar{\lambda}_{i,k+1} + \bar{\rho}_{i,k+1})\right\}, \\
\bar{\Psi}_{i,k+1|k} &= (1+\eta_6)\Omega_{i,k+1|k} + (1+\eta_6^{-1})\hat{x}_{i,k+1|k}\hat{x}_{i,k+1|k}^T.
\end{aligned} \tag{11.18}$$

In addition, $\mathrm{tr}(\Omega_{i,k+1|k+1})$ *can be minimized if* $G_{i,k+1}$ *is chosen as*

$$\begin{aligned}
G_{i,k+1} = {} & \tau_6\bar{\lambda}_{i,k+1}\Omega_{i,k+1|k}C_{i,k+1}^T\Big\{\tau_6\bar{\lambda}_{i,k+1}^2 C_{i,k+1}\Omega_{i,k+1|k}C_{i,k+1}^T + (1+\eta_5^{-1})R_{i,k+1} \\
& + \tau_7\delta_i I + \left[(1+\eta_3^{-1})\check{\bar{\rho}}_{i,k+1} + \tau_8\bar{\rho}_{i,k+1}^2\right]C_{i,k+1}\bar{\Psi}_{i,k+1|k}C_{i,k+1}^T\Big\}^{-1}.
\end{aligned} \tag{11.19}$$

Proof Firstly, noting (11.12)–(11.13), using Lemma 2.2 and considering $\Omega_{i,k} \leq \check{\rho}_{i,k}$, $\Delta\gamma_{i,k}^2 \leq \rho_{i,k}^2$, it is clear to find that

$$P_{i,k+1|k} \leq \tau_1 A_{i,k}P_{i,k|k}A_{i,k}^T + \tau_2 A_{i,k}^d P_{i,k-d|k-d}(A_{i,k}^d)^T + \tau_4\sum_{j=1}^N\sum_{l=1}^N w_{ij}w_{il}\Gamma\mathbb{E}\{x_{j,k}x_{l,k}^T\}$$

$$\times\Gamma^T+\tau_3\bar{\gamma}_{i,k}^2\sum_{j=1}^{N}\sum_{l=1}^{N}w_{ij}w_{il}\Gamma\mathbb{E}\{e_{j,k|k}e_{l,k|k}^T\}\Gamma^T+B_{i,k}Q_{i,k}B_{i,k}^T, \tag{11.20}$$

where τ_i $(i=1,2,3,4)$ are denoted in (11.18) and $\varepsilon_i>0$ $(i=1,2,\ldots,6)$ are scalars. In what follows, the third term of the RHS of (11.20) is solved as

$$\begin{aligned}&\sum_{j=1}^{N}\sum_{l=1}^{N}w_{ij}w_{il}\Gamma\mathbb{E}\{x_{j,k}x_{l,k}^T\}\Gamma^T\\ &\le\frac{1}{2}\sum_{j=1}^{N}\sum_{l=1}^{N}w_{ij}w_{il}\Gamma\mathbb{E}\{x_{j,k}x_{j,k}^T+x_{l,k}x_{l,k}^T\}\Gamma^T\\ &\le w_i(1+\varepsilon_7)\sum_{j=1}^{N}w_{ij}\Gamma P_{j,k|k}\Gamma^T+w_i(1+\varepsilon_7^{-1})\sum_{j=1}^{N}w_{ij}\Gamma\hat{x}_{j,k|k}\hat{x}_{j,k|k}^T\Gamma^T,\end{aligned} \tag{11.21}$$

where w_i is defined in (11.18) and $\varepsilon_7>0$ is a scalar. What's more, it is clear that

$$\sum_{j=1}^{N}\sum_{l=1}^{N}w_{ij}w_{il}\Gamma\mathbb{E}\{e_{j,k|k}e_{l,k|k}^T\}\Gamma^T\le w_i\sum_{j=1}^{N}w_{ij}\Gamma P_{j,k|k}\Gamma^T. \tag{11.22}$$

Together with (11.20)–(11.22), one obtains

$$\begin{aligned}P_{i,k+1|k}\le{}&\tau_1A_{i,k}P_{i,k|k}A_{i,k}^T+\tau_2A_{i,k}^dP_{i,k-d|k-d}(A_{i,k}^d)^T+\tau_5w_i\sum_{j=1}^{N}w_{ij}\Gamma P_{j,k|k}\Gamma^T\\ &+\tau_4(1+\varepsilon_7^{-1})w_i\sum_{j=1}^{N}w_{ij}\Gamma\hat{x}_{j,k|k}\hat{x}_{j,k|k}^T\Gamma^T+B_{i,k}Q_{i,k}B_{i,k}^T\end{aligned} \tag{11.23}$$

with τ_5 being given in (11.18).

Subsequently, the aim is to prove that $P_{i,k+1|k+1}\le\Omega_{i,k+1|k+1}$. It is easy to see from (11.14) and (11.15) that

$$\begin{aligned}P_{i,k+1|k+1}\le{}&\tau_6(I-\bar{\lambda}_{i,k+1}G_{i,k+1}C_{i,k+1})P_{i,k+1|k}(I-\bar{\lambda}_{i,k+1}G_{i,k+1}C_{i,k+1})^T+\tau_7G_{i,k+1}\\ &\times\mathbb{E}\{(\bar{y}_{i,k+1}-y_{i,k+1})(\bar{y}_{i,k+1}-y_{i,k+1})^T\}G_{i,k+1}^T+\big[(1+\eta_3^{-1})\bar{\sigma}_{i,k+1}\\ &+\tau_8\Delta\lambda_{i,k+1}^2\big]G_{i,k+1}C_{i,k+1}\mathbb{E}\{x_{i,k+1}x_{i,k+1}^T\}C_{i,k+1}^TG_{i,k+1}^T\\ &+(1+\eta_5^{-1})G_{i,k+1}R_{i,k+1}G_{i,k+1}^T\end{aligned} \tag{11.24}$$

with τ_i $(i = 6, 7, 8)$ being given in (11.18) and $\eta_i > 0$ $(i = 1, 2, \ldots, 5)$ denoting scalars. Next, let us compute the terms on the RHS of (11.24). According to (11.6), one gets

$$\begin{aligned}&\tau_7 G_{i,k+1}\mathbb{E}\{(\bar{y}_{i,k+1} - y_{i,k+1})(\bar{y}_{i,k+1} - y_{i,k+1})^T\}G_{i,k+1}^T\\ \leq\ &\tau_7\delta_i G_{i,k+1}G_{i,k+1}^T.\end{aligned} \tag{11.25}$$

After that, along the same line with (11.21), one gets

$$\begin{aligned}&\mathbb{E}\{x_{i,k+1}x_{i,k+1}^T\}\\ \leq\ &(1+\eta_6)P_{i,k+1|k} + (1+\eta_6^{-1})\hat{x}_{i,k+1|k}\hat{x}_{i,k+1|k}^T\\ :=\ &\Psi_{i,k+1|k},\end{aligned} \tag{11.26}$$

where $\eta_6 > 0$ denotes a saclar. Thereby, one has

$$\begin{aligned}&G_{i,k+1}C_{i,k+1}\mathbb{E}\{x_{i,k+1}x_{i,k+1}^T\}C_{i,k+1}^T G_{i,k+1}^T\\ \leq\ &G_{i,k+1}C_{i,k+1}\Psi_{i,k+1|k}C_{i,k+1}^T G_{i,k+1}^T.\end{aligned} \tag{11.27}$$

Noticing $\bar{\sigma}_{i,k+1} \leq \check{\bar{\rho}}_{i,k+1}$, $\Delta\lambda_{i,k+1}^2 \leq \bar{\rho}_{i,k+1}^2$, (11.24), (11.25) and (11.27), one derives

$$\begin{aligned}P_{i,k+1|k+1} \leq\ &\tau_6(I - \bar{\lambda}_{i,k+1}G_{i,k+1}C_{i,k+1})P_{i,k+1|k}(I - \bar{\lambda}_{i,k+1}G_{i,k+1}C_{i,k+1})^T\\ &+\tau_7\delta_i G_{i,k+1}G_{i,k+1}^T + \left[\tau_8\bar{\rho}_{i,k+1}^2 + (1+\eta_3^{-1})\check{\bar{\rho}}_{i,k+1}\right]G_{i,k+1}C_{i,k+1}\\ &\times\Psi_{i,k+1|k}C_{i,k+1}^T G_{i,k+1}^T + (1+\eta_5^{-1})G_{i,k+1}R_{i,k+1}G_{i,k+1}^T.\end{aligned} \tag{11.28}$$

Then, it summarizes that $P_{i,k+1|k+1} \leq \Omega_{i,k+1|k+1}$ holds by means of the mathematical induction approach.

Finally, the purpose is to minimize the proposed upper bound matrix $\Omega_{i,k+1|k+1}$ and design a desired estimator gain. In terms of Lemma 2.3, taking the partial derivative of $\mathrm{tr}(\Omega_{i,k+1|k+1})$ regarding $G_{i,k+1}$, one has

$$\begin{aligned}&\frac{\partial \mathrm{tr}(\Omega_{i,k+1|k+1})}{\partial G_{i,k+1}}\\ =\ &-2\bar{\lambda}_{i,k+1}\tau_6(I - \bar{\lambda}_{i,k+1}G_{i,k+1}C_{i,k+1})\Omega_{i,k+1|k}C_{i,k+1}^T + 2G_{i,k+1}\big\{\tau_7\delta_i I\\ &+(1+\eta_5^{-1})R_{i,k+1} + \left[\tau_8\bar{\rho}_{i,k+1}^2 + (1+\eta_3^{-1})\check{\bar{\rho}}_{i,k+1}\right]C_{i,k+1}\bar{\Psi}_{i,k+1|k}C_{i,k+1}^T\big\}.\end{aligned}$$

Let the partial derivative be equal to zero, which generates the result in (11.19).

The main result of Theorem 11.1 can be exhibited in the following EBSE algorithm.

EBSE Algorithm :
Step 1 : Set $k = 0$ and set initial conditions.
Step 2 : Calculate the one-step prediction $\hat{x}_{i,k+1\|k}$ by (11.8).
Step 3 : Derive the value of $\Omega_{i,k+1\|k}$ by (11.16).
Step 4 : Solve the estimator gain matrix $G_{i,k+1}$ by (11.19).
Step 5 : Compute the state estimation $\hat{x}_{i,k+1\|k+1}$ by (11.9).
Step 6 : Obtain $\Omega_{i,k+1\|k+1}$ by (11.17).
Step 7 : Set $k = k + 1$, and go to *Step* 2.

The following corollary is further provided regarding the upper bound of EEC for the deterministic missing probability case, and the proof is omitted here for brevity.

Corollary 11.1 *Consider (11.1)–(11.3) and let the missing probability be deterministic (i.e.,* $\Delta\lambda_{i,k+1} = 0$*). Let* $\varepsilon_i > 0$ $(i = 1, 2, \ldots, 7)$ *and* $\eta_j > 0$ $(j = 1, 3, 5, 6)$ *be scalars. If the matrix equation (11.16) and the following matrix equation:*

$$\begin{aligned}\Omega_{i,k+1|k+1} &= (1+\eta_1)(I - \bar{\lambda}_{i,k+1}G_{i,k+1}C_{i,k+1})\Omega_{i,k+1|k}(I - \bar{\lambda}_{i,k+1}G_{i,k+1}C_{i,k+1})^T \\ &\quad +(1+\eta_1^{-1}+\eta_3+\eta_5)\delta_i G_{i,k+1}G_{i,k+1}^T + (1+\eta_3^{-1})\bar{\rho}_{i,k+1}G_{i,k+1}C_{i,k+1} \\ &\quad \times\bar{\Psi}_{i,k+1|k}C_{i,k+1}^T G_{i,k+1}^T + (1+\eta_5^{-1})G_{i,k+1}R_{i,k+1}G_{i,k+1}^T \end{aligned} \tag{11.29}$$

under the initial condition $\Omega_{i,0|0} \geq P_{i,0|0}$ *have solutions* $\Omega_{i,k+1|k} > 0$ *and* $\Omega_{i,k+1|k+1} > 0$*, then one has* $P_{i,k+1|k+1} \leq \Omega_{i,k+1|k+1}$*. Besides, if the estimator gain matrix is constructed as*

$$\begin{aligned}G_{i,k+1} &= (1+\eta_1)\bar{\lambda}_{i,k+1}\Omega_{i,k+1|k}C_{i,k+1}^T\Big\{(1+\eta_1)\bar{\lambda}_{i,k+1}^2 C_{i,k+1}\Omega_{i,k+1|k}C_{i,k+1}^T \\ &\quad +(1+\eta_1^{-1}+\eta_3+\eta_5)\delta_i I + (1+\eta_5^{-1})R_{i,k+1} \\ &\quad +(1+\eta_3^{-1})\bar{\lambda}_{i,k+1}(1-\bar{\lambda}_{i,k+1})C_{i,k+1}\bar{\Psi}_{i,k+1|k}C_{i,k+1}^T\Big\}^{-1}, \end{aligned} \tag{11.30}$$

then the minimized upper bound matrix $\Omega_{i,k+1|k+1}$ *is determined by*

$$\begin{aligned}\Omega_{i,k+1|k+1} &= (1+\eta_1)\Omega_{i,k+1|k} - (1+\eta_1)^2\bar{\lambda}_{i,k+1}^2\Omega_{i,k+1|k}C_{i,k+1}^T\Big\{(1+\eta_1)\bar{\lambda}_{i,k+1}^2 \\ &\quad \times C_{i,k+1}\Omega_{i,k+1|k}C_{i,k+1}^T + (1+\eta_1^{-1}+\eta_3+\eta_5)\delta_i I + (1+\eta_5^{-1})R_{i,k+1} \\ &\quad +(1+\eta_3^{-1})\bar{\lambda}_{i,k+1}(1-\bar{\lambda}_{i,k+1})C_{i,k+1}\bar{\Psi}_{i,k+1|k}C_{i,k+1}^T\Big\}^{-1} \\ &\quad \times C_{i,k+1}\Omega_{i,k+1|k}. \end{aligned} \tag{11.31}$$

11.3 Monotonicity Analysis

In this section, the relationship between the trace of the upper bound and the deterministic missing probability is discussed and the theoretical proof is provided.

Theorem 11.2 *For each node i with deterministic missing occurrence probability, $\mathrm{tr}(\Omega_{i,k+1|k+1})$ is non-decreasing when $\bar{\lambda}_{i,k+1}$ decreases.*

Proof To simplify the next derivations, define

$$\begin{aligned}\mu_1 &= 1+\eta_1,\ \mu_{2,i} = (1+\eta_1^{-1}+\eta_3+\eta_5)\delta_i,\\ \mu_3 &= 1+\eta_5^{-1},\ \mu_4 = 1+\eta_3^{-1}.\end{aligned}$$

In light of (11.31), taking the partial derivative of $\mathrm{tr}(\Omega_{i,k+1|k+1})$ regarding $\bar{\lambda}_{i,k+1}$ yields

$$\begin{aligned}&\frac{\partial \mathrm{tr}(\Omega_{i,k+1|k+1})}{\partial \bar{\lambda}_{i,k+1}}\\ &\le \mathrm{tr}\Big(-2\mu_1^2\bar{\lambda}_{i,k+1}\Omega_{i,k+1|k}C_{i,k+1}^T\Xi_{i,k+1}^{-1}C_{i,k+1}\Omega_{i,k+1|k}\\ &\quad+\mu_1^2\Omega_{i,k+1|k}C_{i,k+1}^T\Xi_{i,k+1}^{-1}\Big(2\bar{\lambda}_{i,k+1}\Xi_{i,k+1}\\ &\quad-\mu_4\bar{\lambda}_{i,k+1}^2C_{i,k+1}\bar{\Psi}_{i,k+1|k}C_{i,k+1}^T\Big)\Xi_{i,k+1}^{-1}C_{i,k+1}\Omega_{i,k+1|k}\Big),\end{aligned} \tag{11.32}$$

where

$$\begin{aligned}\Xi_{i,k+1} &= \mu_1\bar{\lambda}_{i,k+1}^2C_{i,k+1}\Omega_{i,k+1|k}C_{i,k+1}^T+\mu_{2,i}I+\mu_3R_{i,k+1}\\ &\quad+\mu_4\bar{\lambda}_{i,k+1}(1-\bar{\lambda}_{i,k+1})C_{i,k+1}\bar{\Psi}_{i,k+1|k}C_{i,k+1}^T.\end{aligned}$$

It is easy to summarize that

$$\frac{\partial \mathrm{tr}(\Omega_{i,k+1|k+1})}{\partial \bar{\lambda}_{i,k+1}} \le 0,$$

which finishes the proof.

Remark 11.3 So far, the influences from stochastic coupling and the ETM onto the estimation performance have been investigated. In particular, the available information (such as $\bar{\gamma}_{i,k}$, $\bar{\lambda}_{i,k+1}$, $\rho_{i,k}$, $\bar{\rho}_{i,k+1}$ and δ_i) about the stochastic coupling and the ETM has been explicitly embodied in Theorem 11.1. In addition, the monotonicity relationship between the deterministic missing probability and the trace of the upper bound on the EEC has been adequately discussed in Theorem 11.2.

11.4 An Illustrative Example

In this section, a simulation with comparisons is given to verify the effectiveness of the designed state estimation method under the influences of different types of stochastic coupling.

Consider the delayed complex network (11.1)–(11.3) with following parameters:

$$
\begin{aligned}
A_{1,k} &= \begin{bmatrix} 0.2 & 0.3-0.1\sin(2k) \\ 0.25 & 0.2 \end{bmatrix}, \quad A_{2,k} = \begin{bmatrix} 0.05 & 0.05 \\ 0.04 & 0.06 \end{bmatrix}, \\
A_{3,k} &= \begin{bmatrix} 0.04 & 0.028 \\ 0.025 & 0.05 \end{bmatrix}, \quad A^d_{1,k} = \begin{bmatrix} -0.4 & 0.1 \\ -0.2 & 0.1-0.3\sin(k)\cos(2k) \end{bmatrix}, \\
A^d_{2,k} &= \begin{bmatrix} -0.3 & 0.2 \\ 0 & -0.1 \end{bmatrix}, \quad A^d_{3,k} = \begin{bmatrix} -0.4 & 0.2 \\ 0 & 0.3 \end{bmatrix}, \quad B_{1,k} = \begin{bmatrix} 1.5 & 1.5 \end{bmatrix}^T, \\
B_{2,k} &= \begin{bmatrix} 1.5-0.2\sin(5k) & 1.2 \end{bmatrix}^T, \quad B_{3,k} = \begin{bmatrix} 1.4 & 1.6 \end{bmatrix}^T, \\
C_{1,k} &= \begin{bmatrix} 1.2 & 1.2 \end{bmatrix}, \quad C_{2,k} = \begin{bmatrix} 1.5 & 1 \end{bmatrix}, \\
C_{3,k} &= \begin{bmatrix} 1.6 & 1.9 \end{bmatrix}, \quad d = 2.
\end{aligned}
$$

The inner-coupling matrix is set as $\Gamma = \text{diag}\{0.08, 0.08\}$, and the elements of the coupling configuration matrix are chosen as $w_{ij} = 0.015$ $(i \neq j)$ and $w_{ij} = -0.03$ $(i = j)$.

The initial parameters are set as

$$
\begin{aligned}
& x_{1,0} = \begin{bmatrix} 1.5 & 1.5 \end{bmatrix}^T, \ x_{2,0} = \begin{bmatrix} -0.5 & 2 \end{bmatrix}^T, \ x_{3,0} = \begin{bmatrix} -1 & -2 \end{bmatrix}^T, \\
& x_{i,j} = \begin{bmatrix} 0.2 & 0.2 \end{bmatrix}^T, \ \hat{x}_{i,0} = x_{i,0} - \begin{bmatrix} 1 & 1 \end{bmatrix}^T, \ \hat{x}_{i,j} = \begin{bmatrix} 0.1 & 0.1 \end{bmatrix}^T, \\
& \rho_{i,k} = 0.05, \ \bar{\lambda}_{i,k} = 0.95, \ \bar{\gamma}_{i,k} = 0.95, \ \bar{\rho}_{i,k} = 0.05 \ (i = 1, 2, 3, \ j = -1, -2).
\end{aligned}
$$

Other parameters are given as $Q_{1,k} = Q_{2,k} = Q_{3,k} = 0.3$, $R_{1,k} = R_{2,k} = R_{3,k} = 0.2$, $\varepsilon_1 = \varepsilon_4 = 0.2$, $\varepsilon_2 = \varepsilon_3 = 0.4$, $\varepsilon_5 = \varepsilon_6 = 0.8$, $\varepsilon_7 = 0.9$, $\eta_1 = \eta_6 = 0.6$, $\eta_2 = 0.8$, $\eta_3 = 9.5$, $\eta_4 = 1.2$, $\eta_5 = 0.8$, $\Omega_{1,0|0} = \Omega_{2,0|0} = \Omega_{3,0|0} = 10I_2$ and $\Omega_{i,j|j} = 10I_2$ $(i = 1, 2, 3, \ j = -1, -2)$.

Next, we consider the following cases with various thresholds: Case I: $\delta_1 = \delta_2 = \delta_3 = 0.2$; Case II: $\delta_1 = 7.5$, $\delta_2 = 10.5$ and $\delta_3 = 7.5$. It can be observed from Figs. 11.1, 11.2 and 11.3 that the proposed estimation scheme performs a satisfactory performance.

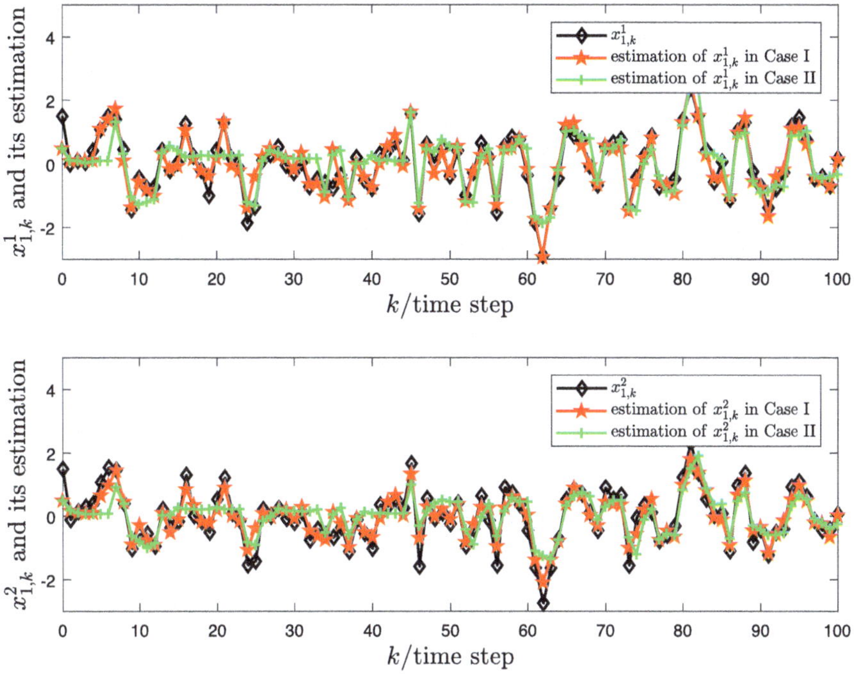

Fig. 11.1 The state $x_{1,k}$ and its estimations

Now, we are ready to discuss the relationship between the missing probability and the trace of this upper bound. We set $\delta_i = 0.2$ $(i = 1, 2, 3)$ and other parameters are similar with mentioned above. In light of Corollary 11.1, the curves of the trace of upper bound $\Omega_{i,k|k}$ are exhibited in Fig. 11.4 under $\bar{\lambda}_{i,k} = 0.7$, $\bar{\lambda}_{i,k} = 0.8$, $\bar{\lambda}_{i,k} = 0.9$ and $\bar{\lambda}_{i,k} = 1$. It is accordant with the theoretical result that $\mathrm{tr}(\Omega_{i,k|k})$ decreases when $\bar{\lambda}_{i,k}$ increases.

11.5 Summary

In this chapter, the robust state estimation problem has been tackled for time-varying coupled complex networks involving ETM and stochastic coupling with UOPs. Firstly, based on available measurements, an optimized upper bound matrix regarding the EEC has been derived and the specific form of the estimator gain has been given by solving the recursive matrix equations. Secondly, a novel state estimation scheme has been presented for time-varying stochastic coupled complex networks in the presence of time delay and missing measurements with UOPs. Moreover, a sufficient condition has been adopted to obtain an optimal upper bound matrix concerning on the EEC. And the theoretical evaluation has been presented to discuss the relationship between the deterministic missing probability and the trace of the upper

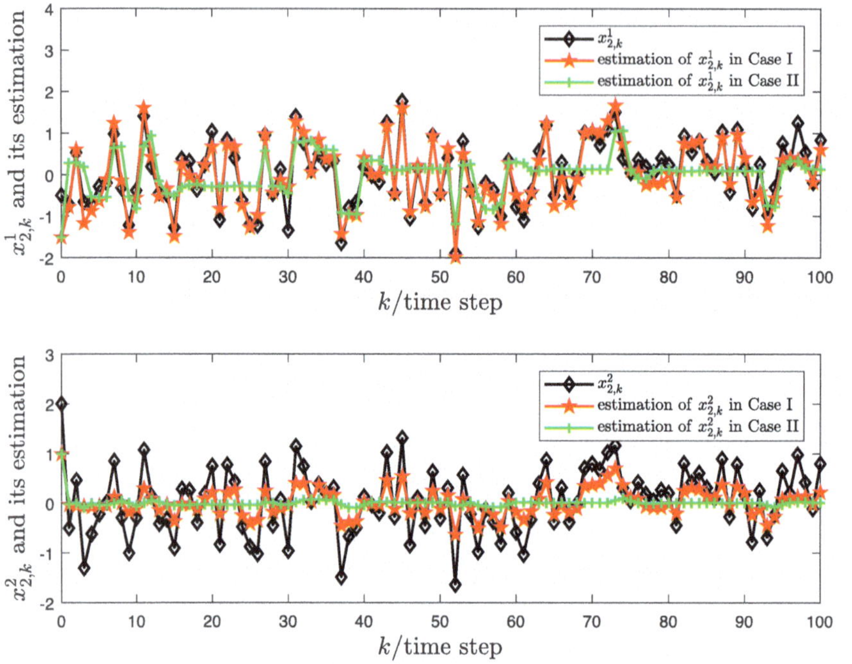

Fig. 11.2 The state $x_{2,k}$ and its estimations

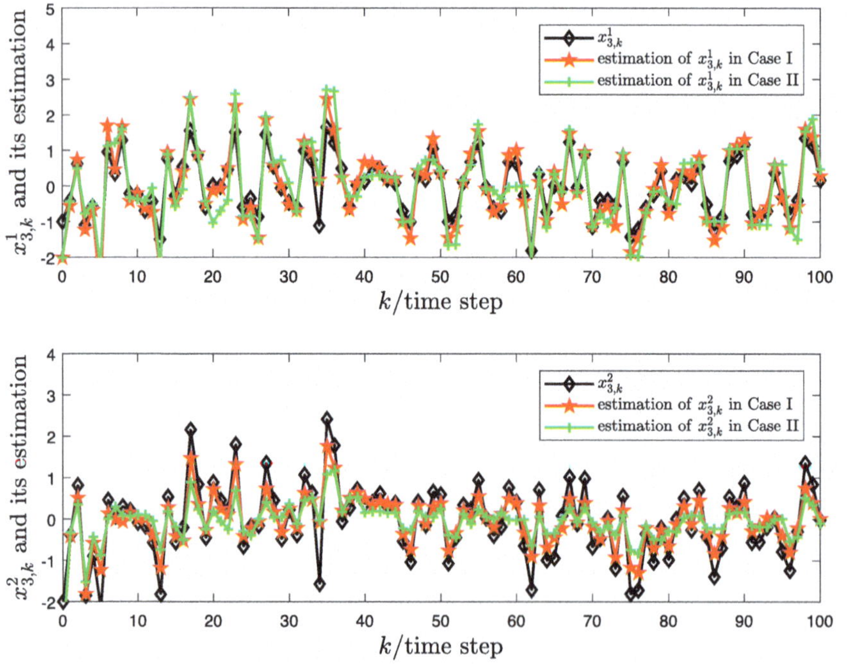

Fig. 11.3 The state $x_{3,k}$ and its estimations

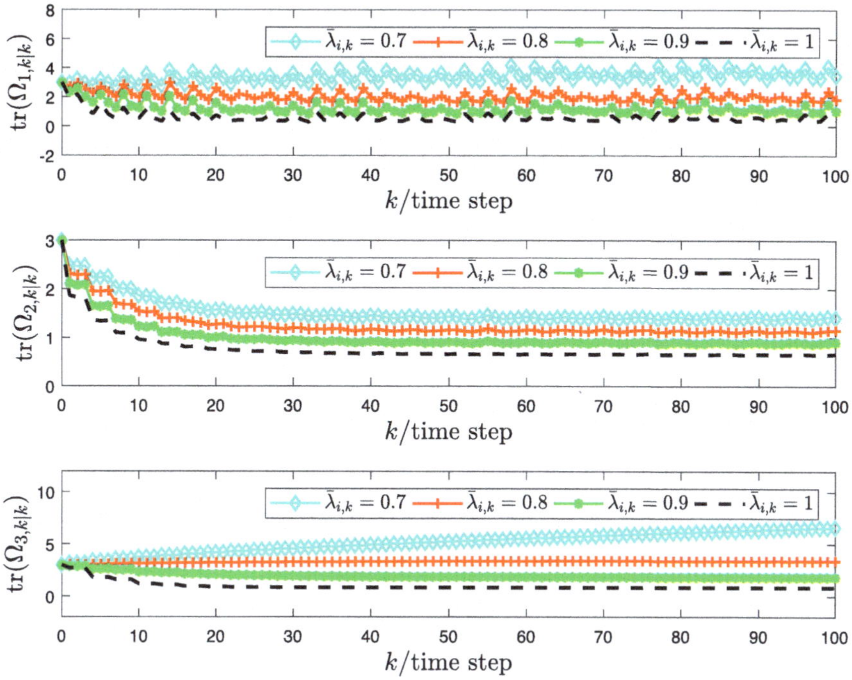

Fig. 11.4 The traces of the upper bound $\Omega_{i,k|k}$ $(i = 1, 2, 3)$ under different occurrence probabilities

bound. Finally, a simulation experiment has been used to testify the effectiveness of the newly proposed state estimation approach for time-varying complex networks under the ETM and missing measurements.

Chapter 12
Event-Based State Estimation for Complex Networks with Fading Observations and Uncertain Switching Topology

In this chapter, the state estimation issue is addressed for time-varying uncertain complex networks involving switching topology, ETM and DPDs. To save network resources and reduce communication cost, the ETM is employed to arrange signal transmission. The binary random variables are adopted to depict the RONs, switching topology and DPDs. The aim is to design a novel state estimation scheme in the context of time-varying uncertain complex networks under the influences of RONs, switching topology and DPDs. Next, the primary results are extended to cope with the state estimation problem for nonlinear complex networks involving ETM, fading measurements and SCS. Specifically, the ETM is applied to regulate signal propagation to enhance communication efficiency, in which the adjustable triggered threshold is used to adjust the transmission frequency. A set of Bernoulli distributed random variables is adopted to model the fading measurements. Subsequently, an event-triggered state estimation scheme is given to tackle the influences from fading measurements and SCS. Moreover, the monotonicity analysis between the fading probability and the proposed estimation scheme is discussed and proved. Finally, the effectiveness of the proposed estimation schemes is testified via simulation experiments.

12.1 Event-Based State Estimation with DPDs and Switching Topology

12.1.1 Problem Formulation

The state estimation problem is handled for time-varying uncertain complex networks involving DPDs, switching topology and RONs. The specific diagram is given in Fig. 12.1, in which there are the relays during the data transmission propagation with long distance.

J. Hu et al., *Variance-Constrained Filtering for Stochastic Complex Systems*, Intelligent Control and Learning Systems 18,
https://doi.org/10.1007/978-981-96-2637-3_12

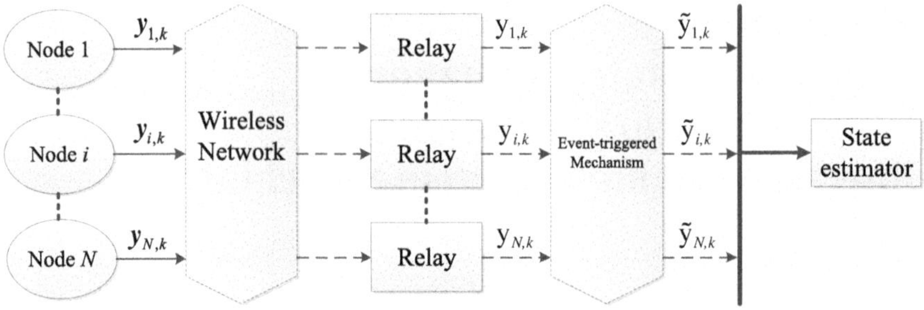

Fig. 12.1 State estimation experiencing DPDs and event-triggered protocol

Consider the following time-varying uncertain complex networks:

$$
\begin{aligned}
\mathbf{x}_{i,k+1} =& (\mathbf{A}_{i,k} + \Delta\mathbf{A}_{i,k})\mathbf{x}_{i,k} + \alpha_{i,k}\mathbf{h}_i(\mathbf{x}_{i,k}) + \sum_{j=1}^{N} \lambda_{ij,k}\varpi_{ij}\bar{\Gamma}\mathbf{x}_{j,k} \\
&+ \mathbf{B}_{i,k}\breve{\omega}_{i,k}, && (12.1) \\
\mathbf{y}_{i,k} =& \mathbf{C}_{i,k}\mathbf{x}_{i,k} + \breve{\nu}_{i,k}, && (12.2) \\
y_{i,k} =& \beta_{i,k}\mathbf{y}_{i,k} + (1 - \beta_{i,k})y_{i,k-1}, && (12.3)
\end{aligned}
$$

where $\mathbf{x}_{i,k} \in \mathbb{R}^n$ is the state with mean $\bar{\mathbf{x}}_{i,0}$ of initial state $\mathbf{x}_{i,0}$, $\mathbf{y}_{i,k} \in \mathbb{R}^m$ is the actual measurement, and $y_{i,k} \in \mathbb{R}^m$ is the measurement received by the relay. The process noise $\breve{\omega}_{i,k} \in \mathbb{R}^p$ and measurement noise $\breve{\nu}_{i,k} \in \mathbb{R}^m$ are with zero-mean and known covariances $\mathscr{W}_{i,k} > 0$ and $\mathscr{V}_{i,k} > 0$, respectively. $\bar{\Gamma}$ represents a given linking matrix and ϖ_{ij} denotes the coupling strength between the i-th node and the j-th node. The norm bounded uncertainty is $\Delta\mathbf{A}_{i,k} = \mathbf{M}_{i,k}F_{i,k}\mathbf{N}_{i,k}$ with $F_{i,k}F_{i,k}^T \leq I$, where $\mathbf{M}_{i,k}$ and $\mathbf{N}_{i,k}$ are pre-defined matrices. $\mathbf{A}_{i,k}$, $\mathbf{B}_{i,k}$ and $\mathbf{C}_{i,k}$ denote given matrices.

The nonlinear function $\mathbf{h}_i(\mathbf{x}_{i,k})$ with $\mathbf{h}_i(0) = 0$ satisfies the following Lipschitz condition:

$$
\|\mathbf{h}_i(\mathbf{x}_{i,k}) - \mathbf{h}_i(\mathbf{z}_{i,k})\| \leq l\|\mathbf{x}_{i,k} - \mathbf{z}_{i,k}\|
$$

with l denoting a pre-determined scalar.

The binary random variables $\alpha_{i,k}$, $\lambda_{ij,k}$ and $\beta_{i,k}$ are employed to describe the RONs, switching topology and DPDs, and the statistical properties regarding these variables are given as

$$
\begin{aligned}
\text{Prob}\{\alpha_{i,k} = 1\} &= \bar{\alpha}_{i,k}, \\
\text{Prob}\{\lambda_{ij,k} = 1\} &= \bar{\lambda}_{ij,k}, \\
\text{Prob}\{\beta_{i,k} = 1\} &= \bar{\beta}_{i,k}, && (12.4)
\end{aligned}
$$

with $\bar{\alpha}_{i,k} \in [0, 1]$, $\bar{\lambda}_{ij,k} \in [0, 1]$ and $\bar{\beta}_{i,k} \in [0, 1]$ representing given constants. Moreover, assume that $\alpha_{i,k}$, $\lambda_{ij,k}$, $\check{\omega}_{i,k}$, $\beta_{i,k}$, $\check{\nu}_{i,k}$ and $\mathbf{x}_{i,0}$ are mutually independent for i, j and k.

Remark 12.1 It should be emphasized that the nonlinearities are widely existed in practical cases and most systems exhibit nonlinear properties influenced by complicated environment, thus it is valuable yet desirable to propose the related nonlinear analysis approaches. However, on account of the limited channel bandwidth, the nonlinearity might occur in a stochastic way, i.e., the so-called RONs, and considerable effort has been endeavored to investigate the estimation problems influenced by the RONs Shen et al. (2011), Huang et al. (2023), Liu et al. (2022). On the other hand, the Bernoulli random variables $\alpha_{i,k}$, $\lambda_{ij,k}$ and $\beta_{i,k}$ are adopted to describe the RONs, switching topology and DPDs. It is noted that this modelling way can clearly depict the above phenomena, which has been widely applied in the existing results Hu et al. (2023), Wang et al. (2019), Wang et al. (2015), Liu et al. (2018).

With the aim of saving network resources, the ETM is utilized to arrange the signal communication. To be specific, the event generator function is characterized as

$$(y_{i,k+p} - y_{i,k_t})^T (y_{i,k+p} - y_{i,k_t}) > \delta_{i,k}, \tag{12.5}$$

where $\delta_{i,k} > 0$ is a known scalar, $y_{i,k+p}$ denotes the current measurement at sampling time $k + p$, and y_{i,k_t} refers to the transmitted measurement at latest triggered time k_t. In terms of (12.5), the released measurement signal $\tilde{y}_{i,k}$ at the sampling time k is given by

$$\tilde{y}_{i,k} = y_{i,k_j}, \quad k \in \{k_j, k_j + 1, \ldots, k_{j+1} - 1\}, \tag{12.6}$$

where $k_0 = 0$.

We define

$$\begin{aligned}
x_{i,k} &= \begin{bmatrix} \mathbf{x}_{i,k} \\ y_{i,k-1} \end{bmatrix}, \quad A_{i,k} = \begin{bmatrix} \mathbf{A}_{i,k} & 0 \\ \beta_{i,k}\mathbf{C}_{i,k} & (1 - \beta_{i,k})I_m \end{bmatrix}, \\
M_{i,k} &= \begin{bmatrix} \mathbf{M}_{i,k} \\ 0 \end{bmatrix}, \quad N_{i,k} = \begin{bmatrix} \mathbf{N}_{i,k} & 0 \end{bmatrix}, \\
\Delta A_{i,k} &= \begin{bmatrix} \Delta\mathbf{A}_{i,k} & 0 \\ 0 & 0 \end{bmatrix}, \quad h_i(x_{i,k}) = \begin{bmatrix} \mathbf{h}_i(\mathbf{x}_{i,k}) \\ 0 \end{bmatrix}, \\
\Gamma &= \begin{bmatrix} \bar{\Gamma} & 0 \\ 0 & 0 \end{bmatrix}, \quad B_{i,k} = \begin{bmatrix} \mathbf{B}_{i,k} & 0 \\ 0 & \beta_{i,k}I_m \end{bmatrix}, \\
\omega_{i,k} &= \begin{bmatrix} \check{\omega}_{i,k} \\ \check{\nu}_{i,k} \end{bmatrix}, \quad \nu_{i,k} = \check{\nu}_{i,k}, \\
C_{i,k} &= \begin{bmatrix} \beta_{i,k}\mathbf{C}_{i,k} & (1 - \beta_{i,k})I_m \end{bmatrix},
\end{aligned}$$

then, the uncertain complex networks can be rewritten as

$$
\begin{aligned}
x_{i,k+1} =&(A_{i,k} + \Delta A_{i,k})x_{i,k} + \alpha_{i,k}h_i(x_{i,k}) + \sum_{j=1}^{N} \lambda_{ij,k}\varpi_{ij}\Gamma x_{j,k} \\
&+ B_{i,k}\omega_{i,k}, \qquad (12.7)\\
y_{i,k} =&C_{i,k}x_{i,k} + \beta_{i,k}v_{i,k}. \qquad (12.8)
\end{aligned}
$$

To facilitate the next derivations, we set

$$
\begin{aligned}
\bar{A}_{i,k} =& \mathbb{E}\{A_{i,k}\} = \begin{bmatrix} \mathbf{A}_{i,k} & 0 \\ \bar{\beta}_{i,k}\mathbf{C}_{i,k} & (1-\bar{\beta}_{i,k})I_m \end{bmatrix},\\
\bar{B}_{i,k} =& \mathbb{E}\{B_{i,k}\} = \begin{bmatrix} \mathbf{B}_{i,k} & 0 \\ 0 & \bar{\beta}_{i,k}I_m \end{bmatrix},\\
\bar{C}_{i,k} =& \mathbb{E}\{C_{i,k}\} = \begin{bmatrix} \bar{\beta}_{i,k}\mathbf{C}_{i,k} & (1-\bar{\beta}_{i,k})I_m \end{bmatrix},\\
\mathbb{E}\{\omega_{i,k}\omega_{i,k}^T\} =& \operatorname{diag}\{\mathscr{W}_{i,k}, \mathscr{V}_{i,k}\} := Q_{i,k},\\
\tilde{A}_{i,k} =& A_{i,k} - \bar{A}_{i,k}\\
=& \begin{bmatrix} 0 & 0 \\ (\beta_{i,k}-\bar{\beta}_{i,k})\mathbf{C}_{i,k} & (\bar{\beta}_{i,k}-\beta_{i,k})I_m \end{bmatrix}\\
=& (\beta_{i,k}-\bar{\beta}_{i,k})\mathscr{A}_{i,k},\\
\tilde{C}_{i,k} =& C_{i,k} - \bar{C}_{i,k}\\
=& \begin{bmatrix} (\beta_{i,k}-\bar{\beta}_{i,k})\mathbf{C}_{i,k} & (\bar{\beta}_{i,k}-\beta_{i,k})I_m \end{bmatrix}\\
=& (\beta_{i,k}-\bar{\beta}_{i,k})\mathscr{C}_{i,k},\\
B_{i,k} =& \begin{bmatrix} \mathbf{B}_{i,k} & 0 \\ 0 & \beta_{i,k}I_m \end{bmatrix} = \Lambda_{i,k}\mathscr{B}_{i,k},\\
\mathscr{A}_{i,k} =& \begin{bmatrix} 0 & 0 \\ \mathbf{C}_{i,k} & -I_m \end{bmatrix}, \ \mathscr{B}_{i,k} = \begin{bmatrix} \mathbf{B}_{i,k} & 0 \\ 0 & I_m \end{bmatrix},\\
\tilde{B}_{i,k} =& B_{i,k} - \bar{B}_{i,k},\\
\mathscr{C}_{i,k} =& \begin{bmatrix} \mathbf{C}_{i,k} & -I_m \end{bmatrix}, \ \Lambda_{i,k} = \begin{bmatrix} I_n & 0 \\ 0 & \beta_{i,k}I_m \end{bmatrix}.
\end{aligned}
$$

It is not difficult to obverse that the expectations of $\tilde{A}_{i,k}$, $\tilde{B}_{i,k}$ and $\tilde{C}_{i,k}$ are zero matrices.

In the sequel, the following time-varying nonlinear estimator is designed as:

$$
\begin{aligned}
\hat{x}_{i,k+1|k} =&\bar{A}_{i,k}\hat{x}_{i,k|k} + \bar{\alpha}_{i,k}h_i(\hat{x}_{i,k|k}) + \sum_{j=1}^{N} \bar{\lambda}_{ij,k}\varpi_{ij}\Gamma\hat{x}_{j,k|k}, \qquad (12.9)\\
\hat{x}_{i,k+1|k+1} =&\hat{x}_{i,k+1|k} + \mathscr{G}_{i,k+1}(\tilde{y}_{i,k+1} - \bar{C}_{i,k+1}\hat{x}_{i,k+1|k}), \qquad (12.10)
\end{aligned}
$$

where $\hat{x}_{i,k|k}$ represents the state estimation, $\hat{x}_{i,k+1|k}$ denotes the prediction estimation of $x_{i,k}$, $\mathscr{G}_{i,k+1} = G_{i,k+1} + \Delta G_{i,k+1}$, and $G_{i,k+1}$ is the estimator gain. $\Delta G_{i,k+1}$ is the estimator gain perturbation satisfying $\Delta G_{i,k+1} = H_{i,k+1}\mathscr{F}_{i,k+1}L_{i,k+1}$, in which $H_{i,k+1}$ and $L_{i,k+1}$ are given matrices, and $\mathscr{F}_{i,k+1}$ satisfies $\mathscr{F}_{i,k+1}\mathscr{F}_{i,k+1}^T \leq I$.

Remark 12.2 Considering the fixed word length, numerical roundoff errors, and so on, it is valuable to consider the estimator gain perturbation when designing the estimator to accurately mirror the reality. Generally speaking, there are primarily three ways to describe the estimator gain perturbation including the multiplicative noise, random variables with statistical properties and the norm bounded parameter uncertainties. To name a few, a resilient filtering scheme has been proposed in Ding et al. (2022) for complex networks, in which a random variable obeying the uniform distribution has been introduced to construct the estimator gain perturbation. In Wang et al. (2022), a resilient filtering problem has been explored for complex networks, in which a multiplicative noise has been utilized to characterize the estimator gain variation. A resilient distributed filtering strategy has been proposed in Li et al. (2021) for multirate systems, where a set of random variables with mean and bounded variance has been introduced to describe the estimator gain perturbation. Unlike the mentioned describing ways, the norm bounded parameter uncertainties can be viewed as a general way. Consequently, the estimator gain perturbations are introduced in this chapter, which are depicted by the norm bounded parameter uncertainties. As will be exhibited later, this phenomenon is effectively addressed and a novel robust state estimation scheme is given to improve the flexibility of the main results.

To facilitate the further derivations, the prediction error and the estimation error are defined as

$$\begin{aligned} e_{i,k+1|k} &= x_{i,k+1} - \hat{x}_{i,k+1|k}, \\ e_{i,k+1|k+1} &= x_{i,k+1} - \hat{x}_{i,k+1|k+1}. \end{aligned}$$

Subsequently, we aim to (1) design the estimator described by (12.9)–(12.10) and search for an upper bound matrix $\Omega_{i,k+1|k+1}$ satisfying

$$P_{i,k+1|k+1} = \mathbb{E}\left\{e_{i,k+1|k+1}e_{i,k+1|k+1}^T\right\} \leq \Omega_{i,k+1|k+1} \tag{12.11}$$

with $\Omega_{i,k+1|k+1} > 0$ being a matrix; (2) design the estimator gain $G_{i,k+1}$ to minimize $\text{tr}(\Omega_{i,k+1|k+1})$.

12.1.2 New Optimized State Estimation Method

In this section, by virtue of the stochastic analysis technique, a locally optimal state estimation strategy is presented for time-varying uncertain complex networks subject to DPDs, switching topology and RONs.

Subtracting (12.9) from (12.7) yields

$$\begin{aligned} e_{i,k+1|k} =&\bar{A}_{i,k}e_{i,k|k} + \tilde{A}_{i,k}x_{i,k} + \Delta A_{i,k}x_{i,k} + \tilde{\alpha}_{i,k}h_i(x_{i,k}) + \bar{\alpha}_{i,k}e^{h_i(x_{i,k})} \\ &+ \sum_{j=1}^{N}\tilde{\lambda}_{ij,k}\varpi_{ij}\Gamma x_{j,k} + \sum_{j=1}^{N}\bar{\lambda}_{ij,k}\varpi_{ij}\Gamma e_{j,k|k} + B_{i,k}\omega_{i,k}, \end{aligned} \tag{12.12}$$

where $\tilde{\alpha}_{i,k} = \alpha_{i,k} - \bar{\alpha}_{i,k}$, $\tilde{\lambda}_{ij,k} = \lambda_{ij,k} - \bar{\lambda}_{ij,k}$ and $e^{h_i(x_{i,k})} = h_i(x_{i,k}) - h_i(\hat{x}_{i,k|k})$. Similarly, it can be obtained that

$$\begin{aligned} e_{i,k+1|k+1} =&(I - \mathscr{G}_{i,k+1}\bar{C}_{i,k+1})e_{i,k+1|k} - \mathscr{G}_{i,k+1}\tilde{C}_{i,k+1}x_{i,k+1} \\ &- \mathscr{G}_{i,k+1}e^{y_{i,k+1}} - \beta_{i,k+1}\mathscr{G}_{i,k+1}\nu_{i,k+1} \end{aligned} \tag{12.13}$$

with $e^{y_{i,k+1}} = \tilde{y}_{i,k+1} - y_{i,k+1}$.

To facilitate subsequent calculations regarding the EEC and its upper bound matrix, the following important lemma is provided.

Lemma 12.1 *For given scalars $\varepsilon_i > 0$ $(i = 1, 2, 3, 4)$, the state covariance matrix $X_{i,k+1} = \mathbb{E}\{x_{i,k+1}x_{i,k+1}^T\}$ satisfies*

$$\begin{aligned} X_{i,k+1} \leq&(1 + \varepsilon_1 + \varepsilon_2)\big[(1 + \varepsilon_4)\bar{A}_{i,k}\bar{X}_{i,k}\bar{A}_{i,k}^T + (1 + \varepsilon_4^{-1})\mathrm{tr}(N_{i,k}\bar{X}_{i,k}N_{i,k}^T)M_{i,k}M_{i,k}^T\big] \\ &+ \bar{\beta}_{i,k}(1 - \bar{\beta}_{i,k})\mathscr{A}_{i,k}\bar{X}_{i,k}\mathscr{A}_{i,k}^T + (\bar{\alpha}_{i,k} + \varepsilon_1^{-1}\bar{\alpha}_{i,k}^2 + \varepsilon_3\bar{\alpha}_{i,k}^2)l^2\mathrm{tr}(\bar{X}_{i,k})I \\ &+ (1 + \varepsilon_2^{-1} + \varepsilon_3^{-1})\Xi_i\sum_{j=1}^{N}\bar{\lambda}_{ij,k}\varpi_{ij}\Gamma\bar{X}_{j,k}\Gamma^T + \sum_{j=1}^{N}\bar{\lambda}_{ij,k}(1 - \bar{\lambda}_{ij,k}) \\ &\times \varpi_{ij}^2\Gamma\bar{X}_{j,k}\Gamma^T + \check{E}_1 \circ (\mathscr{B}_{i,k}Q_{i,k}\mathscr{B}_{i,k}^T) \\ :=&\bar{X}_{i,k+1}, \end{aligned} \tag{12.14}$$

where

$$\Xi_i = \sum_{l=1}^{N}\bar{\lambda}_{il,k}\varpi_{il}, \quad \check{E}_1 = \begin{bmatrix} \check{E}_{n\times n} & \bar{\beta}_{i,k}\check{E}_{n\times m} \\ * & \bar{\beta}_{i,k}\check{E}_{m\times m} \end{bmatrix},$$

$$\check{E}_{m\times n} = \begin{bmatrix} 1 & 1 & \cdots & 1 \\ 1 & 1 & \cdots & 1 \\ \vdots & \vdots & \ddots & \vdots \\ 1 & 1 & \cdots & 1 \end{bmatrix}_{m\times n}.$$

Proof The state covariance matrix $X_{i,k+1}$ can be derived as

$$
\begin{aligned}
X_{i,k+1} =& \mathbb{E}\{(\bar{A}_{i,k} + \Delta A_{i,k})x_{i,k}x_{i,k}^T(\bar{A}_{i,k} + \Delta A_{i,k})^T\} + \bar{\beta}_{i,k}(1 - \bar{\beta}_{i,k})\mathscr{A}_{i,k}X_{i,k}\mathscr{A}_{i,k}^T \\
&+ \bar{\alpha}_{i,k}\mathbb{E}\{h_i(x_{i,k})h_i^T(x_{i,k})\} + \sum_{j=1}^{N}\sum_{l=1}^{N}\bar{\lambda}_{ij,k}\bar{\lambda}_{il,k}\varpi_{ij}\varpi_{il}\Gamma\mathbb{E}\{x_{j,k}x_{l,k}^T\}\Gamma^T \\
&+ \sum_{j=1}^{N}\bar{\lambda}_{ij,k}(1 - \bar{\lambda}_{ij,k})\varpi_{ij}^2\Gamma X_{j,k}\Gamma^T + \check{E}_1 \circ (\mathscr{B}_{i,k}Q_{i,k}\mathscr{B}_{i,k}^T) \\
&+ \mathrm{Sym}(\mathscr{N}_1) + \mathrm{Sym}(\mathscr{N}_2) + \mathrm{Sym}(\mathscr{N}_3),
\end{aligned}
\tag{12.15}
$$

where

$$
\begin{aligned}
\mathscr{N}_1 &= \mathbb{E}\left\{\bar{\alpha}_{i,k}(\bar{A}_{i,k} + \Delta A_{i,k})x_{i,k}h_i^T(x_{i,k})\right\}, \\
\mathscr{N}_2 &= \mathbb{E}\left\{\sum_{j=1}^{N}\bar{\lambda}_{ij,k}\varpi_{ij}\Gamma x_{j,k}x_{i,k}^T(\bar{A}_{i,k} + \Delta A_{i,k})^T\right\}, \\
\mathscr{N}_3 &= \mathbb{E}\left\{\sum_{j=1}^{N}\bar{\lambda}_{ij,k}\varpi_{ij}\bar{\alpha}_{i,k}\Gamma x_{j,k}h_i^T(x_{i,k})\right\}.
\end{aligned}
$$

By virtue of Lemma 2.2, we obtain

$$
\begin{aligned}
\mathrm{Sym}(\mathscr{N}_1) \le& \varepsilon_1\mathbb{E}\{(\bar{A}_{i,k} + \Delta A_{i,k})x_{i,k}x_{i,k}^T(\bar{A}_{i,k} + \Delta A_{i,k})^T\} \\
&+ \varepsilon_1^{-1}\bar{\alpha}_{i,k}^2\mathbb{E}\left\{h_i(x_{i,k})h_i^T(x_{i,k})\right\},
\end{aligned}
\tag{12.16}
$$

$$
\begin{aligned}
\mathrm{Sym}(\mathscr{N}_2) \le& \varepsilon_2\mathbb{E}\{(\bar{A}_{i,k} + \Delta A_{i,k})x_{i,k}x_{i,k}^T(\bar{A}_{i,k} + \Delta A_{i,k})^T\} \\
&+ \varepsilon_2^{-1}\sum_{j=1}^{N}\sum_{l=1}^{N}\bar{\lambda}_{ij,k}\bar{\lambda}_{il,k}\varpi_{ij}\varpi_{il}\Gamma\mathbb{E}\{x_{j,k}x_{l,k}^T\}\Gamma^T,
\end{aligned}
\tag{12.17}
$$

$$
\begin{aligned}
\mathrm{Sym}(\mathscr{N}_3) \le& \varepsilon_3\bar{\alpha}_{i,k}^2\mathbb{E}\{h_i(x_{i,k})h_i^T(x_{i,k})\} \\
&+ \varepsilon_3^{-1}\sum_{j=1}^{N}\sum_{l=1}^{N}\bar{\lambda}_{ij,k}\bar{\lambda}_{il,k}\varpi_{ij}\varpi_{il}\Gamma\mathbb{E}\{x_{j,k}x_{l,k}^T\}\Gamma^T,
\end{aligned}
\tag{12.18}
$$

where $\varepsilon_i > 0$ $(i = 1, 2, 3)$ are given scalars. Thus, it can be inferred that

$$
\begin{aligned}
X_{i,k+1} \le& (1 + \varepsilon_1 + \varepsilon_2)\mathbb{E}\{(\bar{A}_{i,k} + \Delta A_{i,k})x_{i,k}x_{i,k}^T(\bar{A}_{i,k} + \Delta A_{i,k})^T\} \\
&+ \bar{\beta}_{i,k}(1 - \bar{\beta}_{i,k})\mathscr{A}_{i,k}X_{i,k}\mathscr{A}_{i,k}^T + (\bar{\alpha}_{i,k} + \varepsilon_1^{-1}\bar{\alpha}_{i,k}^2 \\
&+ \varepsilon_3\bar{\alpha}_{i,k}^2)\mathbb{E}\{h_i(x_{i,k})h_i^T(x_{i,k})\} + (1 + \varepsilon_2^{-1} + \varepsilon_3^{-1}) \\
&\times \sum_{j=1}^{N}\sum_{l=1}^{N}\bar{\lambda}_{ij,k}\bar{\lambda}_{il,k}\varpi_{ij}\varpi_{il}\Gamma\mathbb{E}\{x_{j,k}x_{l,k}^T\}\Gamma^T + \sum_{j=1}^{N}\bar{\lambda}_{ij,k} \\
&\times (1 - \bar{\lambda}_{ij,k})\varpi_{ij}^2\Gamma X_{j,k}\Gamma^T + \check{E}_1 \circ (\mathscr{B}_{i,k}Q_{i,k}\mathscr{B}_{i,k}^T).
\end{aligned}
\tag{12.19}
$$

In the sequel, the first uncertain term of (12.19) is tackled as

$$\begin{aligned}
&\mathbb{E}\{(\bar{A}_{i,k}+\Delta A_{i,k})x_{i,k}x_{i,k}^T(\bar{A}_{i,k}+\Delta A_{i,k})^T\}\\
\leq&(1+\varepsilon_4)\mathbb{E}\{\bar{A}_{i,k}x_{i,k}x_{i,k}^T\bar{A}_{i,k}^T\}+(1+\varepsilon_4^{-1})\mathbb{E}\{\Delta A_{i,k}x_{i,k}x_{i,k}^T\Delta A_{i,k}^T\}\\
\leq&(1+\varepsilon_4)\bar{A}_{i,k}X_{i,k}\bar{A}_{i,k}^T+(1+\varepsilon_4^{-1})\mathrm{tr}(N_{i,k}X_{i,k}N_{i,k}^T)M_{i,k}M_{i,k}^T
\end{aligned}\tag{12.20}$$

with $\varepsilon_4>0$ denoting a scalar. Next, employing the properties of matrix operations, we have

$$h_i(x_{i,k})h_i^T(x_{i,k})\leq\|h_i(x_{i,k})\|^2I=h_i^T(x_{i,k})h_i(x_{i,k})I.\tag{12.21}$$

Thus, it is obvious to have that

$$\mathbb{E}\{h_i^T(x_{i,k})h_i(x_{i,k})\}\leq l^2\mathbb{E}\{x_{i,k}^Tx_{i,k}\}=l^2\mathrm{tr}(X_{i,k}).\tag{12.22}$$

Moreover, the fourth term of (12.19) is calculated as

$$\begin{aligned}
&\sum_{j=1}^N\sum_{l=1}^N\bar{\lambda}_{ij,k}\bar{\lambda}_{il,k}\varpi_{ij}\varpi_{il}\Gamma\mathbb{E}\{x_{j,k}x_{l,k}^T\}\Gamma^T\\
\leq&\frac{1}{2}\sum_{j=1}^N\sum_{l=1}^N\big\{\bar{\lambda}_{ij,k}\bar{\lambda}_{il,k}\varpi_{ij}\varpi_{il}\Gamma\mathbb{E}\{x_{j,k}x_{j,k}^T\}\Gamma^T\\
&+\bar{\lambda}_{ij,k}\bar{\lambda}_{il,k}\varpi_{il}\varpi_{ij}\Gamma\mathbb{E}\{x_{l,k}x_{l,k}^T\}\Gamma^T\big\}\\
=&\Xi_i\sum_{j=1}^N\varpi_{ij}\bar{\lambda}_{ij,k}\Gamma X_{j,k}\Gamma^T
\end{aligned}\tag{12.23}$$

with Ξ_i given below (12.14).

Finally, together with (12.19)–(12.23), we arrive at

$$\begin{aligned}
X_{i,k+1}\leq&(1+\varepsilon_1+\varepsilon_2)\big[(1+\varepsilon_4)\bar{A}_{i,k}X_{i,k}\bar{A}_{i,k}^T+(1+\varepsilon_4^{-1})\\
&\times\mathrm{tr}(N_{i,k}X_{i,k}N_{i,k}^T)M_{i,k}M_{i,k}^T\big]+\bar{\beta}_{i,k}(1-\bar{\beta}_{i,k})\\
&\times\mathscr{A}_{i,k}X_{i,k}\mathscr{A}_{i,k}^T+(\bar{\alpha}_{i,k}+\varepsilon_1^{-1}\bar{\alpha}_{i,k}^2+\varepsilon_3\bar{\alpha}_{i,k}^2)l^2\mathrm{tr}(X_{i,k})I\\
&+(1+\varepsilon_2^{-1}+\varepsilon_3^{-1})\Xi_i\sum_{j=1}^N\bar{\lambda}_{ij,k}\varpi_{ij}\Gamma X_{j,k}\Gamma^T+\sum_{j=1}^N\bar{\lambda}_{ij,k}\\
&\times(1-\bar{\lambda}_{ij,k})\varpi_{ij}^2\Gamma X_{j,k}\Gamma^T+\check{E}_1\circ(\mathscr{B}_{i,k}Q_{i,k}\mathscr{B}_{i,k}^T),
\end{aligned}\tag{12.24}$$

which completes the proof.

In what follows, consider the definitions of prediction error covariance $P_{i,k+1|k} = \mathbb{E}\{e_{i,k+1|k}e_{i,k+1|k}^T\}$ and EEC $P_{i,k+1|k+1} = \mathbb{E}\{e_{i,k+1|k+1}e_{i,k+1|k+1}^T\}$, according to (12.12) and (12.13), the recursive forms of $P_{i,k+1|k}$ and $P_{i,k+1|k+1}$ can be obtained.

Theorem 12.1 *The prediction error covariance $P_{i,k+1|k}$ is determined by*

$$\begin{aligned}
P_{i,k+1|k} =& \bar{A}_{i,k}P_{i,k|k}\bar{A}_{i,k}^T + \bar{\beta}_{i,k}(1-\bar{\beta}_{i,k})\mathscr{A}_{i,k}\mathbb{E}\{x_{i,k}x_{i,k}^T\}\mathscr{A}_{i,k}^T \\
&+ \Delta A_{i,k}\mathbb{E}\{x_{i,k}x_{i,k}^T\}\Delta A_{i,k}^T + \bar{\alpha}_{i,k}(1-\bar{\alpha}_{i,k}) \\
&\times \mathbb{E}\{h_i(x_{i,k})h_i^T(x_{i,k})\} + \bar{\alpha}_{i,k}^2\mathbb{E}\left\{e^{h_i(x_{i,k})}(e^{h_i(x_{i,k})})^T\right\} \\
&+ \sum_{j=1}^{N}\bar{\lambda}_{ij,k}(1-\bar{\lambda}_{ij,k})\varpi_{ij}^2\Gamma\mathbb{E}\{x_{j,k}x_{j,k}^T\}\Gamma^T + \Xi_i\sum_{j=1}^{N}\bar{\lambda}_{ij,k} \\
&\times \varpi_{ij}\Gamma P_{j,k|k}\Gamma^T + \check{E}_1 \circ (\mathscr{B}_{i,k}Q_{i,k}\mathscr{B}_{i,k}^T) + \mathrm{Sym}(\mathscr{P}_1) \\
&+ \mathrm{Sym}(\mathscr{P}_2) + \mathrm{Sym}(\mathscr{P}_3) + \mathrm{Sym}(\mathscr{P}_4) + \mathrm{Sym}(\mathscr{P}_5) \\
&+ \mathrm{Sym}(\mathscr{P}_6),
\end{aligned} \tag{12.25}$$

where

$$\begin{aligned}
\mathscr{P}_1 =& \mathbb{E}\left\{\bar{A}_{i,k}e_{i,k|k}x_{i,k}^T\Delta A_{i,k}^T\right\}, \\
\mathscr{P}_2 =& \mathbb{E}\left\{\bar{\alpha}_{i,k}\bar{A}_{i,k}e_{i,k|k}(e^{h_i(x_{i,k})})^T\right\}, \\
\mathscr{P}_3 =& \mathbb{E}\left\{\sum_{j=1}^{N}\bar{\lambda}_{ij,k}\varpi_{ij}\Gamma e_{j,k|k}e_{i,k|k}^T\bar{A}_{i,k}^T\right\}, \\
\mathscr{P}_4 =& \mathbb{E}\left\{\bar{\alpha}_{i,k}\Delta A_{i,k}x_{i,k}(e^{h_i(x_{i,k})})^T\right\}, \\
\mathscr{P}_5 =& \mathbb{E}\left\{\sum_{j=1}^{N}\bar{\lambda}_{ij,k}\varpi_{ij}\Gamma e_{j,k|k}x_{i,k}^T\Delta A_{i,k}^T\right\}, \\
\mathscr{P}_6 =& \mathbb{E}\left\{\sum_{j=1}^{N}\bar{\lambda}_{ij,k}\varpi_{ij}\bar{\alpha}_{i,k}\Gamma e_{j,k|k}(e^{h_i(x_{i,k})})^T\right\}.
\end{aligned} \tag{12.26}$$

Proof It is clear to see that (12.25) is correct. Thus, the proof is omitted.

Theorem 12.2 *The EEC $P_{i,k+1|k+1}$ is described as*

$$\begin{aligned}
P_{i,k+1|k+1} =& (I-\mathscr{G}_{i,k+1}\bar{C}_{i,k+1})P_{i,k+1|k}(I-\mathscr{G}_{i,k+1}\bar{C}_{i,k+1})^T \\
&+ \bar{\beta}_{i,k+1}(1-\bar{\beta}_{i,k+1})\mathscr{G}_{i,k+1}\mathscr{C}_{i,k+1}\mathbb{E}\{x_{i,k+1}x_{i,k+1}^T\} \\
&\times \mathscr{C}_{i,k+1}^T\mathscr{G}_{i,k+1}^T + \mathscr{G}_{i,k+1}\mathbb{E}\{e^{y_{i,k+1}}(e^{y_{i,k+1}})^T\}\mathscr{G}_{i,k+1}^T \\
&+ \bar{\beta}_{i,k+1}\mathscr{G}_{i,k+1}\mathscr{V}_{i,k+1}\mathscr{G}_{i,k+1}^T - \mathrm{Sym}(\mathscr{Q}_1) + \mathrm{Sym}(\mathscr{Q}_2),
\end{aligned} \tag{12.27}$$

where

$$\begin{aligned}\mathscr{Q}_1 =&\mathbb{E}\left\{(I-\mathscr{G}_{i,k+1}\bar{C}_{i,k+1})e_{i,k+1|k}(e^{y_{i,k+1}})^T\mathscr{G}_{i,k+1}^T\right\},\\ \mathscr{Q}_2 =&\mathbb{E}\left\{\beta_{i,k+1}\mathscr{G}_{i,k+1}\nu_{i,k+1}(e^{y_{i,k+1}})^T\mathscr{G}_{i,k+1}^T\right\}.\end{aligned} \tag{12.28}$$

Proof Noting $\mathbb{E}\{\nu_{i,k+1}\}=0$ and $\mathbb{E}\{\tilde{C}_{i,k+1}\}=0$, it can be easily obtained that (12.27) is true.

Remark 12.3 In Theorems 12.1–12.2, both the prediction error covariance $P_{i,k+1|k}$ and the EEC $P_{i,k+1|k+1}$ have been derived by the corresponding definitions. Owing to the influence resulting from the estimator gain perturbation, RONs, switching topology, ETM and DPDs, there are some uncertain terms in Theorems 12.1–12.2, such as $\mathscr{P}_i$ $(i=1,2,\ldots,6)$, $\mathscr{Q}_1$, $\mathscr{Q}_2$, $\Delta G_{i,k+1}$ and $\mathbb{E}\{x_{i,k+1}x_{i,k+1}^T\}$ in (12.25) and (12.27). As such, it is difficult to obtain an exact value of the EEC. With the aim to derive the estimator gain in terms of certain criterion, a minimized upper bound matrix of the EEC is given later.

Theorem 12.3 *For* $\eta_l>0$ $(l=1,2,\ldots,9)$, $\theta_{i,k+1}>0$, $\mu_{i,k+1}>0$ *and* $\tau_{i,k+1}>0$, *under the initial condition* $\Omega_{i,0|0}=P_{i,0|0}>0$, *if the following matrix equations:*

$$\begin{aligned}\Omega_{i,k+1|k} =&(1+\eta_1+\eta_2+\eta_3)\bar{A}_{i,k}\Omega_{i,k|k}\bar{A}_{i,k}^T+\bar{\beta}_{i,k}(1-\bar{\beta}_{i,k})\\ &\times\mathscr{A}_{i,k}\bar{X}_{i,k}\mathscr{A}_{i,k}^T+(1+\eta_1^{-1}+\eta_4+\eta_5)\mathrm{tr}(N_{i,k}\bar{X}_{i,k}N_{i,k}^T)\\ &\times M_{i,k}M_{i,k}^T+\bar{\alpha}_{i,k}(1-\bar{\alpha}_{i,k})l^2\mathrm{tr}(\bar{X}_{i,k})I+(1+\eta_2^{-1}\\ &+\eta_4^{-1}+\eta_6)\bar{\alpha}_{i,k}^2l^2\mathrm{tr}(\Omega_{i,k|k})I+\sum_{j=1}^{N}\bar{\lambda}_{ij,k}(1-\bar{\lambda}_{ij,k})\varpi_{ij}^2\\ &\times\Gamma\bar{X}_{j,k}\Gamma^T+(1+\eta_3^{-1}+\eta_5^{-1}+\eta_6^{-1})\Xi_i\sum_{j=1}^{N}\bar{\lambda}_{ij,k}\varpi_{ij}\\ &\times\Gamma\Omega_{j,k|k}\Gamma^T+\check{E}_1\circ(\mathscr{B}_{i,k}Q_{i,k}\mathscr{B}_{i,k}^T)\end{aligned} \tag{12.29}$$

and

$$\begin{aligned}\Omega_{i,k+1|k+1} =&(1+\eta_7)(I-G_{i,k+1}\bar{C}_{i,k+1})\Psi_{i,k+1}^{-1}(I-G_{i,k+1}\bar{C}_{i,k+1})^T\\ &+(1+\eta_7)\theta_{i,k+1}^{-1}H_{i,k+1}H_{i,k+1}^T+\bar{\beta}_{i,k+1}(1-\bar{\beta}_{i,k+1})\\ &\times(1+\eta_9)G_{i,k+1}\mathscr{C}_{i,k+1}\bar{X}_{i,k+1}\mathscr{C}_{i,k+1}^TG_{i,k+1}^T+\bar{\beta}_{i,k+1}\\ &\times(1-\bar{\beta}_{i,k+1})(1+\eta_9^{-1})\mathrm{tr}(L_{i,k+1}\mathscr{C}_{i,k+1}\bar{X}_{i,k+1}\mathscr{C}_{i,k+1}^T\\ &\times L_{i,k+1}^T)H_{i,k+1}H_{i,k+1}^T+(1+\eta_7^{-1}+\eta_8)\delta_{i,k+1}G_{i,k+1}\\ &\times(I-\mu_{i,k+1}L_{i,k+1}^TL_{i,k+1})^{-1}G_{i,k+1}^T+(1+\eta_7^{-1}+\eta_8)\\ &\times\delta_{i,k+1}\mu_{i,k+1}^{-1}H_{i,k+1}H_{i,k+1}^T+(1+\eta_8^{-1})\bar{\beta}_{i,k+1}\\ &\times G_{i,k+1}(\mathscr{V}_{i,k+1}^{-1}-\tau_{i,k+1}L_{i,k+1}^TL_{i,k+1})^{-1}G_{i,k+1}^T\\ &+(1+\eta_8^{-1})\bar{\beta}_{i,k+1}\tau_{i,k+1}^{-1}H_{i,k+1}H_{i,k+1}^T\end{aligned} \tag{12.30}$$

under the following inequality constraints:

$$\begin{aligned}&\theta_{i,k+1}^{-1}I - L_{i,k+1}\bar{C}_{i,k+1}\Omega_{i,k+1|k}\bar{C}_{i,k+1}^{T}L_{i,k+1}^{T} > 0,\\&\mu_{i,k+1}^{-1}I - L_{i,k+1}L_{i,k+1}^{T} > 0,\\&\tau_{i,k+1}^{-1}I - L_{i,k+1}\mathscr{V}_{i,k+1}L_{i,k+1}^{T} > 0,\end{aligned}\tag{12.31}$$

have solutions $\Omega_{i,k+1|k} > 0$ *and* $\Omega_{i,k+1|k+1} > 0$*, where*

$$\bar{\Psi}_{i,k+1} = \Omega_{i,k+1|k}^{-1} - \theta_{i,k+1}\bar{C}_{i,k+1}^{T}L_{i,k+1}^{T}L_{i,k+1}\bar{C}_{i,k+1},$$

then one gets $P_{i,k+1|k+1} \leq \Omega_{i,k+1|k+1}$*. Additionally, if the estimator gain* $G_{i,k+1}$ *is constructed by*

$$\begin{aligned}G_{i,k+1} =&(1+\eta_7)\bar{\Psi}_{i,k+1}^{-1}\bar{C}_{i,k+1}^{T}\big\{(1+\eta_7)\bar{C}_{i,k+1}\bar{\Psi}_{i,k+1}^{-1}\bar{C}_{i,k+1}^{T}\\&+\bar{\beta}_{i,k+1}(1-\bar{\beta}_{i,k+1})(1+\eta_9)\mathscr{C}_{i,k+1}\bar{X}_{i,k+1}\mathscr{C}_{i,k+1}^{T} + (1\\&+\eta_7^{-1}+\eta_8)\delta_{i,k+1}(I-\mu_{i,k+1}L_{i,k+1}^{T}L_{i,k+1})^{-1} + (1\\&+\eta_8^{-1})\bar{\beta}_{i,k+1}(\mathscr{V}_{i,k+1}^{-1} - \tau_{i,k+1}L_{i,k+1}^{T}L_{i,k+1})^{-1}\big\}^{-1},\end{aligned}\tag{12.32}$$

then it can be seen that $\mathrm{tr}(\Omega_{i,k+1|k+1})$ *is minimized.*

Proof Considering the unknown terms in (12.25) and applying Lemma 2.2, we have

$$\begin{aligned}\mathrm{Sym}(\mathscr{P}_1) \leq&\eta_1\bar{A}_{i,k}P_{i,k|k}\bar{A}_{i,k}^{T} + \eta_1^{-1}\Delta A_{i,k}\mathbb{E}\{x_{i,k}x_{i,k}^{T}\}\Delta A_{i,k}^{T},\\\mathrm{Sym}(\mathscr{P}_2) \leq&\eta_2\bar{A}_{i,k}P_{i,k|k}\bar{A}_{i,k}^{T} + \eta_2^{-1}\bar{\alpha}_{i,k}^{2}\mathbb{E}\left\{e^{h_i(x_{i,k})}(e^{h_i(x_{i,k})})^{T}\right\},\\\mathrm{Sym}(\mathscr{P}_3) \leq&\eta_3\bar{A}_{i,k}P_{i,k|k}\bar{A}_{i,k}^{T} + \eta_3^{-1}\Xi_i\sum_{j=1}^{N}\bar{\lambda}_{ij,k}\varpi_{ij}\Gamma P_{j,k|k}\Gamma^{T},\\\mathrm{Sym}(\mathscr{P}_4) \leq&\eta_4\Delta A_{i,k}\mathbb{E}\{x_{i,k}x_{i,k}^{T}\}\Delta A_{i,k}^{T} + \eta_4^{-1}\bar{\alpha}_{i,k}^{2}\mathbb{E}\left\{e^{h_i(x_{i,k})}(e^{h_i(x_{i,k})})^{T}\right\},\\\mathrm{Sym}(\mathscr{P}_5) \leq&\eta_5\Delta A_{i,k}\mathbb{E}\{x_{i,k}x_{i,k}^{T}\}\Delta A_{i,k}^{T} + \eta_5^{-1}\Xi_i\sum_{j=1}^{N}\bar{\lambda}_{ij,k}\varpi_{ij}\Gamma P_{j,k|k}\Gamma^{T},\\\mathrm{Sym}(\mathscr{P}_6) \leq&\eta_6\bar{\alpha}_{i,k}^{2}\mathbb{E}\left\{e^{h_i(x_{i,k})}(e^{h_i(x_{i,k})})^{T}\right\} + \eta_6^{-1}\Xi_i\sum_{j=1}^{N}\bar{\lambda}_{ij,k}\varpi_{ij}\Gamma P_{j,k|k}\Gamma^{T},\end{aligned}\tag{12.33}$$

where $\eta_{l_1} > 0$ ($l_1 = 1, 2, \ldots, 6$) are scalars. Substituting (12.33) into (12.25), one has

$$\begin{aligned}P_{i,k+1|k} \leq&(1+\eta_1+\eta_2+\eta_3)\bar{A}_{i,k}P_{i,k|k}\bar{A}_{i,k}^{T} + \bar{\beta}_{i,k}(1-\bar{\beta}_{i,k})\mathscr{A}_{i,k}\\&\times\mathbb{E}\{x_{i,k}x_{i,k}^{T}\}\mathscr{A}_{i,k}^{T} + (1+\eta_1^{-1}+\eta_4+\eta_5)\Delta A_{i,k}\mathbb{E}\{x_{i,k}x_{i,k}^{T}\}\\&\times\Delta A_{i,k}^{T} + \bar{\alpha}_{i,k}(1-\bar{\alpha}_{i,k})\mathbb{E}\left\{h_i(x_{i,k})h_i^{T}(x_{i,k})\right\}\end{aligned}$$

$$
\begin{aligned}
&+ (1 + \eta_2^{-1} + \eta_4^{-1} + \eta_6)\bar{\alpha}_{i,k}^2 \mathbb{E}\left\{e^{h_i(x_{i,k})}(e^{h_i(x_{i,k})})^T\right\} \\
&+ \sum_{j=1}^{N} \bar{\lambda}_{ij,k}(1 - \bar{\lambda}_{ij,k})\varpi_{ij}^2 \Gamma \mathbb{E}\{x_{j,k}x_{j,k}^T\}\Gamma^T + (1 + \eta_3^{-1} \\
&+ \eta_5^{-1} + \eta_6^{-1})\Xi_i \sum_{j=1}^{N} \bar{\lambda}_{ij,k}\varpi_{ij}\Gamma P_{j,k|k}\Gamma^T \\
&+ \check{E}_1 \circ (\mathscr{B}_{i,k} Q_{i,k} \mathscr{B}_{i,k}^T).
\end{aligned} \tag{12.34}
$$

Next, similar to (12.21), we can obtain

$$
e^{h_i(x_{i,k})}(e^{h_i(x_{i,k})})^T \leq \| e^{h_i(x_{i,k})} \|^2 I = (e^{h_i(x_{i,k})})^T e^{h_i(x_{i,k})} I. \tag{12.35}
$$

Then, we have the following inequality:

$$
\mathbb{E}\{(e^{h_i(x_{i,k})})^T e^{h_i(x_{i,k})}\} \leq l^2 \mathbb{E}\{e_{i,k|k}^T e_{i,k|k}\} = l^2 \mathrm{tr}(P_{i,k|k}). \tag{12.36}
$$

As such, one has

$$
\begin{aligned}
P_{i,k+1|k} \leq &(1 + \eta_1 + \eta_2 + \eta_3)\bar{A}_{i,k} P_{i,k|k} \bar{A}_{i,k}^T + \bar{\beta}_{i,k}(1 - \bar{\beta}_{i,k})\mathscr{A}_{i,k} X_{i,k} \mathscr{A}_{i,k}^T \\
&+ (1 + \eta_1^{-1} + \eta_4 + \eta_5)\mathrm{tr}(N_{i,k} X_{i,k} N_{i,k}^T) M_{i,k} M_{i,k}^T \\
&+ \bar{\alpha}_{i,k}(1 - \bar{\alpha}_{i,k}) l^2 \mathrm{tr}(X_{i,k}) I \\
&+ (1 + \eta_2^{-1} + \eta_4^{-1} + \eta_6)\bar{\alpha}_{i,k}^2 l^2 \mathrm{tr}(P_{i,k|k}) I \\
&+ \sum_{j=1}^{N} \bar{\lambda}_{ij,k}(1 - \bar{\lambda}_{ij,k})\varpi_{ij}^2 \Gamma X_{j,k} \Gamma^T \\
&+ (1 + \eta_3^{-1} + \eta_5^{-1} + \eta_6^{-1})\Xi_i \sum_{j=1}^{N} \bar{\lambda}_{ij,k}\varpi_{ij}\Gamma P_{j,k|k}\Gamma^T \\
&+ \check{E}_1 \circ (\mathscr{B}_{i,k} Q_{i,k} \mathscr{B}_{i,k}^T).
\end{aligned} \tag{12.37}
$$

Subsequently, the aim is to prove $P_{i,k+1|k+1} \leq \Omega_{i,k+1|k+1}$. Employing Lemma 2.2 again, we can obtain

$$
\begin{aligned}
-\mathrm{Sym}(\mathscr{Q}_1) \leq &\eta_7(I - \mathscr{G}_{i,k+1}\bar{C}_{i,k+1})P_{i,k+1|k}(I - \mathscr{G}_{i,k+1}\bar{C}_{i,k+1})^T \\
&+ \eta_7^{-1}\mathscr{G}_{i,k+1}\mathbb{E}\left\{e^{y_{i,k+1}}(e^{y_{i,k+1}})^T\right\}\mathscr{G}_{i,k+1}^T, \\
\mathrm{Sym}(\mathscr{Q}_2) \leq &\eta_8 \mathscr{G}_{i,k+1}\mathbb{E}\left\{e^{y_{i,k+1}}(e^{y_{i,k+1}})^T\right\}\mathscr{G}_{i,k+1}^T \\
&+ \eta_8^{-1}\bar{\beta}_{i,k+1}\mathscr{G}_{i,k+1}\mathscr{V}_{i,k+1}\mathscr{G}_{i,k+1}^T,
\end{aligned} \tag{12.38}
$$

where η_7 and η_8 are positive constants. Substituting (12.38) into (12.27) leads to

$$
\begin{aligned}
P_{i,k+1|k+1} \leq &(1+\eta_7)(I-\mathscr{G}_{i,k+1}\bar{C}_{i,k+1})P_{i,k+1|k}(I-\mathscr{G}_{i,k+1}\bar{C}_{i,k+1})^T \\
&+\bar{\beta}_{i,k+1}(1-\bar{\beta}_{i,k+1})\mathscr{G}_{i,k+1}\mathscr{C}_{i,k+1}\mathbb{E}\{x_{i,k+1}x_{i,k+1}^T\}\mathscr{C}_{i,k+1}^T\mathscr{G}_{i,k+1}^T \\
&+(1+\eta_7^{-1}+\eta_8)\mathscr{G}_{i,k+1}\mathbb{E}\left\{e^{y_{i,k+1}}(e^{y_{i,k+1}})^T\right\}\mathscr{G}_{i,k+1}^T \\
&+(1+\eta_8^{-1})\bar{\beta}_{i,k+1}\mathscr{G}_{i,k+1}\mathscr{V}_{i,k+1}\mathscr{G}_{i,k+1}^T.
\end{aligned} \tag{12.39}
$$

Noting $\mathscr{G}_{i,k+1} = G_{i,k+1} + \Delta G_{i,k+1}$ and employing Lemma 2.1, it is clear that

$$
\begin{aligned}
&(I-\mathscr{G}_{i,k+1}\bar{C}_{i,k+1})P_{i,k+1|k}(I-\mathscr{G}_{i,k+1}\bar{C}_{i,k+1})^T \\
\leq &(I-G_{i,k+1}\bar{C}_{i,k+1})\Psi_{i,k+1}^{-1}(I-G_{i,k+1}\bar{C}_{i,k+1})^T+\theta_{i,k+1}^{-1}H_{i,k+1}H_{i,k+1}^T,
\end{aligned} \tag{12.40}
$$

where $\Psi_{i,k+1} = P_{i,k+1|k}^{-1} - \theta_{i,k+1}\bar{C}_{i,k+1}^T L_{i,k+1}^T L_{i,k+1}\bar{C}_{i,k+1}$. The second unknown term of (12.39) can be given as

$$
\begin{aligned}
&\mathscr{G}_{i,k+1}\mathscr{C}_{i,k+1}\mathbb{E}\{x_{i,k+1}x_{i,k+1}^T\}\mathscr{C}_{i,k+1}^T\mathscr{G}_{i,k+1}^T \\
\leq &(1+\eta_9)G_{i,k+1}\mathscr{C}_{i,k+1}\mathbb{E}\{x_{i,k+1}x_{i,k+1}^T\}\mathscr{C}_{i,k+1}^T G_{i,k+1}^T \\
&+(1+\eta_9^{-1})\Delta G_{i,k+1}\mathscr{C}_{i,k+1}\mathbb{E}\{x_{i,k+1}x_{i,k+1}^T\}\mathscr{C}_{i,k+1}^T\Delta G_{i,k+1}^T \\
\leq &(1+\eta_9)G_{i,k+1}\mathscr{C}_{i,k+1}X_{i,k+1}\mathscr{C}_{i,k+1}^T G_{i,k+1}^T \\
&+(1+\eta_9^{-1})\mathrm{tr}(L_{i,k+1}\mathscr{C}_{i,k+1}X_{i,k+1}\mathscr{C}_{i,k+1}^T L_{i,k+1}^T)H_{i,k+1}H_{i,k+1}^T,
\end{aligned} \tag{12.41}
$$

where $\eta_9 > 0$ is a known scalar. The third unknown term of (12.39) is addressed as

$$
\begin{aligned}
&\mathscr{G}_{i,k+1}\mathbb{E}\left\{e^{y_{i,k+1}}(e^{y_{i,k+1}})^T\right\}\mathscr{G}_{i,k+1}^T \\
\leq &\delta_{i,k+1}G_{i,k+1}(I-\mu_{i,k+1}L_{i,k+1}^T L_{i,k+1})^{-1}G_{i,k+1}^T \\
&+\delta_{i,k+1}\mu_{i,k+1}^{-1}H_{i,k+1}H_{i,k+1}^T.
\end{aligned} \tag{12.42}
$$

Next, the fourth unknown term of (12.39) is derived as

$$
\begin{aligned}
&\mathscr{G}_{i,k+1}\mathscr{V}_{i,k+1}\mathscr{G}_{i,k+1}^T \\
\leq &G_{i,k+1}(\mathscr{V}_{i,k+1}^{-1}-\tau_{i,k+1}L_{i,k+1}^T L_{i,k+1})^{-1}G_{i,k+1}^T+\tau_{i,k+1}^{-1}H_{i,k+1}H_{i,k+1}^T.
\end{aligned} \tag{12.43}
$$

Consequently, substituting (12.40)–(12.43) into (12.39) yields

$$
\begin{aligned}
P_{i,k+1|k+1} \leq &(1+\eta_7)(I-G_{i,k+1}\bar{C}_{i,k+1})\Psi_{i,k+1}^{-1}(I-G_{i,k+1}\bar{C}_{i,k+1})^T \\
&+(1+\eta_7)\theta_{i,k+1}^{-1}H_{i,k+1}H_{i,k+1}^T+\bar{\beta}_{i,k+1}(1-\bar{\beta}_{i,k+1}) \\
&\times(1+\eta_9)G_{i,k+1}\mathscr{C}_{i,k+1}X_{i,k+1}\mathscr{C}_{i,k+1}^T G_{i,k+1}^T+\bar{\beta}_{i,k+1} \\
&\times(1-\bar{\beta}_{i,k+1})(1+\eta_9^{-1})\mathrm{tr}(L_{i,k+1}\mathscr{C}_{i,k+1}X_{i,k+1}\mathscr{C}_{i,k+1}^T \\
&\times L_{i,k+1}^T)H_{i,k+1}H_{i,k+1}^T+(1+\eta_7^{-1}+\eta_8)\delta_{i,k+1}G_{i,k+1} \\
&\times(I-\mu_{i,k+1}L_{i,k+1}^T L_{i,k+1})^{-1}G_{i,k+1}^T+(1+\eta_7^{-1}+\eta_8)
\end{aligned}
$$

$$\times \delta_{i,k+1}\mu_{i,k+1}^{-1}H_{i,k+1}H_{i,k+1}^{T} + (1+\eta_8^{-1})\bar{\beta}_{i,k+1}G_{i,k+1}$$
$$\times (\mathscr{V}_{i,k+1}^{-1} - \tau_{i,k+1}L_{i,k+1}^{T}L_{i,k+1})^{-1}G_{i,k+1}^{T} + (1+\eta_8^{-1})$$
$$\times \bar{\beta}_{i,k+1}\tau_{i,k+1}^{-1}H_{i,k+1}H_{i,k+1}^{T}. \quad (12.44)$$

Considering (12.25), (12.27) and (12.44), by adopting the mathematical induction method, we have $P_{i,k+1|k+1} \leq \Omega_{i,k+1|k+1}$.

Finally, we aim to design a desirable estimator gain. In light of Lemma 2.3, $\frac{\partial \text{tr}(\Omega_{i,k+1|k+1})}{\partial G_{i,k+1}}$ is calculated as follows:

$$\begin{aligned}&\frac{\partial \text{tr}(\Omega_{i,k+1|k+1})}{\partial G_{i,k+1}}\\ =&-2(1+\eta_7)(I-G_{i,k+1}\bar{C}_{i,k+1})\bar{\Psi}_{i,k+1}^{-1}\bar{C}_{i,k+1}^{T} + 2G_{i,k+1}\\ &\times \big\{\bar{\beta}_{i,k+1}(1-\bar{\beta}_{i,k+1})(1+\eta_9)\mathscr{C}_{i,k+1}\bar{X}_{i,k+1}\mathscr{C}_{i,k+1}^{T}\\ &+(1+\eta_7^{-1}+\eta_8)\delta_{i,k+1}(I-\mu_{i,k+1}L_{i,k+1}^{T}L_{i,k+1})^{-1}\\ &+(1+\eta_8^{-1})\bar{\beta}_{i,k+1}(\mathscr{V}_{i,k+1}^{-1}-\tau_{i,k+1}L_{i,k+1}^{T}L_{i,k+1})^{-1}\big\}. \end{aligned} \quad (12.45)$$

Taking $\frac{\partial \text{tr}(\Omega_{i,k+1|k+1})}{\partial G_{i,k+1}} = 0$, and then the estimator gain is obtained as in (12.32), which ends the proof.

Remark 12.4 The distinctive features of the newly designed time-varying event-triggered recursive state estimation approach are summarized as: (1) a new state estimation method involving the prediction and innovation updating steps has certain error correction ability; (2) the available information of switching topology, DPDs and RONs is adequately employed when designing the estimation algorithm; (3) the distributed state estimation scheme is designed without utilizing the state augmentation method, thus decreasing computational burdens; and (4) the state estimation method can be recursively executed, thereby suitable for online calculations.

Remark 12.5 It is essential to select an appropriate optimization index according to different system performance requirements. Note that in this chapter, the optimization index is to minimize $\text{tr}(\Omega_{i,k+1|k+1})$. In particular, the boundedness of $\Omega_{i,k+1|k+1}$ relies on some terms, including system matrices, the bound of the nonlinear disturbances, the noise covariances, and so on. As demonstrated in Li et al. (2018), the related sufficient conditions could be provided to guarantee the boundedness of this upper bound matrix.

12.2 Event-Based State Estimation with Fading Measurements and SCS

12.2.1 Problem Formulation

Consider the following nonlinear stochastic complex networks including N coupled nodes:

$$x_{i,k+1} = f(x_{i,k}) + \sum_{j=1}^{N}(\omega_{ij} + \zeta_{i,k}\Delta\omega_{ij})\Gamma x_{j,k} + B_{i,k}\varpi_{i,k}, \tag{12.46}$$

$$y_{i,k} = \Phi_{i,k}C_{i,k}x_{i,k} + \upsilon_{i,k}, \tag{12.47}$$

where $x_{i,k} \in \mathbb{R}^{n_1}$ is the state and $y_{i,k} \in \mathbb{R}^{n_2}$ is the measurement. The nonlinear function $f(x_{i,k})$ has continuous and bounded second order partial derivatives. $W = [\omega_{ij}]_{N\times N}$ denotes the coupling strength matrix and Γ stands for the inner coupling matrix. $|\Delta\omega_{ij}| \leq \delta_{ij}$ represents the uncertain coupling strength with $\delta_{ij} > 0$ being a scalar. $\varpi_{i,k}$ is the process noise with $\mathbb{E}\{\varpi_{i,k}\} = 0$ and $\text{Var}\{\varpi_{i,k}\} = \mathfrak{Q}_{i,k}$. $\upsilon_{i,k}$ is the measurement noise with $\mathbb{E}\{\upsilon_{i,k}\} = 0$ and $\text{Var}\{\upsilon_{i,k}\} = \mathfrak{R}_{i,k}$. $\zeta_{i,k} \in \mathbb{R}$ is a zero-mean noise sequence with $\mathbb{E}\{\zeta_{i,k}^2\} = 1$. $B_{i,k}$ and $C_{i,k}$ represent known matrices of appropriate dimensions. $\Phi_{i,k} = \text{diag}\{\phi_{i,k}^1, \phi_{i,k}^2, \ldots, \phi_{i,k}^{n_2}\}$ with $\phi_{i,k}^l \in [a_i, b_i]$ $(l = 1, 2, \ldots, n_2)$ satisfies $\mathbb{E}\{\phi_{i,k}^l\} = \bar{\phi}_{i,k}^l$ and $\text{Var}\{\phi_{i,k}^l\} = \hat{\phi}_{i,k}^l$ with $0 \leq a_i \leq b_i \leq 1$.

The ETM is used to reduce the transmission frequency. In particular, the event generator function is described by

$$\Omega(y_{i,k}, \pi_{i,k}) = (y_{i,k} - y_{i,s_t})^T(y_{i,k} - y_{i,s_t}) - \pi_{i,k}, \tag{12.48}$$

where y_{i,s_t} indicates the data information at the latest triggered instant, $\pi_{i,k} = \tau_{i,1}e^{-\tau_{i,2}k} + \tau_{i,3}$ with $\tau_{i,1}$, $\tau_{i,2}$ and $\tau_{i,3}$ being given constants. To be more specific, if $\Omega(y_{i,k}, \pi_{i,k}) > 0$, the current measurement is allowed to transmit to the estimator side. Further, the measurement received by the estimator side is denoted as

$$\tilde{y}_{i,k} = y_{i,s_t}, \quad k \in \{s_t, s_t + 1, \ldots, s_{t+1} - 1\}.$$

Remark 12.6 It is highlighted that the data information would be delivered at each time instant when employing the traditional time-triggered communication scheme, thereby needing high communication resources. To overcome such a difficulty, an ETM described in (12.48) is adopted to manage the transmission case of the data signals, in which three coefficients $\tau_{i,1}$, $\tau_{i,2}$ and $\tau_{i,3}$ are involved in the triggered threshold. In other words, the triggered threshold can be dynamically adjusted according to the practical requirement. Based on (12.48), it can be seen that the transmission frequency of data signals can be properly adjusted. Particularly, when

$\tau_{i,1} = \tau_{i,3} = 0$, the proposed ETM becomes the time-triggered mechanism. Different from the existing approaches, the event generator function given in (12.48) contains adjustment thresholds, thus supplying certain flexibility for addressing the state estimation problem.

Next, the following state estimator is designed as:

$$\hat{x}_{i,k+1|k} = f(\hat{x}_{i,k|k}) + \sum_{j=1}^{N} \omega_{ij} \Gamma \hat{x}_{j,k|k}, \tag{12.49}$$

$$\hat{x}_{i,k+1|k+1} = \hat{x}_{i,k+1|k} + G_{i,k+1} \left(\tilde{y}_{i,k+1} - \bar{\Phi}_{i,k+1} C_{i,k+1} \hat{x}_{i,k+1|k} \right), \tag{12.50}$$

where $\hat{x}_{i,k+1|k}$ is the prediction and $\hat{x}_{i,k|k}$ denotes the state estimation. $\bar{\Phi}_{i,k+1} = \mathbb{E}\left\{\Phi_{i,k+1}\right\}$ and $G_{i,k+1}$ is an estimator gain.

Define the prediction error as $e_{i,k+1|k} = x_{i,k+1} - \hat{x}_{i,k+1|k}$, the estimation error as $e_{i,k+1|k+1} = x_{i,k+1} - \hat{x}_{i,k+1|k+1}$, and their related error covariances as $P_{i,k+1|k} = \mathbb{E}\{e_{i,k+1|k} e_{i,k+1|k}^T\}$ and $P_{i,k+1|k+1} = \mathbb{E}\{e_{i,k+1|k+1} e_{i,k+1|k+1}^T\}$. This chapter is dedicated to constructing an estimator as in (12.49) and (12.50) such that (1) there is a matrix $\Omega_{i,k+1|k+1} > 0$ guaranteeing $P_{i,k+1|k+1} \leq \Omega_{i,k+1|k+1}$; (2) the $G_{i,k+1}$ is designed to guarantee the presence of minimal value of $\mathrm{tr}(\Omega_{i,k+1|k+1})$; and (3) the monotonic analysis between $\mathrm{tr}(\Omega_{i,k+1|k+1})$ and fading probability $\bar{\phi}_{i,k}^{l}$ is discussed and proved.

12.2.2 Design of Optimized State Estimation Scheme

Firstly, the prediction error is calculated as

$$e_{i,k+1|k} = f(x_{i,k}) - f(\hat{x}_{i,k|k}) + \sum_{j=1}^{N} \omega_{ij} \Gamma e_{j,k|k} + \sum_{j=1}^{N} \zeta_{i,k} \Delta\omega_{ij} \Gamma x_{j,k} + B_{i,k} \varpi_{i,k}.$$

By virtue of the Taylor formula, it is clear to see that

$$f(x_{i,k}) = f(\hat{x}_{i,k|k}) + A_{i,k} e_{i,k|k} + o(|e_{i,k|k}^2|),$$

where $A_{i,k} = \frac{\partial f(x_{i,k})}{x_{i,k}}|_{x_{i,k}=\hat{x}_{i,k|k}}$, and $o(|e_{i,k|k}^2|)$ denotes the high-order term which is approximately estimated by $L_{i,k} \mathfrak{M}_{i,k} F_{i,k} e_{i,k|k}$ as in Calafiore (2005) with $L_{i,k}$ and $F_{i,k}$ being given matrices, $\mathfrak{M}_{i,k}$ characterizes the linearization error with $\mathfrak{M}_{i,k} \mathfrak{M}_{i,k}^T \leq I$. Then, it is immediate to obtain that

$$e_{i,k+1|k} = (A_{i,k} + L_{i,k} \mathfrak{M}_{i,k} F_{i,k}) e_{i,k|k} + \sum_{j=1}^{N} \omega_{ij} \Gamma e_{j,k|k}$$

$$+\sum_{j=1}^{N}\zeta_{i,k}\Delta\omega_{ij}\Gamma x_{j,k}+B_{i,k}\varpi_{i,k}. \tag{12.51}$$

Additionally, one obtains

$$\begin{aligned}e_{i,k+1|k+1}=&\left(I-G_{i,k+1}\bar{\Phi}_{i,k+1}C_{i,k+1}\right)e_{i,k+1|k}-G_{i,k+1}\mathscr{Y}_{i,k+1}\\&-G_{i,k+1}\tilde{\Phi}_{i,k+1}C_{i,k+1}x_{i,k+1}-G_{i,k+1}\upsilon_{i,k+1},\end{aligned} \tag{12.52}$$

where $\mathscr{Y}_{i,k+1}=\tilde{y}_{i,k+1}-y_{i,k+1}$ and $\tilde{\Phi}_{i,k+1}=\Phi_{i,k+1}-\bar{\Phi}_{i,k+1}$.

Theorem 12.4 *The prediction error covariance and EEC are computed as*

$$\begin{aligned}P_{i,k+1|k}=&(A_{i,k}+L_{i,k}\mathfrak{M}_{i,k}F_{i,k})P_{i,k|k}(A_{i,k}+L_{i,k}\mathfrak{M}_{i,k}F_{i,k})^T\\&+\sum_{j=1}^{N}\sum_{d=1}^{N}\omega_{ij}\omega_{id}\Gamma\mathbb{E}\left\{e_{j,k|k}e_{d,k|k}^T\right\}\Gamma^T\\&+\sum_{j=1}^{N}\sum_{d=1}^{N}\Delta\omega_{ij}\Delta\omega_{id}\Gamma\mathbb{E}\left\{x_{j,k}x_{d,k}^T\right\}\Gamma^T\\&+B_{i,k}\mathfrak{Q}_{i,k}B_{i,k}^T+\mathfrak{G}_{i,k}+\mathfrak{G}_{i,k}^T\end{aligned} \tag{12.53}$$

and

$$\begin{aligned}P_{i,k+1|k+1}=&(I-G_{i,k+1}\bar{\Phi}_{i,k+1}C_{i,k+1})P_{i,k+1|k}(I-G_{i,k+1}\bar{\Phi}_{i,k+1}C_{i,k+1})^T\\&+G_{i,k+1}\mathbb{E}\left\{\mathscr{Y}_{i,k+1}\mathscr{Y}_{i,k+1}^T\right\}G_{i,k+1}^T-\mathfrak{E}_{i,k+1}-\mathfrak{E}_{i,k+1}^T+\mathfrak{D}_{i,k+1}+\mathfrak{D}_{i,k+1}^T\\&+G_{i,k+1}\mathbb{E}\left\{\tilde{\Phi}_{i,k+1}C_{i,k+1}x_{i,k+1}x_{i,k+1}^TC_{i,k+1}^T\tilde{\Phi}_{i,k+1}^T\right\}G_{i,k+1}^T\\&+\mathfrak{F}_{i,k+1}+\mathfrak{F}_{i,k+1}^T+G_{i,k+1}\mathfrak{R}_{i,k+1}G_{i,k+1}^T,\end{aligned} \tag{12.54}$$

where

$$\begin{aligned}\mathfrak{G}_{i,k}=&(A_{i,k}+L_{i,k}\mathfrak{M}_{i,k}F_{i,k})\sum_{j=1}^{N}\omega_{ij}\mathbb{E}\left\{e_{i,k|k}e_{j,k|k}^T\right\}\Gamma^T,\\\mathfrak{E}_{i,k+1}=&(I-G_{i,k+1}\bar{\Phi}_{i,k+1}C_{i,k+1})\mathbb{E}\left\{e_{i,k+1|k}\mathscr{Y}_{i,k+1}^T\right\}G_{i,k+1}^T,\\\mathfrak{D}_{i,k+1}=&G_{i,k+1}\mathbb{E}\left\{\mathscr{Y}_{i,k+1}x_{i,k+1}^TC_{i,k+1}^T\tilde{\Phi}_{i,k+1}^T\right\}G_{i,k+1}^T,\\\mathfrak{F}_{i,k+1}=&G_{i,k+1}\mathbb{E}\left\{\mathscr{Y}_{i,k+1}\upsilon_{i,k+1}^T\right\}G_{i,k+1}^T.\end{aligned}$$

Proof According to equations (12.51) and (12.52), the prediction error covariance and EEC can be calculated respectively. That is,

$$
\begin{aligned}
P_{i,k+1|k} =& (A_{i,k} + L_{i,k}\mathfrak{M}_{i,k}F_{i,k})P_{i,k|k}(A_{i,k} + L_{i,k}\mathfrak{M}_{i,k}F_{i,k})^T + \sum_{j=1}^{N}\sum_{d=1}^{N}\omega_{ij}\omega_{id} \\
& \times \Gamma\mathbb{E}\left\{e_{j,k|k}e_{d,k|k}^T\right\}\Gamma^T + \sum_{j=1}^{N}\sum_{d=1}^{N}\Delta\omega_{ij}\Delta\omega_{id}\Gamma\mathbb{E}\left\{x_{j,k}x_{d,k}^T\right\}\Gamma^T \\
& + B_{i,k}\mathfrak{Q}_{i,k}B_{i,k}^T + \mathfrak{G}_{i,k} + \mathfrak{G}_{i,k}^T + \mathfrak{G}_{i,k}^1 + \left(\mathfrak{G}_{i,k}^1\right)^T + \mathfrak{G}_{i,k}^2 + \left(\mathfrak{G}_{i,k}^2\right)^T \\
& + \mathfrak{G}_{i,k}^3 + \left(\mathfrak{G}_{i,k}^3\right)^T + \mathfrak{G}_{i,k}^4 + \left(\mathfrak{G}_{i,k}^4\right)^T + \mathfrak{G}_{i,k}^5 + \left(\mathfrak{G}_{i,k}^5\right)^T
\end{aligned}
$$

and

$$
\begin{aligned}
P_{i,k+1|k+1} =& (I - G_{i,k+1}\bar{\Phi}_{i,k+1}C_{i,k+1})P_{i,k+1|k}(I - G_{i,k+1}\bar{\Phi}_{i,k+1}C_{i,k+1})^T \\
& + G_{i,k+1}\mathbb{E}\left\{\mathscr{Y}_{i,k+1}\mathscr{Y}_{i,k+1}^T\right\}G_{i,k+1}^T - \mathfrak{E}_{i,k+1} - \mathfrak{E}_{i,k+1}^T + \mathfrak{D}_{i,k+1} \\
& + \mathfrak{D}_{i,k+1}^T + G_{i,k+1}\mathbb{E}\left\{\tilde{\Phi}_{i,k+1}C_{i,k+1}x_{i,k+1}x_{i,k+1}^T C_{i,k+1}^T\tilde{\Phi}_{i,k+1}^T\right\}G_{i,k+1}^T \\
& + \mathfrak{F}_{i,k+1} + \mathfrak{F}_{i,k+1}^T + G_{i,k+1}\mathfrak{R}_{i,k+1}G_{i,k+1}^T - \mathfrak{F}_{i,k+1}^1 - \left(\mathfrak{F}_{i,k+1}^1\right)^T \\
& - \mathfrak{F}_{i,k+1}^2 - \left(\mathfrak{F}_{i,k+1}^2\right)^T + \mathfrak{F}_{i,k+1}^3 + \left(\mathfrak{F}_{i,k+1}^3\right)^T,
\end{aligned}
$$

where

$$
\begin{aligned}
\mathfrak{G}_{i,k}^1 =& \sum_{j=1}^{N}(A_{i,k} + L_{i,k}\mathfrak{M}_{i,k}F_{i,k})\Delta\omega_{ij}\mathbb{E}\left\{\zeta_{i,k}e_{i,k|k}x_{j,k}^T\right\}\Gamma^T, \\
\mathfrak{G}_{i,k}^2 =& (A_{i,k} + L_{i,k}\mathfrak{M}_{i,k}F_{i,k})\mathbb{E}\left\{e_{i,k|k}\varpi_{i,k}^T\right\}B_{i,k}^T, \\
\mathfrak{G}_{i,k}^3 =& \sum_{j=1}^{N}\sum_{d=1}^{N}\omega_{ij}\Delta\omega_{id}\Gamma\mathbb{E}\left\{\zeta_{i,k}e_{j,k|k}x_{d,k}^T\right\}\Gamma^T, \\
\mathfrak{G}_{i,k}^4 =& \sum_{j=1}^{N}\omega_{ij}\Gamma\mathbb{E}\left\{e_{j,k|k}\varpi_{i,k}^T\right\}B_{i,k}^T, \\
\mathfrak{G}_{i,k}^5 =& \sum_{j=1}^{N}\Delta\omega_{ij}\Gamma\mathbb{E}\left\{\zeta_{i,k}x_{j,k}\varpi_{i,k}^T\right\}B_{i,k}^T, \\
\mathfrak{F}_{i,k+1}^1 =& \left(I - G_{i,k+1}\bar{\Phi}_{i,k+1}C_{i,k+1}\right)\mathbb{E}\left\{e_{i,k+1|k}\upsilon_{i,k+1}^T\right\}G_{i,k+1}^T, \\
\mathfrak{F}_{i,k+1}^2 =& \left(I - G_{i,k+1}\bar{\Phi}_{i,k+1}C_{i,k+1}\right)\mathbb{E}\left\{e_{i,k+1|k}x_{i,k+1}^T C_{i,k+1}^T\tilde{\Phi}_{i,k+1}^T\right\}G_{i,k+1}^T, \\
\mathfrak{F}_{i,k+1}^3 =& G_{i,k+1}\mathbb{E}\left\{\upsilon_{i,k+1}x_{i,k+1}^T C_{i,k+1}^T\tilde{\Phi}_{i,k+1}^T\right\}G_{i,k+1}^T.
\end{aligned}
$$

Due to the fact that $\mathfrak{G}_{i,k}^h = 0$ and $\mathfrak{F}_{i,k}^t = 0$ $(h = 1, 2, 3, 4, 5; t = 1, 2, 3)$, the assertions in the Theorem 12.4 can be obtained and the proof is complete.

Remark 12.7 Until now, the explicit expression forms of the prediction error covariance and EEC have been respectively derived according to the related definitions. Nevertheless, the presence of some factors inevitably leads to the difficulties that the accurate prediction error covariance and the EEC cannot be obtained. Accordingly, an alternative method is to find an upper bound of (12.54) to realize the estimation algorithm design.

Theorem 12.5 *Let $\mu_t > 0$ $(t = 1, 2, \ldots, 6)$ and $\varepsilon_{i,k} > 0$ denote scalars. If the following matrix equations:*

$$\begin{aligned}\Omega_{i,k+1|k} =&(1+\mu_1)\left[A_{i,k}\left(\Omega_{i,k|k}^{-1}-\varepsilon_{i,k}F_{i,k}^TF_{i,k}\right)^{-1}A_{i,k}^T+\varepsilon_{i,k}^{-1}L_{i,k}L_{i,k}^T\right]+B_{i,k}\mathfrak{Q}_{i,k}B_{i,k}^T\\&+(1+\mu_1^{-1})\check{\omega}_{id}\sum_{j=1}^{N}\omega_{ij}\Gamma\Omega_{j,k|k}\Gamma^T+\check{\delta}_{id}\sum_{j=1}^{N}\delta_{ij}\Gamma\ddot{\Theta}_{j,k}\Gamma^T\end{aligned} \tag{12.55}$$

and

$$\begin{aligned}\Omega_{i,k+1|k+1} =&(1+\mu_3)(I-G_{i,k+1}\bar{\Phi}_{i,k+1}C_{i,k+1})\Omega_{i,k+1|k}(I-G_{i,k+1}\bar{\Phi}_{i,k+1}C_{i,k+1})^T\\&+(1+\mu_3^{-1}+\mu_4+\mu_5)\pi_{i,k+1}G_{i,k+1}G_{i,k+1}^T+(1+\mu_5^{-1})G_{i,k+1}\\&\times\mathfrak{R}_{i,k+1}G_{i,k+1}^T+(1+\mu_4^{-1})G_{i,k+1}\Big[\Omega_{i,k+1}\circ(C_{i,k+1}\ddot{\Omega}_{i,k+1}\\&\times C_{i,k+1}^T)\Big]G_{i,k+1}^T\end{aligned} \tag{12.56}$$

under the initial condition $\Omega_{i,0|0} = P_{i,0|0} > 0$ and $\varepsilon_{i,k}^{-1}I - F_{i,k}\Omega_{i,k|k}F_{i,k}^T > 0$ have solutions $\Omega_{i,k+1|k} > 0$ and $\Omega_{i,k+1|k+1} > 0$, then one has $P_{i,k+1|k+1} \leq \Omega_{i,k+1|k+1}$. Besides, the following estimator gain:

$$\begin{aligned}G_{i,k+1} =&(1+\mu_3)\Omega_{i,k+1|k}C_{i,k+1}^T\bar{\Phi}_{i,k+1}^T\Big\{(1+\mu_3)\bar{\Phi}_{i,k+1}C_{i,k+1}\Omega_{i,k+1|k}C_{i,k+1}^T\bar{\Phi}_{i,k+1}^T\\&+(1+\mu_3^{-1}+\mu_4+\mu_5)\pi_{i,k+1}I+(1+\mu_4^{-1})\Omega_{i,k+1}\circ(C_{i,k+1}\ddot{\Omega}_{i,k+1}C_{i,k+1}^T)\\&+(1+\mu_5^{-1})\mathfrak{R}_{i,k+1}\Big\}^{-1}\end{aligned} \tag{12.57}$$

can minimize $\mathrm{tr}(\Omega_{i,k+1|k+1})$, *where*

$$\begin{aligned}\check{\delta}_{id} =&\sum_{d=1}^{N}\delta_{id},\quad \check{\omega}_{id}=\sum_{d=1}^{N}\omega_{id},\\\ddot{\Theta}_{j,k} =&(1+\mu_2)\Omega_{j,k|k}+(1+\mu_2^{-1})\hat{x}_{j,k|k}\hat{x}_{j,k|k}^T,\\\ddot{\Omega}_{i,k+1} =&(1+\mu_6)\Omega_{i,k+1|k}+(1+\mu_6^{-1})\hat{x}_{i,k+1|k}\hat{x}_{i,k+1|k}^T,\\\Omega_{i,k+1} =&\mathrm{diag}\left\{\hat{\phi}_{i,k+1}^1,\hat{\phi}_{i,k+1}^2,\ldots,\hat{\phi}_{i,k+1}^{n_2}\right\}.\end{aligned} \tag{12.58}$$

Proof Adopting Lemma 2.2, we have

$$\begin{aligned}
&\sum_{j=1}^{N}\sum_{d=1}^{N}\omega_{ij}\omega_{id}\Gamma\mathbb{E}\left\{e_{j,k|k}e_{d,k|k}^{T}\right\}\Gamma^{T}\\
\leq&\frac{1}{2}\sum_{j=1}^{N}\sum_{d=1}^{N}\omega_{ij}\omega_{id}\Gamma\mathbb{E}\left\{e_{j,k|k}e_{j,k|k}^{T}+e_{d,k|k}e_{d,k|k}^{T}\right\}\Gamma^{T}\\
=&\check{\omega}_{id}\sum_{j=1}^{N}\omega_{ij}\Gamma P_{j,k|k}\Gamma^{T}
\end{aligned}\tag{12.59}$$

and

$$\begin{aligned}
&\sum_{j=1}^{N}\sum_{d=1}^{N}\Delta\omega_{ij}\Delta\omega_{id}\Gamma\mathbb{E}\left\{x_{j,k}x_{d,k}^{T}\right\}\Gamma^{T}\\
\leq&\frac{1}{2}\sum_{j=1}^{N}\sum_{d=1}^{N}\Delta\omega_{ij}\Delta\omega_{id}\Gamma\mathbb{E}\left\{x_{j,k}x_{j,k}^{T}+x_{d,k}x_{d,k}^{T}\right\}\Gamma^{T}\\
\leq&\check{\delta}_{id}\sum_{j=1}^{N}\delta_{ij}\Gamma\mathbb{E}\left\{x_{j,k}x_{j,k}^{T}\right\}\Gamma^{T},
\end{aligned}\tag{12.60}$$

where $\check{\omega}_{id}$ and $\check{\delta}_{id}$ are given in (12.58).

Applying the same method with (12.60), one has

$$\begin{aligned}
\mathfrak{G}_{i,k}+\mathfrak{G}_{i,k}^{T}\leq&\mu_{1}(A_{i,k}+L_{i,k}\mathfrak{M}_{i,k}F_{i,k})P_{i,k|k}(A_{i,k}+L_{i,k}\mathfrak{M}_{i,k}F_{i,k})^{T}\\
&+\mu_{1}^{-1}\sum_{j=1}^{N}\sum_{d=1}^{N}\omega_{ij}\omega_{id}\Gamma\mathbb{E}\left\{e_{j,k|k}e_{d,k|k}^{T}\right\}\Gamma^{T}
\end{aligned}\tag{12.61}$$

and

$$\mathbb{E}\left\{x_{j,k}x_{j,k}^{T}\right\}\leq(1+\mu_{2})P_{j,k|k}+(1+\mu_{2}^{-1})\hat{x}_{j,k|k}\hat{x}_{j,k|k}^{T}:=\Theta_{j,k}\tag{12.62}$$

with μ_1 and μ_2 being positive constants. Then, we get

$$\sum_{j=1}^{N}\sum_{d=1}^{N}\Delta\omega_{ij}\Delta\omega_{id}\Gamma\mathbb{E}\left\{x_{j,k}x_{d,k}^{T}\right\}\Gamma^{T}\leq\check{\delta}_{id}\sum_{j=1}^{N}\delta_{ij}\Gamma\Theta_{j,k}\Gamma^{T}.\tag{12.63}$$

Considering (12.59)–(12.63), we have

$$
\begin{aligned}
P_{i,k+1|k} \leq &(1+\mu_1)(A_{i,k}+L_{i,k}\mathfrak{M}_{i,k}F_{i,k})P_{i,k|k}(A_{i,k}+L_{i,k}\mathfrak{M}_{i,k}F_{i,k})^T \\
&+B_{i,k}\mathfrak{Q}_{i,k}B_{i,k}^T+(1+\mu_1^{-1})\check{\omega}_{id}\sum_{j=1}^{N}\omega_{ij}\Gamma P_{j,k|k}\Gamma^T \\
&+\check{\delta}_{id}\sum_{j=1}^{N}\delta_{ij}\Gamma\Theta_{j,k}\Gamma^T. \qquad (12.64)
\end{aligned}
$$

Recalling Lemma 2.2, it can be inferred that

$$
\begin{aligned}
-\mathfrak{E}_{i,k+1}-\mathfrak{E}_{i,k+1}^T \leq &\mu_3(I-G_{i,k+1}\bar{\Phi}_{i,k+1}C_{i,k+1})P_{i,k+1|k}(I-G_{i,k+1}\bar{\Phi}_{i,k+1}C_{i,k+1})^T \\
&+\mu_3^{-1}G_{i,k+1}\mathbb{E}\left\{\mathscr{Y}_{i,k+1}\mathscr{Y}_{i,k+1}^T\right\}G_{i,k+1}^T, \qquad (12.65) \\
\mathfrak{D}_{i,k+1}+\mathfrak{D}_{i,k+1}^T \leq &\mu_4^{-1}G_{i,k+1}\mathbb{E}\left\{\tilde{\Phi}_{i,k+1}C_{i,k+1}x_{i,k+1}x_{i,k+1}^TC_{i,k+1}^T\tilde{\Phi}_{i,k+1}^T\right\}G_{i,k+1}^T \\
&+\mu_4G_{i,k+1}\mathbb{E}\left\{\mathscr{Y}_{i,k+1}\mathscr{Y}_{i,k+1}^T\right\}G_{i,k+1}^T \qquad (12.66)
\end{aligned}
$$

and

$$
\begin{aligned}
\mathfrak{F}_{i,k+1}+\mathfrak{F}_{i,k+1}^T \leq &\mu_5G_{i,k+1}\mathbb{E}\left\{\mathscr{Y}_{i,k+1}\mathscr{Y}_{i,k+1}^T\right\}G_{i,k+1}^T \\
&+\mu_5^{-1}G_{i,k+1}\mathfrak{R}_{i,k+1}G_{i,k+1}^T \qquad (12.67)
\end{aligned}
$$

hold, where μ_3, μ_4 and μ_5 are positive constants. Substituting the (12.65)–(12.67) into (12.54) leads to

$$
\begin{aligned}
P_{i,k+1|k+1} \leq &(1+\mu_3)(I-G_{i,k+1}\bar{\Phi}_{i,k+1}C_{i,k+1})P_{i,k+1|k}(I-G_{i,k+1}\bar{\Phi}_{i,k+1}C_{i,k+1})^T \\
&+(1+\mu_3^{-1}+\mu_4+\mu_5)G_{i,k+1}\mathbb{E}\left\{\mathscr{Y}_{i,k+1}\mathscr{Y}_{i,k+1}^T\right\}G_{i,k+1}^T \\
&+(1+\mu_4^{-1})G_{i,k+1}\mathbb{E}\left\{\tilde{\Phi}_{i,k+1}C_{i,k+1}x_{i,k+1}x_{i,k+1}^TC_{i,k+1}^T\tilde{\Phi}_{i,k+1}^T\right\}G_{i,k+1}^T \\
&+(1+\mu_5^{-1})G_{i,k+1}\mathfrak{R}_{i,k+1}G_{i,k+1}^T. \qquad (12.68)
\end{aligned}
$$

It follows from (12.48) that

$$
\mathbb{E}\left\{\mathscr{Y}_{i,k+1}\mathscr{Y}_{i,k+1}^T\right\} \leq \pi_{i,k+1}I. \qquad (12.69)
$$

Similarly, one has

$$
\begin{aligned}
\mathbb{E}\left\{x_{i,k+1}x_{i,k+1}^T\right\} \leq &(1+\mu_6)P_{i,k+1|k}+(1+\mu_6^{-1})\hat{x}_{i,k+1|k}\hat{x}_{i,k+1|k}^T \\
:= &\check{\Omega}_{i,k+1} \qquad (12.70)
\end{aligned}
$$

with $\mu_6 > 0$ denoting a scalar. According to (12.70) and Lemma 2.4, one gets

$$\begin{aligned}&\mathbb{E}\left\{\tilde{\Phi}_{i,k+1}C_{i,k+1}x_{i,k+1}x_{i,k+1}^T C_{i,k+1}^T \tilde{\Phi}_{i,k+1}^T\right\}\\ \leq&\Omega_{i,k+1}\circ(C_{i,k+1}\check{\Omega}_{i,k+1}C_{i,k+1}^T),\end{aligned} \tag{12.71}$$

where $\Omega_{i,k+1}$ is defined in (12.58). Noting (12.69) and (12.71), we have

$$\begin{aligned}P_{i,k+1|k+1}\leq&(1+\mu_3)(I-G_{i,k+1}\bar{\Phi}_{i,k+1}C_{i,k+1})P_{i,k+1|k}(I-G_{i,k+1}\bar{\Phi}_{i,k+1}C_{i,k+1})^T\\&+(1+\mu_3^{-1}+\mu_4+\mu_5)\pi_{i,k+1}G_{i,k+1}G_{i,k+1}^T+(1+\mu_5^{-1})G_{i,k+1}\\&\times\Re_{i,k+1}G_{i,k+1}^T+(1+\mu_4^{-1})G_{i,k+1}\Big[\Omega_{i,k+1}\circ(C_{i,k+1}\check{\Omega}_{i,k+1}\\&\times C_{i,k+1}^T)\Big]G_{i,k+1}^T.\end{aligned} \tag{12.72}$$

Together with (12.55), (12.56), (12.64) and (12.72) and employing the mathematical induction method, one has

$$P_{i,k+1|k+1}\leq\Omega_{i,k+1|k+1}.$$

Finally, we endeavor to derive the estimator gain $G_{i,k+1}$. Setting

$$\begin{aligned}\Xi_{i,k+1}=&(1+\mu_3)\bar{\Phi}_{i,k+1}C_{i,k+1}\Omega_{i,k+1|k}C_{i,k+1}^T\bar{\Phi}_{i,k+1}^T\\&+(1+\mu_5^{-1})\Re_{i,k+1}+(1+\mu_3^{-1}+\mu_4+\mu_5)\pi_{i,k+1}I\\&+(1+\mu_4^{-1})\Big[\Omega_{i,k+1}\circ(C_{i,k+1}\ddot{\Omega}_{i,k+1}C_{i,k+1}^T)\Big]\end{aligned}$$

and completing the square concerning on (12.56), then one has

$$\begin{aligned}\Omega_{i,k+1|k+1}=&\Big[G_{i,k+1}-(1+\mu_3)\Omega_{i,k+1|k}C_{i,k+1}^T\bar{\Phi}_{i,k+1}^T\Xi_{i,k+1}^{-1}\Big]\Xi_{i,k+1}\Big[G_{i,k+1}\\&-(1+\mu_3)\Omega_{i,k+1|k}C_{i,k+1}^T\bar{\Phi}_{i,k+1}^T\Xi_{i,k+1}^{-1}\Big]^T+(1+\mu_3)\Omega_{i,k+1|k}\\&-(1+\mu_3)^2\Omega_{i,k+1|k}C_{i,k+1}^T\bar{\Phi}_{i,k+1}^T\Xi_{i,k+1}^{-1}\bar{\Phi}_{i,k+1}C_{i,k+1}\Omega_{i,k+1|k}.\end{aligned}$$

It is clear that $\mathrm{tr}(\Omega_{i,k+1|k+1})$ can be optimized via the designed estimator gain $G_{i,k+1}$ and the proof is complete.

Remark 12.8 It is important to highlight that the constraint $\varepsilon_{i,k}^{-1}I-F_{i,k}\Omega_{i,k|k}F_{i,k}^T>0$ presented in Theorem 12.5 includes the positive coefficient $\varepsilon_{i,k}$. In the process of the algorithm implementation, $\varepsilon_{i,k}$ can be properly selected to guarantee that $\varepsilon_{i,k}^{-1}I-F_{i,k}\Omega_{i,k|k}F_{i,k}^T>0$ holds, which can be testified in the simulation part.

Based on the above analysis, the following EBSE algorithm is given to facilitate algorithm implementation.

EBSE Algorithm :
Step 1 : Set $k = 0$ and select other initial parameters.
Step 2 : Compute the prediction $\hat{x}_{i,k+1|k}$ via (12.49).
Step 3 : Calculate $\Omega_{i,k+1|k}$ by (12.55).
Step 4 : Solve the estimator gain $G_{i,k+1}$ based on (12.57).
Step 5 : Compute the state estimation $\hat{x}_{i,k+1|k+1}$ by (12.50).
Step 6 : Calculate $\Omega_{i,k+1|k+1}$ via (12.56).
Step 7 : Let $k = k + 1$. Go to *Step* 2.

12.2.3 Monotonicity Analysis

In this section, the influence of fading probability onto the estimation performance is examined. For the fading probability $\bar{\Phi}_{i,k+1} = \text{diag}\{\bar{\phi}^1_{i,k+1}, \bar{\phi}^2_{i,k+1}, \ldots, \bar{\phi}^{n_2}_{i,k+1}\}$, we assume that $\bar{\phi}^l_{i,k+1} = \bar{\phi}_{i,k+1}$ $(l = 1, 2, \ldots, n_2)$.

Theorem 12.6 *If fading probability* $\bar{\phi}_{i,k+1}$ *increases, it concludes that* $\text{tr}(\Omega_{i,k+1|k+1})$ *is non-increasing.*

Proof Taking the partial derivative of $\text{tr}(\Omega_{i,k+1|k+1})$ regarding $\bar{\phi}_{i,k+1}$, one has

$$
\begin{aligned}
&\frac{\text{dtr}(\Omega_{i,k+1|k+1})}{\text{d}\bar{\phi}_{i,k+1}} \\
&= \frac{\text{dtr}}{\text{d}\bar{\phi}_{i,k+1}}\Big((1+\mu_3)\Omega_{i,k+1|k} - (1+\mu_3)^2\bar{\phi}^2_{i,k+1}\Omega_{i,k+1|k}C^T_{i,k+1}\Xi^{-1}_{i,k+1}C_{i,k+1}\Omega_{i,k+1|k}\Big) \\
&= \text{tr}\Big(-2(1+\mu_3)^2\bar{\phi}_{i,k+1}\Omega_{i,k+1|k}C^T_{i,k+1}\Xi^{-1}_{i,k+1}C_{i,k+1}\Omega_{i,k+1|k} \\
&\quad + (1+\mu_3)^2\bar{\phi}^2_{i,k+1}\Omega_{i,k+1|k}C^T_{i,k+1}\Xi^{-1}_{i,k+1}[2(1+\mu_3)\bar{\phi}_{i,k+1}C_{i,k+1}\Omega_{i,k+1|k}C^T_{i,k+1}] \\
&\quad \times \Xi^{-1}_{i,k+1}C_{i,k+1}\Omega_{i,k+1|k}\Big) \\
&\le \text{tr}\Big(-2(1+\mu_3)^2\bar{\phi}_{i,k+1}\Omega_{i,k+1|k}C^T_{i,k+1}\Xi^{-1}_{i,k+1}C_{i,k+1}\Omega_{i,k+1|k} \\
&\quad + (1+\mu_3)^2\bar{\phi}_{i,k+1}\Omega_{i,k+1|k}C^T_{i,k+1}\Xi^{-1}_{i,k+1}[2(1+\mu_3)\bar{\phi}^2_{i,k+1}C_{i,k+1}\Omega_{i,k+1|k}C^T_{i,k+1} \\
&\quad + 2(1+\mu_5^{-1})\Re_{i,k+1} + 2(1+\mu_3^{-1}+\mu_4+\mu_5)\pi_{i,k+1}I \\
&\quad + 2(1+\mu_4^{-1})\Omega_{i,k+1}\circ(C_{i,k+1}\ddot{\Omega}_{i,k+1}C^T_{i,k+1})]\Xi^{-1}_{i,k+1}C_{i,k+1}\Omega_{i,k+1|k}\Big) \\
&= \text{tr}\Big(-2(1+\mu_3)^2\bar{\phi}_{i,k+1}\Omega_{i,k+1|k}C^T_{i,k+1}\Xi^{-1}_{i,k+1}C_{i,k+1}\Omega_{i,k+1|k} \\
&\quad + 2(1+\mu_3)^2\bar{\phi}_{i,k+1}\Omega_{i,k+1|k}C^T_{i,k+1}\Xi^{-1}_{i,k+1}C_{i,k+1}\Omega_{i,k+1|k}\Big) \\
&= 0,
\end{aligned}
$$

which completes the proof of this theorem.

12.3 Two Illustrative Examples

In this section, simulations and comparisons are provided to illustrate the effectiveness of developed state estimation methods.

Example 1 Consider the time-varying uncertain complex networks with switching topology, RONs and DPDs as in (12.1)–(12.3), where the related parameters are given as follows:

$$\begin{aligned}
&\mathbf{A}_{1,k}=\begin{bmatrix}0.95 & 1.24\\ -0.5 & 0.92\sin(5k)\end{bmatrix},\ \mathbf{A}_{2,k}=\begin{bmatrix}-0.65\sin(3k) & -0.8\\ 0.3 & 0.4\end{bmatrix},\\
&\mathbf{A}_{3,k}=\begin{bmatrix}0.38 & -0.7\\ -0.5 & 0.9\end{bmatrix},\ \mathbf{B}_{1,k}=\begin{bmatrix}0.1 & -0.6\end{bmatrix}^T,\\
&\mathbf{B}_{2,k}=\begin{bmatrix}0.3 & -0.9\end{bmatrix}^T,\ \mathbf{B}_{3,k}=\begin{bmatrix}0.1 & -0.6\end{bmatrix}^T,\\
&\mathbf{C}_{1,k}=\begin{bmatrix}0.5 & 1\end{bmatrix},\ \mathbf{C}_{2,k}=\begin{bmatrix}2.4 & 4.5\end{bmatrix},\ \mathbf{C}_{3,k}=\begin{bmatrix}1 & 2\end{bmatrix}.
\end{aligned}$$

We select $\bar{\Gamma}=\text{diag}\{0.08, 0.08\}$. For the coupling strength ϖ_{ij}, select $\varpi_{ij}=-4$ if $i=j$, otherwise, let $\varpi_{ij}=-2$. Others corresponding parameters are set as $\mathbf{M}_{i,k}=H_{i,k}=\begin{bmatrix}0.4 & 0.3\end{bmatrix}^T$, $\mathbf{N}_{i,k}=L_{i,k}=\begin{bmatrix}0.4 & 0.1\end{bmatrix}$ and $F_{i,k}=\mathscr{F}_{i,k}=\sin(3k)$ $(i=1,2,3)$.

The following nonlinear functions are:

$$\mathbf{h}_i(\mathbf{x}_{i,k})=G_i\mathbf{x}_{i,k}+\begin{bmatrix}0.1\sin(\mathbf{x}_{i,k}^1+\mathbf{x}_{i,k}^2)\\ 0.1\mathbf{x}_{i,k}^2\sin(k)\end{bmatrix},$$

where $\mathbf{x}_{i,k}=\begin{bmatrix}\mathbf{x}_{i,k}^1 & \mathbf{x}_{i,k}^2\end{bmatrix}^T$ is the system state of the i-th node and

$$G_1=\begin{bmatrix}0.72 & 0.5\\ 0.48 & 0.5\end{bmatrix},\ G_2=\begin{bmatrix}0.6 & 0.6\\ 0.2 & 0.8\end{bmatrix},\ G_3=\begin{bmatrix}0.2 & 0.3\\ 0.2 & 0.6\end{bmatrix}.$$

In this simulation, let the initial values be $\bar{\mathbf{x}}_{1,0}=\begin{bmatrix}0.5 & 1\end{bmatrix}^T$, $\bar{\mathbf{x}}_{2,0}=\begin{bmatrix}-1 & 0.25\end{bmatrix}^T$, $\bar{\mathbf{x}}_{3,0}=\begin{bmatrix}-0.5 & -0.75\end{bmatrix}^T$, $\hat{\mathbf{x}}_{i,0|0}=\bar{\mathbf{x}}_{i,0}-\begin{bmatrix}1 & 1\end{bmatrix}^T$ with $\hat{\mathbf{x}}_{i,0|0}$ is the estimate of $\mathbf{x}_{i,0}$, $\bar{\alpha}_{i,k}=0.5$, $\bar{\lambda}_{ij,k}=0.95$ $(i=1,2,3)$. Besides, assume that the initial conditions of covariances are $\bar{X}_{1,0}=2I_2$, $\bar{X}_{2,0}=2\bar{X}_{1,0}$, $\bar{X}_{3,0}=3\bar{X}_{1,0}$, $\Omega_{1,0|0}=2I_2$, $\Omega_{2,0|0}=2\Omega_{1,0|0}$ and $\Omega_{3,0|0}=3\Omega_{1,0|0}$. The covariance matrices of noises are $\mathscr{W}_{1,k}=0.1$, $\mathscr{W}_{2,k}=0.15$, $\mathscr{W}_{3,k}=0.2$, $\mathscr{V}_{1,k}=0.2$, $\mathscr{V}_{2,k}=0.1$ and $\mathscr{V}_{3,k}=0.15$. The Lipschitz constant is $l=3$. Select

$$\begin{aligned}
\theta_{i,k+1}=&(1.2\lambda_{\max}(L_{i,k+1}\bar{C}_{i,k+1}\Omega_{i,k+1|k}\bar{C}_{i,k+1}^TL_{i,k+1}^T)+1)^{-1},\\
\mu_{i,k+1}=&(1.2\lambda_{\max}(L_{i,k+1}L_{i,k+1}^T)+1)^{-1},\\
\tau_{i,k+1}=&(1.2\lambda_{\max}(L_{i,k+1}\mathscr{V}_{i,k+1}L_{i,k+1}^T)+1)^{-1}.
\end{aligned}$$

Besides, other parameters are given as $\varepsilon_i = 0.1$ $(i = 1, 2, 3, 4)$, $\eta_i = 0.1$ $(i = 1, 6, 9)$ and $\eta_j = 1$ $(j = 2, 3, 4, 5, 7, 8)$.

To exhibit the merits of the developed state estimation scheme, four cases are taken into account. Firstly, different adjustable thresholds are considered: Case I: $\delta_{1,k} = \delta_{2,k} = \delta_{3,k} = 0.9$; Case II: $\delta_{1,k} = \delta_{2,k} = \delta_{3,k} = 8$. According to Theorem 12.3, the upper bound covariance matrix is calculated and the estimator gain matrix is calculated. The simulation results are presented in Figs. 12.2 and 12.3, which can be seen that the estimation accuracy becomes better along with the decreasing adjustment threshold as in Case I. The reason is due to that more measurement information can be transmitted to the remote estimator side under Case I, thus enhancing the algorithm performance.

Similarly, we consider different occurrence probabilities of DPDs, that is, Case III: the probabilities of DPDs are set as $\bar{\beta}_{1,k} = \bar{\beta}_{2,k} = \bar{\beta}_{3,k} = 0.95$; Case IV: the probabilities of DPDs are set as $\bar{\beta}_{1,k} = \bar{\beta}_{2,k} = \bar{\beta}_{3,k} = 0.05$. Again, the simulation results can be given in Figs. 12.4 and 12.5. For Case IV, i.e., more packets are lost, it is seen that the system state can be roughly tracked. The reason is that the effects of DPDs exist, which should be handled during the design of state estimation scheme. Moreover, it is highlighted that the estimation accuracy in Case IV remains worse compared with Case III as the occurrence probabilities of DPDs become smaller.

In Figs. 12.6 and 12.7, the log(MSE) in Case I and Case II are given with MSE being the mean-square error of the state estimation. Compared with Case II, it is observed that the chances of released information could be increased when the triggered thresholds decrease as in Case I, thus enhancing the estimation accuracy of the designed algorithm. As a result, this fact has been clearly given in Figs. 12.6

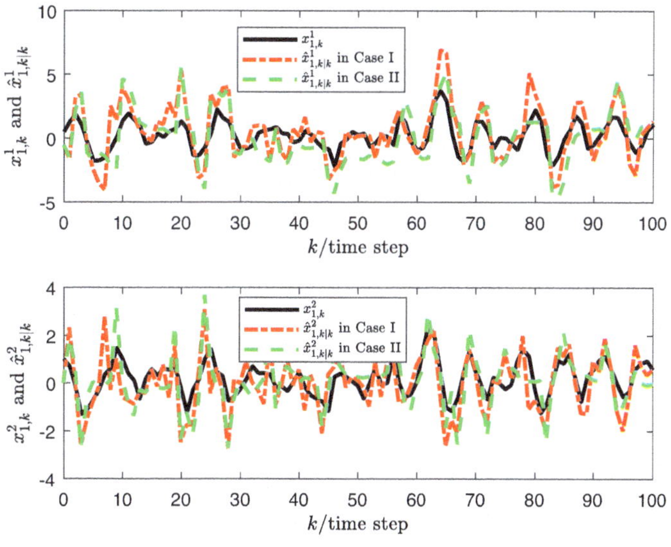

Fig. 12.2 The responses of $\mathbf{x}_{1,k}$ and $\hat{x}_{1,k|k}$

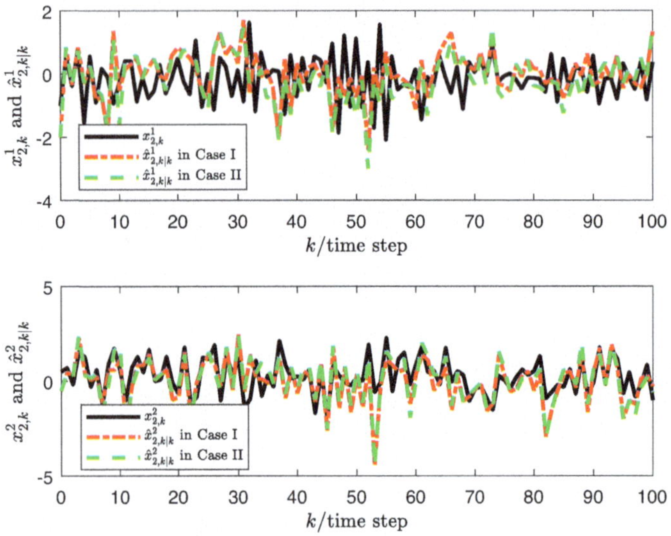

Fig. 12.3 The responses of $\mathbf{x}_{2,k}$ and $\hat{x}_{2,k|k}$

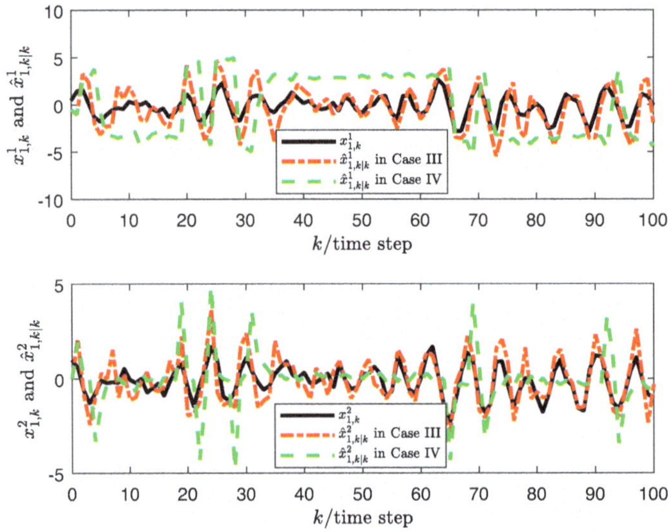

Fig. 12.4 The responses of $\mathbf{x}_{1,k}$ and $\hat{x}_{1,k|k}$

and 12.7. The main reason is due to that more information can be obtained in the estimator side when the triggered thresholds decrease.

With the purpose of evaluating the effects from the degraded measurements, $\log(\mathrm{tr}(\Omega_{i,k|k}))$ is depicted regarding DPDs with different occurrence probabilities in Case III and Case IV, respectively. The simulation results are presented in

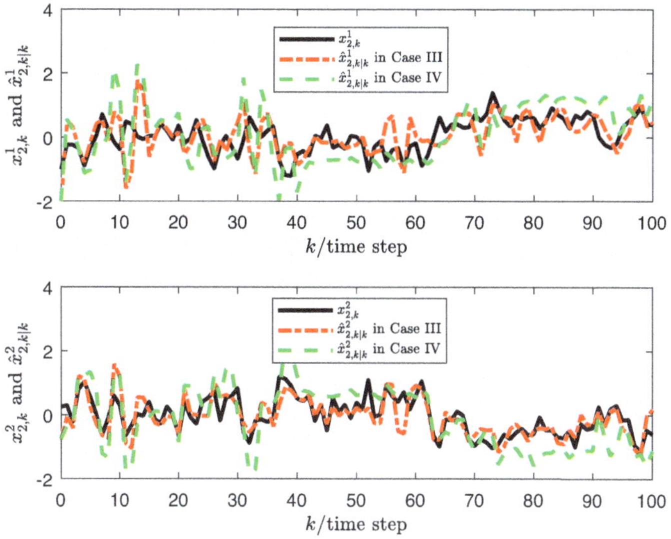

Fig. 12.5 The responses of $\mathbf{x}_{2,k}$ and $\hat{x}_{2,k|k}$

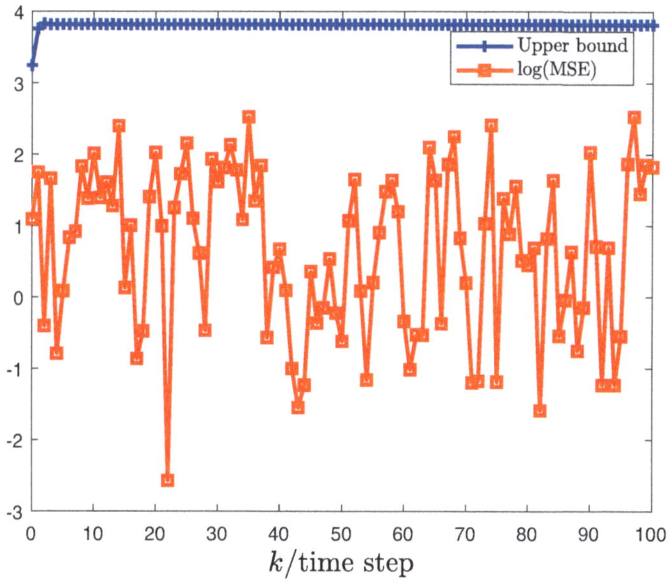

Fig. 12.6 log(MSE) and its upper bound in Case I

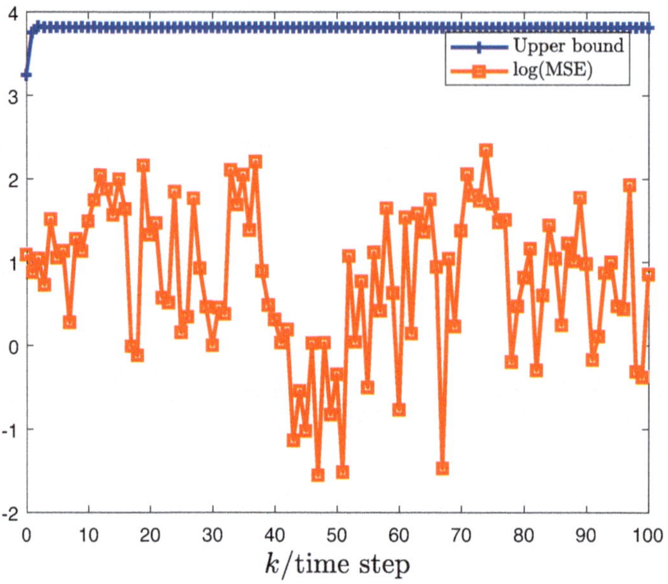

Fig. 12.7 log(MSE) and its upper bound in Case II

Figs. 12.8 and 12.9. It can be seen from Figs. 12.8 and 12.9 that log $(\mathrm{tr}(\Omega_{i,k|k}))$ increases when the occurrence probabilities of DPDs decrease. According to the simulations, it is obvious to testify that the newly proposed time-varying state estimation method possesses a satisfactory performance.

Example 2 Consider the time-varying nonlinear dynamical networks (12.46)–(12.47) with the following coefficients:

$$\begin{aligned}
&B_{1,k} = \left[\, 1.2\ -1.6 \,\right]^T, \quad B_{2,k} = \left[\, -0.8 - 0.1\sin(0.1k)\ 1.6 \,\right]^T, \quad B_{3,k} = \left[\, 0.5\ -2.5 \,\right]^T, \\
&C_{1,k} = \left[\, -2 - 0.1\cos(0.6k)\ 3 \,\right], \quad C_{2,k} = \left[\, -1\ 2 \,\right], \quad C_{3,k} = \left[\, -2\ 1 \,\right], \\
&L_{1,k} = L_{2,k} = L_{3,k} = \mathrm{diag}\{0.5,\ \ 0.5\}, \quad F_{1,k} = F_{2,k} = F_{3,k} = \mathrm{diag}\{0.2,\ \ 0.2\}
\end{aligned}$$

and $x_{i,k} = [x_{i,k}^1\ \ x_{i,k}^2]^T$ is the state vector of the i-th node and the estimation is $\hat{x}_{i,k|k} = [\hat{x}_{i,k|k}^1\ \ \hat{x}_{i,k|k}^2]^T$ $(i = 1, 2, 3)$. The initial values are $x_{1,0} = [0.5\ \ 0.4]^T, x_{2,0} = [0.2\ \ 0.4]^T, x_{3,0} = [0.3\ \ 0.4]^T, \Omega_{1,0|0} = \Omega_{2,0|0} = \Omega_{3,0|0} = 5I_2$. Other parameters are set as $\mu_1 = 0.4,\ \mu_2 = 0.2,\ \mu_3 = 0.6,\ \mu_4 = 0.8,\ \mu_5 = 0.7,\ \mu_6 = 0.2, \mathfrak{Q}_{i,k} = 0.8$, $\mathfrak{R}_{1,k} = \mathfrak{R}_{2,k} = \mathfrak{R}_{3,k} = 0.6,\ \bar{\Phi}_{i,k} = 0.85,\ \hat{\phi}_{i,k} = 0.1,\ \hat{x}_{i,0|0} = x_{i,0} - [1\ \ 1]^T,\ \varepsilon_{i,k} = [1.1\lambda_{\max}(F_{i,k}\Omega_{i,k|k}F_{i,k}^T) + 0.1]^{-1}$. Select the coupling configuration matrix $\Gamma = \mathrm{diag}\{0.04,\ 0.04\}$ and $W = [\omega_{ij}]_{3\times 3}$ with $\omega_{ij} = -0.1$ $(i = j)$ and $\omega_{ij} = 0.05$ $(i \neq j)$. $\Delta\omega_{ij}$ is a random number over the interval $[-0.1, 0.1]$ and $\delta_{ij} = 0.1$. The nonlinear function is set as

$$f(x_{i,k}) = \begin{bmatrix} 0.1x_{i,k}^1 + 0.2x_{i,k}^2 - 0.05\sin(x_{i,k}^1 x_{i,k}^2) \\ -0.1x_{i,k}^1 - 0.1x_{i,k}^2 + 0.06\cos(x_{i,k}^1 x_{i,k}^2) \end{bmatrix}.$$

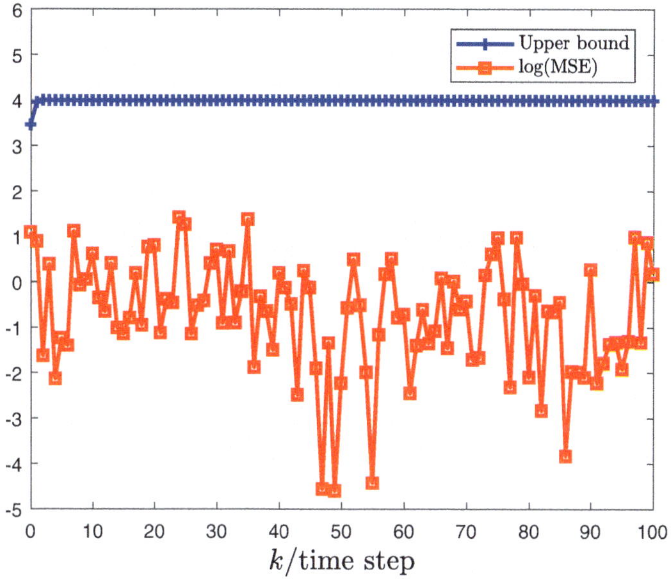

Fig. 12.8 log(MSE) and its upper bound in Case III

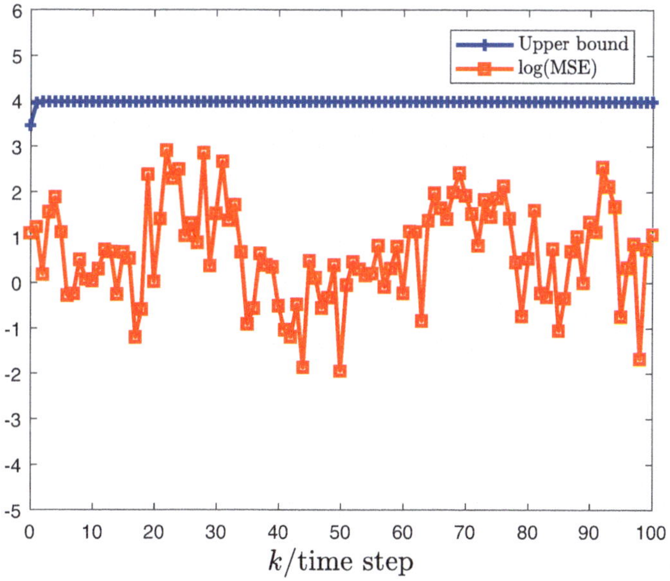

Fig. 12.9 log(MSE) and its upper bound in Case IV

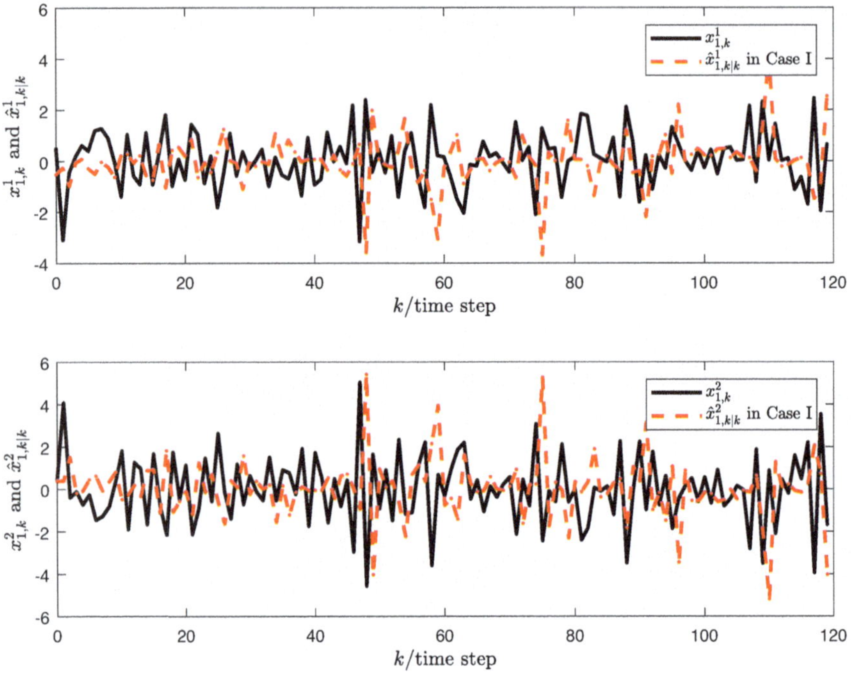

Fig. 12.10 The real state $x_{1,k}$ and the estimation $\hat{x}_{1,k|k}$ in Case I

According to Theorem 12.5, the proposed optimized state estimation issue can be addressed, where the corresponding simulation results are displayed in Figs. 12.10, 12.11, 12.12, 12.13, 12.14, 12.15, 12.16 and 12.17. From Figs. 12.10, 12.11, 12.12, 12.13, 12.14 and 12.15, it is clear that the designed state estimation algorithm performs well under Case I ($\bar{\Phi}_{i,k} = 0.85$) and Case II ($\bar{\Phi}_{i,k} = 0.5$). It can be seen from Fig. 12.16 that log(MSE) of state $x_{i,k}$ is always below the trace of obtained upper bound. In Fig. 12.17, the relationship between $\mathrm{tr}(\Omega_{i,k|k})$ and different fading probabilities is plotted.

12.4 Summary

In this chapter, the event-triggered resilient state estimation issue has been tackled for time-varying uncertain stochastic nonlinear complex networks with switching topology and DPDs firstly. An ETM has been used to attenuate communication burden. By solving the matrix equations, an upper bound of the EEC has been solved and minimized through designing the estimator gain. Subsequently, the optimized state estimation issue has been tackled for nonlinear time-varying complex networks involving ETM, fading measurements and SCS. An upper bound concerning on the EEC has been computed by constructing a proper estimator gain. Furthermore, it has

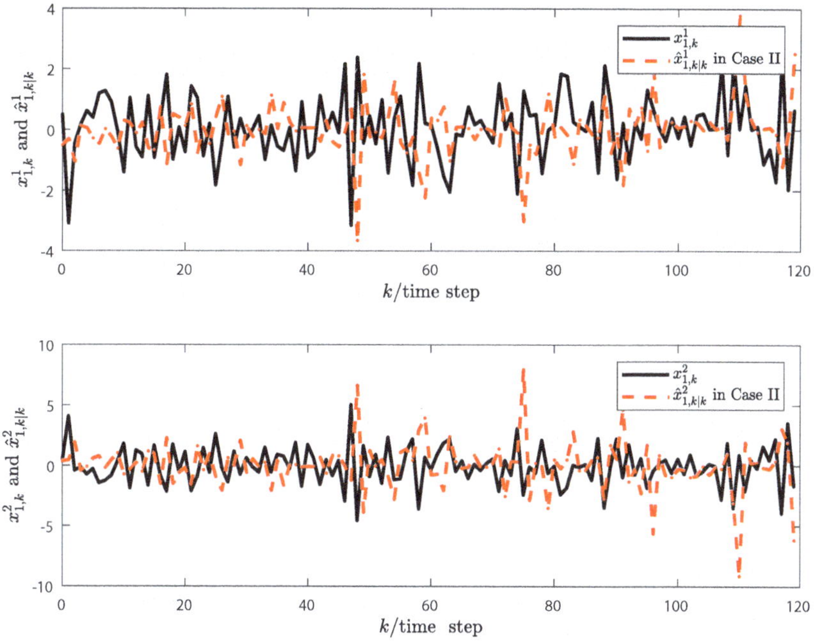

Fig. 12.11 The real state $x_{1,k}$ and the estimation $\hat{x}_{1,k|k}$ in Case II

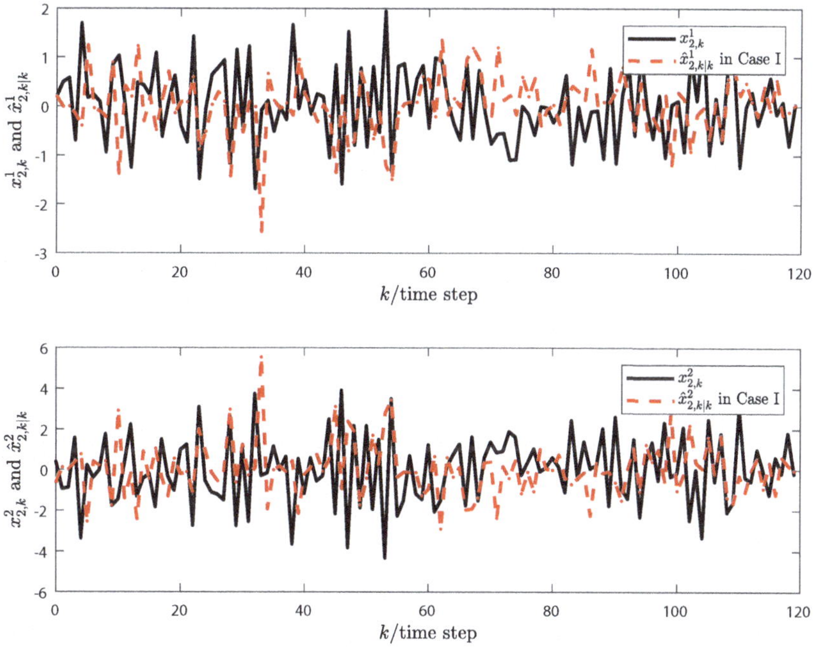

Fig. 12.12 The real state $x_{2,k}$ and the estimation $\hat{x}_{2,k|k}$ in Case I

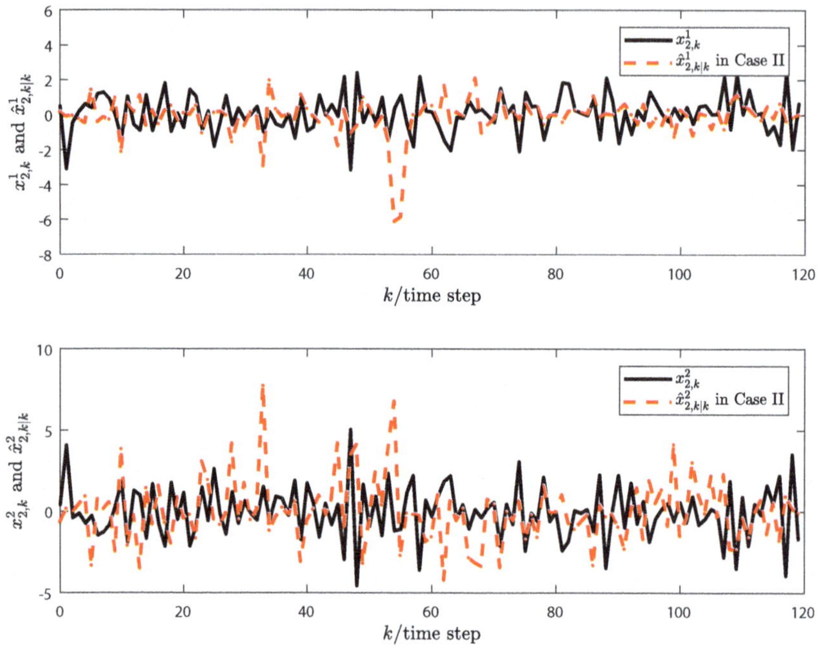

Fig. 12.13 The real state $x_{2,k}$ and the estimation $\hat{x}_{2,k|k}$ in Case II

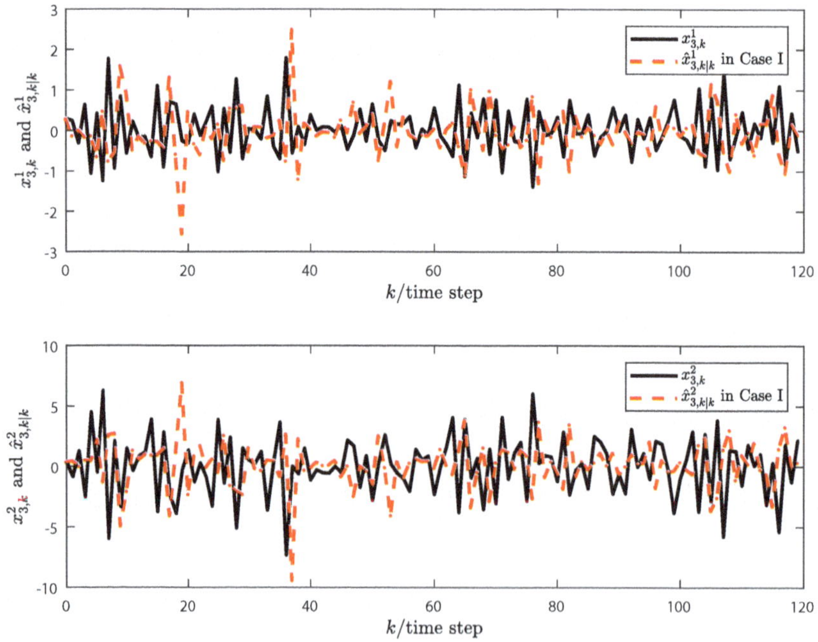

Fig. 12.14 The real state $x_{3,k}$ and the estimation $\hat{x}_{3,k|k}$ in Case I

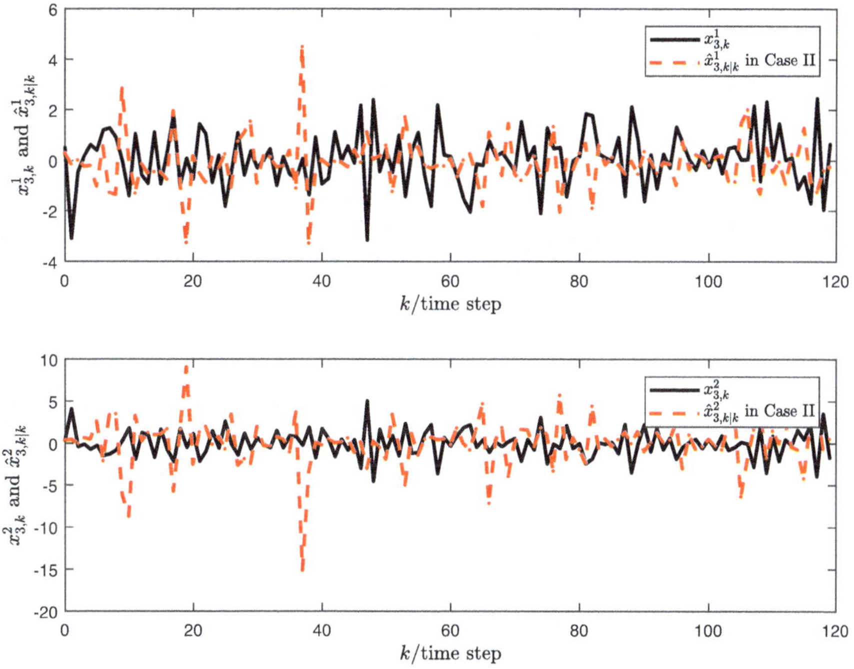

Fig. 12.15 The real state $x_{3,k}$ and the estimation $\hat{x}_{3,k|k}$ in Case II

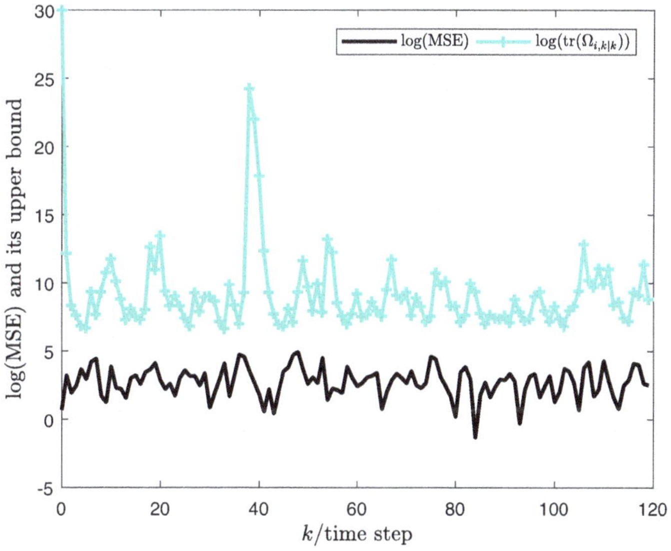

Fig. 12.16 log(MSE) and log(tr($\Omega_{i,k|k}$))

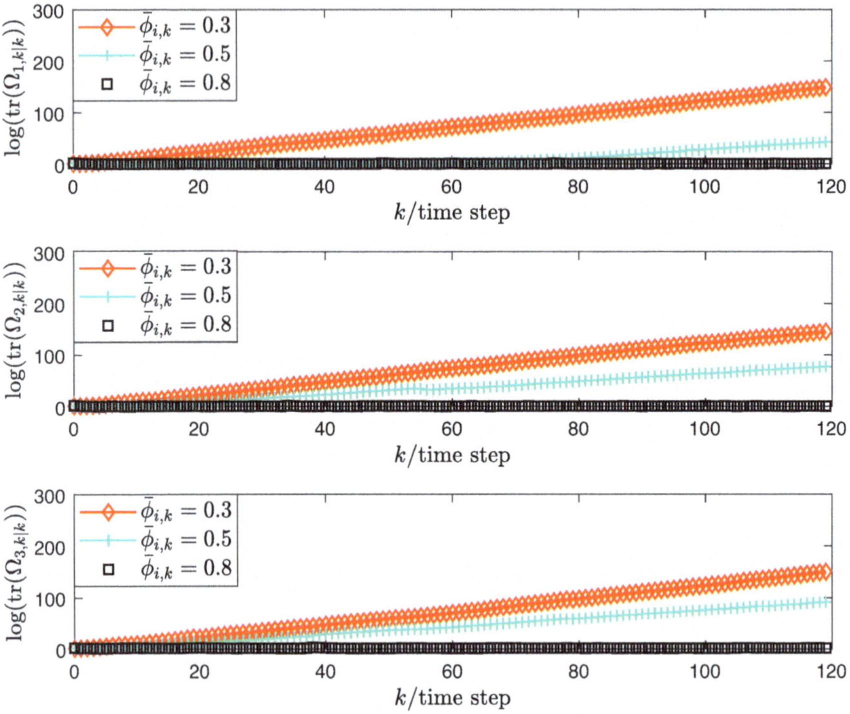

Fig. 12.17 $\log(\mathrm{tr}(\Omega_{i,k|k}))$ regarding different fading probabilities

been proved that the trace of the minimum upper bound is always non-increasing with the increasing fading probability. Finally, some simulation experiments have been provided to validate the usefulness of the proposed state estimation strategy.

References

Calafiore G (2005) Reliable localization using set-valued nonlinear filters. IEEE Trans Syst Man Cybern Part A-Syst Hum 35(2):189–197

Ding D, Liu H, Dong H, Liu H (2022) Resilient filtering of nonlinear complex dynamical networks under randomly occurring faults and hybrid cyber-attacks. IEEE Trans Netw Sci Eng 9(4):2341–2352

Hu J, Hu Z, Caballero-Águila R, Chen C, Fan S, Yi X (2023) Distributed resilient fusion filtering for nonlinear systems with multiple missing measurements via dynamic event-triggered mechanism. Inf Sci 637:118950

Huang C, Mei P, Ding W, Shi Q, Wu Q (2023) State-saturated resilient filtering for nonlinear complex networks under event-triggering protocol. Asian J Control 25(2):1216–1231

Li W, Meng C, Jia Y, Du J (2018) Recursive filtering for complex networks using non-linearly coupled UKF. IET Control Theory Appl 12:549–555

Li Q, Wang Z, Shen B, Liu H, Sheng W (2021) A resilient approach to recursive distributed filtering for multirate systems over sensor networks with time-correlated fading channels. IEEE Trans Signal Inf Proc Netw 7:636–647

Liu H, Wang Z, Shen B, Liu X (2018) Event-triggered H_∞ state estimation for delayed stochastic memristive neural networks with missing measurements: The discrete time case. IEEE Trans Neural Netw Learn Syst 29:3726–3737

Liu S, Zhao X, Tian E, Wei G (2022) Distributed recursive filtering under random access protocols: a multirate strategy. Int J Robust Nonlinear Control 32(12):7132–7148

Shen B, Wang Z, Shu H, Wei G (2011) H_∞ filtering for uncertain time-varying systems with multiple randomly occurred nonlinearities and successive packet dropouts. Int J Robust Nonlinear Control 21(14):1693–1709

Wang S, Fang H, Liu X (2015) Distributed state estimation for stochastic non-linear systems with random delays and packet dropouts. IET Control Theory Appl 9:2657–2665

Wang S, Tian X, Fang H (2019) Event-based state and fault estimation for nonlinear systems with logarithmic quantization and missing measurements. J Frankl Inst-Eng Appl Math 356:4076–4096

Wang S, Wang Z, Dong H, Chen Y (2022) A dynamic event-triggered approach to recursive non-fragile filtering for complex networks with sensor saturations and switching topologies. IEEE Trans Cybern 52(10):11041–11054

Chapter 13
State Estimation for Complex Networks with Uncertain Observations and Coupling Strength

In this chapter, the state estimation issues are solved for time-varying nonlinear complex networks with missing measurements and stochastic inner coupling under variance-constrained framework. Firstly, the multiplicative noise is adopted to describe the stochastic inner coupling and a set of binary random variables with uncertain probabilities is used to characterize the missing measurement. With the aim of avoiding data congestions, the RAP is utilized to arrange the transmission case of sensor nodes. By solving the matrix equations, an optimized upper bound in relation to the EEC is calculated and then the estimator gain is derived while ensuring the minimization value of the trace of this upper bound. In addition, the monotonicity regarding the deterministic occurrence probability and the estimation performance is evaluated. Subsequently, an optimized state estimation issue is studied for nonlinear dynamical networks with RCS, RSDs and ETM. Here, a set of random variables obeying the uniform distribution is utilized to reflect the randomness of the coupling configuration elements. Moreover, a set of binary random variables is adopted to describe the RSDs. For the sake of alleviating communication burden, the ETM is utilized to regulate signal propagations. Based on available measurements, a new state estimation algorithm is established. Furthermore, a criterion is provided to guarantee the boundedness of the estimation error dynamics. Lastly, some illustrative experiments are given to validate the correctness and applicability of the designed state estimation algorithms.

J. Hu et al., *Variance-Constrained Filtering for Stochastic Complex Systems*, Intelligent Control and Learning Systems 18,
https://doi.org/10.1007/978-981-96-2637-3_13

13.1 State Estimation with Missing Measurements and Stochastic Inner Coupling Under RAP

13.1.1 Problem Formulation

Consider the following time-varying nonlinear complex networks with N coupled nodes:

$$\begin{aligned} x_{i,k+1} &= f_i(x_{i,k}) + \sum_{j=1}^{N} w_{ij}(\Gamma + \xi_{i,k}\bar{\Gamma})x_{j,k} + B_{i,k}\varpi_{i,k}, \\ y_{i,k} &= \lambda_{i,k}C_{i,k}x_{i,k} + \nu_{i,k}, \end{aligned} \tag{13.1}$$

where $x_{i,k}$ is the state and $y_{i,k}$ $(i = 1, 2, \ldots, N)$ are the measurements. $W = [w_{ij}]_{N\times N}$ with $w_{ij} \geq 0\,(i \neq j)$ is the coupling configuration matrix. Γ and $\bar{\Gamma}$ represent given linking matrices. $\xi_{i,k}$ is a zero-mean Gaussian noise with unity covariance. The process noise $\varpi_{i,k}$ and the measurement noise $\nu_{i,k}$ are with zero-mean and covariances $Q_{i,k}$ and $R_{i,k}$, respectively. $f_i(x_{i,k})$ is a differentiable nonlinear function and has continuous as well as bounded second order partial derivatives. $B_{i,k}$ and $C_{i,k}$ are pre-defined matrices of suitable dimensions.

In order to depict the missing measurements, the Bernoulli distributed random variable $\lambda_{i,k}$ is introduced here, which satisfies

$$\begin{aligned} \text{Prob}\{\lambda_{i,k} = 1\} &= \bar{\lambda}_{i,k} + \Delta\lambda_{i,k}, \\ \text{Prob}\{\lambda_{i,k} = 0\} &= 1 - \bar{\lambda}_{i,k} - \Delta\lambda_{i,k}, \end{aligned} \tag{13.2}$$

where $\bar{\lambda}_{i,k}$ denotes a known nominal expectation, $|\Delta\lambda_{i,k}| \leq \varrho_{i,k}$ denotes the inaccuracy of the occurrence probability with $\varrho_{i,k}$ being a known scalar. Define the variance of $\lambda_{i,k}$ as $\Omega_{i,k}$. $\varpi_{i,k}$, $\nu_{i,k}$ and $\xi_{i,k}$ are bounded Gaussian noises. Furthermore, suppose that $\varpi_{i,k}$, $\xi_{i,k}$, $\lambda_{i,k}$ and $\nu_{i,k}$ are independent of each other for all i and k.

In the process of signal communication, the RAP is employed to avoid the occurrence of possible data collisions Zou et al. (2018). Let $\gamma_k \in \{1, 2, \ldots, N\}$ denote the chosen sensor at time instant k. Under the RAP, denote the received measurement at the estimator side as $\bar{y}_{i,k}$. It is obvious to find that the updating principle of $\bar{y}_{i,k}$ $(i \in \{1, 2, \ldots, N\})$ under the RAP is modeled by

$$\bar{y}_{i,k} = \begin{cases} y_{i,k}, & \text{if } \gamma_k = i,\ k \geq 0 \\ 0, & \text{otherwise.} \end{cases} \tag{13.3}$$

Further, one has

$$\begin{cases} \bar{y}_{i,k} = \gamma_k^i y_{i,k}, & \text{if } \ k \geq 0 \\ \bar{y}_{i,k} = 0, & \text{otherwise} \end{cases} \tag{13.4}$$

with $\gamma_k^i = \delta(\gamma_k - i)$. $\delta(\cdot)$ is the Kronecker delta function. Next, assume that the occurrence probability of $\gamma_k = i$ ($\forall i \in \{1, 2, \ldots, N\}$) is described by

$$\text{Prob}\{\gamma_k = i\} = p_i, \tag{13.5}$$

where $p_i > 0$ and $\sum_{i=1}^{N} p_i = 1$.

Remark 13.1 As it is well recognized, in order to reduce communication cost and enhance network resource utilization rate, some communication scheduling strategies are employed in the channel to manage signal communication, such as the round-robin protocol, try-once-discard protocol and RAP. Different from the mentioned former two, the RAP involves the probabilistic information, where a stochastic process is adopted to construct the scheduling principle of the RAP Donkers et al. (2012). Specifically, as given in Zou et al. (2018), Balan and Gopalakrishnan (2015), the carrier-sense multiple access protocol as a typical example of RAP has been extensively applied in the industry and effectively addressed the case of constrained communication channel capacities.

To proceed, we denote

$$\begin{aligned}
x_k &= \left[x_{1,k}^T \; x_{2,k}^T \; \cdots \; x_{N,k}^T \right]^T, \\
f(x_k) &= \left[f_1^T(x_{1,k}) \; f_2^T(x_{2,k}) \; \cdots \; f_N^T(x_{N,k}) \right]^T, \\
\xi_k &= \text{diag}\{\xi_{1,k}, \xi_{2,k}, \ldots, \xi_{N,k}\} \otimes I_n, \\
B_k &= \text{diag}\{B_{1,k}, B_{2,k}, \ldots, B_{N,k}\}, \\
\varpi_k &= \left[\varpi_{1,k}^T \; \varpi_{2,k}^T \; \cdots \; \varpi_{N,k}^T \right]^T, \\
C_k &= \text{diag}\{C_{1,k}, C_{2,k}, \ldots, C_{N,k}\}, \\
\nu_k &= \left[\nu_{1,k}^T \; \nu_{2,k}^T \; \cdots \; \nu_{N,k}^T \right]^T, \\
\Lambda_k &= \text{diag}\{\lambda_{1,k}, \lambda_{2,k}, \ldots, \lambda_{N,k}\} \otimes I_p, \\
y_k &= \left[y_{1,k}^T \; y_{2,k}^T \; \cdots \; y_{N,k}^T \right]^T, \\
\bar{y}_k &= \left[\bar{y}_{1,k}^T \; \bar{y}_{2,k}^T \; \cdots \; \bar{y}_{N,k}^T \right]^T, \\
\Delta\Lambda_k &= \text{diag}\{\Delta\lambda_{1,k}, \Delta\lambda_{2,k}, \ldots, \Delta\lambda_{N,k}\} \otimes I_p, \\
\Upsilon_k &= \text{diag}\{\rho_{1,k}, \rho_{2,k}, \ldots, \rho_{N,k}\} \otimes I_p, \\
Q_k &= \text{diag}\{Q_{1,k}, Q_{2,k}, \ldots, Q_{N,k}\}, \\
R_k &= \text{diag}\{R_{1,k}, R_{2,k}, \ldots, R_{N,k}\}.
\end{aligned} \tag{13.6}$$

Then, the system (13.1) is reformulated as

$$\begin{aligned}
x_{k+1} &= f(x_k) + \left[(W \otimes \Gamma) + \xi_k (W \otimes \bar{\Gamma}) \right] x_k + B_k \varpi_k, \\
y_k &= \Lambda_k C_k x_k + \nu_k.
\end{aligned} \tag{13.7}$$

Denoting $\Phi_{\gamma_k} = \text{diag}\{\gamma_k^1 I, \gamma_k^2 I, \ldots, \gamma_k^N I\}$, we design the following state estimator:

$$\hat{x}_{k+1|k} = f(\hat{x}_{k|k}) + (W \otimes \Gamma)\hat{x}_{k|k}, \tag{13.8}$$

$$\hat{x}_{k+1|k+1} = \hat{x}_{k+1|k} + G_{k+1}(\bar{y}_{k+1} - \Phi_{\gamma_{k+1}}\bar{\Lambda}_{k+1}C_{k+1}\hat{x}_{k+1|k}), \tag{13.9}$$

where $\hat{x}_{k+1|k}$ is the one-step prediction, $\hat{x}_{k+1|k+1}$ is the state estimation of x_{k+1}, $\bar{\Lambda}_{k+1} = \text{diag}\{\bar{\lambda}_{1,k+1}, \bar{\lambda}_{2,k+1}, \ldots, \bar{\lambda}_{N,k+1}\} \otimes I_p$, and G_{k+1} is the estimator gain.

Define the one-step prediction error and the estimation error as follows:

$$\begin{aligned} e_{k+1|k} &= x_{k+1} - \hat{x}_{k+1|k}, \\ e_{k+1|k+1} &= x_{k+1} - \hat{x}_{k+1|k+1}. \end{aligned} \tag{13.10}$$

By taking (13.7)–(13.9) into consideration, one obtains

$$e_{k+1|k} = f(x_k) - f(\hat{x}_{k|k}) + (W \otimes \Gamma)e_{k|k} + \xi_k(W \otimes \bar{\Gamma})x_k + B_k\varpi_k \tag{13.11}$$

and

$$\begin{aligned} e_{k+1|k+1} = &(I - G_{k+1}\Phi_{\gamma_{k+1}}\bar{\Lambda}_{k+1}C_{k+1})e_{k+1|k} - G_{k+1}\Phi_{\gamma_{k+1}}\tilde{\Lambda}_{k+1}C_{k+1}x_{k+1} \\ &- G_{k+1}\Phi_{\gamma_{k+1}}\Delta\Lambda_{k+1}C_{k+1}x_{k+1} - G_{k+1}\Phi_{\gamma_{k+1}}v_{k+1}, \end{aligned} \tag{13.12}$$

where $\tilde{\Lambda}_{k+1} = \Lambda_{k+1} - \bar{\Lambda}_{k+1} - \Delta\Lambda_{k+1}$.

Employing the Taylor series expansion method around $\hat{x}_{k|k}$, $f(x_k)$ can be linearized as

$$f(x_k) = f(\hat{x}_{k|k}) + A_k e_{k|k} + o(|e_{k|k}^2|), \tag{13.13}$$

where $A_k = \frac{\partial f(x_k)}{\partial x_k}|_{x_k=\hat{x}_{k|k}}$ and $o(|e_{k|k}^2|)$ is the high-order term of the Taylor series expansion. Next, assume that the high-order term is approximated by the following formulation:

$$o(|e_{k|k}^2|) = L_k F_k H_k e_{k|k}, \tag{13.14}$$

where L_k and H_k are a known problem-dependent scaling matrix and a tuning matrix, respectively. Moreover, F_k characterizing the linearization errors is unknown with $F_k F_k^T \leq I$. Combining (13.11) with (13.13) and (13.14), we obtain

$$e_{k+1|k} = [A_k + L_k F_k H_k + (W \otimes \Gamma)]\, e_{k|k} + \xi_k(W \otimes \bar{\Gamma})x_k + B_k\varpi_k. \tag{13.15}$$

Now, we define the related one-step prediction error covariance and the EEC as

$$\begin{aligned}P_{k+1|k} &= \mathbb{E}\{e_{k+1|k}e_{k+1|k}^T\},\\ P_{k+1|k+1} &= \mathbb{E}\{e_{k+1|k+1}e_{k+1|k+1}^T\}.\end{aligned}$$

Subsequently, we aim to (1) construct a time-varying state estimator (13.8) and (13.9) such that there is a sequence of matrices $\Omega_{k+1|k+1} > 0$ meeting $P_{k+1|k+1} \leq \Omega_{k+1|k+1}$; (2) provide the explicit form of the estimator gain G_{k+1} and minimize $\mathrm{tr}(\Omega_{k+1|k+1})$; and (3) analyze the monotonicity between the deterministic missing probability and the trace of proposed upper bound matrix.

13.1.2 Design of Variance-Constrained State Estimation Algorithm

In this section, by resorting to the stochastic analysis approach, an upper bound of the EEC is solved under the RAP and its trace can be minimized through selecting an appropriate estimator gain matrix. Furthermore, a corollary is shown when $\Delta\lambda_{i,k} = 0$.

Before proceeding, the recursive forms of the one-step prediction error covariance and EEC are derived firstly.

Lemma 13.1 *The one-step prediction error covariance and the EEC have the following recursions:*

$$\begin{aligned}P_{k+1|k} &= [A_k + (W \otimes \Gamma) + L_k F_k H_k] P_{k|k} [A_k + (W \otimes \Gamma) + L_k F_k H_k]^T \\ &\quad + \bar{I} \circ [(W \otimes \bar{\Gamma})\mathbb{E}\{x_k x_k^T\}(W \otimes \bar{\Gamma})^T] + B_k Q_k B_k^T \end{aligned} \tag{13.16}$$

and

$$\begin{aligned}P_{k+1|k+1} &= \mathbb{E}\Big\{(I - G_{k+1}\Phi_{\gamma_{k+1}}\bar{\Lambda}_{k+1}C_{k+1})P_{k+1|k}(I - G_{k+1}\Phi_{\gamma_{k+1}}\bar{\Lambda}_{k+1}C_{k+1})^T\Big\} \\ &\quad + \mathbb{E}\Big\{G_{k+1}\Phi_{\gamma_{k+1}}[\Omega_{k+1} \circ (C_{k+1}\mathbb{E}\{x_{k+1}x_{k+1}^T\}C_{k+1}^T)]\Phi_{\gamma_{k+1}}^T G_{k+1}^T\Big\} \\ &\quad + \mathbb{E}\Big\{G_{k+1}\Phi_{\gamma_{k+1}}\big[\Delta\Lambda_{k+1}C_{k+1}\mathbb{E}\{x_{k+1}x_{k+1}^T\}C_{k+1}^T\Delta\Lambda_{k+1}^T\big]\Phi_{\gamma_{k+1}}^T G_{k+1}^T \\ &\quad + G_{k+1}\Phi_{\gamma_{k+1}}R_{k+1}\Phi_{\gamma_{k+1}}^T G_{k+1}^T\Big\} + \mathcal{B}_{1,k+1} + \mathcal{B}_{1,k+1}^T, \end{aligned} \tag{13.17}$$

where

$$\begin{aligned}\bar{I} &= I_N \otimes 1_{n\times n},\\ \Omega_{k+1} &= \mathrm{diag}\{\Omega_{1,k+1}, \Omega_{2,k+1}, \ldots, \Omega_{N,k+1}\} \otimes 1_{p\times p},\\ \mathcal{B}_{1,k+1} &= -\mathbb{E}\{(I - G_{k+1}\Phi_{\gamma_{k+1}}\bar{\Lambda}_{k+1}C_{k+1})e_{k+1|k}x_{k+1}^T C_{k+1}^T \\ &\quad \times \Delta\Lambda_{k+1}^T \Phi_{\gamma_{k+1}}^T G_{k+1}^T\}. \end{aligned} \tag{13.18}$$

Proof Considering (13.15), it is not difficult to obtain

$$
\begin{aligned}
P_{k+1|k} =& [A_k + L_k F_k H_k + (W \otimes \Gamma)] P_{k|k} [A_k + L_k F_k H_k + (W \otimes \Gamma)]^T \\
&+ B_k Q_k B_k^T + \mathcal{A}_{1,k} + \mathcal{A}_{1,k}^T + \mathcal{A}_{2,k} + \mathcal{A}_{2,k}^T + \mathcal{A}_{3,k} + \mathcal{A}_{3,k}^T \\
&+ \mathbb{E}\{\xi_k (W \otimes \bar{\Gamma}) x_k x_k^T (W \otimes \bar{\Gamma})^T \xi_k^T\},
\end{aligned}
$$

where

$$
\begin{aligned}
\mathcal{A}_{1,k} &= \mathbb{E}\{[A_k + L_k F_k H_k + (W \otimes \Gamma)]\, e_{k|k} x_k^T (W \otimes \bar{\Gamma})^T \xi_k^T\}, \\
\mathcal{A}_{2,k} &= \mathbb{E}\{[A_k + L_k F_k H_k + (W \otimes \Gamma)]\, e_{k|k} \varpi_k^T B_k^T\}, \\
\mathcal{A}_{3,k} &= \mathbb{E}\{\xi_k (W \otimes \bar{\Gamma}) x_k \varpi_k^T B_k^T\}.
\end{aligned}
$$

Note that $\mathbb{E}\{\xi_k\} = 0$, $\mathbb{E}\{\varpi_k\} = 0$, ξ_k and ϖ_k are independent with each other, it is clear that $\mathcal{A}_{i,k} = 0$ $(i = 1, 2, 3)$. Then, it follows from Lemma 2.4 that (13.16) is true.

On the other hand, considering $\mathbb{E}\{\tilde{\Lambda}_{k+1}\} = 0$ and $\mathbb{E}\{\nu_{k+1}\} = 0$, one has

$$
\begin{aligned}
& P_{k+1|k+1} \\
=& \mathbb{E}\Big\{(I - G_{k+1}\Phi_{\gamma_{k+1}} \bar{\Lambda}_{k+1} C_{k+1}) P_{k+1|k} (I - G_{k+1}\Phi_{\gamma_{k+1}} \bar{\Lambda}_{k+1} C_{k+1})^T\Big\} \\
&+ \mathcal{B}_{1,k+1} + \mathcal{B}_{1,k+1}^T + \mathbb{E}\Big\{G_{k+1}\Phi_{\gamma_{k+1}} \tilde{\Lambda}_{k+1} C_{k+1} x_{k+1} x_{k+1}^T C_{k+1}^T \tilde{\Lambda}_{k+1}^T \Phi_{\gamma_{k+1}}^T G_{k+1}^T \\
&+ G_{k+1}\Phi_{\gamma_{k+1}} \big[\Delta\Lambda_{k+1} C_{k+1} \mathbb{E}\{x_{k+1} x_{k+1}^T\} C_{k+1}^T \Delta\Lambda_{k+1}^T\big] \Phi_{\gamma_{k+1}}^T G_{k+1}^T \\
&+ G_{k+1}\Phi_{\gamma_{k+1}} R_{k+1} \Phi_{\gamma_{k+1}}^T G_{k+1}^T\Big\}.
\end{aligned}
$$

Based on Lemma 2.4, (13.17) can be clearly obtained. □

Remark 13.2 Notice that there exist some uncertain terms (e.g. F_k, $\Delta\Lambda_{k+1}$, Ω_{k+1} and $\mathbb{E}\{x_{k+1}x_{k+1}^T\}$) in the one-step prediction error covariance and the EEC, thus we cannot get the precise EEC. For the sake of solving this issue, we aim to find an optimized upper bound regarding the EEC as an alternative scheme to complete the estimator design.

Theorem 13.1 *Letting* $\varepsilon_i > 0$ $(i = 1, 2, 3)$ *and* $\eta_k > 0$ *be scalars, if the following matrix equations:*

$$
\begin{aligned}
\Omega_{k+1|k} =& [A_k + (W \otimes \Gamma)]\,(\Omega_{k|k}^{-1} - \eta_k H_k^T H_k)^{-1}\,[A_k + (W \otimes \Gamma)]^T \\
&+ \eta_k^{-1} L_k L_k^T + B_k Q_k B_k^T + \bar{I} \circ \big[(W \otimes \bar{\Gamma}) \bar{A}_{1,k} (W \otimes \bar{\Gamma})^T\big]
\end{aligned}
\tag{13.19}
$$

$$
\begin{aligned}
\Omega_{k+1|k+1} =& (1 + \varepsilon_2) \sum_{i=1}^{N} p_i \mathcal{K}_{i,k+1} \Omega_{k+1|k} \mathcal{K}_{i,k+1}^T \\
&+ \sum_{i=1}^{N} p_i G_{k+1} \Phi_i \bar{\Omega}_{k+1} \Phi_i^T G_{k+1}^T,
\end{aligned}
\tag{13.20}
$$

under the initial condition $\Omega_{0|0} \geq P_{0|0}$ *and satisfying the following inequality:*

$$\eta_k^{-1} I - H_k \Omega_{k|k} H_k^T > 0, \tag{13.21}$$

have solutions $\Omega_{k+1|k} > 0$ *and* $\Omega_{k+1|k+1} > 0$*, where*

$$\begin{aligned}
\mathcal{K}_{i,k+1} &= I - G_{k+1}\Phi_i \bar{\Lambda}_{k+1} C_{k+1}, \\
\bar{\sigma}_{k+1} &= \mathrm{diag}\{\bar{\sigma}_{1,k+1}, \bar{\sigma}_{2,k+1}, \ldots, \bar{\sigma}_{N,k+1}\} \otimes 1_{p\times p}, \\
\bar{\Lambda}_{1,k} &= (1+\varepsilon_1)\Omega_{k|k} + (1+\varepsilon_1^{-1})\hat{x}_{k|k}\hat{x}_{k|k}^T, \\
\bar{\Lambda}_{2,k+1} &= (1+\varepsilon_3)\Omega_{k+1|k} + (1+\varepsilon_3^{-1})\hat{x}_{k+1|k}\hat{x}_{k+1|k}^T, \\
\bar{\sigma}_{i,k+1} &= \min\left\{\frac{1}{4}, (\bar{\lambda}_{i,k+1} + \varrho_{i,k+1})(1 - \bar{\lambda}_{i,k+1} + \varrho_{i,k+1})\right\}, \quad i = 1, 2, \ldots, N, \\
\bar{\Omega}_{k+1} &= \mathrm{tr}\left(\bar{\sigma}_{k+1} \circ \left(C_{k+1}\bar{\Lambda}_{2,k+1}C_{k+1}^T\right)\right) I + (1+\varepsilon_2^{-1}) \\
&\quad \times \mathrm{tr}(C_{k+1}\bar{\Lambda}_{2,k+1}C_{k+1}^T)\Upsilon_{k+1}\Upsilon_{k+1}^T + R_{k+1},
\end{aligned} \tag{13.22}$$

then $\Omega_{k+1|k+1}$ *is an upper bound of* $P_{k+1|k+1}$. *In addition, if the estimator gain matrix is calculated by*

$$G_{k+1} = (1+\varepsilon_2)\Omega_{k+1|k} C_{k+1}^T \bar{\Lambda}_{k+1}^T \bar{\Phi} \left(\sum_{i=1}^{N} p_i \Phi_i \Pi_{k+1} \Phi_i\right)^{-1} \tag{13.23}$$

with

$$\begin{aligned}
\bar{\Phi} &= \mathrm{diag}\{p_1 I, p_2 I, \ldots, p_N I\}, \\
\Pi_{k+1} &= (1+\varepsilon_2)\bar{\Lambda}_{k+1} C_{k+1} \Omega_{k+1|k} C_{k+1}^T \bar{\Lambda}_{k+1}^T + \bar{\Omega}_{k+1},
\end{aligned} \tag{13.24}$$

then $\mathrm{tr}(\Omega_{k+1|k+1})$ *can be minimized.*

Proof Firstly, noticing $x_k = e_{k|k} + \hat{x}_{k|k}$ and employing Lemma 2.2, we have

$$\begin{aligned}
\mathbb{E}\{x_k x_k^T\} &= \mathbb{E}\{(e_{k|k} + \hat{x}_{k|k})(e_{k|k} + \hat{x}_{k|k})^T\} \\
&\leq (1+\varepsilon_1)P_{k|k} + (1+\varepsilon_1^{-1})\hat{x}_{k|k}\hat{x}_{k|k}^T \\
&:= \Lambda_{1,k}
\end{aligned} \tag{13.25}$$

with $\varepsilon_1 > 0$ being a scalar. Substituting (13.25) into (13.16), one has

$$\begin{aligned}
P_{k+1|k} &\leq [A_k + (W \otimes \Gamma) + L_k F_k H_k] P_{k|k} [A_k + (W \otimes \Gamma) \\
&\quad + L_k F_k H_k]^T + \bar{I} \circ \left[(W \otimes \bar{\Gamma})\Lambda_{1,k}(W \otimes \bar{\Gamma})^T\right] \\
&\quad + B_k Q_k B_k^T.
\end{aligned} \tag{13.26}$$

In light of Lemma 2.2, one has

$$\begin{aligned}\mathcal{B}_{1,k+1}+\mathcal{B}_{1,k+1}^T \le\ & \varepsilon_2\mathbb{E}\big\{(I-G_{k+1}\Phi_{\gamma_{k+1}}\bar{\Lambda}_{k+1}C_{k+1})P_{k+1|k}(I-G_{k+1}\\ &\times\Phi_{\gamma_{k+1}}\bar{\Lambda}_{k+1}C_{k+1})^T\big\}+\varepsilon_2^{-1}\mathbb{E}\{G_{k+1}\Phi_{\gamma_{k+1}}\Delta\Lambda_{k+1}\\ &\times C_{k+1}x_{k+1}x_{k+1}^TC_{k+1}^T\Delta\Lambda_{k+1}^T\Phi_{\gamma_{k+1}}^TG_{k+1}^T\},\end{aligned} \tag{13.27}$$

where $\varepsilon_2>0$ is a constant. Substituting (13.27) into (13.17) leads to

$$\begin{aligned}P_{k+1|k+1}\le\ & \mathbb{E}\big\{(1+\varepsilon_2)(I-G_{k+1}\Phi_{\gamma_{k+1}}\bar{\Lambda}_{k+1}C_{k+1})P_{k+1|k}\\ &\times(I-G_{k+1}\Phi_{\gamma_{k+1}}\bar{\Lambda}_{k+1}C_{k+1})^T\big\}+\mathbb{E}\big\{G_{k+1}\Phi_{\gamma_{k+1}}\\ &\times\big[\Omega_{k+1}\circ(C_{k+1}\mathbb{E}\{x_{k+1}x_{k+1}^T\}C_{k+1}^T)\big]\Phi_{\gamma_{k+1}}^TG_{k+1}^T\big\}\\ &+(1+\varepsilon_2^{-1})\mathbb{E}\Big\{G_{k+1}\Phi_{\gamma_{k+1}}\Delta\Lambda_{k+1}C_{k+1}\mathbb{E}\{x_{k+1}x_{k+1}^T\}\\ &\times C_{k+1}^T\Delta\Lambda_{k+1}^T\Phi_{\gamma_{k+1}}^TG_{k+1}^T+G_{k+1}\Phi_{\gamma_{k+1}}R_{k+1}\Phi_{\gamma_{k+1}}^TG_{k+1}^T\Big\}.\end{aligned} \tag{13.28}$$

Similar to (13.25), it is clear to observe that

$$\begin{aligned}&\mathbb{E}\{x_{k+1}x_{k+1}^T\}\\ &=\mathbb{E}\{(e_{k+1|k}+\hat{x}_{k+1|k})(e_{k+1|k}+\hat{x}_{k+1|k})^T\}\\ &\le(1+\varepsilon_3)P_{k+1|k}+(1+\varepsilon_3^{-1})\hat{x}_{k+1|k}\hat{x}_{k+1|k}^T\\ &:=\Lambda_{2,k+1}\end{aligned} \tag{13.29}$$

with ε_3 being a positive scalar.

Furthermore, noticing $\Delta\Lambda_{k+1}\le\Upsilon_{k+1}$ and $\Omega_{i,k+1}\le\bar{\sigma}_{i,k+1}$, we obtain

$$\begin{aligned}P_{k+1|k+1}\le\ & (1+\varepsilon_2)\mathbb{E}\big\{(I-G_{k+1}\Phi_{\gamma_{k+1}}\bar{\Lambda}_{k+1}C_{k+1})P_{k+1|k}\\ &\times(I-G_{k+1}\Phi_{\gamma_{k+1}}\bar{\Lambda}_{k+1}C_{k+1})^T\big\}+\mathrm{tr}(\bar{\sigma}_{k+1}\circ(C_{k+1}\\ &\times\Lambda_{2,k+1}C_{k+1}^T))\mathbb{E}\big\{G_{k+1}\Phi_{\gamma_{k+1}}\Phi_{\gamma_{k+1}}^TG_{k+1}^T\big\}+\mathbb{E}\big\{(1+\varepsilon_2^{-1})\\ &\times\mathrm{tr}(C_{k+1}\Lambda_{2,k+1}C_{k+1}^T)G_{k+1}\Phi_{\gamma_{k+1}}\Upsilon_{k+1}\Upsilon_{k+1}^T\\ &\times\Phi_{\gamma_{k+1}}^TG_{k+1}^T+G_{k+1}\Phi_{\gamma_{k+1}}R_{k+1}\Phi_{\gamma_{k+1}}^TG_{k+1}^T\big\},\\ =\ &(1+\varepsilon_2)\mathbb{E}\{\mathcal{K}_{\gamma_{k+1},k+1}P_{k+1|k}\mathcal{K}_{\gamma_{k+1},k+1}^T\\ &+G_{k+1}\Phi_{\gamma_{k+1}}\tilde{\Omega}_{k+1}\Phi_{\gamma_{k+1}}^TG_{k+1}^T\big\},\end{aligned} \tag{13.30}$$

where

$$\begin{aligned}\mathcal{K}_{\gamma_{k+1},k+1}&=I-G_{k+1}\Phi_{\gamma_{k+1}}\bar{\Lambda}_{k+1}C_{k+1},\\ \tilde{\Omega}_{k+1}&=\mathrm{tr}\big(\bar{\sigma}_{k+1}\circ\big(C_{k+1}\Lambda_{2,k+1}C_{k+1}^T\big)\big)I+(1+\varepsilon_2^{-1})\\ &\quad\times\mathrm{tr}(C_{k+1}\Lambda_{2,k+1}C_{k+1}^T)\Upsilon_{k+1}\Upsilon_{k+1}^T+R_{k+1},\end{aligned}$$

and $\bar{\sigma}_{k+1}$ is defined as in (13.22).

On the other hand, it follows from the definition of $\Phi_{\gamma_{k+1}}$ that $\Phi_{\gamma_{k+1}} = \sum_{i=1}^{N} \gamma_{k+1}^{i} \Phi_i$. Moreover, it is known that $\mathbb{E}\{\gamma_{k+1}^{i}\} = \sum_{j=1}^{N} p_j \delta(j-i) = p_i$ and

$$\gamma_{k+1}^{i}\gamma_{k+1}^{j} = \begin{cases} 0, & i \neq j \\ \gamma_{k+1}^{i}, & i = j, \end{cases}$$

then, we obtain

$$\begin{aligned}
&\mathbb{E}\{\mathcal{K}_{\gamma_{k+1},k+1} P_{k+1|k} \mathcal{K}_{\gamma_{k+1},k+1}^{T}\} \\
&= \mathbb{E}\left\{\left(I - \sum_{i=1}^{N} \gamma_{k+1}^{i} G_{k+1} \Phi_i \bar{\Lambda}_{k+1} C_{k+1}\right) P_{k+1|k} \left(I - \sum_{i=1}^{N} \gamma_{k+1}^{i} G_{k+1} \Phi_i \bar{\Lambda}_{k+1} C_{k+1}\right)\right\} \\
&= P_{k+1|k} - \mathbb{E}\left\{\sum_{i=1}^{N} \gamma_{k+1}^{i} G_{k+1} \Phi_i \bar{\Lambda}_{k+1} C_{k+1} P_{k+1|k}\right\} \\
&\quad - \mathbb{E}\left\{\sum_{i=1}^{N} \gamma_{k+1}^{i} P_{k+1|k} C_{k+1}^{T} \bar{\Lambda}_{k+1} \Phi_i G_{k+1}^{T}\right\} \\
&\quad + \mathbb{E}\left\{\sum_{i=1}^{N} \gamma_{k+1}^{i} G_{k+1} \Phi_i \bar{\Lambda}_{k+1} C_{k+1} P_{k+1|k} C_{k+1}^{T} \bar{\Lambda}_{k+1} \Phi_i G_{k+1}^{T}\right\} \\
&= \sum_{i=1}^{N} p_i \mathcal{K}_{i,k+1} P_{k+1|k} \mathcal{K}_{i,k+1}^{T}.
\end{aligned} \tag{13.31}$$

Similarly, one has

$$\begin{aligned}
&\mathbb{E}\{G_{k+1} \Phi_{\gamma_{k+1}} \tilde{\Omega}_{k+1} \Phi_{\gamma_{k+1}} G_{k+1}\} \\
&= \mathbb{E}\left\{\sum_{i=1}^{N} \gamma_{k+1}^{i} G_{k+1} \Phi_i \tilde{\Omega}_{k+1} \Phi_i G_{k+1}^{T}\right\} \\
&= \sum_{i=1}^{N} p_i G_{k+1} \Phi_i \tilde{\Omega}_{k+1} \Phi_i G_{k+1}^{T}.
\end{aligned} \tag{13.32}$$

Substituting (13.31) and (13.32) into (13.30) implies

$$\begin{aligned}
P_{k+1|k+1} \leq &(1+\varepsilon_2) \sum_{i=1}^{N} p_i \mathcal{K}_{i,k+1} P_{k+1|k} \mathcal{K}_{i,k+1}^{T} \\
&+ \sum_{i=1}^{N} p_i G_{k+1} \Phi_i \tilde{\Omega}_{k+1} \Phi_i G_{k+1}^{T}.
\end{aligned} \tag{13.33}$$

Applying Lemma 2.1 to (13.26) and noticing (13.19) and (13.20) as well as (13.33), we can conclude that $P_{k+1|k+1} \leq \Omega_{k+1|k+1}$ by resorting to the mathematical induction method.

Finally, based on Lemma 2.3, taking the partial derivative of $\mathrm{tr}(\Omega_{k+1|k+1})$ regarding the estimator gain G_{k+1}, one has

$$\begin{aligned}\frac{\partial \mathrm{tr}(\Omega_{k+1|k+1})}{\partial G_{k+1}} &= -2\sum_{i=1}^{N} p_i(1+\varepsilon_2)\Omega_{k+1|k}C_{k+1}^T\bar{\Lambda}_{k+1}^T\Phi_i^T \\ &\quad +2\sum_{i=1}^{N} p_i G_{k+1}\Phi_i\Pi_{k+1}\Phi_i^T,\end{aligned}$$

where Π_{k+1} is defined as in (13.24). Furthermore, it should be noted that

$$\begin{aligned}\sum_{i=1}^{N} p_i\Phi_i\Pi_{k+1}\Phi_i^T &= \sum_{i=1}^{N} p_i(\Phi_i-\bar{\Phi}+\bar{\Phi})\Pi_{k+1}(\Phi_i-\bar{\Phi}+\bar{\Phi})^T \\ &= \sum_{i=1}^{N} p_i\tilde{\Phi}_i\Pi_{k+1}\tilde{\Phi}_i^T+\bar{\Phi}\Pi_{k+1}\bar{\Phi}^T \\ &\geq \bar{\Phi}\Pi_{k+1}\bar{\Phi}^T \\ &\geq \bar{\Phi}\bar{\Omega}_{k+1}\bar{\Phi}^T \\ &\geq \bar{\Phi}R_{k+1}\bar{\Phi}^T > 0,\end{aligned}$$

where the fact $\sum_{i=1}^{N} p_i\tilde{\Phi}_i = \sum_{i=1}^{N} p_i\left(\Phi_i-\bar{\Phi}\right)=0$ is used, $\bar{\Phi}$ is defined as in (13.24) and $\bar{\Omega}_{k+1}$ is given in (13.22). Therefore, we have $\sum_{i=1}^{N} p_i\Phi_i\Pi_{k+1}\Phi_i^T > 0$. Next, letting the partial derivative denote zero, one has

$$G_{k+1} = (1+\varepsilon_2)\Omega_{k+1|k}C_{k+1}^T\bar{\Lambda}_{k+1}^T\bar{\Phi}\left(\sum_{i=1}^{N} p_i\Phi_i\Pi_{k+1}\Phi_i\right)^{-1}.$$

Therefore, $\mathrm{tr}(\Omega_{k+1|k+1})$ can be minimized by the estimator gain selected in (13.23), which completes the proof now. □

In light of the results in Theorem 13.1, the RAP-based state estimation (RAPBSE) algorithm is given as below.

RAPBSE Algorithm :
Step 1 : Let $k=0$ and choose the initial conditions.
Step 2 : Calculate $\hat{x}_{k+1\|k}$ by (13.8).
Step 3 : Compute the value of $\Omega_{k+1\|k}$ by (13.19).
Step 4 : Get the estimator gain matrix G_{k+1} by (13.23).
Step 5 : Calculate the state estimation $\hat{x}_{k+1\|k+1}$ by (13.9).
Step 6 : Compute the upper bound $\Omega_{k+1\|k+1}$ by (13.20).
Step 7 : Letting $k=k+1$, and go to *Step* 2.

When $\Delta\lambda_{i,k}=0$ and the occurrence probabilities of missing measurements for each sensor node are same, the following corollary is given.

Corollary 13.1 *Suppose that the occurrence probabilities of missing measurements for each sensor node are same and deterministic, i.e.,* $\mathrm{Prob}\{\lambda_{1,k}=1\}=\mathrm{Prob}\{\lambda_{2,k}=1\}=\cdots=\mathrm{Prob}\{\lambda_{N,k}=1\}=\bar{\mu}_k$. *For* $\varepsilon_i>0$ $(i=1,3)$ *and* $\eta_k>0$, *if the recursive matrix equation (13.19) and the following matrix equation:*

$$\begin{aligned}\Omega_{k+1|k+1}=&\sum_{i=1}^{N}p_i(I-\bar{\mu}_{k+1}G_{k+1}\Phi_iC_{k+1})\Omega_{k+1|k}(I-\bar{\mu}_{k+1}G_{k+1}\Phi_iC_{k+1})^T\\&+\sum_{i=1}^{N}p_iG_{k+1}\Phi_i\Big[\bar{\mu}_{k+1}(1-\bar{\mu}_{k+1})\left(I_N\otimes 1_{p\times p}\right)\\&\circ\left(C_{k+1}\bar{\Lambda}_{2,k+1}C_{k+1}^T\right)+R_{k+1}\Big]\Phi_i^TG_{k+1}^T\end{aligned}\tag{13.34}$$

with the initial condition $\Omega_{0|0}\geq P_{0|0}$ *and satisfying*

$$\eta_k^{-1}I-H_k\Omega_{k|k}H_k^T>0$$

have positive-definite solutions $\Omega_{k+1|k}$ *and* $\Omega_{k+1|k+1}$, *then matrix* $\Omega_{k+1|k+1}$ *is an upper bound of* $P_{k+1|k+1}$. *Besides, the trace of the upper bound matrix* $\Omega_{k+1|k+1}$ *can be minimized if the gain matrix is designed by*

$$G_{k+1}=\bar{\mu}_{k+1}\Omega_{k+1|k}C_{k+1}^T\bar{\Phi}\left(\sum_{i=1}^{N}p_i\Phi_i\hat{\Pi}_{k+1}\Phi_i\right)^{-1},\tag{13.35}$$

where

$$\begin{aligned}\hat{\Pi}_{k+1}=&R_{k+1}+\bar{\mu}_{k+1}^2C_{k+1}\Omega_{k+1|k}C_{k+1}^T+\bar{\mu}_{k+1}(1-\bar{\mu}_{k+1})\\&\times\left(I_N\otimes 1_{p\times p}\right)\circ\left(C_{k+1}\bar{\Lambda}_{2,k+1}C_{k+1}^T\right).\end{aligned}$$

Proof Applying the same method as in Theorem 13.1, the proof of this corollary can be obtained directly. □

Remark 13.3 It is highlighted that ε_1, ε_2, ε_3 and η_k have been reflected in Theorem 13.1 and Corollary 13.1, which indeed influence the estimation performance. In the process of the implementation, a feasible way is to choose their values satisfying (13.21) firstly and then adjust them to minimize the trace of the obtained upper bound. Besides, note that the matrices L_k and H_k have been adopted to quantitatively describe the linearization errors induced by the Taylor series expansion. In future, we will investigate other methods to cope with the choice problem of the mentioned scalars and matrices.

13.1.3 Monotonicity Analysis

In this section, the objective is to evaluate how the occurrence probability of missing measurements influences the trace of the obtained upper bound matrix $\Omega_{k+1|k+1}$.

Based on Corollary 13.1 and Theorem 13.1, a theoretical analysis is provided, where the monotonicity between the missing occurrence probability and the trace of this upper bound matrix is exhibited and shown.

Theorem 13.2 *For all sensor nodes, let the occurrence probabilities of missing measurements be same and deterministic. Then,* $\mathrm{tr}(\Omega_{k+1|k+1})$ *is non-increasing as* $\bar{\mu}_{k+1}$ *increases.*

Proof Firstly, substituting (13.35) into (13.34) implies

$$\Omega_{k+1|k+1} = \Omega_{k+1|k} - \bar{\mu}_{k+1}^2 \Omega_{k+1|k}^T C_{k+1}^T \bar{\Phi} \hat{\Xi}_{k+1}^{-1} \bar{\Phi} C_{k+1} \Omega_{k+1|k},$$

where

$$\begin{aligned}\hat{\Xi}_{k+1} = {} & \bar{\mu}_{k+1}^2 \sum_{i=1}^{N} p_i \Phi_i C_{k+1} \Omega_{k+1|k} C_{k+1}^T \Phi_i + \bar{\mu}_{k+1}(1 - \bar{\mu}_{k+1}) \\ & \times \sum_{i=1}^{N} p_i \Phi_i (I_N \otimes 1_{p\times p}) \circ (C_{k+1} \bar{\Lambda}_{2,k+1} C_{k+1}^T) \Phi_i \\ & + \sum_{i=1}^{N} p_i \Phi_i R_{k+1} \Phi_i .\end{aligned}$$

Then, we have

$$\begin{aligned}\frac{\mathrm{d}\,\mathrm{tr}(\Omega_{k+1|k+1})}{\mathrm{d}\bar{\mu}_{k+1}} = {} & \mathrm{tr}\Big(-2\bar{\mu}_{k+1} \Omega_{k+1|k}^T C_{k+1}^T \bar{\Phi} \hat{\Xi}_{k+1}^{-1} \bar{\Phi} C_{k+1} \Omega_{k+1|k} \\ & + \bar{\mu}_{k+1}^2 \Omega_{k+1|k}^T C_{k+1}^T \bar{\Phi} \hat{\Xi}_{k+1}^{-1} S_{k+1} \hat{\Xi}_{k+1}^{-1} \bar{\Phi} C_{k+1} \Omega_{k+1|k} \Big) \\ = {} & \mathrm{tr}\Big(\bar{\mu}_{k+1} \Omega_{k+1|k}^T C_{k+1}^T \bar{\Phi} \hat{\Xi}_{k+1}^{-1} \left(\bar{\mu}_{k+1} S_{k+1} - 2\hat{\Xi}_{k+1} \right) \\ & \times \hat{\Xi}_{k+1}^{-1} \bar{\Phi} C_{k+1} \Omega_{k+1|k} \Big),\end{aligned} \tag{13.36}$$

where

$$\begin{aligned}S_{k+1} = {} & \frac{\mathrm{d}\hat{\Xi}_{k+1}}{\mathrm{d}\bar{\mu}_{k+1}} \\ = {} & 2\bar{\mu}_{k+1} \sum_{i=1}^{N} p_i \Phi_i C_{k+1} \Omega_{k+1|k} C_{k+1}^T \Phi_i \\ & + (1 - 2\bar{\mu}_{k+1}) \sum_{i=1}^{N} p_i \Phi_i \left(I_N \otimes 1_{p\times p} \right) \circ \left(C_{k+1} \bar{\Lambda}_{2,k+1} C_{k+1}^T \right) \Phi_i .\end{aligned}$$

It is easy to see that

$$
\begin{aligned}
&\bar{\mu}_{k+1} S_{k+1} - 2\hat{\Xi}_{k+1} \\
&= 2\bar{\mu}_{k+1}^2 \sum_{i=1}^{N} p_i \Phi_i C_{k+1} \Omega_{k+1|k} C_{k+1}^T \Phi_i + (\bar{\mu}_{k+1} - 2\bar{\mu}_{k+1}^2) \\
&\times \sum_{i=1}^{N} p_i \Phi_i \left(I_N \otimes 1_{p\times p}\right) \circ \left(C_{k+1} \bar{\Lambda}_{2,k+1} C_{k+1}^T\right) \Phi_i - 2\hat{\Xi}_{k+1} \\
&\leq 2\Bigg[\bar{\mu}_{k+1}^2 \sum_{i=1}^{N} p_i \Phi_i C_{k+1} \Omega_{k+1|k} C_{k+1}^T \Phi_i + \bar{\mu}_{k+1} \\
&\times (1 - \bar{\mu}_{k+1}) \sum_{i=1}^{N} p_i \Phi_i \left(I_N \otimes 1_{p\times p}\right) \circ \left(C_{k+1} \bar{\Lambda}_{2,k+1} C_{k+1}^T\right) \Phi_i \Bigg] - 2\hat{\Xi}_{k+1} \\
&= -2 \sum_{i=1}^{N} p_i \Phi_i R_{k+1} \Phi_i \\
&\leq -2\bar{\Phi} R_{k+1} \bar{\Phi} < 0.
\end{aligned}
\tag{13.37}
$$

Secondly, together with (13.36) and (13.37), we have

$$
\begin{aligned}
&\frac{\mathrm{d}\,\mathrm{tr}(\Omega_{k+1|k+1})}{\mathrm{d}\bar{\mu}_{k+1}} \\
&\leq -\mathrm{tr}\left(2\bar{\mu}_{k+1} \Omega_{k+1|k}^T C_{k+1}^T \bar{\Phi} \hat{\Xi}_{k+1}^{-1} \bar{\Phi} R_{k+1} \bar{\Phi} \hat{\Xi}_{k+1}^{-1} \bar{\Phi} C_{k+1} \Omega_{k+1|k}\right).
\end{aligned}
$$

Thus, it is immediate that

$$
\frac{\mathrm{d}\,\mathrm{tr}(\Omega_{k+1|k+1})}{\mathrm{d}\bar{\mu}_{k+1}} \leq 0,
$$

which indicates that $\mathrm{tr}(\Omega_{k+1|k+1})$ is non-increasing when $\bar{\mu}_{k+1}$ increases.

Remark 13.4 So far, a novel state estimation method has been designed for time-varying complex networks subject to RAP, stochastic inner coupling and missing measurements with UOPs. The related information (such as stochastic inner coupling, RAP and missing measurements) has been adequately involved in the estimator gain. In addition, for the sake of evaluating the estimation performance, the monotonicity between the missing probability and the trace of the obtained upper bound has been analyzed from a theoretical viewpoint.

13.2 State Estimation with RSDs and SCS

13.2.1 Problem Formulation

Consider the following class of time-varying nonlinear complex networks with N coupled nodes:

$$\mathbf{x}_{i,k+1} = \mathbf{f}_i(\mathbf{x}_{i,k}) + \sum_{j=1}^{N} \omega_{ij,k}\hat{\Gamma}\mathbf{x}_{j,k} + \mathbf{B}_{i,k}\varpi_{i,k}, \tag{13.38}$$

$$\mathbf{y}_{i,k} = \mathbf{C}_{i,k}\mathbf{x}_{i,k} + \check{\nu}_{i,k}, \tag{13.39}$$

$$\bar{y}_{i,k} = \lambda_{i,k}\mathbf{y}_{i,k} + (1-\lambda_{i,k})\mathbf{y}_{i,k-1}, \tag{13.40}$$

where $\mathbf{x}_{i,k} \in \mathbb{R}^n$ represents the state, $\mathbf{x}_{i,0}$ denotes the initial value with the mean $\bar{\mathbf{x}}_{i,0}$, $\mathbf{y}_{i,k} \in \mathbb{R}^m$ is the measurement, and $\bar{y}_{i,k}$ is the real measurement. The nonlinear function $\mathbf{f}_i(\cdot)$ has continuous and bounded second order partial derivatives. $\varpi_{i,k}$ and $\check{\nu}_{i,k}$ are zero-mean additive noises with covariances $Q_{i,k} > 0$ and $\mathbf{R}_{i,k} > 0$. $W = [\omega_{ij,k}] \in \mathbb{R}^{N\times N}$ is the coupling configuration matrix, where the random variables $\omega_{ij,k}$ obey the uniform distribution over the interval $[a_{ij}, b_{ij}]$. The RSDs are modeled by a set of Bernoulli distributed random variables $\lambda_{i,k} \in \mathbb{R}$. $\hat{\Gamma} = \text{diag}\{\dot{\gamma}_1, \dot{\gamma}_2, \ldots, \dot{\gamma}_n\}$ is the inner coupling matrix, where there is a linking with the j-th state if $\dot{\gamma}_j \neq 0$. $\mathbf{B}_{i,k}$ and $\mathbf{C}_{i,k}$ are bounded time-varying matrices. Moreover, assume that random variables $\omega_{ij,k}$, $\varpi_{i,k}$, $\lambda_{i,k}$, $\check{\nu}_{i,k}$ and $\mathbf{x}_{i,0}$ are independent for each i, j and k.

Note that random variable $\omega_{ij,k}$ obeys the uniform distribution over the interval $[a_{ij}, b_{ij}]$, which satisfies the following statistical properties:

$$\mathbb{E}\{\omega_{ij,k}\} = \bar{\omega}_{ij} = \frac{a_{ij}+b_{ij}}{2}, \quad \text{Var}\{\omega_{ij,k}\} = \Omega_{ij}^2 = \frac{(b_{ij}-a_{ij})^2}{12}.$$

The uncertain occurrence probability of RSDs for each node is described by

$$\begin{aligned} \text{Prob}\{\lambda_{i,k} = 1\} &= \bar{\lambda}_{i,k} + \Delta\lambda_{i,k}, \\ \text{Prob}\{\lambda_{i,k} = 0\} &= 1 - (\bar{\lambda}_{i,k} + \Delta\lambda_{i,k}), \end{aligned} \tag{13.41}$$

where $\bar{\lambda}_{i,k}$ is the nominal expectation of random variable $\lambda_{i,k}$, the probability uncertainty is described by $|\Delta\lambda_{i,k}| \leq \varrho_{i,k}$ with $\varrho_{i,k}$ denoting a given bound.

Remark 13.5 Similar to Hu et al. (2019), on account of some inevitable disturbances including implementation errors, inaccuracy statistical test results and so on, the uncertain occurrence probability is introduced in this chapter to mirror the practical engineering reality. To be more specific, the RSDs are depicted in (13.40) via an array of random variables $\lambda_{i,k}$ $(i = 1, 2, \ldots, N)$, in which the occurrence probabilities are consisted of the nominal expectation $\bar{\lambda}_{i,k}$ and probability uncertainty $\Delta\lambda_{i,k}$ together. Particularly, it is pointed out that the traditional RSDs can be attained via setting $\Delta\lambda_{i,k} = 0$.

Before proceeding, we denote

$$x_{i,k} = \begin{bmatrix} \mathbf{x}_{i,k} \\ \mathbf{x}_{i,k-1} \end{bmatrix}, \quad f_i(x_{i,k}) = \begin{bmatrix} \mathbf{f}_i(\mathbf{x}_{i,k}) \\ 0 \end{bmatrix}, \quad \bar{I} = \begin{bmatrix} 0 & 0 \\ I & 0 \end{bmatrix},$$

$$B_{i,k} = \begin{bmatrix} \mathbf{B}_{i,k} \\ 0 \end{bmatrix}, \quad C_{i,k} = \begin{bmatrix} \mathbf{C}_{i,k} & 0 \\ 0 & \mathbf{C}_{i,k-1} \end{bmatrix}, \quad \nu_{i,k} = \begin{bmatrix} \check{\nu}_{i,k} \\ \check{\nu}_{i,k-1} \end{bmatrix},$$

$$\Gamma = \begin{bmatrix} \hat{\Gamma} & 0 \\ 0 & 0 \end{bmatrix}, \quad \Upsilon_{i,k} = [\lambda_{i,k} I \quad (1 - \lambda_{i,k}) I].$$

Then, one has

$$x_{i,k+1} = f_i(x_{i,k}) + \bar{I} x_{i,k} + \sum_{j=1}^{N} \omega_{ij,k} \Gamma x_{j,k} + B_{i,k} \varpi_{i,k}, \tag{13.42}$$

$$\bar{y}_{i,k} = \Upsilon_{i,k}(C_{i,k} x_{i,k} + \nu_{i,k}). \tag{13.43}$$

Here, $\nu_{i,k}$ is the one-step correlated measurement noise of (13.42) and (13.43) with zero mean and covariance

$$\mathbb{E}\{\nu_{i,k} \nu_{i,l}^T\} = R_{i,k} \delta(k-l) + R^i_{k,k-1} \delta(k-l-1) + R^i_{k,k+1} \delta(k-l+1), \tag{13.44}$$

where

$$R_{i,k} = \begin{bmatrix} \mathbf{R}_{i,k} & 0 \\ 0 & \mathbf{R}_{i,k-1} \end{bmatrix}, \quad R^i_{k,k-1} = \begin{bmatrix} 0 & 0 \\ \mathbf{R}_{i,k-1} & 0 \end{bmatrix}, \quad R^i_{k,k+1} = \begin{bmatrix} 0 & \mathbf{R}_{i,k} \\ 0 & 0 \end{bmatrix}.$$

With the purpose of alleviating the communication burden, the ETM is employed to arrange signal transmission. Thus, for the i-th node, the triggered instants $0 = k^i_0 < k^i_1 < \cdots < k^i_l < \cdots$ can be determined by

$$k^i_{l+1} = \min\{k \in \mathbb{N} | k > k^i_l, \ \ \phi(\bar{y}_{i,k}, \bar{y}_{i,k^i_l}, \theta_i) > 0\}, \tag{13.45}$$

where $\phi(\bar{y}_{i,k}, \bar{y}_{i,k^i_l}, \theta_i)$ is the event generator function described by

$$\phi(\bar{y}_{i,k}, \bar{y}_{i,k^i_l}, \theta_i) = (\bar{y}_{i,k} - \bar{y}_{i,k^i_l})^T (\bar{y}_{i,k} - \bar{y}_{i,k^i_l}) - \theta_i. \tag{13.46}$$

Here, k^i_l is the latest triggered time step of the i-th node, and the scalar $\theta_i > 0$ is an adjustable threshold. Next, it follows from (13.45) and (13.46) that the released signal $y_{i,k}$ for i-th node is modeled as:

$$y_{i,k} = \bar{y}_{k^i_l}, \quad k \in \{k^i_l, k^i_l + 1, \ldots, k^i_{l+1} - 1\}. \tag{13.47}$$

Remark 13.6 Compared with the existing complex dynamical network models, the system model could characterize the RCS and RSDs under time-varying framework, which can be regarded as a more general form of Liang et al. (2014). Besides, in comparison to Li et al. (2017), the ETM is introduced to enhance transmission efficiency and reduce communication cost. In fact, some other communication scheduling strategies have been proposed to regulate signal propagations, such as round-robin protocol, RAP and try-once-discard protocol Bu et al. (2019), Dong et al. (2016), Zhang et al. (2019), Zou et al. (2017). Unlike the communication protocols mentioned above, the chief difference lies in that the released data signal in (13.47) is updated via different rules. According to the practical requirements, the proper communication protocols can be chosen to manage the signal transmission.

For the i-th node, the following state estimator is designed:

$$\hat{x}_{i,k+1|k} = f_i(\hat{x}_{i,k|k}) + \bar{I}\hat{x}_{i,k|k} + \sum_{j=1}^{N} \bar{\omega}_{ij}\Gamma\hat{x}_{j,k|k}, \tag{13.48}$$

$$\hat{x}_{i,k+1|k+1} = \hat{x}_{i,k+1|k} + G_{i,k+1}(y_{i,k+1} - \bar{\Upsilon}_{i,k+1}C_{i,k+1}\hat{x}_{i,k+1|k}), \tag{13.49}$$

where $\hat{x}_{i,k+1|k}$ depicts the prediction, $\hat{x}_{i,k|k}$ represents the estimation with the initial value $\hat{x}_{i,0|0} = [\bar{\mathbf{x}}_{i,0}^T \quad 0^T]^T$, $\bar{\Upsilon}_{i,k+1} = [\bar{\lambda}_{i,k+1}I \quad (1-\bar{\lambda}_{i,k+1})I]$, and $G_{i,k+1}$ represents the estimator gain.

Next, let $e_{i,k+1|k} = x_{i,k+1} - \hat{x}_{i,k+1|k}$ denote the prediction error. Set the estimation error characterized by $e_{i,k+1|k+1} = x_{i,k+1} - \hat{x}_{i,k+1|k+1}$. Accordingly, it is easy to see that

$$\begin{aligned} e_{i,k+1|k} = {} & f_i(x_{i,k}) - f_i(\hat{x}_{i,k|k}) + \bar{I}e_{i,k|k} + \sum_{j=1}^{N}(\omega_{ij,k} - \bar{\omega}_{ij})\Gamma x_{j,k} \\ & + \sum_{j=1}^{N} \bar{\omega}_{ij}\Gamma(x_{j,k} - \hat{x}_{j,k|k}) + B_{i,k}\varpi_{i,k}. \end{aligned} \tag{13.50}$$

Adopting the Taylor series expansion method regarding $\hat{x}_{i,k|k}$, one obtains

$$f_i(x_{i,k}) = f_i(\hat{x}_{i,k|k}) + \bar{A}_{i,k}e_{i,k|k} + o(|e_{i,k|k}^2|), \tag{13.51}$$

where $\bar{A}_{i,k} = \frac{\partial f_i(x_{i,k})}{\partial x_{i,k}}|_{x_{i,k}=\hat{x}_{i,k|k}}$ and $o(|e_{i,k|k}^2|)$ stand for the high-order terms. As in Kai et al. (2010), the terms $o(|e_{i,k|k}^2|)$ are described by $o(|e_{i,k|k}^2|) = L_{i,k}\aleph_{i,k}H_{i,k}e_{i,k|k}$, where $L_{i,k} \in \mathbb{R}^{2n\times 2n}$ are known problem-dependent scaling matrices, $H_{i,k} \in \mathbb{R}^{2n\times 2n}$ are known tuning matrices and $\aleph_{i,k} \in \mathbb{R}^{2n\times 2n}$ depicts the resultant linearization errors with $\aleph_{i,k}\aleph_{i,k}^T \leq I$. Based on (13.50) and (13.51), the prediction error is calculated as

$$\begin{aligned} e_{i,k+1|k} = {} & (A_{i,k} + L_{i,k}\aleph_{i,k}H_{i,k})e_{i,k|k} + \sum_{j=1}^{N}(\omega_{ij,k} - \bar{\omega}_{ij})\Gamma x_{j,k} \\ & + \sum_{j=1}^{N} \bar{\omega}_{ij}\Gamma(x_{j,k} - \hat{x}_{j,k|k}) + B_{i,k}\varpi_{i,k}, \end{aligned} \tag{13.52}$$

where $A_{i,k} = \bar{A}_{i,k} + \bar{I}$.

Similarly, the estimation error is computed as

$$\begin{aligned} e_{i,k+1|k+1} = & (I - G_{i,k+1}\bar{\Upsilon}_{i,k+1}C_{i,k+1})e_{i,k+1|k} - G_{i,k+1}(y_{i,k+1} - \bar{y}_{i,k+1}) \\ & -G_{i,k+1}\tilde{\Upsilon}_{i,k+1}C_{i,k+1}x_{i,k+1} - G_{i,k+1}\check{\Upsilon}_{i,k+1}C_{i,k+1}x_{i,k+1} \\ & -G_{i,k+1}\Upsilon_{i,k+1}\nu_{i,k+1}, \end{aligned} \tag{13.53}$$

where

$$\begin{aligned} \tilde{\Upsilon}_{i,k+1} &= \Upsilon_{i,k+1} - \hat{\Upsilon}_{i,k+1} = \tilde{\lambda}_{i,k+1}\begin{bmatrix} I & -I \end{bmatrix}, \\ \hat{\Upsilon}_{i,k+1} &= \begin{bmatrix} (\bar{\lambda}_{i,k+1} + \Delta\lambda_{i,k+1})I & (1 - \bar{\lambda}_{i,k+1} - \Delta\lambda_{i,k+1})I \end{bmatrix}, \\ \check{\Upsilon}_{i,k+1} &= \begin{bmatrix} \Delta\lambda_{i,k+1}I & -\Delta\lambda_{i,k+1}I \end{bmatrix} \end{aligned}$$

with $\tilde{\lambda}_{i,k+1} = \lambda_{i,k+1} - (\bar{\lambda}_{i,k+1} + \Delta\lambda_{i,k+1})$. Then, it follows from $\mathbb{E}\{\tilde{\lambda}_{i,k+1}\} = 0$ that $\mathbb{E}\{\tilde{\Upsilon}_{i,k+1}\} = 0$.

In what follows, the prediction error covariance and EEC are defined as $P_{i,k+1|k} = \mathbb{E}\{e_{i,k+1|k}e_{i,k+1|k}^T\}$ and $P_{i,k+1|k+1} = \mathbb{E}\{e_{i,k+1|k+1}e_{i,k+1|k+1}^T\}$, respectively. Next, we aim to construct a novel estimator of (13.48) and (13.49) such that an optimal upper bound is derived for $P_{i,k+1|k+1}$ and the explicit expression form of the estimator gain $G_{i,k+1}$ is given. Moreover, a sufficient condition is shown to prove the exponential boundedness regarding the estimation error dynamics in the mean-square sense.

13.2.2 Design of State Estimation Method

In this section, the prediction error covariance and EEC are derived by resorting to the related definitions. In the sequel, the uncertain/cross terms are addressed and the upper bound matrix of the EEC is calculated. Thirdly, a desirable estimator gain matrix is given via minimizing this upper bound.

Lemma 13.2 *The prediction error covariance $P_{i,k+1|k}$ is characterized as*

$$\begin{aligned} & P_{i,k+1|k} \\ & = (A_{i,k} + L_{i,k}\aleph_{i,k}H_{i,k})P_{i,k|k}(A_{i,k} + L_{i,k}\aleph_{i,k}H_{i,k})^T \\ & + \sum_{j=1}^{N} \bar{\omega}_{ij}\mathbb{E}\{(A_{i,k} + L_{i,k}\aleph_{i,k}H_{i,k})e_{i,k|k}e_{j,k|k}^T\Gamma^T + \Gamma e_{j,k|k}e_{i,k|k}^T(A_{i,k} + L_{i,k}\aleph_{i,k}H_{i,k})^T\} \\ & + \sum_{j=1}^{N} \Omega_{ij}^2\Gamma\mathbb{E}\{x_{j,k}x_{j,k}^T\}\Gamma^T + \sum_{j=1}^{N}\sum_{l=1}^{N} \bar{\omega}_{ij}\bar{\omega}_{il}\Gamma\mathbb{E}\{e_{j,k|k}e_{l,k|k}^T\}\Gamma^T \\ & + B_{i,k}Q_{i,k}B_{i,k}^T. \end{aligned} \tag{13.54}$$

Proof Note that $\mathbb{E}\{(\omega_{ij,k}-\bar{\omega}_{ij})(\omega_{il,k}-\bar{\omega}_{il})\}=0 \quad (j\neq l)$ and $\mathbb{E}\{(\omega_{ij,k}-\bar{\omega}_{ij})(\omega_{ij,k}-\bar{\omega}_{ij})\}=\Omega_{ij}^2$, then it can be concluded that (13.54) holds. □

Lemma 13.3 *The EEC $P_{i,k+1|k+1}$ is computed as*

$$
\begin{aligned}
&P_{i,k+1|k+1}\\
&=(I-G_{i,k+1}\bar{\Upsilon}_{i,k+1}C_{i,k+1})P_{i,k+1|k}(I-G_{i,k+1}\bar{\Upsilon}_{i,k+1}C_{i,k+1})^T\\
&\quad+G_{i,k+1}\mathbb{E}\{(y_{i,k+1}-\bar{y}_{i,k+1})(y_{i,k+1}-\bar{y}_{i,k+1})^T\}G_{i,k+1}^T-\mathscr{A}_{k+1}-\mathscr{A}_{k+1}^T\\
&\quad-\mathscr{B}_{k+1}-\mathscr{B}_{k+1}^T+\mathscr{C}_{k+1}+\mathscr{C}_{k+1}^T+\mathscr{D}_{k+1}+\mathscr{D}_{k+1}^T+\mathscr{E}_{k+1}+\mathscr{E}_{k+1}^T\\
&\quad+G_{i,k+1}\mathbb{E}\{\tilde{\Upsilon}_{i,k+1}C_{i,k+1}x_{i,k+1}x_{i,k+1}^TC_{i,k+1}^T\tilde{\Upsilon}_{i,k+1}^T\}G_{i,k+1}^T\\
&\quad+G_{i,k+1}\check{\Upsilon}_{i,k+1}C_{i,k+1}\mathbb{E}\{x_{i,k+1}x_{i,k+1}^T\}C_{i,k+1}^T\check{\Upsilon}_{i,k+1}^TG_{i,k+1}^T\\
&\quad+G_{i,k+1}\mathbb{E}\{\Upsilon_{i,k+1}\nu_{i,k+1}\nu_{i,k+1}^T\Upsilon_{i,k+1}^T\}G_{i,k+1}^T,
\end{aligned}
\tag{13.55}
$$

where

$$
\begin{aligned}
\mathscr{A}_{k+1}&=(I-G_{i,k+1}\bar{\Upsilon}_{i,k+1}C_{i,k+1})\mathbb{E}\{e_{i,k+1|k}(y_{i,k+1}-\bar{y}_{i,k+1})^T\}G_{i,k+1}^T,\\
\mathscr{B}_{k+1}&=(I-G_{i,k+1}\bar{\Upsilon}_{i,k+1}C_{i,k+1})\mathbb{E}\{e_{i,k+1|k}x_{i,k+1}^T\}C_{i,k+1}^T\check{\Upsilon}_{i,k+1}^TG_{i,k+1}^T,\\
\mathscr{C}_{k+1}&=G_{i,k+1}\mathbb{E}\{(y_{i,k+1}-\bar{y}_{i,k+1})x_{i,k+1}^T\}C_{i,k+1}^T\tilde{\Upsilon}_{i,k+1}^TG_{i,k+1}^T,\\
\mathscr{D}_{k+1}&=G_{i,k+1}\mathbb{E}\{(y_{i,k+1}-\bar{y}_{i,k+1})x_{i,k+1}^T\}C_{i,k+1}^T\check{\Upsilon}_{i,k+1}^TG_{i,k+1}^T,\\
\mathscr{E}_{k+1}&=G_{i,k+1}\mathbb{E}\{(y_{i,k+1}-\bar{y}_{i,k+1})\nu_{i,k+1}^T\Upsilon_{i,k+1}^T\}G_{i,k+1}^T.
\end{aligned}
\tag{13.56}
$$

Proof In light of the definition of the EEC, one gets

$$
\begin{aligned}
&P_{i,k+1|k+1}\\
&=\mathbb{E}\left\{e_{i,k+1|k+1}e_{i,k+1|k+1}^T\right\}\\
&=(I-G_{i,k+1}\bar{\Upsilon}_{i,k+1}C_{i,k+1})P_{k+1|k}(I-G_{i,k+1}\bar{\Upsilon}_{i,k+1}C_{i,k+1})^T-\mathscr{A}_{k+1}\\
&\quad-\mathscr{A}_{k+1}^T-\mathscr{A}_{1,k+1}-\mathscr{A}_{1,k+1}^T-\mathscr{B}_{k+1}-\mathscr{B}_{k+1}^T-\mathscr{A}_{2,k+1}-\mathscr{A}_{2,k+1}^T\\
&\quad+G_{i,k+1}\mathbb{E}\{(y_{i,k+1}-\bar{y}_{i,k+1})(y_{i,k+1}-\bar{y}_{i,k+1})^T\}G_{i,k+1}^T\\
&\quad+\mathscr{C}_{k+1}+\mathscr{C}_{k+1}^T+\mathscr{D}_{k+1}+\mathscr{D}_{k+1}^T+\mathscr{E}_{k+1}+\mathscr{E}_{k+1}^T\\
&\quad+G_{i,k+1}\mathbb{E}\{\tilde{\Upsilon}_{i,k+1}C_{i,k+1}x_{i,k+1}x_{i,k+1}^TC_{i,k+1}^T\tilde{\Upsilon}_{i,k+1}^T\}G_{i,k+1}^T\\
&\quad+\mathscr{A}_{3,k+1}+\mathscr{A}_{3,k+1}^T+\mathscr{A}_{4,k+1}+\mathscr{A}_{4,k+1}^T\\
&\quad+G_{i,k+1}\check{\Upsilon}_{i,k+1}C_{i,k+1}\mathbb{E}\{x_{i,k+1}x_{i,k+1}^T\}C_{i,k+1}^T\check{\Upsilon}_{i,k+1}^TG_{i,k+1}^T+\mathscr{A}_{5,k+1}+\mathscr{A}_{5,k+1}^T\\
&\quad+G_{i,k+1}\mathbb{E}\{\Upsilon_{i,k+1}\nu_{i,k+1}\nu_{i,k+1}^T\Upsilon_{i,k+1}^T\}G_{i,k+1}^T,
\end{aligned}
\tag{13.57}
$$

where

$$
\begin{aligned}
\mathscr{A}_{1,k+1}&=(I-G_{i,k+1}\bar{\Upsilon}_{i,k+1}C_{i,k+1})\mathbb{E}\{e_{i,k+1|k}x_{i,k+1}^TC_{i,k+1}^T\tilde{\Upsilon}_{i,k+1}^T\}G_{i,k+1}^T,\\
\mathscr{A}_{2,k+1}&=(I-G_{i,k+1}\bar{\Upsilon}_{i,k+1}C_{i,k+1})\mathbb{E}\{e_{i,k+1|k}\nu_{i,k+1}^T\Upsilon_{i,k+1}^T\}G_{i,k+1}^T,\\
\mathscr{A}_{3,k+1}&=G_{i,k+1}\mathbb{E}\{\tilde{\Upsilon}_{i,k+1}C_{i,k+1}x_{i,k+1}x_{i,k+1}^T\}C_{i,k+1}^T\check{\Upsilon}_{i,k+1}^TG_{i,k+1}^T,
\end{aligned}
$$

$$\mathscr{A}_{4,k+1} = G_{i,k+1}\mathbb{E}\{\tilde{\Upsilon}_{i,k+1}C_{i,k+1}x_{i,k+1}v_{i,k+1}^T\Upsilon_{i,k+1}^T\}G_{i,k+1}^T,$$
$$\mathscr{A}_{5,k+1} = G_{i,k+1}\mathbb{E}\{\check{\Upsilon}_{i,k+1}C_{i,k+1}x_{i,k+1}v_{i,k+1}^T\Upsilon_{i,k+1}^T\}G_{i,k+1}^T.$$

It can be summarized that $\mathscr{A}_{i,k+1} = 0$ $(i = 1, 2, \ldots, 5)$, which completes this lemma.

Remark 13.7 It is crucial to note that it is impossible to calculate the exact EEC because of the existence of some unknown terms (e.g. $\aleph_{i,k+1}$, $\mathbb{E}\{x_{i,k+1}x_{i,k+1}^T\}$ and $\check{\Upsilon}_{i,k+1}$) in (13.54) and (13.55), in which those terms are resulted from the linearization errors, RSDs and ETM. In order to overcome this challenge, we adopt an alternative scheme to complete the estimator design, i.e., calculating an optimized upper bound of the EEC and then deriving the estimator gain through minimizing the trace of this upper bound.

According to the theoretical results given in Lemmas 13.2-13.3, the recursion of an optimized upper bound concerning on the EEC is given.

Theorem 13.3 *Given some constants $\gamma_k > 0$ and $\varepsilon_i > 0$ $(i = 1, 2, \ldots, 8)$, assume that the following recursive matrix equations:*

$$\begin{aligned}\Omega_{i,k+1|k} &= (1+\bar{\omega}_i)A_{i,k}(\Omega_{i,k|k}^{-1} - \gamma_k H_{i,k}^T H_{i,k})^{-1}A_{i,k}^T + (1+\bar{\omega}_i)\gamma_k^{-1}L_{i,k}L_{i,k}^T \\ &\quad + \sum_{j=1}^{N}\left[\bar{\omega}_{ij} + \bar{\omega}_i\bar{\omega}_{ij} + (1+\varepsilon_1)\Omega_{ij}^2\right]\Gamma\Omega_{j,k|k}\Gamma^T \\ &\quad + (1+\varepsilon_1^{-1})\sum_{j=1}^{N}\Omega_{ij}^2\Gamma\hat{x}_{j,k|k}\hat{x}_{j,k|k}^T\Gamma^T + B_{i,k}Q_{i,k}B_{i,k}^T,\end{aligned} \tag{13.58}$$

$$\begin{aligned}\Omega_{i,k+1|k+1} &= (1+\varepsilon_2+\varepsilon_3)(I - G_{i,k+1}\bar{\Upsilon}_{i,k+1}C_{i,k+1})\Omega_{i,k+1|k}(I - G_{i,k+1}\bar{\Upsilon}_{i,k+1}C_{i,k+1})^T \\ &\quad + (1+\varepsilon_2^{-1}+\varepsilon_4+\varepsilon_5+\varepsilon_6)\theta_i G_{i,k+1}G_{i,k+1}^T \\ &\quad + (1+\varepsilon_4^{-1})\check{\rho}_{i,k+1}G_{i,k+1}\bar{H}C_{i,k+1}\Psi_{i,k+1|k}C_{i,k+1}^T\bar{H}^T G_{i,k+1}^T \\ &\quad + (1+\varepsilon_3^{-1}+\varepsilon_5^{-1})\varrho_{i,k+1}^2 G_{i,k+1}\bar{H}C_{i,k+1}\Psi_{i,k+1|k}C_{i,k+1}^T\bar{H}^T G_{i,k+1}^T \\ &\quad + (1+\varepsilon_6^{-1})\check{\rho}_{i,k+1}G_{i,k+1}\bar{H}R_{i,k+1}\bar{H}^T G_{i,k+1}^T \\ &\quad + (1+\varepsilon_6^{-1})(1+\varepsilon_8)G_{i,k+1}\bar{\Upsilon}_{i,k+1}R_{i,k+1}\bar{\Upsilon}_{i,k+1}^T G_{i,k+1}^T \\ &\quad + (1+\varepsilon_6^{-1})(1+\varepsilon_8^{-1})\varrho_{i,k+1}^2 G_{i,k+1}\bar{H}R_{i,k+1}\bar{H}^T G_{i,k+1}^T\end{aligned} \tag{13.59}$$

under the initial condition $\Omega_{i,0|0} \geq P_{i,0|0}$ and the following constraint:

$$\gamma_k^{-1}I - H_{i,k}\Omega_{i,k|k}H_{i,k}^T > 0 \tag{13.60}$$

have solutions $\Omega_{i,k+1|k} > 0$ *and* $\Omega_{i,k+1|k+1} > 0$, *where*

$$\bar{H} = [I \quad -I],\ \bar{\omega}_i = \sum_{j=1}^{N} \bar{\omega}_{ij},$$

$$\check{\rho}_{i,k+1} = \min\left\{\frac{1}{4}, (\bar{\lambda}_{i,k+1} + \varrho_{i,k+1})(1 + \varrho_{i,k+1} - \bar{\lambda}_{i,k+1})\right\},$$

$$\bar{\Psi}_{i,k+1|k} = (1 + \varepsilon_7)\Omega_{i,k+1|k} + (1 + \varepsilon_7^{-1})\hat{x}_{i,k+1|k}\hat{x}_{i,k+1|k}^T.$$

Then, it is shown that

$$P_{i,k+1|k+1} \le \Omega_{i,k+1|k+1}. \tag{13.61}$$

Additionally, if the following estimator gain is adopted:

$$\begin{aligned}
&G_{i,k+1} \\
&= (1 + \varepsilon_2 + \varepsilon_3)\Omega_{i,k+1|k} C_{i,k+1}^T \bar{\Upsilon}_{i,k+1}^T \\
&\times \Big\{(1 + \varepsilon_2 + \varepsilon_3)\bar{\Upsilon}_{i,k+1} C_{i,k+1} \Omega_{i,k+1|k} C_{i,k+1}^T \bar{\Upsilon}_{i,k+1}^T \\
&+ (1 + \varepsilon_2^{-1} + \varepsilon_4 + \varepsilon_5 + \varepsilon_6)\theta_i I + (1 + \varepsilon_4^{-1})\check{\rho}_{i,k+1} \bar{H} C_{i,k+1} \bar{\Psi}_{i,k+1|k} C_{i,k+1}^T \bar{H}^T \\
&+ (1 + \varepsilon_3^{-1} + \varepsilon_5^{-1})\varrho_{i,k+1}^2 \bar{H} C_{i,k+1} \bar{\Psi}_{i,k+1|k} C_{i,k+1}^T \bar{H}^T + (1 + \varepsilon_6^{-1})\check{\rho}_{i,k+1} \bar{H} R_{i,k+1} \bar{H}^T \\
&+ (1 + \varepsilon_6^{-1})(1 + \varepsilon_8)\bar{\Upsilon}_{i,k+1} R_{i,k+1} \bar{\Upsilon}_{i,k+1}^T \\
&+ (1 + \varepsilon_6^{-1})(1 + \varepsilon_8^{-1})\varrho_{i,k+1}^2 \bar{H} R_{i,k+1} \bar{H}^T \Big\}^{-1},
\end{aligned} \tag{13.62}$$

then the trace of the obtained upper bound matrix can be minimized.

Proof Firstly, by virtue of Lemma 2.2, the second and the fourth unknown terms on the right side of (13.54) can be handled as

$$\begin{aligned}
&\sum_{j=1}^{N} \bar{\omega}_{ij} \mathbb{E}\{(A_{i,k} + L_{i,k}\aleph_{i,k} H_{i,k}) e_{i,k|k} e_{j,k|k}^T \Gamma^T + \Gamma e_{j,k|k} e_{i,k|k}^T (A_{i,k} + L_{i,k}\aleph_{i,k} H_{i,k})^T\} \\
&\le \sum_{j=1}^{N} \bar{\omega}_{ij} \mathbb{E}\{(A_{i,k} + L_{i,k}\aleph_{i,k} H_{i,k}) e_{i,k|k} e_{i,k|k}^T (A_{i,k} + L_{i,k}\aleph_{i,k} H_{i,k})^T + \Gamma e_{j,k|k} e_{j,k|k}^T \Gamma^T\} \\
&= \bar{\omega}_i (A_{i,k} + L_{i,k}\aleph_{i,k} H_{i,k}) P_{i,k|k} (A_{i,k} + L_{i,k}\aleph_{i,k} H_{i,k})^T + \sum_{j=1}^{N} \bar{\omega}_{ij} \Gamma P_{j,k|k} \Gamma^T
\end{aligned} \tag{13.63}$$

and

$$\begin{aligned}
&\sum_{j=1}^{N}\sum_{l=1}^{N}\bar{\omega}_{ij}\bar{\omega}_{il}\Gamma\mathbb{E}\{e_{j,k|k}e_{l,k|k}^{T}\}\Gamma^{T}\\
&\le\frac{1}{2}\sum_{j=1}^{N}\sum_{l=1}^{N}\bar{\omega}_{ij}\bar{\omega}_{il}\Gamma\mathbb{E}\{e_{j,k|k}e_{j,k|k}^{T}+e_{l,k|k}e_{l,k|k}^{T}\}\Gamma^{T}\\
&=\bar{\omega}_{i}\sum_{j=1}^{N}\bar{\omega}_{ij}\Gamma P_{j,k|k}\Gamma^{T},
\end{aligned}\tag{13.64}$$

where $\bar{\omega}_i$ is defined below (13.60). In what follows, the third unknown term of (13.54) is tackled as

$$\begin{aligned}
&\sum_{j=1}^{N}\Omega_{ij}^{2}\Gamma\mathbb{E}\{x_{j,k}x_{j,k}^{T}\}\Gamma^{T}\\
=&\sum_{j=1}^{N}\Omega_{ij}^{2}\Gamma\mathbb{E}\{(e_{j,k|k}+\hat{x}_{j,k|k})(e_{j,k|k}+\hat{x}_{j,k|k})^{T}\}\Gamma^{T}\\
\le&(1+\varepsilon_{1})\sum_{j=1}^{N}\Omega_{ij}^{2}\Gamma P_{j,k|k}\Gamma^{T}+(1+\varepsilon_{1}^{-1})\sum_{j=1}^{N}\Omega_{ij}^{2}\Gamma\hat{x}_{j,k|k}\hat{x}_{j,k|k}^{T}\Gamma^{T},
\end{aligned}\tag{13.65}$$

where $\varepsilon_1 > 0$ is a scalar. Substituting (13.63)–(13.65) into (13.54) results in

$$\begin{aligned}
P_{i,k+1|k}\le&(1+\bar{\omega}_{i})(A_{i,k}+L_{i,k}\aleph_{i,k}H_{i,k})P_{i,k|k}(A_{i,k}+L_{i,k}\aleph_{i,k}H_{i,k})^{T}\\
&+\sum_{j=1}^{N}\left[\bar{\omega}_{ij}+\bar{\omega}_{i}\bar{\omega}_{ij}+(1+\varepsilon_{1})\Omega_{ij}^{2}\right]\Gamma P_{j,k|k}\Gamma^{T}\\
&+(1+\varepsilon_{1}^{-1})\sum_{j=1}^{N}\Omega_{ij}^{2}\Gamma\hat{x}_{j,k|k}\hat{x}_{j,k|k}^{T}\Gamma^{T}+B_{i,k}Q_{i,k}B_{i,k}^{T}.
\end{aligned}\tag{13.66}$$

Based on Lemma 2.2, it follows that

$$\begin{aligned}
-\mathscr{A}_{k+1}-\mathscr{A}_{k+1}^{T}\le&\varepsilon_{2}(I-G_{i,k+1}\bar{\Upsilon}_{i,k+1}C_{i,k+1})P_{i,k+1|k}(I-G_{i,k+1}\bar{\Upsilon}_{i,k+1}C_{i,k+1})^{T}\\
&+\varepsilon_{2}^{-1}G_{i,k+1}\mathbb{E}\{(y_{i,k+1}-\bar{y}_{i,k+1})(y_{i,k+1}-\bar{y}_{i,k+1})^{T}\}\\
&\times G_{i,k+1}^{T},
\end{aligned}\tag{13.67}$$

$$\begin{aligned}
-\mathscr{B}_{k+1}-\mathscr{B}_{k+1}^{T}\le&\varepsilon_{3}(I-G_{i,k+1}\bar{\Upsilon}_{i,k+1}C_{i,k+1})P_{i,k+1|k}(I-G_{i,k+1}\bar{\Upsilon}_{i,k+1}C_{i,k+1})^{T}\\
&+\varepsilon_{3}^{-1}G_{i,k+1}\check{\Upsilon}_{i,k+1}C_{i,k+1}\mathbb{E}\{x_{i,k+1}x_{i,k+1}^{T}\}\\
&\times C_{i,k+1}^{T}\check{\Upsilon}_{i,k+1}^{T}G_{i,k+1}^{T},
\end{aligned}\tag{13.68}$$

$$\begin{aligned}\mathscr{C}_{k+1}+\mathscr{C}_{k+1}^T \leq & \varepsilon_4 G_{i,k+1}\mathbb{E}\{(y_{i,k+1}-\bar{y}_{i,k+1})(y_{i,k+1}-\bar{y}_{i,k+1})^T\}G_{i,k+1}^T \\ & +\varepsilon_4^{-1}G_{i,k+1}\mathbb{E}\{\tilde{\Upsilon}_{i,k+1}C_{i,k+1}x_{i,k+1}x_{i,k+1}^T C_{i,k+1}^T\tilde{\Upsilon}_{i,k+1}^T\} \\ & \times G_{i,k+1}^T, \end{aligned} \tag{13.69}$$

$$\begin{aligned}\mathscr{D}_{k+1}+\mathscr{D}_{k+1}^T \leq & \varepsilon_5 G_{i,k+1}\mathbb{E}\{(y_{i,k+1}-\bar{y}_{i,k+1})(y_{i,k+1}-\bar{y}_{i,k+1})^T\}G_{i,k+1}^T \\ & +\varepsilon_5^{-1}G_{i,k+1}\check{\Upsilon}_{i,k+1}C_{i,k+1}\mathbb{E}\{x_{i,k+1}x_{i,k+1}^T\} \\ & \times C_{i,k+1}^T\check{\Upsilon}_{i,k+1}^T G_{i,k+1}^T, \end{aligned} \tag{13.70}$$

$$\begin{aligned}\mathscr{E}_{k+1}+\mathscr{E}_{k+1}^T \leq & \varepsilon_6 G_{i,k+1}\mathbb{E}\{(y_{i,k+1}-\bar{y}_{i,k+1})(y_{i,k+1}-\bar{y}_{i,k+1})^T\}G_{i,k+1}^T \\ & +\varepsilon_6^{-1}G_{i,k+1}\mathbb{E}\{\Upsilon_{i,k+1}\nu_{i,k+1}\nu_{i,k+1}^T\Upsilon_{i,k+1}^T\}G_{i,k+1}^T \end{aligned} \tag{13.71}$$

with $\varepsilon_i > 0\ (i = 2, 3, \ldots, 6)$ being scalars. Substituting (13.67)–(13.71) into (13.55) leads to

$$\begin{aligned} & P_{i,k+1|k+1} \\ \leq & (1+\varepsilon_2+\varepsilon_3)(I-G_{i,k+1}\bar{\Upsilon}_{i,k+1}C_{i,k+1})P_{i,k+1|k}(I-G_{i,k+1}\bar{\Upsilon}_{i,k+1}C_{i,k+1})^T \\ & +(1+\varepsilon_2^{-1}+\varepsilon_4+\varepsilon_5+\varepsilon_6)G_{i,k+1}\mathbb{E}\{(y_{i,k+1}-\bar{y}_{i,k+1})(y_{i,k+1}-\bar{y}_{i,k+1})^T\}G_{i,k+1}^T \\ & +(1+\varepsilon_4^{-1})G_{i,k+1}\mathbb{E}\{\tilde{\Upsilon}_{i,k+1}C_{i,k+1}x_{i,k+1}x_{i,k+1}^T C_{i,k+1}^T\tilde{\Upsilon}_{i,k+1}^T\}G_{i,k+1}^T \\ & +(1+\varepsilon_3^{-1}+\varepsilon_5^{-1})G_{i,k+1}\check{\Upsilon}_{i,k+1}C_{i,k+1}\mathbb{E}\{x_{i,k+1}x_{i,k+1}^T\}C_{i,k+1}^T\check{\Upsilon}_{i,k+1}^T G_{i,k+1}^T \\ & +(1+\varepsilon_6^{-1})G_{i,k+1}\mathbb{E}\{\Upsilon_{i,k+1}\nu_{i,k+1}\nu_{i,k+1}^T\Upsilon_{i,k+1}^T\}G_{i,k+1}^T. \end{aligned} \tag{13.72}$$

Subsequently, noting the event-triggered condition (13.46), one obtains

$$\begin{aligned} & (1+\varepsilon_2^{-1}+\varepsilon_4+\varepsilon_5+\varepsilon_6)G_{i,k+1}\mathbb{E}\{(y_{i,k+1}-\bar{y}_{i,k+1})(y_{i,k+1}-\bar{y}_{i,k+1})^T\}G_{i,k+1}^T \\ & \leq (1+\varepsilon_2^{-1}+\varepsilon_4+\varepsilon_5+\varepsilon_6)\theta_i G_{i,k+1}G_{i,k+1}^T. \end{aligned} \tag{13.73}$$

Noting the formulation of $\tilde{\Upsilon}_{i,k+1}$, we have

$$\begin{aligned} & \mathbb{E}\{\tilde{\Upsilon}_{i,k+1}C_{i,k+1}x_{i,k+1}x_{i,k+1}^T C_{i,k+1}^T\tilde{\Upsilon}_{i,k+1}^T\} \\ = & \bar{\sigma}_{i,k+1}\bar{H}C_{i,k+1}\mathbb{E}\{x_{i,k+1}x_{i,k+1}^T\}C_{i,k+1}^T\bar{H}^T, \end{aligned} \tag{13.74}$$

where $\bar{\sigma}_{i,k+1} = (\bar{\lambda}_{i,k+1}+\Delta\lambda_{i,k+1})[1-(\bar{\lambda}_{i,k+1}+\Delta\lambda_{i,k+1})]$. It indicates that $\bar{\sigma}_{i,k+1} \leq \check{\rho}_{i,k+1}$.

On the other hand, for a real number $\varepsilon_7 > 0$, similar to the analysis of (13.65), one has

$$\begin{aligned}\mathbb{E}\{x_{i,k+1}x_{i,k+1}^T\} = & \mathbb{E}\{(e_{i,k+1|k}+\hat{x}_{i,k+1|k})(e_{i,k+1|k}+\hat{x}_{i,k+1|k})^T\} \\ \leq & \mathbb{E}\{(1+\varepsilon_7)e_{i,k+1|k}e_{i,k+1|k}^T+(1+\varepsilon_7^{-1})\hat{x}_{i,k+1|k}\hat{x}_{i,k+1|k}^T\} \\ = & (1+\varepsilon_7)P_{i,k+1|k}+(1+\varepsilon_7^{-1})\hat{x}_{i,k+1|k}\hat{x}_{i,k+1|k}^T \\ := & \Psi_{i,k+1|k}. \end{aligned} \tag{13.75}$$

Thus, it can be inferred from (13.74) and (13.75) that

$$\begin{aligned}&(1+\varepsilon_4^{-1})G_{i,k+1}\mathbb{E}\{\tilde{\Upsilon}_{i,k+1}C_{i,k+1}x_{i,k+1}x_{i,k+1}^T C_{i,k+1}^T\tilde{\Upsilon}_{i,k+1}^T\}G_{i,k+1}^T\\&\leq(1+\varepsilon_4^{-1})\breve{\rho}_{i,k+1}G_{i,k+1}\bar{H}C_{i,k+1}\Psi_{i,k+1|k}C_{i,k+1}^T\bar{H}^T G_{i,k+1}^T.\end{aligned}\tag{13.76}$$

Now, the fourth unknown term of the right side of (13.72) is addressed as

$$\begin{aligned}&(1+\varepsilon_3^{-1}+\varepsilon_5^{-1})G_{i,k+1}\check{\Upsilon}_{i,k+1}C_{i,k+1}\mathbb{E}\{x_{i,k+1}x_{i,k+1}^T\}C_{i,k+1}^T\check{\Upsilon}_{i,k+1}^T G_{i,k+1}^T\\&\leq(1+\varepsilon_3^{-1}+\varepsilon_5^{-1})\Delta\lambda_{i,k+1}^2 G_{i,k+1}\bar{H}C_{i,k+1}\Psi_{i,k+1|k}C_{i,k+1}^T\bar{H}^T G_{i,k+1}^T\\&\leq(1+\varepsilon_3^{-1}+\varepsilon_5^{-1})\varrho_{i,k+1}^2 G_{i,k+1}\bar{H}C_{i,k+1}\Psi_{i,k+1|k}C_{i,k+1}^T\bar{H}^T G_{i,k+1}^T.\end{aligned}\tag{13.77}$$

It follows from $\Upsilon_{i,k+1}=\tilde{\Upsilon}_{i,k+1}+\bar{\Upsilon}_{i,k+1}+\check{\Upsilon}_{i,k+1}$ that

$$\begin{aligned}&G_{i,k+1}\mathbb{E}\{\Upsilon_{i,k+1}\nu_{i,k+1}\nu_{i,k+1}^T\Upsilon_{i,k+1}^T\}G_{i,k+1}^T\\&\leq\bar{\sigma}_{i,k+1}G_{i,k+1}H_{i,k+1}R_{i,k+1}H_{i,k+1}^T G_{i,k+1}^T+(1+\varepsilon_8)G_{i,k+1}\bar{\Upsilon}_{i,k+1}R_{i,k+1}\bar{\Upsilon}_{i,k+1}^T G_{i,k+1}^T\\&\quad+(1+\varepsilon_8^{-1})G_{i,k+1}\check{\Upsilon}_{i,k+1}R_{i,k+1}\check{\Upsilon}_{i,k+1}^T G_{i,k+1}^T\\&\leq\breve{\rho}_{i,k+1}G_{i,k+1}\bar{H}R_{i,k+1}\bar{H}^T G_{i,k+1}^T+(1+\varepsilon_8)G_{i,k+1}\bar{\Upsilon}_{i,k+1}R_{i,k+1}\bar{\Upsilon}_{i,k+1}^T G_{i,k+1}^T\\&\quad+(1+\varepsilon_8^{-1})\varrho_{i,k+1}^2 G_{i,k+1}\bar{H}R_{i,k+1}\bar{H}^T G_{i,k+1}^T\end{aligned}\tag{13.78}$$

with $\varepsilon_8>0$ being a real scalar. Combining with (13.73)–(13.78), one obtains

$$\begin{aligned}P_{i,k+1|k+1}\leq&(1+\varepsilon_2+\varepsilon_3)(I-G_{i,k+1}\bar{\Upsilon}_{i,k+1}C_{i,k+1})P_{i,k+1|k}(I-G_{i,k+1}\bar{\Upsilon}_{i,k+1}C_{i,k+1})^T\\&+(1+\varepsilon_2^{-1}+\varepsilon_4+\varepsilon_5+\varepsilon_6)\theta_i G_{i,k+1}G_{i,k+1}^T\\&+(1+\varepsilon_4^{-1})\breve{\rho}_{i,k+1}G_{i,k+1}\bar{H}C_{i,k+1}\Psi_{i,k+1|k}C_{i,k+1}^T\bar{H}^T G_{i,k+1}^T\\&+(1+\varepsilon_3^{-1}+\varepsilon_5^{-1})\varrho_{i,k+1}^2 G_{i,k+1}\bar{H}C_{i,k+1}\Psi_{i,k+1|k}C_{i,k+1}^T\bar{H}^T G_{i,k+1}^T\\&+(1+\varepsilon_6^{-1})\breve{\rho}_{i,k+1}G_{i,k+1}\bar{H}R_{i,k+1}\bar{H}^T G_{i,k+1}^T\\&+(1+\varepsilon_6^{-1})(1+\varepsilon_8)G_{i,k+1}\bar{\Upsilon}_{i,k+1}R_{i,k+1}\bar{\Upsilon}_{i,k+1}^T G_{i,k+1}^T\\&+(1+\varepsilon_6^{-1})(1+\varepsilon_8^{-1})\varrho_{i,k+1}^2 G_{i,k+1}\bar{H}R_{i,k+1}\bar{H}^T G_{i,k+1}^T.\end{aligned}\tag{13.79}$$

Noticing (13.58), (13.59), (13.66), (13.79) and adopting the mathematical induction method, it is verified that $P_{i,k+1|k+1}\leq\Omega_{i,k+1|k+1}$ is true for all k.

Next, the partial derivative of $\mathrm{tr}(\Omega_{i,k+1|k+1})$ regarding $G_{i,k+1}$ is calculated as

$$\begin{aligned}&\frac{\partial\mathrm{tr}(\Omega_{i,k+1|k+1})}{\partial G_{i,k+1}}\\&=-2(1+\varepsilon_2+\varepsilon_3)(I-G_{i,k+1}\bar{\Upsilon}_{i,k+1}C_{i,k+1})\Omega_{i,k+1|k}C_{i,k+1}^T\bar{\Upsilon}_{i,k+1}^T\\&+2G_{i,k+1}\Big\{(1+\varepsilon_2^{-1}+\varepsilon_4+\varepsilon_5+\varepsilon_6)\theta_i I+(1+\varepsilon_6^{-1})\breve{\rho}_{i,k+1}\bar{H}R_{i,k+1}\bar{H}^T\end{aligned}$$

$$+\left[(1+\varepsilon_3^{-1}+\varepsilon_5^{-1})\varrho_{i,k+1}^2+(1+\varepsilon_4^{-1})\check{\rho}_{i,k+1}\right]\bar{H}C_{i,k+1}\Psi_{i,k+1|k}C_{i,k+1}^T\bar{H}^T$$
$$+(1+\varepsilon_6^{-1})(1+\varepsilon_8)\bar{\Upsilon}_{i,k+1}R_{i,k+1}\bar{\Upsilon}_{i,k+1}^T$$
$$+(1+\varepsilon_6^{-1})(1+\varepsilon_8^{-1})\varrho_{i,k+1}^2\bar{H}R_{i,k+1}\bar{H}^T\Big\}. \quad (13.80)$$

Then, let the result be zero and the desired estimator gain is designed as in (13.62).

Based on Theorem 13.3, the following corollary can be given when the occurrence probability of RSDs is deterministic (i.e., $\Delta\lambda_{i,k}=0$).

Corollary 13.2 *Consider the discrete time-varying nonlinear dynamical networks with RSDs under deterministic occurrence probability* $\bar{\lambda}_{i,k}$. *For scalars* $\gamma_k>0$ *and* $\varepsilon_i>0\,(i=1,2,4,6)$, *assume that the equation (13.58) and the following recursion:*

$$\begin{aligned}\Omega_{i,k+1|k+1}=&(1+\varepsilon_2)(I-G_{i,k+1}\bar{\Upsilon}_{i,k+1}C_{i,k+1})\Omega_{i,k+1|k}(I-G_{i,k+1}\bar{\Upsilon}_{i,k+1}C_{i,k+1})^T\\&+(1+\varepsilon_2^{-1}+\varepsilon_4+\varepsilon_6)\theta_iG_{i,k+1}G_{i,k+1}^T\\&+(1+\varepsilon_4^{-1})\bar{\lambda}_{i,k+1}(1-\bar{\lambda}_{i,k+1})G_{i,k+1}\bar{H}C_{i,k+1}\bar{\Psi}_{i,k+1|k}C_{i,k+1}^T\bar{H}^TG_{i,k+1}^T\\&+(1+\varepsilon_6^{-1})\bar{\lambda}_{i,k+1}(1-\bar{\lambda}_{i,k+1})G_{i,k+1}\bar{H}R_{i,k+1}\bar{H}^TG_{i,k+1}^T\\&+(1+\varepsilon_6^{-1})G_{i,k+1}\bar{\Upsilon}_{i,k+1}R_{i,k+1}\bar{\Upsilon}_{i,k+1}^TG_{i,k+1}^T\end{aligned} \quad (13.81)$$

with the initial condition $\Omega_{i,0|0}\geq P_{i,0|0}$ *and under the constraint (13.60) have solutions* $\Omega_{i,k+1|k}>0$ *and* $\Omega_{i,k+1|k+1}>0$ *for all* $k\geq 0$. *Then one has* $P_{i,k+1|k+1}\leq\Omega_{i,k+1|k+1}$. *Besides, if the estimator parameter matrix is adopted as*

$$\begin{aligned}G_{i,k+1}=&(1+\varepsilon_2)\Omega_{i,k+1|k}C_{i,k+1}^T\bar{\Upsilon}_{i,k+1}^T\Big\{(1+\varepsilon_2)\bar{\Upsilon}_{i,k+1}C_{i,k+1}\Omega_{i,k+1|k}C_{i,k+1}^T\bar{\Upsilon}_{i,k+1}^T\\&+(1+\varepsilon_2^{-1}+\varepsilon_4+\varepsilon_6)\theta_iI\\&+(1+\varepsilon_4^{-1})\bar{\lambda}_{i,k+1}(1-\bar{\lambda}_{i,k+1})\bar{H}C_{i,k+1}\bar{\Psi}_{i,k+1|k}C_{i,k+1}^T\bar{H}^T\\&+(1+\varepsilon_6^{-1})\bar{\lambda}_{i,k+1}(1-\bar{\lambda}_{i,k+1})\bar{H}R_{i,k+1}\bar{H}^T\\&+(1+\varepsilon_6^{-1})\bar{\Upsilon}_{i,k+1}R_{i,k+1}\bar{\Upsilon}_{i,k+1}^T\Big\}^{-1},\end{aligned} \quad (13.82)$$

it concludes that $\mathrm{tr}(\Omega_{i,k+1|k+1})$ *is minimized.*

Remark 13.8 By virtue of the state augmented method, the state estimation algorithms have been reported in Wang et al. (2016), Hu et al. (2016) for dynamical networks and the estimator gain has been derived. In this chapter, without adopting the state augmented method, a distributed state estimation scheme is proposed for dynamical complex networks in the presence of RCS and RSDs under ETM, in which the estimator gain is derived for each node. It is crucial to highlight that the calculation burden can be effectively decreased due to the fact that there is no need to compute the EEC of the augmentation state. In addition, the designed estimation strategy is of recursive manner, which is suitable for distributed calculations and implementations.

Remark 13.9 It is essential to emphasize that the developed estimation scheme can effectively address the influences of RSDs, RCS and ETM. Particularly, it can be seen that the parameter γ_k is reflected in Theorem 13.3, thus affecting the feasibility of main results in Theorem 13.3. In the implementation process, the parameter γ_k can be properly selected according to (13.60) and then adjusted to enhance the performance of the estimation algorithm.

13.2.3 Boundedness Analysis

In this section, the boundedness analysis of the estimation error dynamics is provided. Firstly, the following useful lemma is introduced here.

Lemma 13.4 *Özbek et al. (2010) For stochastic process ξ_k, assume that there is a stochastic process $V_k(\xi_k)$, real numbers $\underline{\nu}$, $\bar{\nu}$, $\mu > 0$ and $0 < \kappa \leq 1$ satisfying*

$$\underline{\nu}\|\xi_k\|^2 \leq V_k(\xi_k) \leq \bar{\nu}\|\xi_k\|^2 \tag{13.83}$$

and

$$\mathbb{E}\{V_k(\xi_k)|\xi_{k-1}\} \leq (1-\kappa)V_{k-1}(\xi_{k-1}) + \mu. \tag{13.84}$$

Then, one has

$$\mathbb{E}\{\|\xi_k\|^2\} \leq \frac{\bar{\nu}}{\underline{\nu}}\mathbb{E}\{\|\xi_0\|^2\}(1-\kappa)^k + \frac{\mu}{\underline{\nu}}\sum_{i=1}^{k-1}(1-\kappa)^i. \tag{13.85}$$

Before discussing the boundedness of the estimation error dynamics, the following assumption is provided.

Assumption 13.1 For each $1 \leq i \leq N$, $\tau > 0$ and $k \geq 0$, there are positive real numbers $\bar{a}_i$, $\bar{b}_i$, $\bar{c}_i$, $\underline{c}_i$, $\bar{l}_i$, $\bar{\lambda}_i$, $\underline{\lambda}_i$, $\bar{h}_i$, $\underline{\delta}_1$, $\bar{\delta}_1$, $\underline{\delta}_2$, $\vartheta_{i,k+1}$ and $\bar{\delta}_2$ such that

$$\begin{aligned}
&\|A_{i,k}\| \leq \bar{a}_i, \quad \|B_{i,k}\| \leq \bar{b}_i, \quad \underline{c}_i \leq \|C_{i,k}\| \leq \bar{c}_i,\\
&\|L_{i,k}\| \leq \bar{l}_i, \quad \|H_{i,k}\| \leq \bar{h}_i, \quad \underline{\lambda}_i \leq \|\bar{\lambda}_{i,k}\| \leq \bar{\lambda}_i,\\
&\underline{\delta}_1 I \leq C_{i,k}C_{i,k}^T \leq \bar{\delta}_1 I, \quad \underline{\delta}_2 I \leq \Omega_{i,k+1|k} \leq \bar{\delta}_2 I,\\
&\mathrm{tr}(\bar{\Psi}_{i,k+1|k}) \leq \vartheta_{i,k+1},\\
&\rho = (1+\tau)\bar{k}_2^2(\bar{a} + \|\overline{W} \otimes \Gamma\|)^2 + (1+\tau^{-1})\bar{h}^2\bar{k}_2^2\bar{l}^2 < 1,
\end{aligned}$$

where

$$\overline{W} = [\bar{\omega}_{ij}] \in \mathbb{R}^{N\times N},\ \bar{a} = \max_i\{\bar{a}_i\},\ \bar{l} = \max_i\{\bar{l}_i\},$$

$$\bar{h} = \max_i\{\bar{h}_i\},\ \bar{k}_2 = \max_i\{\bar{k}_{2,i}\},\ \bar{k}_{2,i} = 1 + \frac{\bar{\delta}_2\bar{c}_i^{-2}k_{1,i}^2}{\underline{\delta}_1\underline{\delta}_2 k_{2,i}^2},$$

$$k_{1,i} = \max\left\{\sqrt{\bar{\lambda}_i^2 + (1-\bar{\lambda}_i)^2}, \sqrt{\underline{\lambda}_i^2 + (1-\underline{\lambda}_i)^2}\right\},$$

$$k_{2,i} = \min\left\{\sqrt{\bar{\lambda}_i^2 + (1-\bar{\lambda}_i)^2}, \sqrt{\underline{\lambda}_i^2 + (1-\underline{\lambda}_i)^2}\right\}. \tag{13.86}$$

Next, substituting (13.52) into (13.53) results in

$$e_{i,k+1|k+1} = M_{i,k+1}F_{i,k}e_{i,k|k} + M_{i,k+1}\textstyle\sum_{j=1}^N \bar{\omega}_{ij}\Gamma e_{j,k|k} + p_{i,k+1} + q_{i,k+1} \tag{13.87}$$

with the notations

$$\begin{aligned}
M_{i,k+1} &= I - G_{i,k+1}\bar{\Upsilon}_{i,k+1}C_{i,k+1},\ F_{i,k} = A_{i,k} + L_{i,k}\aleph_{i,k}H_{i,k},\\
p_{i,k+1} &= -G_{i,k+1}(y_{i,k+1} - \bar{y}_{i,k+1}) - G_{i,k+1}(\tilde{\Upsilon}_{i,k+1} + \check{\Upsilon}_{i,k+1})C_{i,k+1}x_{i,k+1}\\
&\quad -G_{i,k+1}\Upsilon_{i,k+1}v_{i,k+1},\\
q_{i,k+1} &= M_{i,k+1}\sum_{j=1}^N(\omega_{ij,k} - \bar{\omega}_{ij})\Gamma x_{j,k} + M_{i,k+1}B_{i,k}\varpi_{i,k}.
\end{aligned}$$

To facilitate the subsequent formulation, let

$$\begin{aligned}
e_{k+1|k+1} &= [e_{1,k+1|k+1}^T\ \ e_{2,k+1|k+1}^T\ \ \cdots\ \ e_{N,k+1|k+1}^T]^T,\\
p_{k+1} &= [p_{1,k+1}^T\ \ p_{2,k+1}^T\ \ \cdots\ \ p_{N,k+1}^T]^T,\\
q_{k+1} &= [q_{1,k+1}^T\ \ q_{2,k+1}^T\ \ \cdots\ \ q_{N,k+1}^T]^T,\\
A_k &= \mathrm{diag}\{A_{1,k}, A_{2,k}, \ldots, A_{N,k}\},\\
L_k &= \mathrm{diag}\{L_{1,k}, L_{2,k}, \ldots, L_{N,k}\},\\
H_k &= \mathrm{diag}\{H_{1,k}, H_{2,k}, \ldots, H_{N,k}\},\\
\aleph_k &= \mathrm{diag}\{\aleph_{1,k}, \aleph_{2,k}, \ldots, \aleph_{N,k}\},\\
M_{k+1} &= \mathrm{diag}\{M_{1,k+1}, M_{2,k+1}, \ldots, M_{N,k+1}\}.
\end{aligned}$$

From the above notations and applying the Kronecker product, the estimation error $e_{k+1|k+1}$ is modeled as

$$e_{k+1|k+1} = (M_{k+1}\bar{A}_k + M_{k+1}L_k\aleph_k H_k)e_{k|k} + p_{k+1} + q_{k+1}, \tag{13.88}$$

where $\bar{A}_k = A_k + \overline{W} \otimes \Gamma$.

Now, we endeavor to give a sufficient condition that can guarantee the exponential boundedness of the estimation error under the mean-square sense.

Theorem 13.4 *Under Assumption 13.1, it can be testified that the estimation error (13.88) is exponentially mean-square bounded if $e_{i,0|0}$ is bounded.*

Proof From (13.62), we have

$$\begin{aligned}\|G_{i,k+1}\| &\leq \|(1+\varepsilon_2+\varepsilon_3)\Omega_{i,k+1|k}C_{i,k+1}^T\bar{\Upsilon}_{i,k+1}^T\\ &\quad\times\left[(1+\varepsilon_2+\varepsilon_3)\bar{\Upsilon}_{i,k+1}C_{i,k+1}\Omega_{i,k+1|k}C_{i,k+1}^T\bar{\Upsilon}_{i,k+1}^T\right]^{-1}\|\\ &< \frac{\bar{\delta}_2\bar{c}_i k_{1,i}}{\underline{\delta}_1\underline{\delta}_2 k_{2,i}^2} := \bar{k}_{1,i}. \end{aligned} \tag{13.89}$$

Similarly, it indicates that

$$\|M_{i,k+1}\| = \|I - G_{i,k+1}\bar{\Upsilon}_{i,k+1}C_{i,k+1}\| \leq 1 + \frac{\bar{\delta}_2\bar{c}_i^2 k_{1,i}^2}{\underline{\delta}_1\underline{\delta}_2 k_{2,i}^2} := \bar{k}_{2,i}. \tag{13.90}$$

On the other hand, one obtains

$$\begin{aligned}&\mathbb{E}\{p_{i,k+1}^T p_{i,k+1}\}\\ &\leq \theta_i(1+\varsigma_1+\varsigma_2)\bar{k}_{1,i}^2 + 2(1+\varsigma_1^{-1}+\varsigma_3)\bar{k}_{1,i}^2\bar{c}_i^2\vartheta_{i,k}\check{\rho}_{i,k+1}\\ &+2(1+\varsigma_1^{-1}+\varsigma_3)\varrho_{i,k+1}^2\bar{k}_{1,i}^2\bar{c}_i^2\vartheta_{i,k}\\ &+(1+\varsigma_2^{-1}+\varsigma_3^{-1})(1+\varsigma_4+\varsigma_5)\bar{k}_{1,i}^2\check{\rho}_{i,k+1}\mathrm{tr}\left(\bar{H}R_{i,k}\bar{H}^T\right)\\ &+(1+\varsigma_2^{-1}+\varsigma_3^{-1})(1+\varsigma_4^{-1}+\varsigma_6)\bar{k}_{1,i}^2\mathrm{tr}\left(R_{i,k+1}\right)[\bar{\lambda}_{i,k+1}^2+(1-\bar{\lambda}_{i,k+1})^2]\\ &+2(1+\varsigma_2^{-1}+\varsigma_3^{-1})(1+\varsigma_5^{-1}+\varsigma_6^{-1})\varrho_{i,k+1}^2\bar{k}_{1,i}^2\mathrm{tr}\left(R_{i,k+1}\right)\\ &:= \bar{p}_i \end{aligned} \tag{13.91}$$

and

$$\begin{aligned}&\mathbb{E}\{q_{i,k+1}^T q_{i,k+1}\}\\ &\leq \sum_{j=1}^N \Omega_{ij}^2\bar{k}_{2,i}^2\|\Gamma\|^2\mathrm{tr}\left(\mathbb{E}\{x_{j,k}x_{j,k}^T\}\right) + \bar{k}_{2,i}^2\bar{b}_i^2\mathrm{tr}(Q_{i,k})\\ &\leq \sum_{j=1}^N \Omega_{ij}^2\bar{k}_{2,i}^2\|\Gamma\|^2\vartheta_{j,k} + \bar{k}_{2,i}^2\bar{b}_i^2\mathrm{tr}(Q_{i,k})\\ &:= \bar{q}_i \end{aligned} \tag{13.92}$$

with $\varsigma_i > 0$ $(i = 1, 2, \ldots, 6)$ being scalars. Accordingly, we have

$$\begin{aligned} p_{k+1}^T p_{k+1} &= \sum_{i=1}^N p_{i,k+1}^T p_{i,k+1} \leq \sum_{i=1}^N \bar{p}_i,\\ q_{k+1}^T q_{k+1} &= \sum_{i=1}^N q_{i,k+1}^T q_{i,k+1} \leq \sum_{i=1}^N \bar{q}_i. \end{aligned} \tag{13.93}$$

Next, define the following recursive matrix equation:

$$\begin{aligned}\Xi_{k+1} =&(1+\tau)M_{k+1}\bar{A}_k\Xi_k\bar{A}_k^T M_{k+1}^T \\ &+ (1+\tau^{-1})\|\Xi_k\|\bar{h}^2 M_{k+1}L_kL_k^T M_{k+1}^T + \eta I,\end{aligned} \tag{13.94}$$

where the initial condition $\Xi_0 = \eta I$ and η is a positive scalar. Then, it can be easily seen that

$$\begin{aligned}\|\Xi_{k+1}\| \le \|\Xi_k\|\left[(1+\tau)\|M_{k+1}\|^2\|\bar{A}_k\|^2 + (1+\tau^{-1})\bar{h}^2\|M_{k+1}L_k\|^2\right] + \eta \\ \le \rho\|\Xi_k\| + \eta.\end{aligned} \tag{13.95}$$

By iteration, we obtain

$$\|\Xi_k\| \le \rho^k\|\Xi_0\| + \eta\sum_{i=0}^{k-1}\rho^i. \tag{13.96}$$

It can be derived from $\rho < 1$ that

$$\|\Xi_k\| \le \|\Xi_0\| + \eta\sum_{i=0}^{+\infty}\rho^i = \|\Xi_0\| + \frac{\eta}{1-\rho}. \tag{13.97}$$

Furthermore, it is obvious that

$$\Xi_{k+1} \ge \eta I. \tag{13.98}$$

In view of (13.97) and (13.98), it is summarized that there are positive scalars $\bar{\xi}$ and $\underline{\xi}$ such that $\underline{\xi} I \le \Xi_k \le \bar{\xi} I$ is true for all $k \ge 0$.

Adopting the matrix inversion lemma to Ξ_{k+1}, one obtains

$$\begin{aligned}&(M_{k+1}\bar{A}_k + M_{k+1}L_k\aleph_k H_k)^T\Xi_{k+1}^{-1}(M_{k+1}\bar{A}_k + M_{k+1}L_k\aleph_k H_k) - \Xi_k^{-1} \\ &\le (M_{k+1}\bar{A}_k + M_{k+1}L_k\aleph_k H_k)^T[(M_{k+1}\bar{A}_k + M_{k+1}L_k\aleph_k H_k)\Xi_k(M_{k+1}\bar{A}_k \\ &+ M_{k+1}L_k\aleph_k H_k)^T + \eta I]^{-1}(M_{k+1}\bar{A}_k + M_{k+1}L_k\aleph_k H_k) - \Xi_k^{-1} \\ &= -[\Xi_k + \frac{1}{\eta}\Xi_k(M_{k+1}\bar{A}_k + M_{k+1}L_k\aleph_k H_k)^T(M_{k+1}\bar{A}_k + M_{k+1}L_k\aleph_k H_k)\Xi_k]^{-1} \\ &= -[I + \frac{1}{\eta}(M_{k+1}\bar{A}_k + M_{k+1}L_k\aleph_k H_k)^T(M_{k+1}\bar{A}_k + M_{k+1}L_k\aleph_k H_k)\Xi_k]^{-1}\Xi_k^{-1} \\ &\le -\left[1 + \frac{\bar{k}_2^2(\bar{a} + \|\overline{W}\otimes\Gamma\| + \bar{h}\bar{l})^2}{\eta}\right]^{-1}\Xi_k^{-1} \\ &:= -\delta_0\Xi_k^{-1}.\end{aligned}$$

Since $\delta_0 < 1$, then a scalar $\kappa > 0$ can be found, which satisfies $\delta = (1+\kappa)(1-\delta_0) < 1$. Next, considering the Lyapunov function $V_k(e_{k|k}) = e_{k|k}^T \Xi_k^{-1} e_{k|k}$, one has

$$\begin{aligned}
&\mathbb{E}\{V_{k+1}(e_{k+1|k+1})|e_{k|k}\} - (1+\kappa)V_k(e_{k|k}) \\
&\leq (1+\kappa)\mathbb{E}\{e_{k|k}^T[(M_{k+1}\bar{A}_k + M_{k+1}L_k\aleph_k H_k)^T \Xi_{k+1}^{-1} \\
&\times (M_{k+1}\bar{A}_k + M_{k+1}L_k\aleph_k H_k) - \Xi_k^{-1}]e_{k|k}|e_{k|k}\} \\
&+ (1+\kappa^{-1}+\kappa_1)\mathbb{E}\{p_{k+1}^T \Xi_{k+1}^{-1} p_{k+1}\} + (1+\kappa_1^{-1})\mathbb{E}\{q_{k+1}^T \Xi_{k+1}^{-1} q_{k+1}\} \\
&\leq -\delta_0(1+\kappa)V_k(e_{k|k}) + \left[(1+\kappa^{-1}+\kappa_1)\sum_{i=1}^{N}\bar{p}_i + (1+\kappa_1^{-1})\sum_{i=1}^{N}\bar{q}_i\right]\underline{\xi}^{-1} \\
&:= -\delta_0(1+\kappa)V_k(e_{k|k}) + \mu.
\end{aligned}$$

Then, it indicates that

$$\mathbb{E}\{V_{k+1}(e_{k+1|k+1})\} \leq \delta V_k(e_{k|k}) + \mu,$$

which results in

$$\mathbb{E}\left\{\|e_{k|k}\|\right\} \leq \frac{\bar{\xi}}{\underline{\xi}}\|e_{0|0}\|^2\delta^k + \mu\bar{\xi}\sum_{i=0}^{+\infty}\delta^i = \frac{\bar{\xi}}{\underline{\xi}}\|e_{0|0}\|^2\delta^k + \frac{\mu\bar{\xi}}{1-\delta},$$

thus ending the proof of this theorem. □

Up to now, a novel optimized event-triggered estimation (OETE) algorithm has been presented. In light of Theorem 13.3, the OETE algorithm is exhibited here.

OETE Algorithm:
Step 1 : Propose the initial values and related parameters. Assume $k = 0$.
Step 2 : Obtain the prediction estimation $\hat{x}_{i,k+1|k}$ based on (13.48).
Step 3 : Calculate the upper bound $\Omega_{i,k+1|k}$ via the equation (13.58).
Step 4 : Compute the estimator gain $G_{i,k+1}$ in terms of the equation (13.62).
Step 5 : Calculate the state estimation $\hat{x}_{i,k+1|k+1}$ via the updating equation (13.49).
Step 6 : Compute the optimized upper bound $\Omega_{i,k+1|k+1}$ in view of the equation (13.59).
Step 7 : Letting $k = k+1$, and go back to *Step* 2.

13.3 Two Illustrative Examples

In this section, the usefulness of the designed state estimation methods is evaluated by two simulation examples.

Example 13.1 Consider the complex networks (13.1) with nonlinear function as

$$f_i(x_{i,k}) = M_{i,k}x_{i,k} + \begin{bmatrix} 0.1\sin(x_{i,k}^1 + x_{i,k}^2) \\ 0.1(x_{i,k}^2)^2\cos(k) \end{bmatrix},$$

where $x_{i,k} = \left[x_{i,k}^1 \; x_{i,k}^2 \right]^T$ is the state of the i-th node at time k and

$$M_{1,k} = \begin{bmatrix} 0.1 + 0.2\sin(0.5\pi k) & 0.5 \\ 0.65 & 0.7 \end{bmatrix},$$
$$M_{2,k} = \begin{bmatrix} 0.3 & 0.6 \\ 0.6 & 0.3 + 0.2\cos(3k) \end{bmatrix}, \quad M_{3,k} = \begin{bmatrix} 0.3 & 0.5 \\ 0.6 & 0.5 \end{bmatrix}.$$

Other system parameters are selected as

$$B_{1,k} = \left[0.55 \; 0.7 \right]^T, \quad B_{2,k} = \left[0.9 + 0.1\sin(2k) \; 0.6 \right]^T,$$
$$B_{3,k} = \left[0.7 \; 0.6 \right]^T, \quad C_{1,k} = \left[1.3 \; 1.5 \right],$$
$$C_{2,k} = \left[1.6 \; 1.5 \right], \quad C_{3,k} = \left[1.2 \; 1.3 + 0.1\cos(7k) \right].$$

The inner-coupling matrices are given as $\Gamma = \mathrm{diag}\{0.9, 0.9\}$, $\bar{\Gamma} = \mathrm{diag}\{0.5, 0.5\}$, and the elements of the coupling configuration matrix are $w_{ij} = 0.12$ $(i \neq j)$ and $w_{ij} = -0.16$ $(i = j)$. The matrices L_k and H_k are set as

$$L_k = \mathrm{diag}\{0.01, 0.02, 0.01, 0.02, 0.01, 0.02\},$$
$$H_k = \mathrm{diag}\{0.01, 0.01, 0.01, 0.01, 0.01, 0.02\}.$$

Other parameters are selected as $\bar{\lambda}_{1,k} = 0.75$, $\bar{\lambda}_{2,k} = \bar{\lambda}_{3,k} = 0.85$, $\varrho_{j,k} = 0.1$ $(j = 1, 2, 3)$, $Q_{1,k} = 0.1$, $Q_{2,k} = 0.3$, $Q_{3,k} = 0.2$, $R_{1,k} = 0.2$, $R_{2,k} = 0.1$, $R_{3,k} = 0.15$, $\varepsilon_1 = 1$, $\varepsilon_2 = 0.05$, $\varepsilon_3 = 0.5$ and $\eta_k = 1/\left(1.2\lambda_{\max}(H_k \Omega_{k+1|k} H_k^T) + 1\right)$.

The initial conditions are given as

$$x_0 = \left[x_{1,0}^T \; x_{2,0}^T \; x_{3,0}^T \right]^T = \left[5 \; 2 \; 3 \; 5 \; -1 \; -3 \right]^T,$$
$$\hat{x}_{0|0} = \left[\hat{x}_{1,0|0}^T \; \hat{x}_{2,0|0}^T \; \hat{x}_{3,0|0}^T \right]^T = \left[-4 \; 2 \; 2 \; -6 \; -4 \; -4 \right]^T, \quad \Omega_{0|0} = 360 I_6.$$

Next, let MSE represent the mean-square error for the estimation of the x_k. To compare the estimation performance under different situations, we consider the following two cases, i.e.,

Case I: $\mathrm{Prob}\{\gamma_k = 1\} = 0.3$, $\mathrm{Prob}\{\gamma_k = 2\} = 0.3$ and $\mathrm{Prob}\{\gamma_k = 3\} = 0.4$;

Case II: $\mathrm{Prob}\{\gamma_k = 1\} = 0.1$, $\mathrm{Prob}\{\gamma_k = 2\} = 0.8$ and $\mathrm{Prob}\{\gamma_k = 3\} = 0.1$.

Now, the estimation algorithm can be carried out and the related simulation results are plotted in Figs. 13.1, 13.2, 13.3, 13.4 and 13.5. Among the simulation, Figs. 13.1, 13.2 and 13.3 depict the states $x_{i,k}$ and their estimations $\hat{x}_{i,k|k}$, where $\hat{x}_{i,k|k}$ can be computed by $\hat{x}_{i,k|k} = N_i \hat{x}_{k|k}$ with $N_1 = \left[I_n \; 0 \; 0 \right]$, $N_2 = \left[0 \; I_n \; 0 \right]$ and $N_3 = \left[0 \; 0 \; I_n \right]$. The MSEs and the trace of their upper bound under logarithmic transformation are depicted in Fig. 13.4, which exhibits that the MSEs indeed remain below their upper bounds. Besides, Fig. 13.5 depicts the sensor nodes being chosen to access the network channel. From the simulations, it is observed that the established estimation scheme has a satisfactory performance despite considering the missing measure-

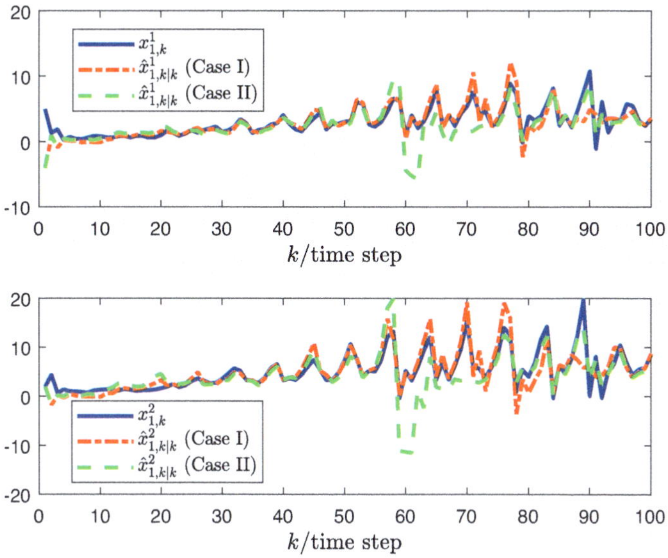

Fig. 13.1 The real state $x_{1,k}$ and its estimation $\hat{x}_{1,k|k}$

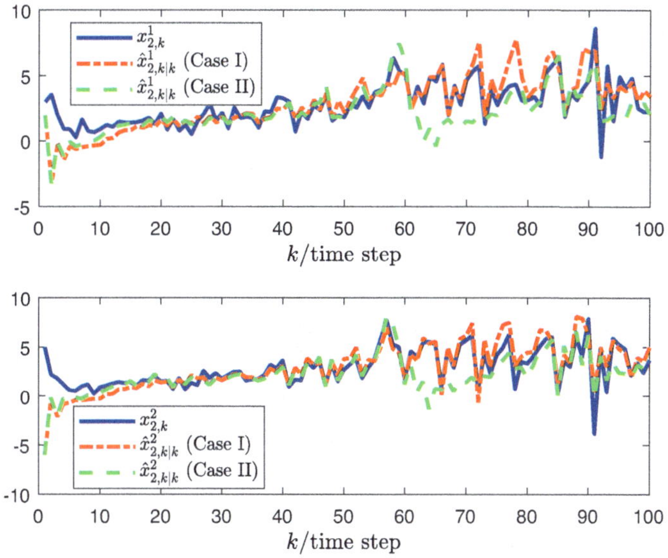

Fig. 13.2 The real state $x_{2,k}$ and its estimation $\hat{x}_{2,k|k}$

ments and RAP. On the other hand, consider the parameters in above simulation and set $\Omega_{0|0} = 20I_6$. In light of Corollary 13.1, the curves of $\mathrm{tr}(\Omega_{k+1|k+1})$ are plotted in Fig. 13.6, which further indicates the analysis results in Theorem 13.2.

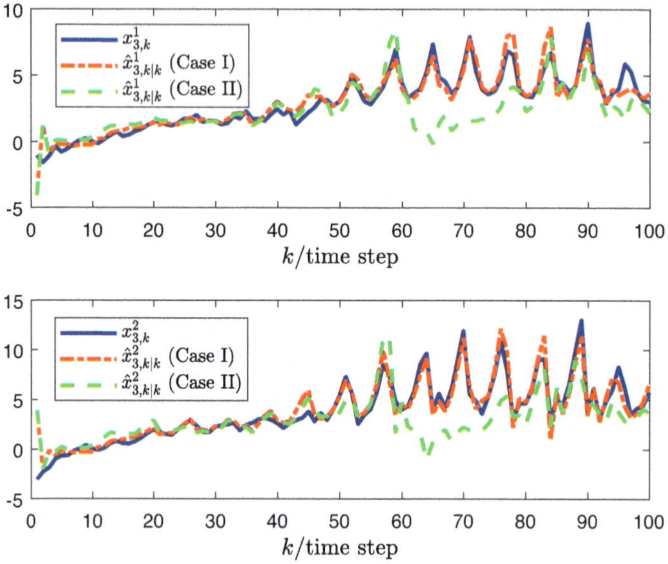

Fig. 13.3 The real state $x_{3,k}$ and its estimation $\hat{x}_{3,k|k}$

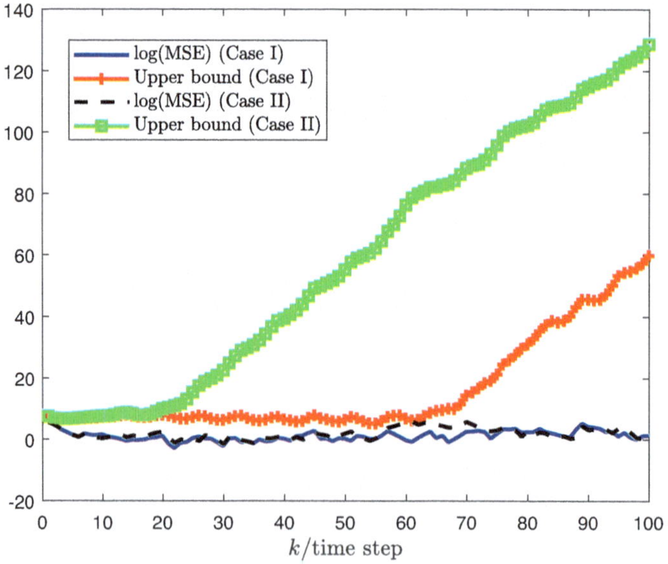

Fig. 13.4 MSE and their upper bounds

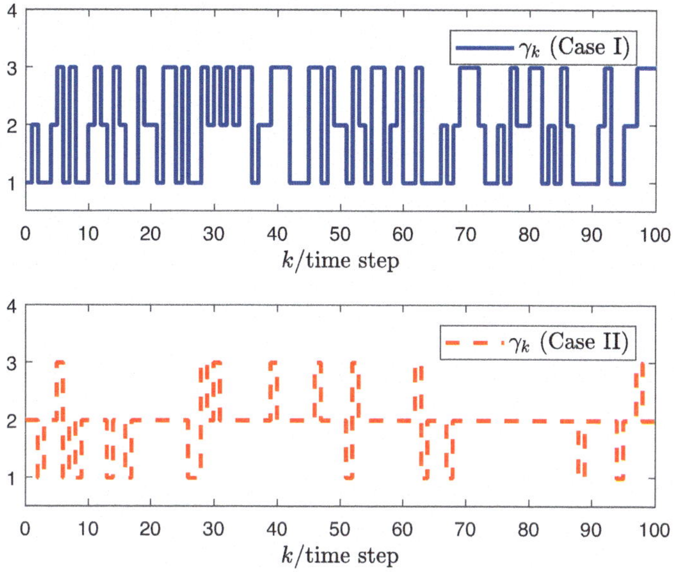

Fig. 13.5 Selected sensor nodes at each time step

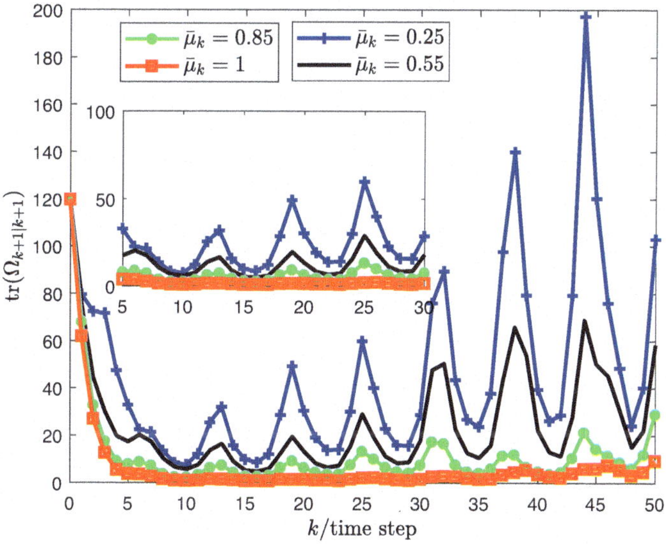

Fig. 13.6 The trace of $\Omega_{k+1|k+1}$ with different missing probabilities

Example 13.2 In the following simulation, assume that the elements $\omega_{ij,k}$ obey the following distribution:

$$\omega_{ij,k} \sim \begin{cases} U(-0.23, -0.17), & i = j \\ U(0.07, 0.13), & i \neq j \end{cases}$$

and $\hat{\Gamma} = \text{diag}\{0.6, 0.6\}$. The additive noises $\varpi_{i,k}$ $(i = 1, 2, 3)$ and $\check{\nu}_{i,k}$ are zero mean with covariances $Q_{1,k} = \mathbf{R}_{2,k} = 0.1$, $Q_{2,k} = 0.3$, $Q_{3,k} = \mathbf{R}_{1,k} = \mathbf{R}_{3,k} = 0.2$. Other system matrices are set as

$$\mathbf{B}_{1,k} = \begin{bmatrix} 1.3 - 0.1\sin(3k) \\ 1.5 \end{bmatrix}, \quad \mathbf{B}_{2,k} = \begin{bmatrix} 0.5 \\ 0.7 \end{bmatrix}, \quad \mathbf{B}_{3,k} = \begin{bmatrix} 0.8 \\ 1.3 \end{bmatrix},$$
$$\mathbf{C}_{1,k} = [1.8\ 1.5], \quad \mathbf{C}_{2,k} = [1.6\ 1.7], \quad \mathbf{C}_{3,k} = [1.1\ 1.3].$$

The following nonlinear functions are chosen as:

$$\mathbf{f}_i(\mathbf{x}_{i,k}) = M_i \mathbf{x}_{i,k} + \begin{bmatrix} 0.1\sin(\mathbf{x}_{i,k}^1 + \mathbf{x}_{i,k}^2) \\ 0.5\mathbf{x}_{i,k}^2 \sin(k) \end{bmatrix},$$

where $\mathbf{x}_{i,k} = \begin{bmatrix} \mathbf{x}_{i,k}^1 & \mathbf{x}_{i,k}^2 \end{bmatrix}^T$ is the system state of the i-th node and

$$M_1 = \begin{bmatrix} 0.42\ 0.5 \\ 0.28\ 0.5 \end{bmatrix}, \quad M_2 = \begin{bmatrix} 0.5\ \ 0.6 \\ -0.1\ 0.8 \end{bmatrix}, \quad M_3 = \begin{bmatrix} 0.6\ 0.5 \\ 0.2\ 0.6 \end{bmatrix}.$$

In the simulation, set $\bar{\lambda}_{s,k} = 0.75$, $\varrho_{s,k} = 0.05$ $(s = 1, 2, 3)$, $\varepsilon_i = 0.1$ $(i = 2, 3, 7)$ and $\varepsilon_j = 1$ $(j = 1, 4, 5, 6, 8)$. The initial states are selected as $\mathbf{x}_{1,0} = [0.1\ \ 0.2]^T$, $\mathbf{x}_{2,0} = [-0.2\ \ 0.1]^T$, $\mathbf{x}_{3,0} = [-0.1\ \ -0.15]^T$ and $\hat{x}_{i,0|0} = [\bar{\mathbf{x}}_{i,0}^T\ \ 0^T]^T$ with $\bar{\mathbf{x}}_{i,0} = \mathbf{x}_{i,0} - [1\ \ 1]^T$ $(i = 1, 2, 3)$. Moreover, select the parameter $\gamma_k = \left[1.5\lambda_{\max}(H_{i,k}^T \Omega_{i,k|k} H_{i,k}) + 1\right]^{-1}$ satisfying the inequality (13.60), and choose $\Omega_{i,0|0} = 2.5I_4$, $L_{i,k} = 0.01I_4$ and $H_{i,k} = 0.01I_4$.

In the following simulation experiment, two scenarios with different adjustable thresholds are considered: Case I: $\theta_1 = 1, \theta_2 = 1$ and $\theta_3 = 1$; Case II: $\theta_1 = 10, \theta_2 = 10$ and $\theta_3 = 10$. Specifically, Figs. 13.7, 13.8 and 13.9 depict the state trajectories and their estimates under different adjustable thresholds. Compared with the two cases, it is seen that the tracking performance in Case I is better than the case in Case II, which are consistent with the simulations given in Figs. 13.10 and 13.11. Furthermore, the MSE and the related upper bounds under two cases are given in Figs. 13.12 and 13.13, which further illustrate the effectiveness of the theoretical results in Theorem 13.3. Assume that the adjustable threshold is zero and occurrence probability is deterministic, i.e., $\theta_i = 0$ and $\Delta\lambda_{i,k} = 0$. Then, the curves of $\text{tr}(\Omega_{i,k|k})$ are plotted in Fig. 13.14 under $\bar{\lambda}_{i,k} = 0.85,\ 0.9,\ 0.95$ and 1, respectively, which can be observed that $\text{tr}(\Omega_{i,k|k})$ decreases along with the increasing the occurrence probability $\bar{\lambda}_{i,k}$ of RSDs.

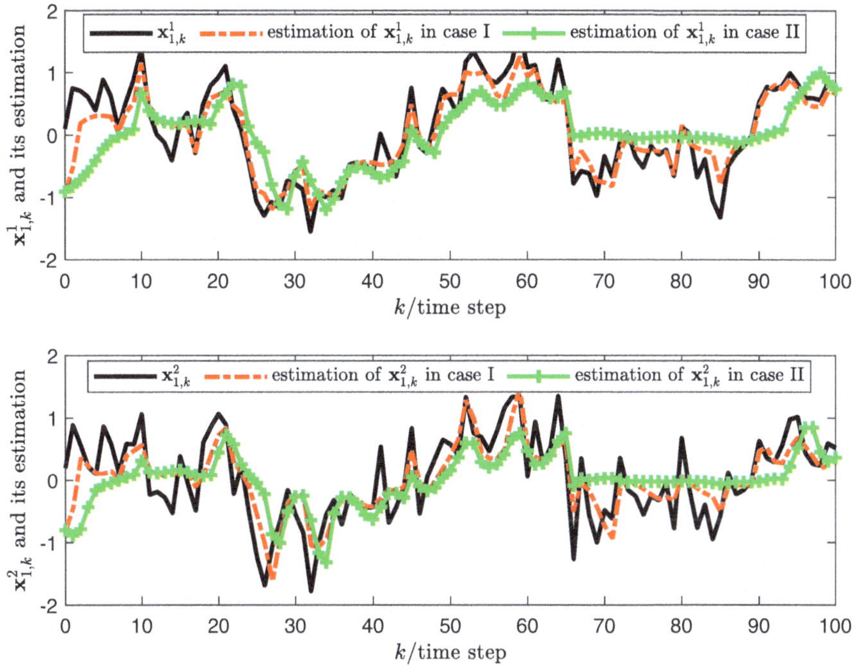

Fig. 13.7 The responses of $\mathbf{x}_{1,k}$ and $\hat{\mathbf{x}}_{1,k|k}$

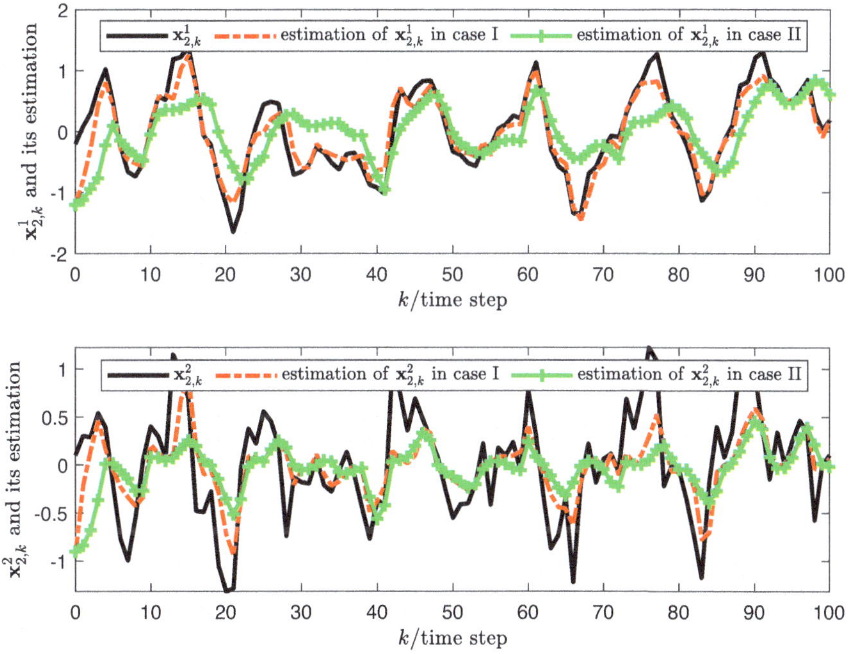

Fig. 13.8 The responses of $\mathbf{x}_{2,k}$ and $\hat{\mathbf{x}}_{2,k|k}$

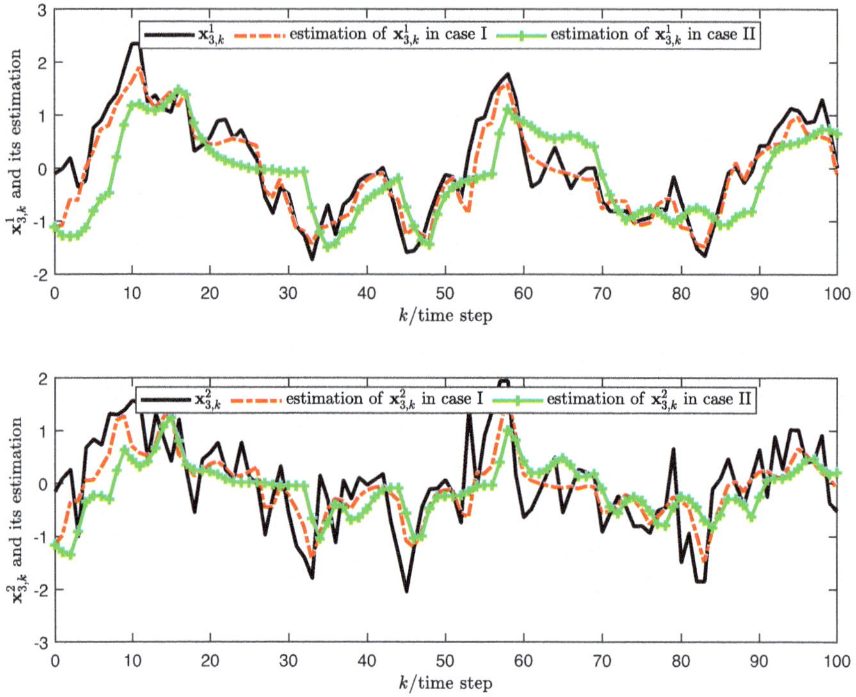

Fig. 13.9 The responses of $\mathbf{x}_{3,k}$ and $\hat{\mathbf{x}}_{3,k|k}$

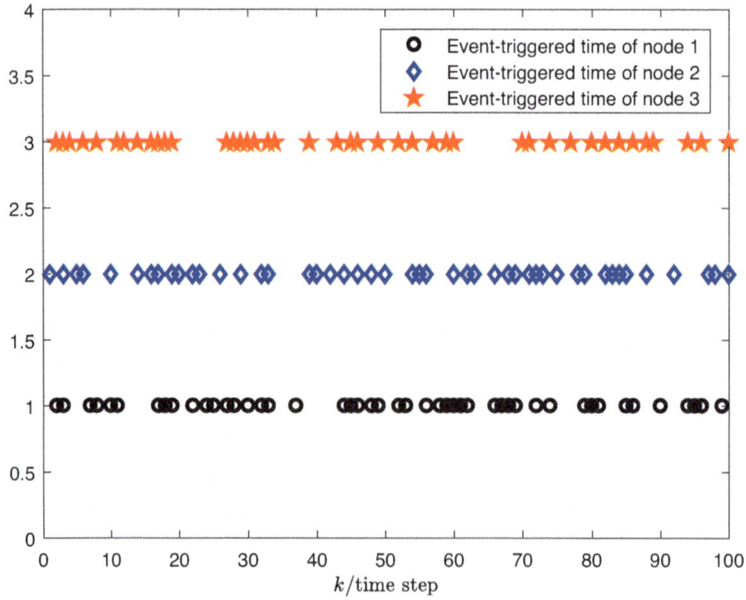

Fig. 13.10 The event-triggered time step (Case I)

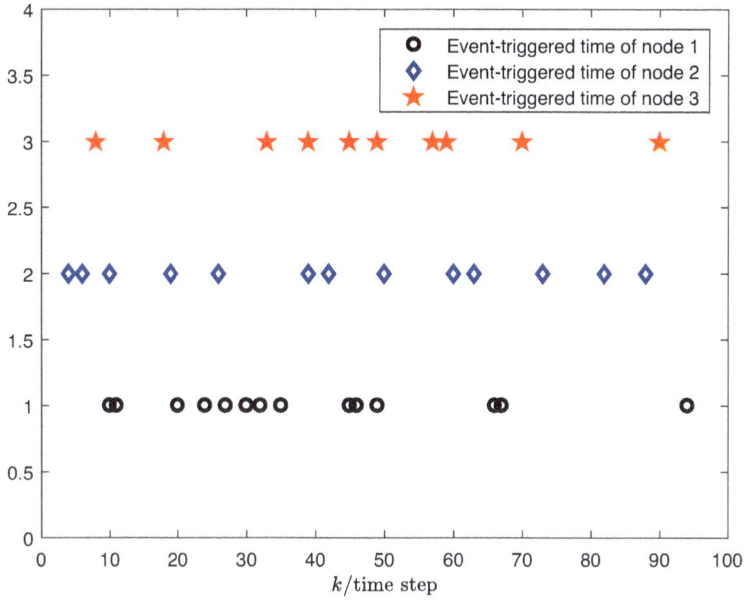

Fig. 13.11 The event-triggered time step (Case II)

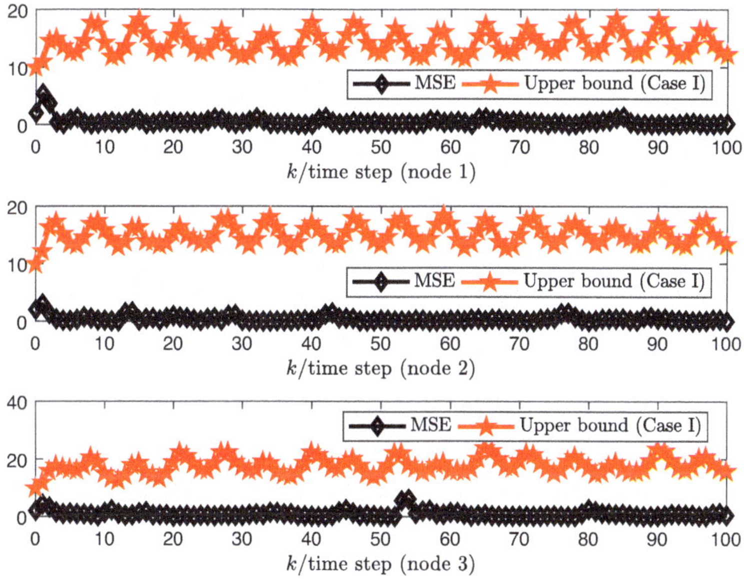

Fig. 13.12 MSE and corresponding upper bounds (Case I)

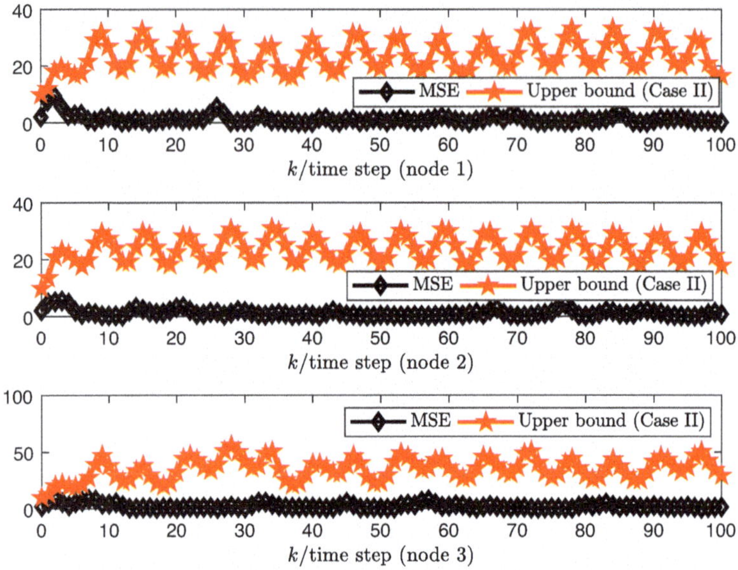

Fig. 13.13 MSE and corresponding upper bounds (Case II)

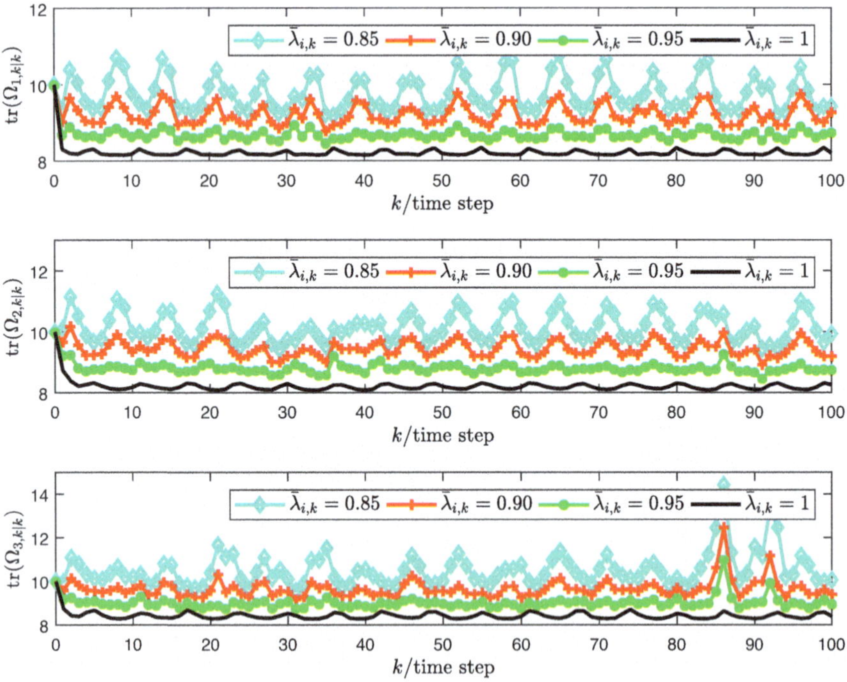

Fig. 13.14 $\mathrm{tr}(\Omega_{i,k|k})$ under different $\bar{\lambda}_{i,k}$

13.4 Summary

In this chapter, a new protocol-based state estimation strategy has been proposed for nonlinear complex networks involving stochastic inner coupling and missing measurements with UOPs. With the aim of avoiding undesirable data collisions, the RAP has been employed to regulate data signals. In particular, an optimized upper bound of the EEC has been solved, and the desired estimator gain has been derived. Furthermore, the monotonicity between the deterministic missing probability and the trace of this upper bound has been evaluated. Secondly, a novel state estimation strategy has been established for time-varying nonlinear complex networks in the presence ETM, RSDs and RCS. The ETM for each node has been used to improve communication efficiency. Next, a new time-varying estimator for each node has been designed and the theoretical analysis of the boundedness of the estimation error has been discussed. Finally, the feasibility of the designed state estimation algorithms has been demonstrated by providing some simulation experiments with comparisons.

References

Balan AI, Gopalakrishnan VT (2015) Analysis of carrier sense multiple access protocols for channels supporting multi-packet reception. IET Commun 9(4):468–475

Bu X, Dong H, Han F, Hou N, Li G (2019) Distributed filtering for time-varying systems over sensor networks with randomly switching topologies under the round-robin protocol. Neurocomputing 346:58–64

Dong H, Wang Z, Shen B, Ding D (2016) Variance-constrained H_∞ control for a class of nonlinear stochastic discrete time-varying systems: the event-triggered design. Automatica 72:28–36

Donkers MCF, Heemels WPMH, Bernardini D, Bemporad A, Shneer V (2012) Stability analysis of stochastic networked control systems. Automatica 48(5):917–925

Hu J, Wang Z, Liu S, Gao H (2016) A variance-constrained approach to recursive state estimation for time-varying complex networks with missing measurements. Automatica 64:155–162

Hu J, Zhang P, Kao Y, Liu H, Chen D (2019) Sliding mode control for Markovian jump repeated scalar nonlinear systems with packet dropouts: the uncertain occurrence probabilities case. Appl Math Comput 362:124574

Kai X, Wei C, Liu L (2010) Robust extended Kalman filtering for nonlinear systems with stochastic uncertainties. IEEE Trans Syst Man Cybern Part A-Syst Hum 40(2):399–405

Li W, Jia Y, Du J (2017) Recursive state estimation for complex networks with random coupling strength. Neurocomputing 219:1–8

Liang J, Wang Z, Liu Y, Liu X (2014) State estimation for two-dimensional complex networks with randomly occurring nonlinearities and randomly varying sensor delays. Int J Robust Nonlinear Control 24(1):18–38

Özbek L, Babacan EK, Efe M (2010) Stochastic stability of the discrete-time constrained extended Kalman filter. Turk J Electr Eng Comput Sci 18(2):211–223

Wang L, Wang Z, Huang T, Wei G (2016) An event-triggered approach to state estimation for a class of complex networks with mixed time delays and nonlinearities. IEEE Trans Cybern 46(11):2497–2508

Zhang H, Hu J, Liu H, Yu X, Liu F (2019) Recursive state estimation for time-varying complex networks subject to missing measurements and stochastic inner coupling under random access protocol. Neurocomputing 346:48–57

Zou L, Wang Z, Han QL, Zhou D (2017) Ultimate boundedness control for networked systems with try-once-discard protocol and uniform quantization effects. IEEE Trans Autom Control 62(12):6582–6588

Zou L, Wang Z, Han QL, Zhou DH (2018) Recursive filtering for time-varying systems with random access protocol. IEEE Trans Autom Control 64(2):720–727

Chapter 14
Conclusions and Future Work

In this chapter, the main conclusions of this book are given and the potential research directions are pointed out accordingly.

14.1 Conclusions

The major objective of this book has been focused on the variance-constrained optimized filtering and its applications for time-varying nonlinear dynamical systems. Firstly, the variance-constrained optimized filtering algorithms have been designed for time-varying nonlinear dynamical systems with incomplete information. Secondly, the proposed filtering methods have been extended to handle the joint state and fault estimation problems for time-varying nonlinear dynamical systems in the presence of incomplete measurements under the variance-constrained sense. Besides, the proposed variance-constrained filtering and state estimation algorithms have been applied to deal with the state estimation problems for time-varying complex dynamical networks involving incomplete measurements, where the performance evaluation criteria have been given for the developed optimal filtering/estimation algorithms. The specific contents with respect to each chapter are listed as follows:

- In Chap. 2, the robust variance-constrained filtering problem has been discussed for a class of time-varying networked systems influenced by stochastic nonlinearity, ROQ and ROUs. It is worth mentioning that we have proposed a novel optimal variance-constrained filtering algorithm and desirable sufficient criterion to guarantee the boundedness of the filtering error dynamics. To be more specific, the ROUs and random signal quantization have been firstly characterized by some mutually independent random variables obeying the Bernoulli distribution. A locally optimal variance-constrained filtering scheme has been proposed, where a minimum upper bound of the FEC has been given and the expected time-varying filter gain has been obtained by solving the matrix equations. Furthermore, the

J. Hu et al., *Variance-Constrained Filtering for Stochastic Complex Systems*, Intelligent Control and Learning Systems 18,
https://doi.org/10.1007/978-981-96-2637-3_14

performance analysis concerning the developed filtering scheme has been discussed, where a sufficient criterion has been provided to guarantee the exponential mean-square boundedness of the filtering error dynamics. At last, some comparative simulations have been utilized to validate the feasibility and validity of the obtained variance-constrained filtering algorithm.

- In Chap. 3, the resilient variance-constrained filtering scheme has been proposed for time-varying nonlinear stochastic systems, in which the stochastic uncertainties, filter gain perturbations as well as incomplete measurements have been taken into account. Firstly, the stochastic uncertainties have been discussed and characterized by a known bounded matrix and a set of multiplicative noise sequences. Secondly, a unified framework has been established to depict the phenomena of incomplete measurements by employing a Kronecker-δ function, where the ROQ and missing measurements have been considered and depicted simultaneously. Moreover, a set of multiplicative noises has been introduced to model and reflect the filter gain perturbations in order to enhance the robustness of the presented filtering algorithm. By resorting to the available information, a new time-varying nonlinear filter has been designed and the expression form of filter gain matrix has been characterized by solving the recursive matrix difference equations. To be more specific, during the derivations, the nonlinear function has been linearized and the resultant linearization errors have been approximatively estimated. In terms of the matrix theory and stochastic analysis technique, a variance-constrained filtering approach has been given such that, in the presence of stochastic uncertainties, filter gain perturbations and incomplete measurements, a minimal upper bound on the FEC has been given by appropriately selecting a filter gain matrix. At last, the validity of the designed non-fragile filtering algorithm has been shown by some simulations.
- With the purpose of saving network resources, the proposed variance-constrained filtering schemes have been further extended and the event-triggering condition has been introduced in Chap. 4. Here, the effects from the stochastic uncertainties and SPDs onto the filtering performance have been examined. Firstly, an event-based resilient filtering issue has been handled for time-varying stochastic systems in the presence of stochastic uncertainties. According to the definitions of one-step prediction error covariance and FEC, the desired recursions have been derived. Next, an optimal upper bound concerning on the FEC has been solved via selecting an appropriate filter gain matrix. It has been further indicated that the proposed recursive filtering scheme is applicable for online implementations and the network resources can be further saved because of the adding of ETM. Secondly, the designed method has been further extended to deal with the event-triggered resilient filtering issue for time-varying systems affected by SPDs and stochastic uncertainties, where the related information of the stochastic uncertainties and SPDs has been clearly reflected in the proposed optimal nonlinear filtering approaches. Finally, the simulation experiments have been utilized to validate the usefulness of the developed event-triggered resilient filtering algorithms.
- In Chap. 5, the proposed event-based filtering methods in Chap. 4 have been further extended to tackle the multiple missing measurements, where the occurrence probabilities can be uncertain. Firstly, an ETM has been employed to decrease

the communication burden, which can schedule the information transmissions between the sensor and the remote filter. The merits of the proposed filtering scheme lie in that it is robust against the filter gain perturbations and then the resilient properties of the proposed filtering algorithm have been ensured. Moreover, the RONs have been characterized and reflected in main results. Through the solutions to the Riccati-like difference equations, an upper bound for the FEC has been guaranteed and such an upper bound matrix has been minimized by constructing the filter gain. Secondly, the above results have been further extended, where we have addressed the event-triggered filtering issue for nonlinear stochastic systems subject to multiple missing measurements. Particularly, the multiple missing measurements have been considered and the modeling way of the UOPs has been given via the nominal mean and known upper bound of the probabilities. Furthermore, the ETM has been employed to decrease undesirable data transmission and save network resources. By taking the available probability information concerning on the multiple missing measurements into account, a novel filtering scheme with compensation ability has been proposed. To be more specific, the recursive equations of the one-step prediction error covariance and the FEC have been derived. Because of the presence of some unknown terms, the upper bound of the FEC has been derived and then minimized by using the stochastic analysis technique. It should be noticed that the developed recursive filtering algorithm is suitable for online applications. Finally, the feasibility and correctness of the designed event-triggered filtering approaches have been illustrated by some practical simulations.

- Based on the proposed variance-constrained filtering approaches, we continue to handle the joint state and fault estimation problems for several classes of time-varying nonlinear stochastic systems, where additive fault and dynamical fault have been considered. In Chap. 6, the co-design problem of the joint state and fault estimator has been solved for time-varying stochastic systems involving RONs and RODAs, in which two sets of mutually independent random variables governed by Bernoulli distribution with known conditional probabilities have been given to depict the behaviors of RONs and RODAs, respectively. The major effort has been devoted to designing a novel compensation scheme and revealing the effects from both RONs and RODAs onto the filtering algorithm accuracy. In particular, a variance-constrained filtering algorithm within the minimum variance sense has been given and the desired filter gain has been provided by resorting to the solutions to two Riccati-like difference matrix equations. Finally, the effectiveness of the designed fault estimation strategy has been shown via an illustrative simulation.
- In Chap. 7, the designed variance-constrained state estimation method in Chap. 6 has been further extended to address the joint estimation problem with ROUs, stochastic nonlinearities and packet dropouts. The mutually independent random variables have been adopted to describe the ROUs and packet dropouts. Moreover, the stochastic nonlinearities have been considered and the main statistical information regarding the stochastic nonlinearities has been reflected in the main results. The main effort has been devoted to constructing a new fault estimation scheme under the influences of ROUs, stochastic nonlinearities and packet dropouts. Moreover, the monotonicity between the trace of this optimal

upper bound and the missing probability has been discussed in detail. Finally, the validity of the established joint estimation compensation method has been testified by a simulation experiment.

- In Chap. 8, we have further addressed the joint state and fault estimation issue for time-varying nonlinear stochastic systems in the presence of ROFs and sensor saturations. Here, a set of Bernoulli distributed random variables has been adopted to model the ROFs and the signum function has been introduced to characterize the sensor saturation by taking the practical constraints of the measurements into consideration. After that, much effort has been dedicated to designing a locally optimal estimator to realize the simultaneous estimation of the state and fault. In particular, the upper bound of the EEC has been minimized by deriving a proper estimator gain. It has been revealed that the established estimation scheme can be online executed. What's more, the boundedness of the estimation error dynamics in the mean-square sense has been proved by providing the sufficient condition. At last, some simulation examples have been adopted to exhibit the usefulness of the designed fault estimation scheme.
- Next, the presented variance-constrained filtering/estimation methods have been employed to address the state estimation problem for time-varying complex networks. In Chap. 9, the recursive state estimation issue has been solved for discrete time-varying coupled stochastic complex networks influenced by missing measurements. An array of random variables obeying given probability has been utilized to model the missing measurements for each sensor. In order to handle the nonlinearity, the Taylor series expansion method has been used to address the nonlinearity and the corresponding high-order terms with respect to the linearization errors have also been estimated. Based on the available information, a new state estimator has been designed by utilizing the variance-constrained method. By solving the Riccati-like difference equations, the estimator gain with expression form has been designed on the premise of minimizing the upper bound of the EEC. Finally, an illustrative example has been given to verify the validity of the presented state estimation algorithm.
- In Chap. 10, the proposed recursive state estimation problems in Chap. 9 have been further extended to address the state estimation problem for time-varying complex networks subject to uncertain inner coupling and signal quantization. According to the practical requirements, the coupling strengths have been always allowed to be varying in pre-determined intervals. Meanwhile, the measurements have been quantized due to limited network bandwidth. After that, a state estimation scheme has been proposed under the variance-constrained sense and the locally optimized upper bound of the EEC has been built by constructing a desirable estimator gain. In addition, the boundedness of the developed estimation scheme has been evaluated. Lastly, the numerical experiments have been provided to validate the usefulness of the newly designed estimation scheme.
- In order to decrease the communication burden and improve the transmission efficiency, the ETM has been utilized in Chap. 11 to tackle the state estimation problem for time-varying complex networks under the variance-constraint index. Firstly, we have handled the event-triggered filtering issue for time-varying delayed

complex networks subject to stochastic coupling and missing measurements with UOPs. By resorting to two sets of mutually independent random variables obeying Bernoulli distribution, the stochastic coupling and missing measurements have been characterized simultaneously. The ETM has been used to further schedule the measurements to improve the transmission efficiency. After that, a new state estimation strategy has been developed for conducted time-varying complex networks. Furthermore, the monotonicity evaluation between the trace of the obtained upper bound of the EEC and the deterministic missing probability has been discussed. Finally, some simulations with comparisons have been provided to testify the correctness and applicability of the developed robust state estimation algorithm.

- In Chap. 12, the recursive state estimation issues have been addressed for time-varying uncertain complex networks with RONs, switching topology and packet dropouts. Firstly, the ETM has been applied to regulate data transmissions and the phenomena of RONs, switching topology and packet dropouts have been depicted by means of some binary random variables. By fully taking the switching topology, packet dropouts, RONs and ETM into account, a novel state estimation scheme has been proposed for addressed time-varying dynamical complex networks. Specifically, a minimal upper bound concerning on the EEC has been given via constructing a desirable estimator gain. Secondly, the fading measurements have been characterized by a set of random variables with given probability distribution. A recursive event-triggered state estimation approach has been provided and the monotonicity relationship between the trace of the resulting upper bound of the EEC and the fading probability has been discussed from a theoretical perspective. At last, some illustrative experiments have been employed to clarify the effectiveness of the designed state estimation methods.
- In Chap. 13, the state estimation issue has been handled for time-varying nonlinear complex networks under communication scheduling strategies in the variance-constrained framework. Firstly, a sequence of Bernoulli distributed random variables has been adopted to model the missing measurements under the case of the uncertain probabilities. For the sake of improving the transmission efficiency, the RAP has been utilized to determine the node transmission cases. Through solving the recursive matrix equations, an optimized upper bound concerning on the EEC has been derived and then the desirable filter gain has been constructed on the premise of guaranteeing the minimization of the proposed upper bound. Furthermore, the monotonicity regarding the missing probability and the algorithm performance has been analyzed from a theoretical viewpoint. Secondly, the main results have been extended to handle the optimized state estimation issue for nonlinear dynamical complex networks affected by the RCS, RSDs and ETM. Here, an array of random variables satisfying uniform distribution over certain interval has been used to mirror the randomness of the elements of coupling configuration. The RSDs have been depicted by a set of random variables obeying the Bernoulli distribution. Based on the available measurements influenced by RSDs and random coupling strengths, a new robust state estimation

scheme has been proposed for time-varying nonlinear complex networks. Additionally, a sufficient criterion has been provided to prove the boundedness of the estimation error dynamics. Lastly, two simulation experiments with comparisons have been used to illustrate the validity of the presented state estimation algorithms.

14.2 Future Work

Related topics in relation to the future research work are given below:

- The addressed nonlinearities possess some limitations to some extent. The variance-constrained filtering and estimation issues for more general nonlinear systems with incomplete information would be one of the future research directions.
- Another future research topic is to discuss the variance-constrained filtering and estimation problems for more dynamical complex systems with other scheduling protocols, such as the weighted try-once-discard protocol, FlexRay protocol, amplified-and-forward relay, decode-and-forward relay, token bucket protocol, and so on.
- In order to adapt the digital communication requirement, the encoding-decoding mechanism is necessary to be introduced in the process of data transmission. Moreover, taking into account the constrained bit rate, the encoding-decoding with the constrained bit rate is essential to be considered in the transmission process. Thus, the filter design methods under the encoding-decoding mechanism with the constrained bit rate are desirable for time-varying dynamical systems.
- With the purpose of ensuring the data quality and security, the binary encoding scheme is essential to be introduced. However, owing to the presence of the channel noise, the bit error might appear in the transmission process. Accordingly, it is valuable to study the filtering/estimation problems under bit error.
- From the perspective of sensor energy constraint, it is valuable to develop a variance-constrained filtering algorithm for time-varying dynamical systems under the sensor energy harvesting mechanism.
- When the system states are constrained and own memory characteristics, the variance-constrained filter design is desired for fractional-order state-saturated systems with incomplete information.
- An additional trend for future research topic is to apply the proposed methods to handle more filtering issues in the context of time-varying dynamical systems with incomplete information, such as CKF, UKF, adaptive filtering, particle filtering, and so on.

Index

J. Hu et al., *Variance-Constrained Filtering for Stochastic Complex Systems*, Intelligent Control and Learning Systems 18,
https://doi.org/10.1007/978-981-96-2637-3